Society of
Exploration
Geophysicists

SEISMIC WAVE
ATTENUATION

Edited by
David H. Johnston and M. Nafi Toksöz

Series Editor
Franklyn K. Levin

Geophysics reprint series
No. 2

Library of Congress Catalog Card Number: 81-50381

Society of Exploration Geophysicists
P.O. Box 3098
Tulsa, Oklahoma 74101

Printed in the United States of America

ISBN 0-931830-16-8
    0-931830-00-1

CONTENTS

Preface

Seismic waves propagating through the earth are attenuated by the conversion of some fraction of the elastic energy to heat. Using terminology analogous to that used for the elastic properties that control seismic velocities, attenuation properties are characterized as anelastic properties. Attenuation data complement other physical measurements for characterizing rock properties. In seismic studies, attenuation data can at least double the information obtained from velocities alone.

An understanding of the attenuative properties of the earth has two major motivations. First, seismic wave amplitudes are reduced as waves propagate through an anelastic medium, and this reduction is generally frequency dependent. Second, attenuation characteristics reveal much information, such as lithology, physical state, and degree of saturation of rocks.

The phenomenon of attenuation is much more complex than the elastic aspects of seismic wave propagation. Both laboratory and field measurements are difficult to make. The mechanisms contributing to attenuation are numerous, and small changes in some conditions can affect attenuation significantly. However, sensitivities to some parameters, such as fluid saturation, make the measurement and understanding of attenuation highly important for many applications. The realization of the need and promise for specific data and models has prompted a strong resurgence of interest and research concerning attenuation in the fields of both seismology and rock physics.

Laboratory measurements of attenuation in rock samples under varying pressures, temperatures, strain amplitudes, frequencies, and saturation conditions are presently being carried out. Detailed theoretical modeling of processes that may be responsible for attenuation is being undertaken. Measurements of attenuation in the earth using direct and refracted compressional and shear waves, surface waves, reflection seismograms, vertical seismic profiling, and full-wave acoustic well logs are being explored intensively. The net

result of these field, laboratory, and theoretical studies will be a rapid expansion of our knowledge concerning the attenuation of seismic waves in the earth's crust. We have undertaken the editing of this volume to help the broad-range research effort gain a better understanding of attenuation and its applications to seismic exploration problems.

Attenuation is such an extensive subject that a comprehensive treatment of all relevant material would require many volumes. In order to provide a coherent coverage of the subject matter in one volume, we have done three things. First, we have emphasized the material most relevant to exploration geophysics. As a result, most of the papers compiled here deal with sedimentary rocks, the effects of fluids, and the pressure ranges encountered in shallow crustal layers. A hard choice had to be made among many good papers in order to keep the size of this volume manageable. The many important papers dealing with attenuation under mantle conditions have been exluded, as have some articles of historic significance (in order to make room for the more recent results). Because of the breadth of the subject matter, attenuation literature is widely dispersed. Thus, as a second measure, we sought the most relevant articles published not only in GEOPHYSICS but in other journals as well. As a result, the formats and styles of the articles included in this volume vary. In spite of these variations, we feel that the continuity of these articles is evident. As a third step, we have written some additional introductory and explanatory sections. The introductory chapter concerning definitions and terminology and the article (in chapter 2) presenting a summary of attenuation data are two such examples.

This volume contains 33 articles grouped into five chapters. Chapter 1 contains a review of the definitions of attenuation. Chapter 2, with 12 articles, is devoted to laboratory measurement techniques and data. At the end of chapter 2, these data are summarized and recent developments are presented. Chapter 3 deals with mechanisms of attenuation in dry and saturated rocks. The nine articles in this chapter cover various mechanisms and use many data from articles in chapter 2 for testing theoretical formulations. Chapter 4 covers field techniques for the measurement of attenuation in situ. The six articles in chapter 4 describe measurement techniques and data for sedimentary formations, soil, and the earth's crust. The final chapter (chapter 5) is devoted to the effects of attenuation on the seismic pulse. Pulse broadening because of attenuation is a major phenomenon we face in exploration seismology. As a matter of format, references made in the introductory material to articles appearing in this volume are in italics.

We would like to express our thanks to the authors of the articles included here, and to the various journals in which they were published for their permission to reproduce the papers. We thank Eleanor Davis for typing, and Bryan Brown for editing the introductory material, chapter 1, and the summary of attenuation data. We are deeply grateful to the SEG Publications Committee and to the Series Editor, Dr. F. K. Levin, for their support and boundless aid in the collection and publication of the material in this volume.

M. Nafi Toksöz
David H. Johnston

Chapter 1
Definitions and terminology

David H. Johnston and M. Nafi Toksöz

The elastic properties of rock are uniquely defined by elastic moduli and/or P- and S-wave velocities. Generally accepted definitions and units for these two parameters make their use commonplace. The attenuative properties of rocks, however, are specified by a wide range of measures. In order to compare attenuation data properly from different sources, it is important to present definitions of the different measures and to show how they are related to one another.

The most commonly used measures of attenuation found in the literature are the attenuation coefficient α which is the exponential decay constant of the amplitude of a plane wave traveling in a homogeneous medium; the quality factor Q and its inverse Q^{-1}, sometimes called the internal friction or dissipation factor; and the logarithmic decrement δ. These quantities are related as follows:

$$\frac{1}{Q} = \frac{\alpha v}{\pi f} = \frac{\delta}{\pi} , \qquad (1)$$

where v is the velocity and f is the frequency. Since both velocity and attenuation are associated with a particular mode of wave propagation, one experimental technique may yield an extensional wave velocity controlled by the Young's modulus and a dissipation factor denoted Q_E^{-1}, while another may determine the P-wave velocity and Q_p^{-1}. In general, these results are not equivalent.

In the following sections, the relationships expressed in equation (1) are derived and conventional units and conversion factors discussed. Finally, equations relating the attenuation measures of different modes of wave propagation are presented.

Attenuation coefficient and logarithmic decrement

For plane waves propagating in a homogeneous medium, the amplitude is given by

$$A(x,t) = A_o e^{i(kx - \omega t)} , \qquad (2)$$

where ω is the angular frequency and k is the wavenumber. Attenuation may be introduced mathematically by allowing either the frequency or wavenumber to be complex. In the latter case,

$$k = k_r + i\alpha \qquad (3)$$

so that

$$A(x,t) = A_o e^{-\alpha x} e^{i(k_R x - \omega t)} ,$$

where α is the attenuation coefficient in units of inverse length and the phase velocity is

$$v = \frac{\omega}{k_R} . \qquad (4)$$

Attenuation may also be defined in terms of inverse time by allowing ω to be complex.

Letting the attenuation be determined by

$$A(x) = A_o e^{-\alpha x} , \qquad (5)$$

α may be written as

$$\alpha = -\frac{1}{A(x)} \frac{dA(x)}{dx} = -\frac{d}{dx} \ln A(x) .$$

For two different positions, x_1 and x_2 ($x_1 < x_2$), with respective amplitudes $A(x_1)$ and $A(x_2)$,

1

$$\alpha = \frac{1}{x_2 - x_1} \ln\left[\frac{A(x_1)}{A(x_2)}\right] \qquad (6)$$

in nepers/unit length (or simply inverse length). Alternatively,

$$\alpha = \frac{1}{x_2 - x_1} \cdot 20 \log\left[\frac{A(x_1)}{A(x_2)}\right]$$

in units of dB/unit length. Conversion of units is accomplished as follows: α (dB/unit length) = 8.686 α (nepers/unit length).

For an oscillating system in free decay, the definition of the logarithmic decrement δ follows from equation (6) and is

$$\delta = \ln\left[\frac{A_1}{A_2}\right] = \alpha\lambda = \frac{\alpha v}{f}, \qquad (7)$$

where A_1 and A_2 are the amplitudes of two consecutive cycles, v is the velocity, f is the frequency, and λ is the wavelength.

Equivalent to δ is the attenuation measure with units of dB/wavelength (or dB/period if α is defined in units of inverse time). This measure is commonly referred to as an attenuation coefficient and its conversion to δ is

$$\delta = \frac{\alpha \, (dB/\lambda)}{8.686}.$$

Quality factor (Q)

The most common measures of attenuation are the dimensionless quality factor Q and its inverse Q^{-1}. As an intrinsic property of rock, Q is a ratio of stored energy to dissipated energy. O'Connell and Budiansky (1978) discussed in detail various definitions of Q and their relationships to the viscoelastic constitutive equations for a given material.

Intrinsic Q may differ under some conditions from Q values derived from processes such as wave propagation. Yet these processes are valuable tools for measuring the anelastic response of a rock. The various definitions of Q presented in this chapter are equivalent to intrinsic Q if losses are assumed to be small ($Q > 10$). Fortunately, under most conditions of interest in geophysics, the small-dissipation assumption is valid.

Intrinsic Q may be defined as

$$Q = \frac{\omega E}{-dE/dt} = \frac{2\pi W}{\Delta W}, \qquad (8)$$

where E is the instantaneous energy in the system, dE/dt is the rate of energy loss, W is the elastic energy stored at maximum stress and strain, and ΔW is the energy loss (per cycle) of a harmonic excitation.

For nearly elastic or low-loss linear solids, an alternative definition may be found from the stress-strain relations. Given a sinusoidally varying stress, the strain response will also be sinusoidal. The two are related by the appropriate elastic modulus M and the phase lag ϕ of strain behind stress. Allowing M to be complex where $M = M_R + iM_I$, it can be shown that

$$\frac{1}{Q} = \frac{M_I}{M_R} = \tan \phi \sim \phi \qquad (9)$$

(White, 1965). ϕ is analogous to the loss tangent in electromagnetic (EM) problems. Equation (9) results in an elliptical stress-strain curve under sinusoidal loading. The equivalence of the definitions of Q given in equations (8) and (9) arises from the fact that the area of the ellipse is proportional to the energy loss per cycle and the energy stored is given by the total area under the loading curve.

For lossy material the equation relating Q with the attenuation coefficient of equation (1) must include a second-order term, since the stored energy depends upon the derivative of the complex modulus with respect to frequency as well as the value of the modulus (O'Connell and Budiansky, 1978). As discussed by *Hamilton* (*1972*) in chapter 4,

$$\frac{1}{Q} = \frac{\alpha V}{\pi f - \dfrac{\alpha^2 V^2}{4\pi f}}. \qquad (10)$$

However, under the low-loss assumption, i.e., when $M_I \ll M_R$, the $\alpha^2 V^2/4\pi f$ term is negligible and may be dropped.

Finally, for the dynamic systems most commonly used to measure attenuation, Q may be defined in terms of a resonance-peak bandwidth

$$Q = \frac{f_r}{\Delta f}, \qquad (11)$$

where Δf is the frequency width between half-power (3 dB in amplitude) points about a resonance peak at f_r on a power-frequency plot. Equation (11) is equivalent to equation (8) for low-loss materials as will now be shown.

We first consider the transient decay of free oscillations. The equation of motion including a damping term is

$$\ddot{x}(t) + \Gamma \dot{x}(t) + \omega_0^2 x(t) = 0 , \quad (12)$$

where ω_0 is the natural frequency of the undamped system and is equal to $(K/M)^{1/2}$, K is the restoring force, and M is the mass. Γ is the damping constant per unit mass. This equation has a solution of the form

$$x(t) = e^{-t/2\tau} \cos (\omega_1 t + \theta) , \quad (13)$$

where $\tau = 1/\Gamma$ and $\omega_1^2 = \omega_0^2 - (\Gamma/2)^2$. Of course a more general solution results from the superposition of linearly independent solutions. For weakly damped systems we may assume $e^{-t/2\tau} = e^{-\Gamma t/2}$ to be relatively constant during one cycle of oscillation. We may then write the time derivative of x as

$$\dot{x}(t) = -\omega_1 e^{-\Gamma t/2} \sin (\omega_1 t + \theta) . \quad (14)$$

Total energy (kinetic plus potential) is then

$$E = \frac{1}{2} M \dot{x}^2 + \frac{1}{2} M \omega_0^2 x^2 . \quad (15)$$

This can be shown from equation (14) to be

$$E = E_0 e^{-\Gamma t} , \quad (16)$$

where $E_0 = M(\omega_1^2 + \omega_0^2)/2$. If we set $\Gamma = \omega_1/Q$, then equation (14) is a solution to

$$\frac{dE}{dt} = \frac{-\omega_1 E}{Q} ,$$

rewritten as

$$Q = \frac{\omega_1 E}{-dE/dt} , \quad (17)$$

which is the definition given for Q in equation (8).

We now consider steady-state oscillation under a harmonic driving force of the form

$$F(t) = F_0 \cos \omega t .$$

The equation of motion,

$$\ddot{x}(t) + \Gamma \dot{x}(t) + \omega_0^2 x(t) = \frac{F_0}{M} \cos \omega t , \quad (18)$$

has the steady-state solution

$$x(t) = A \sin \omega t + B \cos \omega t , \quad (19)$$

where (Crawford, 1968)

$$A = \frac{F_0}{M} \frac{\Gamma \omega}{[(\omega_0^2 - \omega^2)^2 + \Gamma^2 \omega^2]} , \quad (20)$$

and

$$B = \frac{F_0}{M} \frac{(\omega_0^2 - \omega^2)}{[(\omega_0^2 - \omega^2)^2 + \Gamma^2 \omega^2]} . \quad (21)$$

The instantaneous power delivered to the system by the driving force is

$$P(t) = F_0 \cos \omega t [\dot{x}(t)] \quad (22)$$

$$= F_0 \cos \omega t (\omega A \cos \omega t - \omega B \sin \omega t) .$$

The time average input power is determined from the average over one cycle. Thus

$$\bar{P} = F_0 \omega A \langle \cos^2 \omega t \rangle - F_0 \omega B \langle \cos \omega t \sin \omega t \rangle$$

$$= \frac{1}{2} F_0 \omega A . \quad (23)$$

We see from equation (23) that only the velocity component of x that is in phase with the driving force (displacement 90 degrees out of phase) contributes to the average input power. For steady-state oscillations, the time average input power is equal to the power dissipated by attenuation. It can be shown that the average power loss is

$$P_\ell = M\Gamma \langle \dot{x}^2 \rangle$$

$$= \frac{1}{2} M\Gamma \omega^2 (A^2 + B^2) \quad (24)$$

and is in fact equivalent to equation (23). This is not to say that the instantaneous power input and the instantaneous power loss are equal, but rather that the average power input and loss are equal over a cycle.

We shall now derive the frequency bandwidth definition of Q. From equations (23) and (20), it can be shown that

$$\bar{P} = \frac{1}{2} \frac{F_0^2}{M} \frac{\Gamma \omega^2}{(\omega_0^2 - \omega^2)^2 + \Gamma^2 \omega^2} . \quad (25)$$

The maximum average power will occur when the denominator of equation (25) is at a minimum. This resonance occurs when $\omega = \omega_0$ so that $P_{max} = F_0^2 / 2M\Gamma$. Equation (25) can be rewritten as

$$\bar{P} = P_{max} \frac{\Gamma^2 \omega^2}{(\omega_0^2 - \omega^2)^2 + \Gamma^2 \omega^2} . \quad (26)$$

Values of ω for which the power is one-half maximum are found when $\omega^2 = \omega_0^2 \pm \Gamma\omega$ with the two realistic solutions giving

$$\omega = \sqrt{\omega_0^2 + \frac{1}{4}\Gamma^2} \pm \frac{\Gamma}{2} . \quad (27)$$

The full-frequency width of the half-power units is simply

$$\Delta\omega = \Gamma . \quad (28)$$

If $\Gamma = \omega_1 / Q$ as defined previously, then

$$\frac{\Delta\omega}{\omega_1} = \frac{\Delta f}{f_1} = \frac{1}{Q} . \quad (29)$$

Equation (29) is valid if the assumption of weak damping is made. That is, $\omega_0 \approx \omega_1 \approx \omega$. Again, for most rocks, the low-loss assumption is appropriate. Also, since it has been determined that the free decay time $\tau = 1/\Gamma$, therefore, $\tau\Delta\omega = 1$. Thus the Q values obtained from free decay and resonance bandwidth are equivalent.

The same analysis and result obtained in equation (29) also holds true for half-energy points. From equation (15), it can be seen that

$$E = E_0 \frac{\frac{1}{2}\Gamma^2 (\omega^2 + \omega_0^2)}{(\omega_0^2 - \omega^2)^2 + \Gamma^2 \omega^2} . \quad (30)$$

Again it may be shown that the values of ω for which $E = E_0 / 2$ are:

$$\omega^2 = \omega_0^2 \pm \Gamma\omega ,$$

or

$$\omega = \sqrt{\omega_0^2 + \frac{1}{4}\Gamma^2} \pm \frac{\Gamma}{2} . \quad (31)$$

Q for modes of wave propagation

Attenuation can be measured in a variety of ways, including quasi-static techniques (stress-strain phase lag) and dynamic methods using compressional and shear waves. Different Q values are obtained for different methods. For example, the use of extensional stress-strain phase lag results in the loss parameter Q_E^{-1}, that is, the anelasticity of the Young's modulus. The bulk loss Q_K^{-1} can be obtained from hydrostatic stress and dilational strain. Compressional and shear waves provide values for Q_P^{-1} and Q_S^{-1}. These quantities can be related to one another using the definition of Q given in equation (9) rewritten here

$$Q_M^{-1} = M_I / M_R , \quad (32)$$

where M_I and M_R are the imaginary and real parts of the modulus M relating stress and strain. For extensional stress, M is the Young's modulus E; for hydrostatic stress, M is the bulk modulus K; for shear or torsional stress, M is the shear modulus μ; and for P-waves, $M = K + 4\mu/3$. Appropriate substitution of these moduli into equation (32) leads to three equations relating the attenuation measures of different stress states (*Winkler and Nur, 1979*):

$$\frac{(1-\nu)(1-2\nu)}{Q_P} = \frac{1+\nu}{Q_E} - \frac{2\nu(2-\nu)}{Q_S} ,$$

$$\frac{3}{Q_E} = \frac{1-2\nu}{Q_K} + \frac{2(\nu+1)}{Q_S} , \quad (33)$$

$$\frac{1+\nu}{Q_K} = \frac{3(1-\nu)}{Q_P} - \frac{2(1-2\nu)}{Q_S} ,$$

where ν is the Poisson's ratio that must be determined. For a purely elastic solid, the velocities are related by

$$V_P^2 = \frac{V_S^2 (4 V_P^2 - V_E^2)}{3 V_S^2 - V_E^2} ,$$

$$V_E^2 = \frac{V_S^2 (3 V_P^2 - 4 V_S^2)}{V_P^2 - V_S^2} , \quad (34)$$

and

$$V_K^2 = V_P^2 - \frac{4}{3}V_S^2 \ ,$$

and the Poisson's ratio is

$$\nu = \frac{V_P^2 - 2V_S^2}{2(V_P^2 - V_S^2)} = \frac{V_E^2 - 2V_S^2}{2V_S^2} \ . \qquad (35)$$

The relationships expressed in equation (33) are crucial for comparing Q values obtained through different experimental methods. Also, decomposition of Q values into pure compressional and shear losses has proved to be a useful diagnostic aid in the analysis of attenuation mechanisms.

Summary

The common measures of attenuation are defined and related to one another as follows:

$$\frac{1}{Q} = \frac{M_I}{M_R} = \frac{\Delta f}{f} = \frac{\delta}{\pi} = \frac{\alpha V}{\pi f} = \frac{\alpha(\mathrm{dB}/\lambda)}{8.686\,\pi} \ ,$$
$$(36)$$

where

Q = quality factor,
α = attenuation coefficient,
δ = logarithmic decrement,
Δf = resonance width,

M_R, M_I = real and imaginary parts, respectively, of complex modulus M,
V = velocity,
f = frequency,

as long as the low-loss assumption is valid and it is recognized that each mode of wave propagation has a distinct attenuation measure associated with it.

As a final reminder about units, the attenuation coefficient may be expressed in nepers/unit length (or simply inverse length) or in dB/unit length. The relationship between the two is given by $\alpha(\mathrm{dB}/\text{unit length}) = 8.686\alpha(\text{nepers}/\text{unit length})$, and as noted in equation (36), $\alpha(\mathrm{dB}/\lambda) = 8.686\,\pi/Q$.

REFERENCES

Crawford, F.S., 1968, Waves, Berkeley Physics Course, v. 3: New York, McGraw-Hill Book Co., Inc., p. 102–108.

Hamilton, E.L., 1972, Compressional-wave attenuation in marine sediments: Geophysics, v. 37, p. 620–646.

O'Connell, R.J., and Budiansky, B., 1978, Measures of dissipation in viscoelastic media: Geophys. Res. Lett., v. 5, p. 5–8.

White, J.E., 1965, Seismic waves: Radiation, transmission, and attenuation: New York, McGraw-Hill Book Co., Inc.

Winkler, K., and Nur, A., 1979, Pore fluids and seismic attenuation in rocks: Geophys. Res. Lett., v. 6, p. 1–4.

Chapter 2
Laboratory measurements of attenuation

The accurate measurement of attenuation is a difficult task and seriously limits the utilization of anelastic rock properties. Both in the laboratory and field, seismic wave amplitudes are strongly affected by geometrical spreading, reflections, and scattering in addition to intrinsic damping. Thus, in many cases, to obtain the true intrinsic attenuation it is necessary to correct for these other effects. This correction process can be a formidable task. In this introduction, we briefly describe the experimental techniques commonly used in laboratory measurements. Following the papers, a summary of current attenuation data is presented.

The methods generally used for measuring attenuation in the laboratory can be separated into the following categories (Zener, 1948; Kolsky, 1953; Schreiber et al, 1973):

- Free vibration,
- Forced vibration,
- Wave propagation,
- Observation of stress-strain curves.

Each method can be further split into subgroups. The choice of an experimental method is based largely on the frequency range of interest, the actual values of attenuation, and the physical conditions under which the sample will be studied. For example, the use of stress-strain curves will provide information at frequencies below 1 Hz, while resonance vibrations measure the attenuative properties in the range of 100 Hz–100 kHz, and wave-propagation experiments are commonly restricted to the ultrasonic range of 100 kHz or higher. Furthermore, each method determines a different attenuation or loss parameter. The definitions of, and relationships among, these parameters (see chapter 1) must be specified before any meaningful comparisons can be made.

Free vibration method

In this technique the attenuation is found by the amplitude decay of successive cycles of free vibrations. We define the logarithmic decrement δ as follows:

$$\delta = \frac{\ln (A_1 / A_2)}{(t_1 - t_2) f} = \ln (A_\lambda / A_0), Q^{-1} = \frac{\delta}{\pi},$$

(1)

where A_1 and A_2 are the amplitudes at times t_1 and t_2 and f is the natural free-vibration frequency of the system. The alternative definition is in terms of the amplitude A_λ, one wavelength from a starting amplitude A_0.

The free vibration method has been implemented in several ways. The first is a torsion pendulum (*Peselnick and Outerbridge, 1961*) in which a rod of rock is suspended vertically so that a mass with a large moment of inertia can be attached to its lower end. If the mass is given a "kick" and the system is allowed to vibrate freely, the free-vibration frequency is a function of the rock properties and the moment of inertia of the mass. The rate of decay of the amplitude of these oscillations is attributable to energy loss in the rock if other losses can be made small. One important application of this method is in the study of the elastic constant and attenuation of metals and composite materials at high temperatures and very low (≈ 1 Hz) frequencies (Kê, 1947; Kingery, 1959; Jackson, 1969). However, it is difficult to fabricate long cylindrical rock samples with the uniform cross-sections needed for this type of experiment. Furthermore, the initial applied stress must be low so that the sample does not fracture, Q and the elastic moduli are independent of amplitude, and stress inhomogeneities in the sample are minimized.

Vibration may also be excited by piezoelectric, electromagnetic (EM), and electrostatic effects. These methods become necessary for short samples that have resonant frequencies in the kilohertz range. In each case (for rocks) one must apply a conductive coating or a transducer, and therefore corrections to the resonant frequency must be made. While the effects of these corrections are minimal for attenuation measurements, they are crucial for the determination of elastic parameters. As with all resonant techniques, the determination of the specific loss parameter depends upon the mode of vibration excited.

Forced vibration method

One of the more common methods for measuring attenuation is the forced longitudinal, flexural, or torsional vibration of long bars. This method is based on the phenomenon of standing waves. For longitudinal and torsional waves the velocity of the wave is given by

$$V = \lambda f = \frac{2\ell f}{n}, \quad n = 1, 2, 3, \ldots, \quad (2)$$

where f is the resonant frequency of mode n and ℓ is the length of the sample. The expression for flexural waves is more complicated because the nodes do not occur at the quarter points.

Reasonably exact solutions of the three-dimensional wave equation exist for cylindrical and rectangular geometries; therefore, bars of those shapes are most suited for resonance experiments. Based on solutions of this sort, Spinner and Tefft (1961) derived relationships between the Young's modulus and resonant frequency for longitudinal and flexural vibrations and between the shear modulus and resonant frequency for torsional vibrations. These relationships are summarized by Schreiber et al (1973).

For a continuously driven system, Q may be found from the width of the resonance amplitude peak. Defining Δf to be the frequency range between those values for which amplitude is 3 dB lower than the resonance peak amplitude,

$$Q = \frac{f_n}{\Delta f}. \quad (3)$$

If the system can be driven easily, this method may be used for low Q materials with an accuracy of about 1 to 5 percent. Alternatively, the system may be first driven at resonance. Then, when the driving force is turned off, the subsequent decay of amplitude with time can be used to calculate Q.

Again, the sample may be driven by electrostatic, EM, or piezoelectric methods. Because of its weakness, the electrostatic drive is better suited for high-Q materials. The EM drive is useful if the sample is electrically conducting. Otherwise, conducting plates must be attached to the specimen ends and one must correct for the end-loading effects. Operation over a wide frequency range is possible, especially if the sample is purposely end-loaded to lower its resonance frequency (Tittmann, 1977). For piezoelectric excitation (Quimby, 1925), piezoelectric transducers are cemented to the specimen and the combination is made to resonate. Corrections must be made for the impedance and mechanical Q of the transducer, but highly efficient, low-loss ceramic transducers are available, making this method attractive for both low and relatively high-Q samples.

While resonance methods are easily implemented and Q values are easily determined over a wide frequency range, the physical conditions under which these types of experiments may be performed without concern for extraneous losses are limited. In particular, it is necessary to consider radiation losses into the surrounding medium.

In torsional vibration only shear waves are involved. Since coupling between the shear waves in the solid and in the surrounding medium (e.g., air, gas) is poor, little energy is lost. For longitudinal and flexural waves, however, damping along the cylindrical surface and radiation from the ends can be substantial. Browne and Pattison (1957) analyzed this problem in great detail. Corrections for these effects are negligible for low-Q materials in ambient air conditions. For high-Q materials such as metals it is advised to perform resonance experiments under vacuum. However, even for rocks, losses when the sample is subjected to pressure can be high when compared to the intrinsic attenuation. For these types of experiments to be performed successfully, helium is commonly used as the pressure medium and radiation losses must be determined and subtracted from the measured values.

The effects of a jacket on a sample are worthy of consideration. A jacket will alter the resonant frequency and the observed attenuation in a rock sample. These changes can be calculated using Rayleigh's principle (Gardner et al, 1964) and are generally small. For some porous rocks, however, the jacket will penetrate the sample; thus the results obtained will represent an average of the sample and jacket.

Wave propagation

The use of wave propagation experiments in the lower ultrasonic frequency range to determine attenuation in the laboratory is of particular interest since the loss parameters involved closely parallel those measured in field experiments. However, these types of techniques are fraught with experimental and interpretational difficulties. The methods assume the amplitude of the seismic wave (generally considered to be a plane wave) decays exponentially with distance or time and that one can correct for losses other than the intrinsic attenuation. The extraneous losses include beam spreading, coupling losses, diffraction losses, and wedging effects.

At low frequencies (<1 MHz) the effect of beam spreading can become significant. While direct corrections can be made, their validity is based on the assumption that plane or spherical waves are present. The extent of the plane-wave region, dependent upon the size of the transducer and wavelength of the seismic wave, must be determined. It is easier, in most cases, to design the experiment so that spreading losses are minimal when compared to the intrinsic attenuation.

At long wavelengths (relative to the transducer diameter) diffraction losses may become important. Because of beam spreading, sidewall reflections and mode conversions may occur, interfering with the directly propagating wave. This interference can be readily observed in the pulse-echo technique as a nonexponential decay in amplitude. While it is possible in many cases to design the experiment so that the sample diameter/transducer diameter ratio is large, this is not always possible and diffraction may cause problems even with pulse transmission techniques. Truell et al (1969) treated diffraction theoretically, but generally this problem must be tackled empirically.

Energy loss can occur in the transducer itself, at the bond between transducer and sample, and in the electronic measurement system. Transducer properties are generally known, and one can choose a material with a much higher mechanical Q than the sample. However, the other losses are impossible to calculate theoretically and must be determined empirically (Truell et al, 1969) or eliminated by the experimental design.

Finally, if the ends of the sample are not exactly parallel, nonexponential losses owing to phase variations over the surface of the transducer may result. That is, a plane wave is not reflected or transmitted in phase by a nonparallel or "wedged" boundary. Again, this effect is a most important consideration for the pulse-echo technique but, unlike diffraction, is more pronounced at higher frequencies. Similar errors can occur if the thickness of the bond between transducer and sample is nonuniform or if, as is the case with single crystals, elastic constants vary slightly within the ultrasonic beam. This problem has been analyzed by Truell and Oates (1963), who showed that the allowed deviation from parallelism is inversely dependent upon both the Q of the sample and the frequency.

In general, wave propagation experiments can be classified according to whether they make use of pulse-echo or through-transmission methods. Also the type of excitation may either be a pulse, which provides an attenuation value averaged over a relatively broad frequency band, or a tone burst, which is strictly band-limited.

For the pulse-echo technique, the attenuation is found by observing the amplitude decay of multiple reflections from a free surface. Of course, the decay must be assumed to be exponential. If the electronic measuring system is linear and the amplitudes can be directly determined, then the attenuation coefficient can be calculated from

$$\alpha = \frac{1}{2\ell} \ln \left[A(1) / A(2) \right] , \qquad (4)$$

where ℓ is the sample length, $A(1)$ is the amplitude at one echo, and $A(2)$ is the amplitude at the next echo.

Most of the time the attenuation coefficient is found from a calibrated exponential decay curve that is superimposed on the echo images as they are observed on an oscilloscope. The curve is generated by allowing a capacitor to discharge through a known resistor so that the R-C time constant can be evaluated. Measurements obtained in this way are given in terms of inverse time. The essential advantage of this method over using the absolute amplitudes is that nonexponential behavior, if present, is easily seen. However, the technique does not work well for highly attenuating samples. For such samples the pulse comparator method, where a reference pulse of the same frequency as that applied to the transducer is sent through the same electronics, must be employed. If this comparator pulse can be accurately attenuated, then the relative amplitude loss between the two echoes can be determined.

Pulse-echo techniques are generally used for high-Q samples and are popular for determining Q values for single crystals. An important consideration of this method, however, is that the reflection at the free boundary is loss free. This assumption limits the method's usefulness in high-pressure studies where energy will be lost into the pressure medium. The technique has, however, been used successfully on some fine-grained limestones at atmospheric pressure (Peselnick and Zietz, 1959). In that particular study, beam spreading, diffraction losses, and incomplete reflection at the free surface were deduced to be negligible when compared to the intrinsic attenuation.

The through-transmission method is best suited for use with jacketed samples in pressure vessels. And, with the use of spectral ratios, many of the problems that plague the pulse-echo method can be, at least conceptually if not practically, eliminated.

Transmission experiments can be categorized in terms of sample size and the locations of transmitter and receiver transducers. In most cases the transducers are located at opposite ends of the sample. If the sample diameter is larger than the length, sidewall reflections are minimized. Alternatively, the guided-wave method, for which a cylindrical rod acts as a waveguide (McSkimin, 1956), may be used. The primary distinction between these two methods lies in the types of corrections that must be made. A third arrangement involves the use of large blocks of sample material upon which the transmitter and receiver are used to measure amplitude loss as a function of distance (Watson and Wuenschel, 1973). In this case, if the transducer diameters are small and the distances between them are large with respect to the wavelength, simple inverse-length beam spreading may be assumed.

Data for the first two arrangements described above may be analyzed in terms of echoes. While the two arrangements are equally accurate, the guided-wave method requires extensive sample preparation and calculations to determine the effect of mode conversions at the sidewall. These calculations not only depend upon the sample's elastic properties but also on the attenuation. Mode conversions are minimized when the sample

length-to-diameter ratio is small, but one must then account for diffraction fields.

For large samples the attenuation may be found, although with some uncertainty, from the amplitude decay of a particular peak in a wavetrain as a function of transmitter-receiver separation. Of course, it must be assumed that the coupling between transducer and sample is the same in each case and that spreading has been accurately accounted for.

Attenuation may also be found from pulse-type waveforms using a predictive waveform. If the input pulse is known or the waveforms at greater distances are normalized to a starting point, then the match of a synthetic waveform for a variable Q material with the actual waveform will yield an attenuation value (Watson and Wuenschel, 1973).

Spectral ratios, by far the most common technique used in seismology, allow for the elimination of many of the problems associated with wave-propagation methods. This technique relies on the fact that high frequencies are preferentially attenuated relative to low frequencies. In general, the spectral amplitude of the propagating wave may be expressed (Ward and Toksöz, 1971) as

$$A(f,x) = GA_r(f) \exp(-t^*f) , \quad (5)$$

where G includes geometrical spreading and transmission or reflection coefficients, A_r is the receiver response, and

$$t^* = \pi \int_{path} \frac{dx}{QV} \quad (6)$$

for frequency-independent Q. This expression is also valid for slowly varying Q in the frequency band of interest. For spectral amplitudes obtained at two points for a common source,

$$\ln \frac{A_1(f)}{A_2(f)} = (t^*_2 - t^*_1)f$$

$$+ \ln \frac{G_1}{G_2} + \ln \frac{A_{r_1}(f)}{A_{r_2}(f)} . \quad (7)$$

If the receiver responses are either equal or unequal but known, equation (7) may be reduced so that the slope of a straight line fitted to the log of the spectral ratio-versus-frequency function will yield the differential attenuation $t_2^* - t_1^*$ between the two receivers. A modification of this technique has been successfully applied to laboratory samples utilizing a high-Q reference standard (*Toksöz et al, 1979*).

Another technique for determining attenuation utilizing wave propagation methods employs the change in pulse shape, usually characterized by the rise time τ as it travels through the rock. Gladwin and Stacey (1974) suggest the following empirical relationship:

$$\tau = \tau_0 + c \int_0^t Q^{-1} dt , \qquad (8)$$

where t is the traveltime, τ_0 is the initial rise time at $t = 0$, and c is a constant. This finding contradicts *Ricker's* (*1953*) observation using explosion seismology that $\tau \propto t^{1/2}$. Ramana and Rao (1974), following the theoretical work of Knopoff (1956), characterize pulse shape by its width Δt. Thus,

$$Q^{-1} t = \Delta t , \qquad (9)$$

where t is again the traveltime.

The theoretical basis for this technique is discussed in several papers in chapter 5. Present use of this method, however, is limited to massive rock or samples for which several lengths are available.

Observation of stress-strain curves

Energy loss may be measured directly from hysteresis in stress-strain curves obtained from loading-unloading cycles well off the resonance frequency of the sample (*Gordon and Davis, 1968; McKavanagh and Stacey, 1974; Brennan and Stacey, 1977;* Pelselnick et al, 1979). Specifically, the area between the loading and unloading paths on the stress-strain curve is the energy lost ΔW in that cycle. The relative attenuation may be found by dividing ΔW by W, the maximum work done, i.e., the area under the loading path during loading or by measuring the phase lag of strain behind stress. Conceptually, this type of experiment offers great promise for obtaining Q values for small samples under high pressure and temperature conditions at seismic frequencies and over a wide range of strain amplitudes. Furthermore, by analyzing the shape of the stress-strain curve, the linearity or nonlinearity of the operative attenuation mechanism can be inferred.

Ideally, the experiment should produce pure shear and dilatational stresses. Reported applications of this technique, however, are limited

to either torsional (shear) or uniaxial stress geometries. Another limitation of the method is the ability to measure small dynamic strains. However, coaxial capacitance transducers and phase-lock amplifiers have been successfully employed for radial strain determinations. Furthermore, the response of the system seems to depend critically on the driving stress function. Very low distortion sine waves must be used if information regarding the linearity of the attenuation mechanism is to extracted. Harmonic distortion in the driving function can lead to spurious cusps in the stress-strain curve that can appear to be diagnostic of a nonlinear mechanism.

The papers included in this chapter were chosen to represent many of the experimental methods described above. In cases where investigators have published several papers presenting attenuation data, we have usually chosen the work that best illustrates the technique used. The first paper by *Peselnick and Outerbridge* (*1961*) describes a torsional pendulum device and also discusses the frequency dependence of attenuation in Solenhofen limestone. An extensive study of attenuation and velocity in saturated and partially saturated porous media utilizing both resonant bar and pulse methods is presented in the second paper by *Wyllie et al* (*1962*). Continuing this work, *Gardner et al* (*1964*) measured attenuation and velocity as a function of both confining pressure and fluid saturation using resonant bars at about 30 kHz in Berea sandstone and some unconsolidated sands. *Gordon and Davis* (*1968*) describe both the composite piezoelectric resonator method and the system driven off resonance with attenuation measured from stress-strain hysteresis. Attenuation is measured as a function of strain amplitude and confining pressure.

Spetzler and Anderson (*1968*) presented some very interesting data on velocity and attenuation in an ice-brine system illustrating the effects of phase changes. The technique employed is a resonant bar. In the paper by *Tittmann* (*1977*), laboratory results concerning highly outgassed rocks as a function of confining pressure are discussed in relation to Q models of the crust. A unique resonance apparatus employing end loading to lower the resonance frequency for flexural mode vibrations is described. Tittmann is presently using the method for saturated sedimentary rocks with great success.

The next two papers discuss the use of stress-strain hysteresis loops to obtain attenuation. *McKavanagh and Stacey* (*1974*) present an experimental technique utilizing a capacitance displacement transducer and cyclic axial loading. The presence of cusps in the hysteresis loops at strain amplitudes of 10^{-5} suggest a nonlinear attenuation mechanism at these high amplitudes. However, *Brennan and Stacey* (*1977*), using a more sensitive torsion experiment, report purely elliptical loops at strain amplitudes of 10^{-6} indicating a linear mechanism. Peselnick et al (1979) and Liu and Peselnick (1979) present similar data and also conclude that the attenuation mechanism is linear at low strain amplitudes.

In the paper by *Toksöz et al* (*1979*), a modification of the spectral ratio method is used for determining the attenuation of ultrasonic waves as a function of pressure for both dry and saturated rocks.

Winkler et al (*1979*), utilizing a resonant bar technique, investigate the strain amplitude dependence of attenuation in terms of a friction versus fluid flow mechanism. Finally, *Winkler and Nur* (*1979*) present very interesting data on the effects of fluid saturation that may prove to be an important diagnostic tool for petroleum exploration.

Interest in the experimental determination of attenuation has grown dramatically in the past few years as is evident from the publication dates of the papers in this chapter. It would be difficult to include all of the recent work done in the field; thus at the end of this chapter we have attempted to summarize the major results of the laboratory work done to date.

REFERENCES

Brennan, B.J., and Stacey, F.D., 1977, Frequency dependence of elasticity of rock—Test of seismic velocity dispersion: Nature, v. 268, p. 220–222.

Browne, M.T., and Pattison, J.R., 1957, The damping effect of surrounding gases on a cylinder in longitudinal oscillations: Brit. J. Appl. Phys., v. 8, p. 452–456.

Gardner, G.H.F., Wyllie, M.R.J., and Droschak, D.M., 1964, Effects of pressure and fluid saturation on the attenuation of elastic waves in sands: J. Petr. Tech., February, p. 189–198.

Gladwin, M.T., and Stacey, F.D., 1974, Anelastic degradation of acoustic pulses in rock: Phys. Earth Plan. Int., v. 8, p. 332–336.

Gordon, R.B., and Davis, L.A., 1968, Velocity and attenuation of seismic waves in imperfectly elastic rock: J. Geophys. Res., v. 73, p. 3917–3935.

Jackson, D.D., 1969, Grain boundary relaxations and the attenuation of seismic waves: Ph.D. thesis, M.I.T.

Kê, T.S., 1947, Experimental evidence on the viscoelastic behavior of grain boundaries in metals: Phys. Rev., v. 25, p. 533.

Kingery, W.D., 1959, Property measurements at high temperature: New York, Wiley.

Knopoff, L., 1956, The seismic pulse in materials possessing solid friction, 1, Plane waves: SSA Bull., v. 46, p. 175–183.

Kolsky, H., 1953, Stress waves in solids: London, Oxford Univ. Press; also 1963, New York, Dover Publications.

Liu, H.P., and Peselnick, L., 1979, Mechanical hysteresis loops of an anelastic solid and the determination of rock attenuation properties: Geophys. Res. Lett., v. 6, p. 545–548.

McKavanagh, B., and Stacey, F.D., 1974, Mechanical hysteresis in rocks at low strain amplitudes and seismic frequencies: Phys. Earth Plan. Int., v. 8, p. 246–250.

McSkimin, H.J., 1956, Propagation of longitudinal waves and shear waves in cylindrical rods at high frequencies: J. Acoust. Soc. Am., v. 28, p. 484–494.

Peselnick, L., Liu, H.P., and Harper, K.R., 1979, Observations of details of hysteresis loops in Westerly granite: Geophys. Res. Lett., v. 6, p. 693–696.

Peselnick, L., and Outerbridge, W.F., 1961, Internal friction in shear and shear modulus of Solenhofen limestone over a frequency range of 10^7 cycles per second: J. Geophys. Res., v. 66, p. 581, 588.

Peselnick, L., and Zietz, I., 1959, Internal friction of fine grained limestones at ultrasonic frequencies: Geophysics, v. 24, p. 285–296.

Quimby, S.L., 1925, On the experimental determination of the viscosity of vibrating solids: Phys. Rev., v. 25, p. 558–573.

Ramana, Y.V., and Rao, M.V.M.S., 1974, *Q* by pulse broadening in rocks under pressure: Phys. Earth Plan. Int., v. 8, p. 337–341.

Ricker, N., 1953, The form and laws of propagation of seismic wavelets: Geophysics, v. 18, p. 10–40.

Schreiber, E., Anderson, O.L., and Soga, N., 1973, Elastic constants and their measurement: New York, McGraw-Hill Book Co., Inc.

Spetzler, H., and Anderson, D.L., 1968, The effect of temperature and partial melting on velocity and attenuation in a simple binary system: J. Geophys. Res., v. 73, p. 6051–6060.

Spinner, S., and Tefft, W.E., 1961, Method for determining mechanical resonance frequencies and for calculating elastic moduli from these frequencies: Am. Soc. Test. Mat. Proc., v. 61, p. 1221–1238.

Tittmann, B.R., 1977, Internal friction measurements and their implications in seismic *Q* structure models of the crust: AGU Geophys. monogr. 20, The earth's crust, p. 197–213.

Toksöz, M.N., Johnston, D.H., and Timur, A., 1979, Attenuation of seismic waves in dry and saturated rocks: I. Laboratory measurements: Geophysics, v. 44, p. 681–690.

Truell, R., Elbaum, C., and Chick, B., 1969, Ultrasonic methods in solid state physics: New York, Academic Press.

Truell, R., and Oates, W., 1963, Effect of lack of parallelism in sample faces on measurement on ultrasonic attenuation: J. Acoust. Soc. Am., v. 35, p. 1382.

Ward, R.W., and Toksöz, M.N., 1971, Causes of regional variation of magnitude: SSA Bull., v. 61, p. 649–670.

Watson, T.H., and Wuenschel, P.C., 1973, An experimental study of attenuation in fluid saturated porous media, compressional waves and interfacial waves: Presented at the 43rd Annual International SEG Meeting, October 24, in Mexico City.

Winkler, K., and Nur, A., 1979, Pore fluids and seismic attenuation in rocks: Geophys. Res. Lett., v. 6, p. 1–4.

Winkler, K., Nur, A., and Gladwin, M., 1979, Friction and seismic attenuation in rocks: Nature, v. 277, p. 528–531.

Wyllie, R.J., Gardner, G.H.F., and Gregory, A.R., 1962, Studies of elastic wave attenuation in porous media: Geophysics, v. 27, p. 569–589.

Zener, C., 1948, Elasticity and anelasticity of metals: Chicago, Univ. of Chicago Press.

Reprinted from Journal of Geophysical Research, v. 66, p. 581–588.

Internal Friction in Shear and Shear Modulus of Solenhofen Limestone over a Frequency Range of 10^7 Cycles per Second

Louis Peselnick and W. F. Outerbridge

U. S. Geological Survey, Washington 25, D. C.

Abstract. The internal friction in shear and modulus of rigidity of dry Solenhofen limestone has been investigated over a frequency range from 4 cps to 10 Mc/s at room temperature. The results found are: (1) The rigidity modulus is constant ($U = 2.64 \times 10^{11}$ dynes/cm²) to within ±2 per cent over the total frequency range covered, provided that the samples have the same density. (2) The shear internal friction (as measured by the logarithmic decrement) in the cycle-per-second frequency range is about a factor of 5 lower than the internal friction in the megacycle frequency range; the logarithmic decrement at 4 cps = 3.4×10^{-3}, the logarithmic decrement at 10^7 cps = 17×10^{-3}. (3) The shear internal friction in the infrasonic frequency range increases by 18 per cent with the application of a 7.2-kg/cm² static axial tensile stress, but no large change in the internal friction occurs for axial compressive stresses of the same magnitude. (4) The shear internal friction is strain-dependent even for strains as small as 10^{-6}, a static axial tensile stress being superposed on the dynamic torsional stress.

Introduction. Previous work on internal friction of Solenhofen limestone [*Peselnick and Zietz*, 1959] has shown a linear dependence of the shear and dilatational absorption coefficient with frequency in the range 3 to 10 Mc/s and no variation in the modulus of rigidity with frequency within 4 per cent. The purpose of the present work is to extend the frequency range of the measurements of the shear internal friction and the elastic modulus for Solenhofen limestone. The frequency range below 500 cps is of particular interest in geophysics because the higher-frequency components of stress waves are severely attenuated in relatively short distances.

A torsion pendulum was constructed that enabled measurements to be taken of internal friction and the rigidity modulus of coherent rocks at frequencies of the order of 1 to 10 cps, and a resonance device was constructed on the basis of a design by *Kê* [1957] for measurements of the elastic constants and internal friction in rocks at frequencies from 8000 to 30,000 cps. Elastic properties and the internal friction at ultrasonic frequencies were determined with the apparatus described by *Peselnick and Zietz* [1959].

Experimental apparatus and procedure. The torsion pendulum method for obtaining the rigidity modulus and the internal friction of materials has been discussed in detail by *Cottell,*

Entwhistle, and Thompson [1948] and by *Jensen* [1952].

Figure 1 is a mechanical drawing of the torsion pendulum. The test specimen (1) has a diameter of 1/16 to 1/4 inch and is about 6 inches long. One end of the specimen is clamped by the collet (2), which is press-fit and bolted to the frame. The other end is connected by means of another collet to the inertia members, which consist of a pair of weights (3), a threaded bar (4), and a centerpiece (5). The center pin (6) and pivot bearing (7) ensure that the oscillations are entirely in the torsional mode, and permit the pendulum to be used in an inverted position without danger of the specimen's being broken by severe bending. Nontorsional modes of vibration in the specimen occur whenever the apparatus is operated without a pivot bearing. The contribution to the internal friction that these extraneous vibrations make is quite variable, frequently amounting to as much as 10^{-4}. Therefore it is desirable to operate with a bearing. The pivot bearing (7) is a spring-loaded sapphire that can be positioned both horizontally and vertically to compensate for any tendency of the center pin to be forced against the side of the conical surface. Three pins (8) mounted in a supporting frame fit into an annular groove in the bottom of the center-piece (9), and an adjustable spring clamp (10)

13

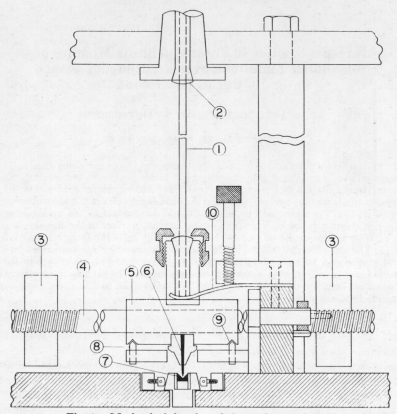

Fig. 1.   Mechanical drawing of the torsion pendulum.

holds the centerpiece against the mounting pins so that the pendulum can be securely clamped. By this system of clamping, the apparatus can be inverted without damage to the sample, pivot bearing, or needle.

The differences between our pendulum and those of Cottell and Jensen are: (a) An ionization tube is used for detecting the oscillations of the torsion pendulum through changes in electrical capacity. (b) A sapphire V jewel bearing and a carpet sewing needle serve as the center bearing and pin, respectively. (c) Three movable pins are used to adjust the position of the center bearing for minimum friction. (d) Miniature collet lathe chucks serve as specimen grips. (e) The inertia member consists of two equal weights placed equidistant from the axis of the pendulum. (f) An electronic counter measures the period of the oscillations. (g) An oscilloscope camera is used for photographing the decay curve displayed on an oscilloscope. (h) The sample, which supports the inertia member, may be changed to be in either uniaxial

tension or compression by inverting the pendulum. A perspective drawing of the apparatus is shown in Figure 2.

The pendulum rests on a 1-inch-thick marble slab, which in turn rests on a cemented construction of cinder blocks. External noise, predominantly vibrations from automobile traffic, prevented us from obtaining data for strains much less than 10^{-6} cm/cm. Operation of the torsion pendulum in a vibration-free location is therefore important to the making of small strain-amplitude measurements.

The apparatus is enclosed in a bell jar and operated from atmospheric pressure to 100 microns of mercury. A wire cage over the bell jar protects the experimenters against implosions and also serves to shield the detector probe from external 60-cps magnetic fields. Oscillations of the pendulum are started by passing current through the air-core coil C placed near one of the soft-iron weights (Fig. 2, W_2), thereby causing an initial deflection. The decay of the free oscillations is recorded

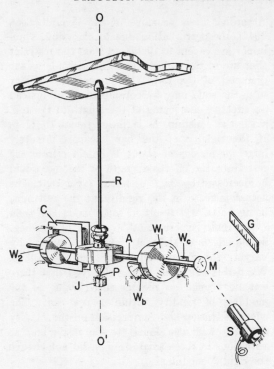

Fig. 2. Perspective drawing of the torsion pendulum. *OO'*, axis of torsional oscillation; *R*, specimen having length *L* between chucks; *C*, air core coil; *W₁* and *W₂*, soft-iron inertia weights; *A*, stainless-steel threaded inertia arm; *J*, sapphire V jewel bearing; *P*, carpet sewing needle; *M*, mirror; *S*, light source; *G*, scale.

after the current is turned off. Except for the soft-iron weights and wire cage, permanent magnets and magnetic materials were avoided in order to reduce losses from magnetic damping. The frequency of the oscillations is varied by symmetrically changing the positions of the inertia elements W_1 and W_2 with respect to the axis of torsion and also by substituting other weights of different mass.

The decay curve is recorded by converting the mechanical oscillations to an equivalent electrical output by means of a commercially available ionization transducer [*Lion,* 1956]. The device works on the principle that a small change in capacity introduced on the electrodes of a gas tube in the presence of an r-f field will result in a d-c voltage on those electrodes which is directly proportional to this change in capacity. As shown in Figure 3, the components C_1 and C_2 represent two physical capacitances with W_1 as the reference potential. These capacitances

C_1 and C_2 are connected in series and are applied to the ionization transducer. Mechanical oscillation of the weight W results in a differential change in capacity that is linearly recorded by the ionization transducer.

The output of the ionization gage is amplified by a differential amplifier (Tektronix, type 122). The amplified signal is then connected to an oscilloscope and an electronic frequency counter. The decay of the mechanical oscillations presented on the screen is photographed, and the period of the oscillations is determined by the counter, using ten period averages.

Deflections on the oscilloscope were calibrated in terms of the surface strain of the specimen as determined by an optical lever arrangement. The detecting equipment was checked for linearity by observing the oscilloscope deflection against the deflection on the galvanometer scale of the optical system. For the range of strains used, the oscilloscope output is found to be proportional to the strain. The computed strain was the maximum strain or surface strain of the specimen equal to $r\theta/L$, where r is the radius of the specimen, θ is the angle of twist in radians, and L is the length of the specimen.

The data in the kilocycle region were obtained with conventional bar resonance techniques [*Kê,* 1957]. The apparatus consists essentially of two phonograph pickup cartridges from which a sample in the form of a long, thin rod is suspended by fine wires placed close to the ends of the rod. For high-loss specimens such as Solenhofen limestone, no change in the experimental value of the logarithmic decrement could be detected when the specimen was supported

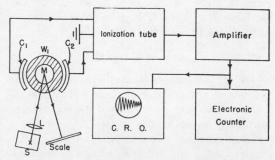

Fig. 3. Block diagram of detecting equipment. *L*, lens; *S*, light source; C_1 and C_2, capacitor plates; W_1, one of the soft-iron inertia weights serving as a ground for the capacitor system; *M*, mirror.

either at the resonant nodes or at the ends. One cartridge serves as a driver, the other as the detector. The elastic constant (s) and internal friction are determined from the resonant frequencies and the two frequencies corresponding to the half-power points on either side of a given resonant frequency.

For both the bar resonance and torsion pendulum methods, measurements were made with the equipment and specimens in an evacuated atmosphere. The amount of water saturation in the samples for these measurements was consequently negligible.

Calculations and accuracy of the torsion pendulum low-frequency experimental method. The torsion pendulum, treated as a lumped system having inertia I, restoring torque constant K, and internal friction constant R, has for its equation of motion of free torsional oscillations

$$I\ddot{\theta} + R\dot{\theta} + K\theta = 0$$

where θ is the angle of twist of the rod. The solution for $R/(2I) \ll K/I$ is

$$\theta \simeq \theta_0 \cos \omega_0 t \exp(-Rt/2I) \qquad (1)$$

where

$$\omega_0^2 = K/I = (2\pi f_0)^2 \qquad (2)$$

The decay term, $\exp(-Rt/2I)$, may be written as $\exp(-\delta f_0 t)$, the logarithmic decrement δ frequently being used as the measure of internal friction. For losses as large as $\delta = 0.01$ (which were never exceeded in these measurements) the ratio of $R/(2I)$ to K/I is about 10^{-5}, so that the approximate solution is permissible.

The restoring torque constant K for a solid circular shaft of diameter d, length L, and modulus of rigidity U is given from the theory of elasticity [*Southwell*, 1941] as

$$K = \pi\, d^4 U/32L \qquad (3)$$

where it is assumed that the applied torque consists of circumferential stresses that increase linearly from zero at the axis of the rod to some maximum value at the surface. This is not a valid assumption when the ends of the rod are clamped, or if the rod is in tension or compression, so that an estimate of the error involved is required.

The effect of the end restraints is to stiffen the rod torsionally at and near the restraints.

According to a principle of St. Venant [*Love*, 1944], the perturbation due to clamping is unimportant except at distances from the restraint that are of the same order of magnitude as the diameter of the rod. The increase in rigidity near the ends may be considered to be instead an effective shortening of the length of the rod, since the constant K is proportional to U/L. If the rigidity of the bar becomes infinitely large for regions of about 1 diameter from the restraints, the effective length of the rod would be decreased by $2d/L$ or about 7 per cent. The actual increase in the rigidity at the restraint, being finite, will result in some value less than 7 per cent. Experimental evidence of the negligible effects of the restraints was obtained by varying the torque adjustment on the collets. When this was done, it was observed that there was no change on the measured value of the modulus of rigidity within 1 per cent. The effect of tension and compression on the rigidity modulus was also found to be negligible, as variations in these axial stresses did not change the rigidity modulus.

Returning to the simple pendulum, the modulus of rigidity U can be determined from the measurements of the rod dimensions, the moment of inertia I, and the resonant frequency f_0 by combining equations 2 and 3:

$$U = 128\pi I L f_0^2/d^4 \qquad (4)$$

The logarithmic decrement δ is determined by measuring the time Δt required for the amplitude to decay to θ_0/e. Substituting $\theta = \theta_0/e$ into the decay expression

$$\theta = \theta_0 \exp(-\delta f_0\, \Delta t)$$

gives

$$\delta = (f_0\, \Delta t)^{-1}$$

The validity of the lumped-components assumption depends on the proper design of the apparatus. The moment of inertia of the frame of the pendulum with respect to the axis OO' is very much greater than the moment of inertia of members 3, 4, and 5 of Figure 1. The very small moment of inertia of the specimen and the collet chucks can be neglected, because their dimensions and mass are small in comparison with those of the other members of the pendulum. At the same time, since the ratio

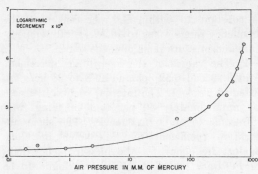

Fig. 4. Logarithmic decrement vs. air pressure for duraluminum.

of the diameter to length of the specimen is small with respect to the diameter/length ratios of the other members, the shear strain of the specimen rod will be much greater than the shear strain in other members, all other rigidities in the system being of the same order of magnitude.

The decay of the oscillations of the pendulum is a result not only of the torsional (shear) internal friction of the sample but also of the external sources of friction such as air damping, losses at the supports or grips of the specimen, losses in the center pin bearing, losses due to nontorsional vibrations in the specimen, and losses due to the motion of the frame of the pendulum. These external losses must be determined, or reduced to a value much smaller than the internal friction of the specimen.

Air losses were determined by operating the pendulum at atmospheric pressure down to a pressure of 100 microns of mercury. Figure 4 shows the internal friction expressed in logarithmic decrement (δ) vs. air pressure for a duraluminum specimen. The air damping is taken as the difference in the internal friction between the low-pressure limit and the internal friction at 1 atmosphere, in this case about 2.1×10^{-4}. The strain amplitude was maintained at a value of less than 10^{-4} cm/cm.

The torque adjustment of the grips on Solenhofen limestone, duraluminum, and steel drill rod was varied, and the losses were found to be independent of this adjustment above 5 foot-pounds. Although this assures constant loss at the grips for any particular specimen, it does not ensure zero losses at the grips. It is assumed that grip losses are not very different for different specimens.

The bearing losses were determined by taking the difference between the internal friction of the duraluminum specimen under tension with the center pin in a cup of mercury and the internal friction of the same specimen with the center pin in the sapphire bearing. With the center pin in the mercury cup, flexural modes of vibration which were initially predominant were attenuated more rapidly than the torsional vibrations, so that a minute or so after the pendulum is set into oscillation nearly all the vibrations are in the pure torsional mode. The bearing friction was found to be 2×10^{-5}.

The parts of the frame are securely bolted together. Since the moment of inertia of the frame is of the order of 100 times that of the pendulum, it is unlikely that the motion of the pendulum will affect or be affected by the frame.

The uncertainty of the absolute value of internal friction in Solenhofen limestone can be estimated from the measurements in Solenhofen limestone and duraluminum. The indicated value of the logarithmic decrement for Solenhofen limestone, excluding bearing and air losses, which will be called δ_s, is about 37×10^{-4}. The same quantity for duraluminum, δ_d, is 4.1×10^{-4}. Both δ_s and δ_d are composed of the true logarithmic decrement of the specimen plus a decrement, δ_x, representing unaccounted-for external losses in the apparatus, i.e., external losses other than those due to air damping and bearing friction. The true logarithmic decrement representing the internal friction for Solenhofen limestone and duraluminum is the indicated value minus the quantity δ_x:

$$\text{true} \quad \delta_s = 37 \times 10^{-4} - \delta_x \qquad (5)$$

$$\text{true} \quad \delta_d = 4.1 \times 10^{-4} - \delta_x \qquad (6)$$

Because the unknowns are positive quantities, the values of δ_x is in the range $0 < \delta_x < 4.1 \times 10^{-4}$. This allows the true logarithmic decrement for Solenhofen limestone to be written as

$$\text{true} \quad \delta_s = (35.0 \pm 2.0) \times 10^{-4}$$

which represents an uncertainty of about ± 6 per cent. The precision of δ_s was approximately ± 5 per cent.

A series of measurements taken on a duraluminum rod shows that the rigidity modulus can be measured to a precision of 0.5 per cent. The accuracy of the rigidity modulus is determined by considering the uncertainties in the quantities on the right side of equation 4. The moment of inertia I is calculated from measurements of weight and dimension to better than 0.1 per cent; L is measured to the nearest half millimeter and is accurate to 0.36 per cent; the frequency f_0 is measured to 1 part in a million; the diameter d is measured with a micrometer, and an average of ten or so measurements is used. The accuracy of the measurement, or constancy, or both, of d effectively controls the accuracy of the measurement of the rigidity modulus. Since the error in the value of d is ± 0.5 per cent, this contributes ± 2 per cent error to the value of U. If all the variables are taken with their minimum accuracy, the absolute value of U could be in error by about ± 2.5, the end effect notwithstanding.

Preparation of samples. Solenhofen limestone rods having a diameter of 3/16 inch and a length of 7 inches were made by first coring limestone rods 1/4 inch or more in diameter. The core was then placed in a centerless grinder and ground until the diameter reached the desired dimension. Close tolerance on the constancy of diameter is important for accuracy of the shear rigidity determination, as was mentioned above.

Results. The results for Solenhofen limestone at frequencies from 4 cps to 10 Mc/s are listed in Table 1.

1. The modulus of rigidity of samples of Solenhofen limestone having the same density

is constant to within ± 2 per cent over the frequency range 4 cps to 10 Mc/s, the temperature and pressure remaining the same.

2. The values of the logarithmic decrement δ at 3.89 and 6.52 cps and at 8 to 28 kc/s are listed in Table 1. The internal friction increases by approximately 50 per cent in going from infrasonic to audio frequencies. The shear absorption coefficient for Solenhofen limestone against frequency for the high-frequency data of *Peselnick and Zietz* [1959] was found to be linear with frequency and may be represented approximately by $\alpha_s = 5 \times 10^{-7} f$, where α_s is in decibels per centimeter and f is in cycles per second.

The logarithmic decrement δ is related to the shear absorption coefficient by

$$\alpha_s = 8.686 \, \delta f / C \qquad (7)$$

where C is shear velocity in centimeters per second, so that the high-frequency data can be expressed as

$$\delta = 5 \times 10^{-7} C / 8.686 \qquad (8)$$

or about 17×10^3. Therefore the internal friction is increased at most by a factor of 5 from the infrasonic to the ultrasonic frequencies.

3. Figure 5, curve I, gives the experimentally determined values of internal friction as measured by the logarithmic decrement for Solenhofen limestone sample B at 3.89 cps, where the abscissa is the lapse of time in days of the sample under a tensile stress. The maximum shear

TABLE 1.   Results on Solenhofen Limestone

Sample	Density, g/cm²	Frequency, cps	Rigidity Modulus, dynes/cm²	Shear Velocity, km/s	Logarithmic Decrement, δ	Remarks
B	2.67	3.89	2.58×10^{11}	3.1	3.4–4.1×10^{-3}	(a)
A	2.67	6.52	2.59	3.1	4×10^{-3}	(b)
C	2.67	8,200	2.66	3.15	5	(c)
C	2.67	16,400	2.66	3.15	5	(c)
B	2.67	9,500	2.68	3.17	5	(c)
B	2.67	18,600	2.61	3.12	5	(c)
B	2.67	28,500	2.69	3.18	$\cdots$	(c)
S-25	2.66	10×10^6	2.6	3.1	$\cdots$	(d)
S-1	2.59	9×10^6	2.2	2.9	17	(d)

(a) 7.2 kg/cm² axial tension; maximum shear strain $\sim 7 \times 10^{-6}$ (precision in $\delta = \pm 5$ per cent).
(b) 4.3 kg/cm² axial compression; maximum shear strain $\sim 5 \times 10^{-6}$ (precision in $\delta = \pm 5$ per cent).
(c) Bar resonance technique (precision in $\delta = \pm 10$ per cent).
(d) Pulse-echo technique (precision in $\delta = \pm 15$ per cent).

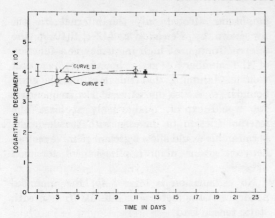

Fig. 5. Static axial loading effect on the internal friction of Solenhofen limestone. Curve I shows that the internal friction increases with the application of a 7.2-kg/cm² axial tensile stress to sample B. Curve II shows that the internal friction is either constant or is slightly decreasing with the application of a 4.3-kg/cm² axial compressive stress to sample A. The frequency and maximum shear strain for both samples were about 4 cps and 5 to 7×10^{-6} cm/cm, respectively.

strain amplitude for each measurement was 7×10^{-6} cm/cm. A uniaxial tensile stress of 7.2 kg/cm² was present, owing to the inertia weights. The internal friction was initially 3.4×10^{-3}. Measurements taken as a function of time showed rapid increase of internal friction 3 and 4 days after the original measurement, followed by a much slower increase thereafter. A limiting value of internal friction was reached after about 11 days. The total increase in internal friction for the 11-day period, with a 7.2-kg/cm² tensile stress, amounted to 18 per cent. There was an indication during this period of a slight decrease in the rigidity modulus under tensile load, but the change was smaller than the error of the measurement.

After 11 days the specimen was removed from the torsion pendulum for 1 day. A subsequent measurement the following day (dashed portion of curve I) showed no large change in the value of internal friction just before removal.

A new sample, A, under a 4.3-kg/cm² compressive stress, showed the internal friction to be either constant or slightly decreasing during a 2-week period (see curve II, Fig. 5).

4. Figure 6, curves I and II, shows the internal friction plotted against the maximum shear strain under a tensile stress of 7.2 kg/cm² for

the Solenhofen limestone specimen B. Curve II, a repeat run, was taken 3 days after curve I. The internal friction for both curves increases approximately linearly with strain. The internal friction is strain-amplitude-dependent, even for strains as low as 10^{-6} cm/cm. However, curve II is approximately 7 per cent higher than curve I, which is a consequence either of the time dependence of internal friction under axial tension or of the previously applied large strains, or of both. No observable change in the rigidity modulus from 2.58×10^{11} could be detected as the strain was varied.

Discussion. The shear velocity for dry Solenhofen limestone at 25°C and under the aforementioned axial stresses as calculated from the rigidity at 4 cps is the same as the velocity determined in the megacycle frequency range, provided that the samples have the same density. This result is important to seismic prospecting when it is considered that most laboratory velocity measurements of rocks are made using either the resonance method or the pulse-echo technique, and that the frequencies employed for these tests are much greater than the frequencies in the seismic range.

These results support the validity of using high-frequency elastic data for seismic application, at least for homogeneous and well-compacted rocks. This in no way rules out the effects of pressure and temperature on the

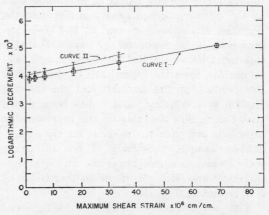

Fig. 6. Internal friction versus maximum shear strain for Solenhofen limestone. The internal friction increases linearly with strain. Curve II is a repeat run taken 3 days after curve I and shows a constant increase in internal friction of about 7 per cent.

velocity, which *Hughes and Cross* [1951] and *Birch and Bancroft* [1938a and b], to mention but a few, have shown to be of considerable importance.

The lower velocity of sample S-1, compared with the velocities of the other samples, is a consequence of lower rigidity for this sample, since one would expect, solely on the basis of a lower density, an increase in the velocity ($C^2 = U/\rho$), where ρ is the density.

The shear internal friction, δ, increases at most by a factor of 5 from infrasonic to ultrasonic frequencies. The assumption of a linear dependence of the absorption coefficient, α_s, with frequency, however, is wholly justified for a small frequency range. The justification is even better at the lower range of frequencies where the frequency has to be increased by a factor of several thousand in order to increase the internal friction by 50 per cent. A phenomenological model has been proposed by *Knopoff and MacDonald* [1960] having a frictional force proportional to the absolute magnitude of the displacement. This results in an absorption coefficient which depends linearly on frequency. For the frequency range of seismic prospecting (of the order of 1 to 500 cps), one can be especially confident of the solid-hysteresis type of behavior of Solenhofen limestone near atmospheric pressure and temperature.

The factor of 5 of internal friction at the ultrasonic and infrasonic frequencies may be too high, because the density of the sample used at the ultrasonic frequencies was less than the densities of the samples used at the lower frequencies. It is reasonable to expect that the internal friction for the less-dense sample will be greater, from the general rule of thumb that the product of the modulus and the internal friction is a constant; and it has been shown that the rigidity modulus for the lower-density sample is less than the modulus for the higher-density sample. Scattering of the ultrasonic wave, which becomes important at higher frequencies, may also contribute to this factor of 5 in the internal friction.

The low-frequency results of static axial tension and compression show relatively no effect on the rigidity or shear velocity but appreciable effect on the internal friction for static axial tension. Similarly, the low-frequency results of the variation of the dynamic strain

amplitude affected only the internal friction. As shown by *Peselnick and Zietz* [1959], the internal friction at high frequency is a function of the number of grain boundaries per unit volume. Assuming that this loss (at the grain boundaries) is also operative at low frequencies, one would expect, qualitatively at least, the internal friction to increase with tensile loads, because this would allow a greater relative motion between grains upon application of torsional stress.

No explanation is offered for the time dependence of the increase of internal friction with static tensile load.

Acknowledgment. We wish to express our appreciation to Mr. Rudolph Raspet of the U. S. Geological Survey for his assistance in the mechanical design of the torsion pendulum.

Publication of this paper is authorized by the Director, U. S. Geological Survey.

REFERENCES

Birch, Francis, and Dennison Bancroft, The effect of pressure on the rigidity of rocks, part I, *J. Geol.*, 46, 59–87, 1938a.

Birch, Francis, and Dennison Bancroft, The effect of pressure on the rigidity of rocks, part II, *J. Geol.*, 46, 113–141, 1938b.

Cottell, G. A., K. M. Entwhistle, and F. C. Thompson, The measurement of the damping capacity of metals in torsional vibration, *J. Inst. Metals*, 74, 374–424, 1948.

Hughes, D. S., and J. H. Cross, Elastic wave velocities at high pressures and temperatures, *Geophys.*, 16, 577–593, 1951.

Jensen, J. W., A torsion pendulum of improved design for measuring damping capacity, *Rev. Sci. Instr.*, 23, p. 397, 1952.

Ke, T. S., A study on the acoustic internal friction of iron vibrating transversely in a steady magnetic field by piezoelectric crystal plates, *Scientia Sinica*, 6, (2), 237–245, 1957.

Knopoff, L., and G. J. F. MacDonald, Models for acoustic loss in solids, *J. Geophys. Research.* 65, 2191–2197, 1960.

Lion, K. S., Mechanic-electric transducer, *Rev. Sci. Instr.*, 27, (4), 222–225, 1956.

Love, A. E. H., *A Treatise on the Mathematical Theory of Elasticity*, Dover Publications, New York, 4th ed., p. 132, 1944.

Peselnick, L., and I. Zietz, Internal friction of fine grained limestones at ultrasonic frequencies: *Geophysics*, 24, (2), 285–296, 1959.

Southwell, R. V., *An Introduction to the Theory of Elasticity for Engineers and Physicists*, Oxford University Press, 2d ed., p. 153 and footnote, 1941.

(Manuscript received November 29, 1960.)

Reprinted from Geophysics, v. 27, p. 569–589.

STUDIES OF ELASTIC WAVE ATTENUATION IN POROUS MEDIA*

M. R. J. WYLLIE,† G. H. F. GARDNER,† AND A. R. GREGORY†

Elastic wave attenuation in porous media is due in part to the relative motion of the liquid and the solid. Biot's theory expresses this component in terms of permeability, fluid viscosity, frequency, and the elastic constants of the material. Experiments were performed to measure attenuation in the frequency range $f < 20,000$ cps by a resonant bar method; attempts to measure attenuation at very high frequencies gave more equivocal results. Alundum bars were used to test the validity of the theory, for with these the loss not due to fluid motion is relatively small. Experiments were also made with natural specimens of rock. These showed that when not subjected to compacting pressure both the velocities and decrements of specimens were affected chemically and physically by the presence of liquid pore saturants.

It is concluded that Biot's theory seems generally applicable to the determination of the fluid-solid or "sloshing" losses in resonated porous media. There is still some doubt about the applicability of the theory in the case of measurements made by pulse techniques.

The use of attenuation measurements as a logging technique, possibly to estimate permeability, is also discussed.

NOMENCLATURE

Symbol	Quantity	Units
K	permeability	cm² (1 darcy$\sim$10⁻⁸ cm²)
ϕ	porosity	—
ρ_F	fluid density	gm/cc
μ	fluid viscosity	poise
E	structure factor ⎫ shape parameters in Biot's theory; see Biot	—
δ	geometrical factor ⎭ (1956), Wyllie et al (1961)	—
M_2	mass of fluid in saturated rock	gm
M_1	mass of dry rock	gm
S_w	saturation of wetting fluid	—
f_c	critical frequency $\mu\phi/2\pi\rho_F K$	cps
f	frequency	cps
Δ	logarithmic decrement	—
σ	Poisson's ratio	—
κ	$\delta(f/f_c)^{1/2}$	—
F	formation factor	—

INTRODUCTION

A number of measurements have been made of the attenuation of elastic waves in natural porous media. Knopoff and Macdonald (1958) reviewed the information then available to them. More recently, the results of two additional laboratory studies have appeared (Peselnick and Zietz, 1959; Peselnick and Outerbridge, 1960).

The available data do not lend themselves to empirical generalization or to comparison with theoretical expectations because no systematic attempts have been made to measure attenuation

---

* Presented at the 31st Annual SEG Meeting, Denver, Colorado, November 9, 1961. Manuscript received by the Editor January 19, 1962.

† Gulf Research and Development Co., Pittsburgh, Pennsylvania.

21

as a function of those parameters which might reasonably be assumed to influence the dissipation of the energy of elastic waves in rocks. This is not to say that existing information is of little value. This is far from being the case. A number of tentative conclusions may be reached on the basis of published data. Of these we deem the most important to be:

(1) Attenuation is affected by compacting pressure with an increase in pressure leading to a decrease in attenuation (Birch and Bancroft, 1938).

(2) The logarithmic decrement at zero differential pressure and very low water saturation may be the sum of two components, one of which is independent of frequency and one of which is proportional to frequency (Born, 1941).

(3) The logarithmic decrement in subsurface formations is variable but is possibly in the range 0.001 to 0.10 for frequencies below 20,000 cps. Data from several sources are given in Table 1. The possible upper limit is based on a statement by Blizard (1959) that the performance of CVL equipment indicates attenuations in situ to be less than 2 db/ft.

(4) At the frequencies characteristically used in seismic recording, the logarithmic decrement per cycle seems to be very nearly constant, i.e., the losses are mainly of the solid friction or "jostling" type (Table 1). Nevertheless, the accuracy of the field data does not exclude the possibility that a truer relationship might be, for example, of the form $\Delta = A + Bf$, where Δ = logarithmic decrement, f = frequency, and A and B are constants. Such a relationship is compatible with Born's measurements made on a moist sandstone.

(5) The theory of Biot (1956) for the propagation of elastic waves in fluid-saturated porous solids seems to provide an admirable framework on which to display the results of systematic measurements of attenuation. Biot's theory does not cover jostling losses but enables losses which result from the relative movement of fluid and solid ("sloshing losses") to be estimated.

(6) Rayleigh scattering and other losses which

may occur if very high-frequency energy is propagated through porous media may be excluded from consideration if attention is directed only to the frequency range zero to approximately 10 mc, i.e., the frequency range used in seismic prospecting, continuous velocity logging and many laboratory experiments.

It is realized that the validity of each of the above six items is not firmly established. The items are listed because they constituted the basis of the working hypothesis used to plan the experiments discussed below.

One main purpose of the present work was to obtain experimental data which could be used to check available theory and, particularly, the theory of Biot.

EXPERIMENTAL CONSIDERATIONS

Methods of measuring logarithmic decrement in rock samples fall into three main groups: (a) torsion pendulum methods, (b) resonant bar methods and (c) ultrasonic pulse methods. Each method is usually only suitable for making measurements over a comparatively small range of frequencies. Essentially, the first two are rather low-frequency methods and the third a rather high-frequency one. Since the adjectives defining frequency have real significance only in terms of the properties of the specimen being measured, it is perhaps preferable to note that:

method (a) is generally limited to frequencies less than 100 cps,

(b) is generally limited to frequencies less than about 30,000 cps, and

(c) is generally limited to frequencies in excess of about 50,000 cps.

The experimental problems involved in the use of these methods have been frequently discussed. The description of Niblett and Wilks (1960) is succinct.

In the present work, we have not had occasion to use method (a) but methods (b) and (c) have been rather extensively examined. Method (b), that of the resonant bar, has a number of important advantages. These are the relative simplicity of the apparatus required and the fact that longitudinal and torsional vibrations are easily generated. The drawbacks to the method lie in the interpretation of the so-called "bar" values which

Table 1. A selection of logarithmic decrements measured in the laboratory and in the field.

Source	Nature of Rock	Water Saturation %	Frequency cps	Logarithmic Decrement	Nature of Decrement	Reference
Laboratory	Amherst Sandstone	0	0–4,000	0.06	Extensional bar	Born (1941)
Laboratory	Amherst Sandstone	3.3	0–2,000	$0.057 + 0.2 \times 10^{-3}$ f	Extensional bar	Born (1941)
Field	Various, Oklahoma	100	45	0.041	Bulk Dilatational	Kendall (1941)
Field	Various, Oklahoma	100	37	0.023	Bulk Dilatational	Kendall (1941)
Field	Pierre Shale 250–750 ft	100	0–600	0.097	Bulk Dilatational	McDonal et al (1958)
Field	Pierre Shale 4,000 ft	100	~50	0.032 ± 0.016	Bulk Dilatational	McDonal et al (1958)
Field	Various, Oklahoma, Texas	100	20–50	0.013	Bulk Dilatational	Bardeen (1960)
Laboratory	Tennessee Marble	0	0–20,000	0.007	Extensional bar	Present work
Laboratory	Solenhofen Limestone	0	3–10 Mc	0.017	Bulk Shear	Peselnick and Zeitz (1959)
Laboratory	Solenhofen Limestone	0	3.59	0.005	Bulk Shear	Peselnick and Outerbridge (1960)

are obtained in terms of the required bulk properties. This drawback was discussed by Born (1941). Considerable progress in overcoming this difficulty has been made (Gardner, 1962). In our view, the most serious drawback to the resonant bar method lies in the inability of the technique to yield direct data for dilatational waves, and the greatest experimental difficulty is in obtaining reliable data when the specimen being vibrated is jacketed. Jacketed specimens are essential if net compacting pressures are to be applied. We have noted elsewhere our view (Wyllie, Gardner, and Gregory, 1961) that velocity measurements made on uncompacted samples of sedimentary rock are not representative of subsurface conditions. The same view, if anything a fortiori, we feel is applicable to attenuation measurements. This is also reflected in Point 1 of the introduction. We are, as yet, not confident of our ability to measure reliable decrements on specimens unchanged except for the addition of a jacket (or to allow for the presence of the jacket by appropriate calculation). We believe that the difficulties we have experienced are not insuperable. Recent results we have obtained indicate that these difficulties have been overcome. This belief is shared by Geertsma (1960). Geertsma gives no details of his technique.

Several ultrasonic methods have been explored. The most promising one is based on the use of a gated pulse and is in all essentials similar to that of Peselnick and Zietz (1958). Using this technique we have been able to use jacketed specimens and to measure attenuation as a function of net compacting pressure. (Results are qualitatively similar to those reported by Birch and Bancroft, 1938.) Nevertheless, the difficulty of accounting properly for geometric and other errors has not yet been overcome; hence we do not place quantitative reliance on decrements that we have measured. Indeed, our experience leads us to believe that even the data presented by Peselnick and Zietz are probably only qualitatively correct; they are likely to err on the high side. This may be one reason for the discrepancy between the shear decrements in Solenhofen limestone measured at very high frequencies and at a low frequency (Table 1). This discrepancy is the more marked when it is realized that the velocities measured at the same high and low frequencies agreed within the precision of the measurements. Measurements we have made of the torsional decrement of Solenhofen limestone agree with the low-frequency data of Peselnick and Outerbridge.

It was found in a study of elastic wave velocities in porous media (Wyllie, Gregory, and Gardner, 1958) that certain types of synthetic alundum cores were relatively free from pressure-induced velocity changes and analogous to natural rocks under high compacting pressure. Alundum cores can be obtained over a wide range of permeabilities and in several porosities. They are also reasonably uniform in their physical properties. Most important, they are almost inert chemically. This latter property is of primary importance in an experimental program and sharply distinguishes alundum from most natural sandstones. This essential difference between the properties of alundum and sandstone, particularly in relation to interaction with water, may be seen by comparing Figures 5 and 6 with Figures 7 and 8. The differences are further reviewed below. It will suffice to note here that in order to avoid complications which result from the apparent chemical interaction between liquids such as water and the solid constituents of many sandstones, the following experimental approach was adopted.

(1) Use the resonant bar method and alundum specimens to measure sloshing losses (Wyllie, Gardner, and Gregory, 1961) and compare these losses with those computed from the Biot theory, using as parameters in the theory the porosity, permeability, structure factor, and elastic constants of the cores and the viscosity and density of the contained fluid.

(2) If the validity of the Biot theory is confirmed, calculate the attenuation caused by sloshing losses in any core of natural rock. In order to make this calculation applicable to a rock under high compacting pressure (in the absence of marked chemical interaction between rock frame and pore saturant), it is necessary only to measure the parameters required by Biot's equation, and particularly the dilatational and shear velocities, under appropriate conditions of compaction. This is relatively easy to do.

It must be pointed out again that the solid friction or jostling losses in the frame can be measured only experimentally; there is no extant theory enabling them to be computed. The actual decrement of a core in situ will be the sum of the unknown jostling loss and the computed sloshing loss. Only if jostling losses under compaction in situ are generally small by comparison with sloshing losses will this backdoor method be of any immediate practical use. Some calculated sloshing decrements are given in Table 2.

Since the jostling decrement may be relatively independent of frequency (at least for rocks under pressure) while the sloshing decrement, according to Biot's theory, is a function of frequency, it is possible that at the frequencies used in continuous velocity logging (circa 25,000 cps), the latter losses will always dominate. On the other hand, at the low recording frequencies used in seismic prospecting, it is probable that jostling losses are the only ones of significance.

APPARATUS

The apparatus used to make measurements of resonant frequency and decrement was similar to that used by Born (1941). It was found convenient to cement a small, flat, ceramic permanent magnet to one end face of the cylindrical specimen; then by the use of an external electromagnet which could be turned through 90 degrees, either longitudinal or torsional vibrations could be induced. Bi-morphs cemented to the specimen at the end opposite to the magnet acted as pickups. The most satisfactory of the suspensions tried consisted of a fine wire equilateral triangle held in a circular ring so that the wire could be pulled taut around the central nodal circle of the rod. The fundamental mode of vibration was used to obtain all the data discussed here. The principal refinement incorporated in the equipment was to house it in a rectangular metal box sealed on its top face by a transparent, pliable film of plastic. The specimen itself could be picked up by two wires passing through small holes in the film and thus weighed at any time without removing it from the box. This feature, in conjunction with facilities to pass air through the box so as to control evaporation, was found almost essential if reliable measurements were to be made of decrement as a function of air/liquid saturation.

An apparatus to measure longitudinal and shear velocities in jacketed specimens capable of being subjected to controlled differences (up to 10,000 psi) between frame pressure and pore pressure is shown sectioned in Figure 1. This apparatus is based on the critical angle method discussed by Schneider and Burton (1949) and, more recently, by Sabbarao and Ramchandra Rao (1957). The specimens used were copper-jacketed with film 0.010 inch thick. Agreement between dilatational and shear velocities measured by the critical angle method, and by direct pulses generated by quartz crystals cut to accentuate the required modes, was reasonably satisfactory. Some comparisons are shown in Table 3. It is necessary to use an apparatus of the critical angle type because the unequivocal measurement of shear velocities by a pulse method is not always possible. The amplitude minima corresponding to the critical angles for total reflection of longitudinal and shear waves were generally well defined. The method, however, seems inherently to lack the precision which characterizes the measure-

Table 2. Velocity and viscous absorption data for laboratory samples.

Sample	Porosity, percent	Saturant	Pressure,* psi	V_d, ft/sec	V_s, ft/sec	Poisson's Ratio	Saturant Density, gm/cm²	Saturant Viscosity, cp at 20 °C	Permeability, millidarcys	f	f_c	f/f_c	Log. Decr. P-wave	Log. Decr. S-wave
Berea	19.0	Air	6,000	12,050	7,120	.231	0.00129	0.0183	150	25 kc	409 kc	0.061	.005	.0125
Berea	19.0	Oil	9,000	12,570	7,500	.224	0.75	1.5	150	25 kc	197 kc	0.127	.010	.038
Berea	19.0	Brine	6,000	13,120	8,630	.118	1.14	1.1	200	20 kc	156 kc	0.128	.010	.034
Berea	19.4	Water	6,000	13,190	7,980	.209	1.0	1.0	150	25 kc	239 mc	1.05×10^{-4}	$.52 \times 10^{-5}$	3.3×10^{-6}
Berea	18.8	Glycerol	8,000	14,000	9,250	.109	1.26	1,490	200	25 kc	189 mc	1.32×10^{-4}	—	4.38×10^{-5}
Berea	18.8	Glycerol	10,000	14,080	8,400	.224	1.26	1,490	200	10 kc	100 kc	0.100	.025	.097
Berea	19.8	CCl₄	6,000	12,800	7,700	.214	1.50	0.969	200	2.5 kc	17.8 kc	0.140	.139	.244
Berea	19.3	Mercury	7,000	10,810	7,050	.130	13.6	1.55	200	2.5 kc	17.4 kc	0.144	—	—
Berea	18.9	Mercury	5,000	11,130	6,890	.189	13.6	1.55	200					
Fox Hills	7.4	Air	6,000	14,600	8,250	.265	0.00129	0.0183	32.5	25 kc	735 kc	0.034	.0014	.0023
Fox Hills	7.4	Oil	6,000	15,000	8,650	.249	0.75	1.5	32.5					
Tensleep	15.3	Air	6,000	12,900	7,920	.197	0.00129	0.0183	220	10 kc	112 kc	.089	.006	.018
Tensleep	15.3	Water	6,000	13,800	8,150	.231	1.0	1.0	220	25 kc	133 kc	1.88×10^{-4}	1.86×10^{-5}	4.71×10^{-5}
Tensleep	15.3	Glycerol	6,000	14,500	8,400	.247	1.26	1,490	220					
Oriskany	3.3	Air	7,000	18,400	11,400	.185	0.00129	0.0183	.02	25 kc	267 mc	0.94×10^{-4}	4.4×10^{-6}	3.8×10^{-6}
Oriskany	3.3	Water	7,000	17,200	10,550	.199	1.0	1.0	.02					
Jelm	23.5	Air	4,500	11,570	7,100	.199	0.00129	0.0183	1,140	7.5 kc	66 kc	0.114	.013	.03
Jelm	23.5	Oil	4,500	12,050	7,250	.216	0.75	1.5	1,140					
Teapot	29.7	Air	4,600	10,000	6,120	.200	0.00129	0.0183	1,900	7.5 kc	50 kc	0.150	.043	.054
Teapot	29.7	Oil	7,000	11,000	6,190	.269	0.75	1.5	1,900					
Alundum 7919	25.7	Air	8,000	19,000	10,600	.274	0.00129	0.0183	448	12 kc	92 kc	0.130	.024	.033
Alundum 7919	25.7	Water	8,000	19,400	10,650	.284	1.0	1.0	448	25 kc	109 mc	2.3×10^{-4}	4.4×10^{-5}	7.12×10^{-5}
Alundum 7919	25.7	Glycerol	8,000	20,500	10,700	.313	1.26	1,490	448	7.5 kc				
Alundum 7919	25.7	CCl₄	6,000	19,200	10,700	.275	1.5	0.969	448	7.5 kc	60 kc	0.125	.0076	.046
Alundum 7919	25.7	Mercury	5,000	17,140	10,300	.217	13.6	1.55	448	1.5 kc	10.5 kc	0.143	.150	.246
Alundum Aggregate	~38	Air	5,000	6,520	4,250	.130	0.00129	0.0183	—	—	—	—	—	—
Alundum Aggregate	~38	Water	5,000	7,780	4,350	.272	1.0	1.0	—	—	—	—	—	—
Alundum Aggregate	~38	Brine	5,000	8,150	4,570	.270	1.14	1.1	—	—	—	—	—	—

* Differential pressure between rock frame & fluid saturant.

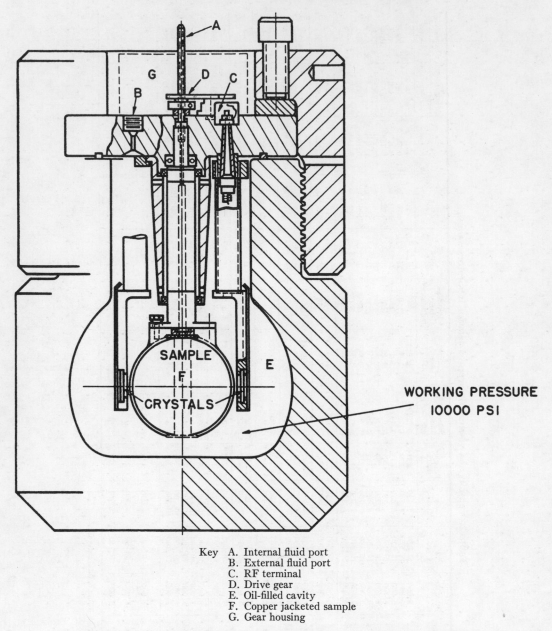

Key A. Internal fluid port
       B. External fluid port
       C. RF terminal
       D. Drive gear
       E. Oil-filled cavity
       F. Copper jacketed sample
       G. Gear housing

Fɪɢ. 1. Cross-section of sonic pressure chamber used for critical angle measurements.

ment of P-wave velocity by conventional pulse techniques. Two principal difficulties were experienced. The first, that of interference by internal reflections in the cell, was overcome by using pulsed energy and not continuous waves as advocated by the Indian workers. The second stems from the nonexistence of liquids which have low velocities, particularly when subjected to pressures of 5,000 psi and above.[1] Thus, it was not possible to measure the critical angle of reflec-

---

[1] The velocity of silicone oil, for example, increases from 3,320 ft/sec at zero psi to 3,890 ft/sec at 5,000 psi and 4,350 ft/sec at 10,000 psi (100 centistoke oil at 24° C).

Table 3. Comparisons of dilatational and shear wave velocities derived
from the critical angle and single pulse methods.

Sample	Porosity percent	Saturant	Critical Angle				Pulse				Percent Difference	
			Uniform Pressure, psi	V_d ft/sec	V_s, ft/sec	σ	Axial Pressure, psi	V_d, ft/sec	V_s, ft/sec	σ	from Pulse Method Value, V_d	from Critical Angle Method Value, V_s
Berea	19.0	Air	6,000	12,050	7,120	.231	4,250	11,200	7,050	.174	+ 7.6	−1.0
Tensleep	15.3	Water	6,000	15,900	8,470	.302	5,000	13,800	8,120	.235	+15.2	−4.1
Fox Hills	7.4	Air	6,000	14,600	8,250	.265	7,500	14,070	8,670	.193	+ 3.8	+5.1
Fox Hills	7.4	Oil	6,000	15,000	8,650	.249	6,500	15,660	8,360	.300	− 4.2	−3.3
Solenhofen Limestone	3.5	Air	9,000	18,600	9,670	.314	5,000	18,260	9,9:0	.288	+ 1.9	+2.9
Alundum 7919	25.7	Air	8,000	18,290	10,500	.253	6,500	18,970	10,740	.264	− 3.6	+2.3
Alundum 7919	25.7	Water	8,000	19,450	10,630	.287	6,500	19,300	10,690	.277	+ 0.8	+0.6
Alundum 7919	25.7	Glycerol	8,000	20,900	10,800	.317	6,500	20,150	10,690	.304	+ 3.4	−1.0

Notes:
(1) Comparisons could not be made at the same pressures, but it is assumed that velocities were near terminal values in each case.
(2) Exactly the same specimens were not used for different methods.
(3) It is probable that V_d is more reliable using pulse method and V_s is more reliable using critical angle method.

tion for shear waves in a packing of quartz grains of 35 percent porosity under a net compacting pressure of 5,000 psi. The shear velocity is presumably less than 3,890 ft/sec. This conclusion tends to be confirmed by the fact that the shear velocity through a similar packing of grains of aluminum oxide could be measured. The velocity of aluminum oxide is some 50 percent greater than that of quartz.

Figures 2 and 3 show typical curves illustrating the effect of compacting pressure on shear and dilatational velocities. Also plotted are Poisson ratios calculated from the velocities measured. It will be observed that only under rather high compacting pressures does the Poisson ratio reach a limiting value. Note, however, that the total increase in Poisson ratio caused by pressure is only about 10–15 percent.

In Figure 4 the terminal shear velocities in water-saturated sandstones, i.e. the rather constant velocities measured under high compacting pressures, have been plotted as a function of

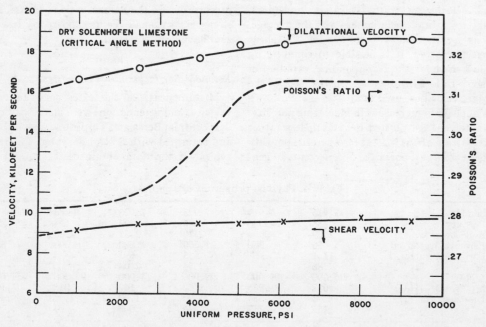

FIG. 2. Dilatational and shear-wave velocities as a function of pressure for dry Solenhofen limestone.

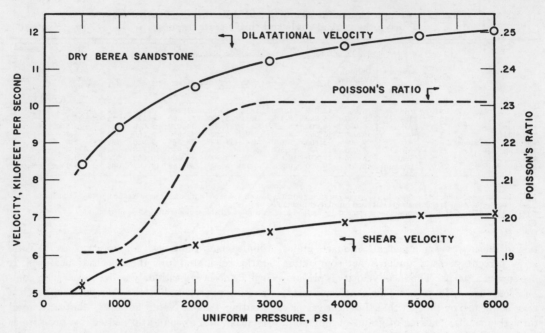

Fig. 3. Dilatational and shear-wave velocities as a function of pressure for dry Berea sandstone.

porosity. Also plotted are field data taken from Vogel (1952) logs, some measurements made by Bayuk (1960) on dry specimens under a pressure of 14,223 psi and the time-average plot of dilatational velocity in water-saturated sandstones. Using this average relationship and the average, also as a function of porosity, for the shear velocities plotted, it is possible to compute by classical methods the corresponding variation of Poisson's ratio with porosity. On the basis of the average velocities used, the average Poisson's ratio appears to increase with increasing porosity. The total change plotted is nevertheless rather small — from about 0.19 at zero percent porosity to about 0.27 at a porosity of 35 percent. In general, it seems that at any porosity the effect of pressure is to increase Poisson's ratio; and at any pressure an increase in porosity effects a similar increase. As shown in Table 4, at a compacting pressure of zero the Poisson ratio of natural sandstones is very small. Values apparently less than zero have been observed.

RESULTS

Resonant Bar Experiments: Sandstone Rods

Measurements of the extensional and shear velocity, and attenuation, were made with cylinders cut from Berea and Torpedo sandstone blocks and saturated with Soltrol or water. Figure 5 is typical of the data obtained; it shows $1/V^2_{S_w}$

Table 4. Physical properties of samples.

Sample	7,915	7,919*	7,923	7,928	Berea	Torpedo
Porosity, ϕ	.258	.266	.306	.392	.194	.230
Bulk Density	2.89	2.84	2.70	2.32	2.08	2.03
Permeability, K md	70	350	1,500	4,000	300	570
Formation Factor, F	11.1	—	8.3	5.3	—	—
ϕF	2.86	—	2.54	2.08	—	—
$f_c(H_2O)$	526,000	109,000	29,000	14,000	92,200	57,600
M_2/M_1+M_2 (H_2O)	.082	.086	1.02	.147	.085	.102
Diameter, cm	1.90	2.54	1.90	1.90	2.54	1.90
σ	.152	.117	.128	.068	.04	.04

* This sample had been subjected to an axial compression greater than 5,000 psi many times.

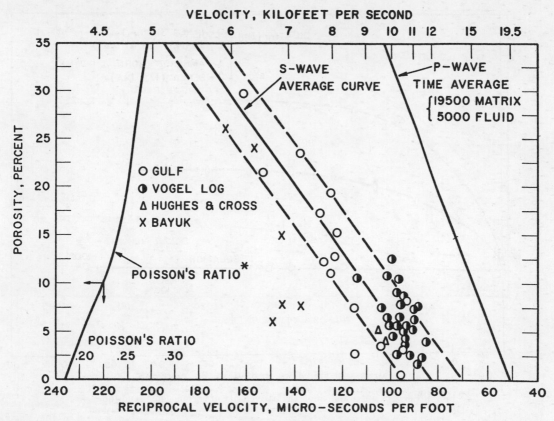

FIG. 4. Shear wave terminal velocities, P-wave time average and poisson's ratio shown as a function of porosity for water saturated sedimentary rocks.

* Derived from V_d & V_s taken from time average & S-wave average curves respectively.

versus $S_w M_2/M_1$. Here, V_{S_w} is the extensional velocity, M_1 the mass of the dry rod, M_2 the mass of the fluid required to saturate the pore volume of the rod, and S_w the fraction of the pore volume filled with the fluid. For comparison, Born and Owen's (1935) data for Amherst sandstone are also plotted.

If the addition of fluid changes the velocity only because of the concomitant change in bulk density, the velocity at saturation S_w can be predicted. Thus,

$$\frac{1}{V^2_{S_w}} = \frac{1}{V_0^2} \left[1 + S_w M_2/M_1 \right]. \qquad (1)$$

In equation (1), V_0 is the dry velocity.

A study of Figure 5 leads to the conclusion that three effects are often operative. The first, which may be primarily chemical in origin, is associated with the initial addition of fluid to the rod and gives a rapid increase in $1/V^2_{S_w}$. This effect was noted by Born and Owen and attributed by them to water weakening the intergranular cement and thereby lowering the Young's modulus of their bar. The magnitude of this effect depends on the fluid added. Thus, water may cause a decrease in velocity of over 50 percent whereas Soltrol (purified by passage through a column of silica gel) may cause a decrease of less than two percent.

The second effect is evident at intermediate saturations. The change in $1/V^2_{S_w}$ with $S_w M_2/M_1$ is linear. The slope of this line is in accord with equation (1) if $1/V_0^2$ is taken to be the value of $1/V^2_{S_w}$ when extrapolated linearly to $S_w=0$. The tacit assumption underlying the extrapolation is that the fluid when first added affects the mechanical properties of the frame but that further additions merely serve to load it.

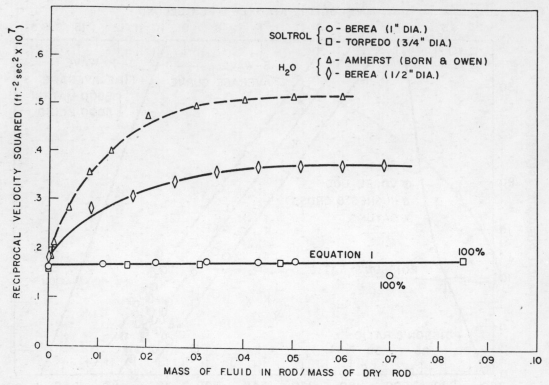

FIG. 5. Effect of water and oil on the bar velocity of uncompacted sandstones.

The third effect, like the first, is not always manifest. When the rod is fully saturated with fluid, its velocity may appear to be greater or less than that obtained by linearly extrapolating $1/V^2_{S_w}$ from intermediate saturations. It appears that the presence of free water on the surface of the rod may cause an additional lowering of the velocity, whereas the operation of saturating the rod under vacuum may increase the velocity.

Figure 6 shows the corresponding data for shear velocities. The addition of water causes a substantial decrease in the modulus of rigidity; the addition of purified Soltrol causes a comparatively minor change which may have as its origin the physical adsorption of the oil in micro-cracks in the rock frame. It may be noted that Poisson's ratio for both sandstones is very small and not greatly changed by the addition of water. A small Poisson's ratio seems to be characteristic of natural sedimentary rocks not subjected to compacting pressure. Physical properties of the specimens and fluids are included in Tables 4 and 5.

The attenuations of the extensional and shear

waves are closely related to the velocities. The initial decrease in velocity with the addition of a fluid is accompanied by a proportional increase in attenuation. In the intermediate saturation range, where the elastic moduli are constant and the velocity decrease is caused by mass loading, the attenuations increase owing to the relative motion of the fluid and the solid. For fully saturated sandstone rods, the attenuations, like the velocities, have been found to be somewhat erratic, probably because of the technique used to achieve complete saturation.

Effect of pressure

The experiments with uncompacted sandstone rods verified that the nature of the bond between the grains can have an important effect on both velocity and attenuation. Indeed, for a chemically reactive saturant like water, the variation in the elastic moduli of the frame is much more important than the change in bulk density. Even for Soltrol, which is a straight-chain hydrocarbon and thus relatively inert chemically, the effects are of

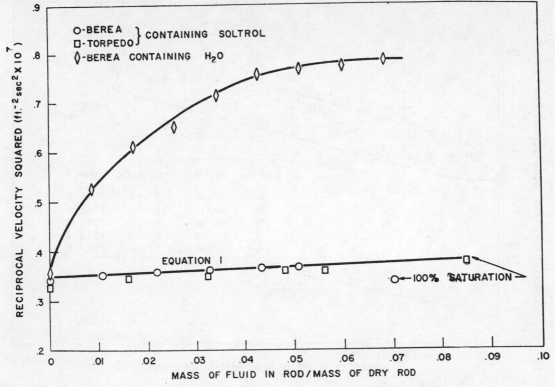

Fig. 6. Effect of water and oil on the shear velocity of uncompacted sandstones.

about equal size. There is evidence to suggest that at high differential pressures the elastic moduli of the frame are more constant and may even increase upon the addition of water. Thus, the change in dilatational velocity when a dry rock is completely water saturated and subjected to high compacting pressure can usually be accounted for solely by mass loading as discussed by the present authors (1961). This indicates that under pressure the change in bulk modulus of the frame caused by wetting it is small. On the other hand, the data given in Table 2 indicate that at a high enough pressure the shear velocity often *increases* with the addition of a fluid. This indicates that the rigidity modulus of the frame is

greater in the wet state than in the dry. If the frame is not appreciably influenced at high pressure by the addition of fluids, it is reasonable to expect that the attenuation of the waves by the frame is also not appreciably different when wet than when dry. This conjecture was tested by experiment. Measurements made by the gated *RF* pulse method at 1 mcps gave a logarithmic decrement for dilatational waves of less than 0.05 for Berea sandstone subjected to a compacting pressure of 2,500 psi both dry and wet with distilled water. This value includes the geometrical and reflection losses inherent in the method, and the actual decrement when these are subtracted is probably less than 0.02. Thus, not only velocities but also attenuations measured for natural rocks under no compacting stress are unlikely to be representative of rocks deeply buried in the ground.

Alundum Rods

A large change in the elastic moduli of the frame upon the addition of fluids was avoided by the use

Table 5. Physical properties of fluids.

Fluid	Density at 25°C. gm/cc	Viscosity at 25°C. cp	$\dfrac{\mu}{2\pi\rho}$
Water	0.9970	0.8937	.14266
Soltrol	0.7523	1.45	.30675
FC-75	1.760	1.44	.13021

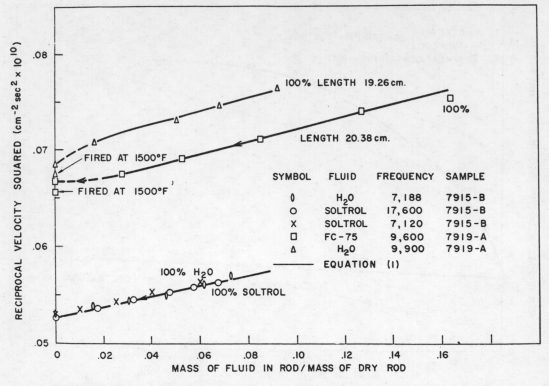

FIG. 7. Variation of shear velocity of alundum rods with saturation in the low-frequency range.

of alundum. Only the density, viscosity, and compressibility of the fluid are significant variables and hence, the measurements can be compared readily with results from Biot's theory. Four grades of alundum were used and the physical properties are listed in Table 4. However, it was found that alundum became less inert after some saturants had been removed by evaporation. This behavior could be eliminated by firing the rods at 1,500°C.

Figure 7 shows shear velocity data obtained in the low-frequency range ($f/f_c < 0.15$). It may be noted that the variation of velocity with saturation is entirely accounted for by mass loading since the velocity measured at zero saturation is here the same as that obtained by extrapolating measurements made at intermediate saturations. The assumption that the elastic moduli of the frame remain unaltered seems justified. The reproducibility of the data is also satisfactory. Figure 8 shows the corresponding data for the extension mode of vibration. Again the phenomena observed with sandstone rods, and attributed to changes in the elastic moduli of the frame, are largely absent.

The logarithmic decrements for both types of waves are shown in Figure 9. The measurements proved reasonably straightforward to make if care was taken to ensure that the resonant peak for the rod did not overlap the peak for the crystal pickups or some overtone of the flexural modes of vibration. If this precaution is not taken, results obtained are often curious.

The dry losses in all cases are small. Since the elastic moduli of the frame remain constant when a liquid is added to the rod, it seems reasonable to assume that the attenuation of energy by the frame will also remain constant. If this is true, the increase in attenuation with saturation is entirely caused by the relative motion of fluid and solid and may be derived from Biot's theory. In the low-frequency range the decrement for shear waves is given by

$$\Delta_{\text{shear}} = \pi \left[\frac{M_2}{M_1 + M_2} \right] \left[\frac{f}{f_c} \right]. \qquad (2)$$

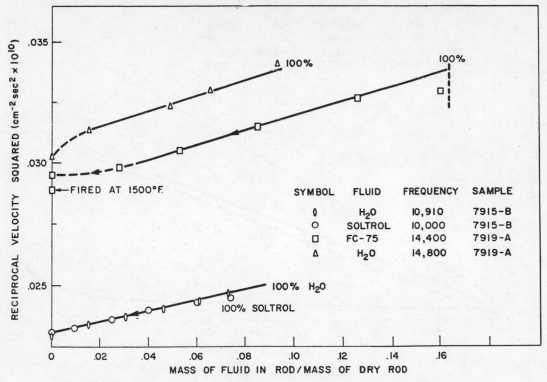

Fig. 8. Variation of bar velocity of alundum rods with saturation in the low-frequency range.

The formula for the decrement of the extensional waves is more complicated and is discussed by Gardner (1962). The measured torsional decrements are in satisfactory agreement with equation (2); an example is plotted in Figure 9. Exemplary calculations of logarithmic decrements appear in the Appendix.

It is remarkable that the logarithmic decrement is almost proportional to the fluid saturation. Equation (2) implies that f_c is, therefore, almost independent of saturation. Furthermore, the same loss was measured for saturations obtained by adding fluid and saturations obtained by evaporating fluid. These observations suggest that the critical frequency, f_c, at partial saturation is not related to the usual relative permeability of the fluid since such relative permeabilities are generally proportional to saturation raised to the third or fourth power and depend also on the method of obtaining a particular saturation. Recent measurements of the relative permeability characteristics of the specimen cast doubt on this conclusion. The relative permeability was found to be an almost linear function of saturation with

ho hysteresis. The formula for $f_c(S)$, the critical frequency at saturation S, may be written

$$f_c(S) = \frac{\mu\phi S}{2\pi\rho_F K(S)}. \qquad (3)$$

Therefore, the critical frequency is independent of saturation only if $K(S)$ is proportional to S. Thus, the appropriate relative permeability is equal to S. This conclusion is consistent with the facts that alundum has pores which are almost all the same size, and that the surface tension forces are very small compared with the viscous and inertial forces at the frequencies employed. For materials with a wider range of pore sizes, the decrement may deviate from a linear relation with the fluid saturation.

Data obtained in a higher frequency range ($f/f_c > 0.15$) are illustrated in Figures 10, 11, 12, and 13. It may be observed in Figures 10 and 11 that dispersion occurs at these frequencies. At a fluid content above 25 percent, both velocities deviate from the straight line given by equation (1). This deviation is in accordance with Biot's

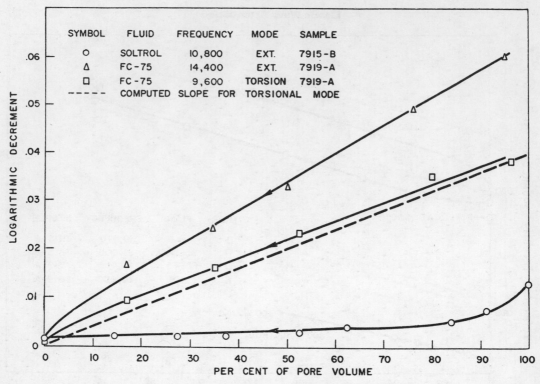

FIG. 9. Variation of torsional and extensional decrements of alundum rods with saturation in the low-frequency range.

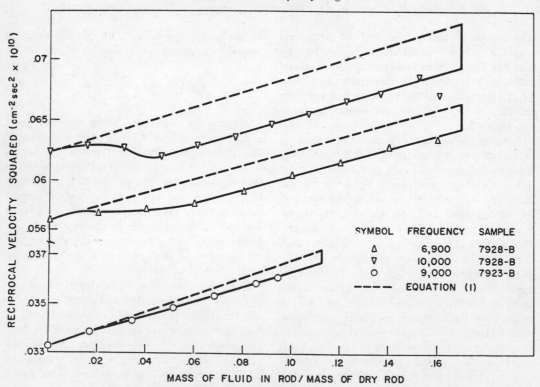

FIG. 10. Change in bar velocity on the addition of water to alundum rods in the higher-frequency range $f/f_c > 0.15$.

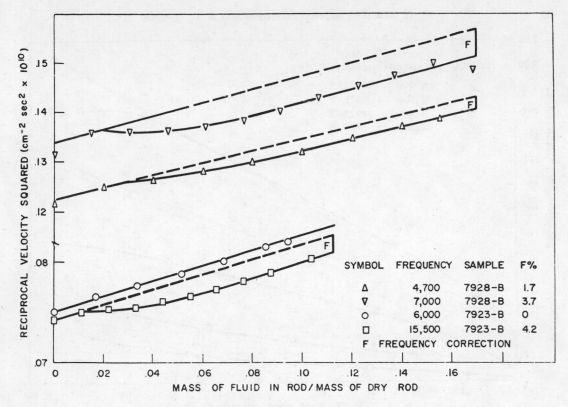

FIG. 11. Change in shear velocity on the addition of water to alundum rods in the
higher-frequency range $f/f_c > 0.15$.

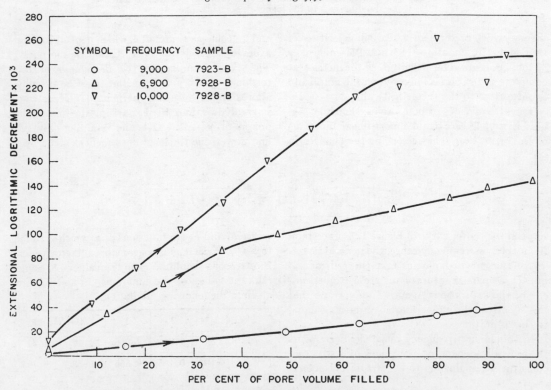

FIG. 12. Change in extensional logarithmic decrements on the addition of water to
alundum rods in the higher-frequency range.

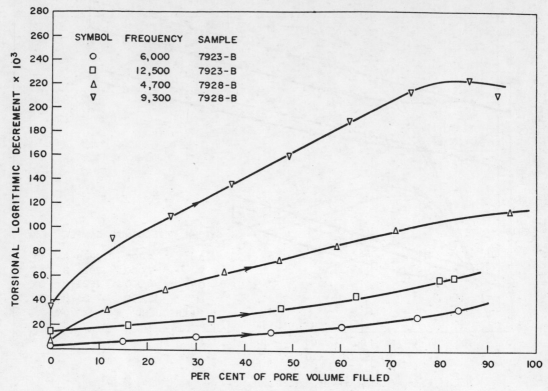

FIG. 13. Change in torsional logarithmic decrements on the addition of water to alundum rods in the higher-frequency range.

theory, which predicts an increase in velocity over that given by equation (1) when Poiseuille flow no longer occurs. The magnitude of the increase at complete saturation is indicated on the graph and is quite close to the theoretical value.

The attenuations of both waves, shown in Figures 12 and 13, are again almost linear functions of the saturation. The reason for the changes in slope in the center curve of Figure 12 is uncertain. As in the low-frequency range, the general linearity of the curves indicates that the critical frequency at partial saturation is almost independent of the saturation. Again we may argue that the effective relative permeability must be equal to the saturation.

In this higher frequency range the decrements theoretically depend on the values of a structure constant, E, analogous to a "tortuosity factor,"

and a geometrical factor, δ, which depends on the average shape of the cross-section of the pores. The exact formula for the decrement of the torsional waves, given by Biot, may be usefully approximated by using the fact that M_2 is considerably less than M_1. Expanding the formula in powers of $M_2/(M_1+M_2)$, and retaining only the first power, the torsional decrement is given by

$$\Delta = \pi \left[\frac{M_2 \; -}{M_1 \; + \; M_2} \right]\left[\frac{f}{f_c} \right] \cdot \frac{F_r}{F_r{}^2 + \left\{ E(f/f_c) + F_i \right\}^2}, \qquad (4)$$

where F_r and F_i are functions of κ, which is equal to $\delta(f/f_c)^{1/2}$. Here, δ is a geometrical factor which Biot reasons should be near the value $\sqrt{8}$. Some relevant values of the functions, F_r and F_i, are given in the following tabulation.

κ	1	2	3	4	5
F_r	1.0	1.01	1.065	1.17	1.32
F_i	0.04	0.165	0.355	0.58	0.795

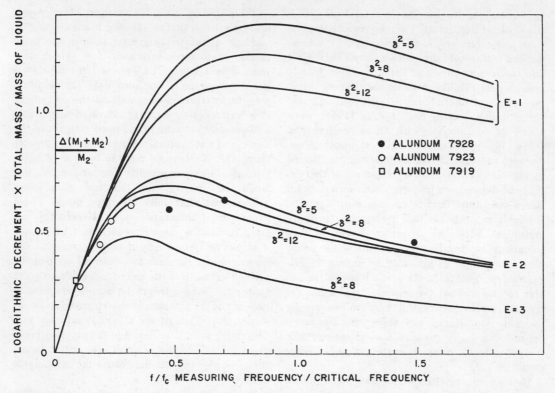

Fig. 14. Comparison of measured torsional decrements of fully saturated alundum rods with the Biot theory.

In Figure 14 the group of terms, $\Delta(M_1+M_2)/M_2$, is plotted as a function of f/f_c for some fixed values of E and δ. The solid curves may be obtained from equation (4). The isolated points show experimental data for fully saturated rods. These data were obtained by submerging the rods in the fluid and removing the trapped air by means of a vacuum pump.

Figure 14 shows that a value of E close to two gives the best agreement between theory and experiment. This supports the view that the structure constant, E, is approximately equal to ϕF (see Table 4). However, the accuracy of the data is not great enough to pick out a particular value of δ.

It was observed that the decrements obtained at partial saturations did not extrapolate linearly to the decrements obtained at 100 percent saturation in the higher frequency range, whereas they appeared to do so in the low-frequency range. Indeed, the decrement for a fully saturated rod was less than when the rod contained about 20 percent gas saturation. This may be caused by the dependence of E on saturation. If, at partial saturation, E has a value closer to one than two, then the observed behavior could be accounted for. However, there are not sufficient data available to test this theory.

DISCUSSION

The attenuation of waves in fluid-saturated porous material may be regarded as the sum of the loss caused by fluid motion and the loss caused by the solid framework, i.e. sloshing and jostling losses. The jostling loss may in turn be regarded as composed of two parts, a dry loss of the solid friction type and a viscous loss resulting from the chemical and physical effect of the fluid on the cementing material of the solid and perhaps within cracks of the grains themselves. The fragmentary results available indicate that when the rock frame is compressed as it would be in situ, the jostling loss is decreased. Whether this decrease typically renders the jostling loss negligible by comparison with sloshing losses measured at frequencies used in continuous velocity logging is uncertain. These frequencies are in the range 10^4 to 2×10^4 cps. It would be convenient to assume that at logging frequencies the total decrement is the sloshing decrement. This convenient assumption, unfortunately, is not sustained by a comparison of the field decrements shown in Table 1 and those calculated and listed in Table 2. Only in high permeability rocks do the sloshing decrements appear likely to exceed the jostling ones.[2] Unfortunately, also, Born's data (1941) when taken in conjunction with those of Born and Owen (1935) show that in Amherst sandstone under zero pressure the frame decrement is affected by addition of water and a component of that decrement depends on frequency. Our own data for Berea sandstone tend to confirm Born's results. While it is probable that compacting pressure minimizes both the frequency-dependent and frequency-independent components of the frame decrement, there is presently no certain knowledge on the point. On the whole it seems reasonable to assume that the dependence of frame decrement on frequency will vary widely among rocks in situ. Rocks may show zero frequency dependence under pressure only if entirely free from hydratable minerals in the vicinity of grain contacts.

If the evidence presented above for alundum rods is accepted as confirmation of the main tenets of Biot's theory, it becomes possible to compute the decrement of P-waves of rocks in situ. This was done for the sandstones listed in Table 2 (see Appendix). From a practical standpoint it would be more useful to log decrements in situ at two or more frequencies and then to use Biot's theory to calculate the fluid mobility (K/μ) in the rock pores. The logging of shear decrements would seem particularly desirable. The decrements would not alone suffice to calculate mobilities, but the other parameters required in Biot's equation may be reasonably assumed or read from other logs. As already pointed out, this possible method of using decrements to determine the productivity of formations in situ must await confirmation that frame decrements at logging frequencies can be assumed to be negligible in comparison with sloshing decrements, or frequency-independent.

It is rather widely believed, largely on the basis

of "skip-cycling" (see, for example, Tixier, Alger, and Doh, 1959) that attenuation in gas-containing sands is qualitatively greater than in the same sands when water-saturated, i.e. skip-cycling tends to be seen on CVL's above but not below a gas-water contact. Our own data for alundum may be interpreted to show that the decrement of a water-wet sand partly saturated with water and gas may be greater than the decrement of the same sand water-saturated provided f/f_c is greater than 0.15. This would tend to be true of high-porosity, high-permeability formations. At low frequencies, partly saturated sands have lower attenuations than fully saturated sands. These experimental observations are not necessarily related to what is sometimes seen on CVL's since a skip-cycle occurs only if each recorder of the logging device responds to a different cycle of the arrival signal. If both recorders respond to the same cycle, even if this cycle, because of attenuation, is not the one usually used for triggering, no skips occur. Thus, if the attenuation in the gas-containing sand is critical and skip-cycling takes place, either a higher or lower attenuation below the gas-water contact may cause the skipping to cease.[3]

We have dwelt at some length on the problem of the attenuation which occurs in gas-sands in situ because of the practical possibilities which are opened if Biot's theories can be used to interpret decrements derived from appropriate in-situ measurements. In spite of the experimental difficulties which becloud the picture, we believe there is now strong evidence to support the correctness of Biot's theory when applied to resonating bars. It is in its application to the interpretation of measurements made by pulse techniques that we feel some uncertainty still exists. This uncertainty stems from our observation (which is supported by Hughes and Cross, 1951) that the shear velocity through liquid-saturated specimens, measured by a pulse technique, is often greater than the velocity through the same specimen when dry. The data are shown in Table 2. Biot's theory states unequivocally that shear velocities through liquid-saturated specimens must always be less than through the same speci-

---

[2] Note that the viscosity of water decreases by about one percent for each 1°F increase in temperature. Thus, at reservoir temperatures sloshing losses in a particular rock will be correspondingly higher than those measured at atmospheric temperature.

[3] It is possible that the attenuation is principally caused by reflection of the signal at the interface between the water- and gas-containing portions of the rock and is not primarily a phenomenon of the type discussed in this paper.

mens when dry. Thus, our observation must mean, if Biot's theory is applicable, that the rigidity of the frame of a specimen *under pressure* is often *increased* by the presence of a liquid in its pores to an extent which offsets the loading effect of the added liquid. This is, of course, possible since little is known concerning the nature of the interaction of liquids and solids. Nevertheless, the effect seems qualitatively surprisingly different from that shown in Figures 5 and 6 when no pressure is applied to the frame.

These apparent discrepancies between theory and experiment, as well as other considerations arising out of the results reported, are currently being investigated. At this stage, therefore it seems wise to suggest that the validity of Biot's theory be restricted to resonating systems.

CONCLUSIONS

In the light of work so far carried out we believe that the following conclusions are justified. These conclusions may be conveniently divided into three parts: those that apply at low frequency, where $f/f_c < 0.15$; those that apply at higher frequencies, where $f/f_c > 0.15$; and those that are independent of frequency.

Low Frequency

(1) Biot's theory appears to give quantitative expression to the logarithmic decrement which defines the losses caused by the mutual interaction between the frame of a porous body and a liquid which saturates its pores (sloshing decrement).

(2) At partial saturations of a liquid in a porous body, the logarithmic decrement may increase as the first power of the liquid saturation, when this saturation is expressed as a fraction of the total pore-volume.

(3) The total logarithmic decrement of a porous body containing a liquid is the sum of the frame decrement and the frequency-dependent sloshing decrement. In many natural rocks containing water it appears that (at atmospheric pressure) the frame decrement is itself a dichotomy, one component being of the solid friction-type and frequency-independent and one component being frequency-dependent.

(4) In rocks in situ it appears probable that

the frame decrement and the sloshing decrement are of similar magnitude. Only when permeabilities exceed about 500 md (for viscosity = 1 cp) is the latter decrement, either shear or dilatational, likely to exceed the former significantly. At reservoir temperatures and concomitant lower liquid viscosities, this permeability will be proportionately reduced.

High Frequency

(1) Pulse measurements suggest that certain liquids may increase the rigidity of compacted rocks.

(2) In some cases, the logarithmic decrement of a porous body may be lower when fully saturated with a liquid than when partially saturated.

All Frequencies

(1) The effect of compacting pressure is to reduce the frame decrement; its effect on sloshing decrement is smaller and results only from pressure-induced changes in porosity, permeability, and elastic constants.

(2) Natural rocks when uncompacted appear to be characterized by Poisson's ratios, as these are usually computed, which approach zero. As compacting pressure increases at constant porosity, Poisson's ratio increases; a similar increase is observed as porosity increases at constant compacting pressure.

APPENDIX

Calculation of Shear Decrement

In the low-frequency range the decrement of plane shear waves, Δ_s, may be obtained from

$$\Delta_s = \pi \frac{M_2}{M_1 + M_2} \frac{f}{f_c}, \quad f/f_c < 0.15. \quad (5)$$

Example: Water-saturated Berea Sandstone

$$\mu = .01 \text{ poise}, \quad \rho_F = 1.0 \text{ gm/cc}, \quad \phi = 0.19,$$

$$K = 200 \text{ millidarcys}$$

$$= 2(.987) \, 10^{-9} \text{ cm}^2.$$

Hence,

$$f_c = \phi\mu/2\pi\rho_F K$$
$$= 153{,}225 \text{ cps.}$$

The grain density of the sandstone is 2.65 gm/cc.

Hence,

$$M_2/(M_1 + M_2) = 0.0814.$$

The shear decrement at 20,000 cps is therefore given by

$$\Delta_s = \pi(.0814)(.13)$$
$$= .033.$$

Calculation of Dilatational Decrement

In the low-frequency range the decrement of plane dilatational waves may be obtained from

$$\Delta_d = \pi \frac{M_1 + M_2}{M^2} \left| (Z_1 - 1)(Z_2 - 1) \right| \cdot (\sigma_{11}\sigma_{22} - \sigma_{12}{}^2) \frac{f}{f_c},$$
$$\text{for } \frac{f}{f_c} < 0.15. \tag{6}$$

Here, Z_1 and Z_2 are the roots of the quadratic equation

$$(\sigma_{11}\sigma_{22} - \sigma_{12}{}^2)Z^2 - (\sigma_{22}\gamma_{11} + \sigma_{11}\gamma_{22}$$
$$- 2\sigma_{12}\gamma_{12})Z + (\gamma_{11}\gamma_{22} - \gamma_{12}{}^2) = 0. \tag{7}$$

The σ_{ij} and γ_{ij} are elastic and density coefficients which may be given numerical values as illustrated in the example below.

Example: Water-saturated Berea Sandstone

$$\phi = .19, \quad \rho_F = 1.0 \text{ gm/cc},$$
$$\rho_R = \text{grain density} = 2.65 \text{ gm/cc},$$

E taken equal to $3(\cong\phi F)$. From these data the density coefficients may be computed. We have:

$$\rho_{22} = E\phi\rho_F \text{ (by definition of } E) = 0.570,$$
$$\rho_{11} = (1 - \phi)\rho_R + \phi(E - 1)\rho_F = 2.526,$$
$$\rho_{12} = -(E - 1)\phi\rho_F = -0.380,$$
$$\rho = \rho_{11} + 2\rho_{12} + \rho_{22} = 2.34.$$

The γ_{ij} are defined equal to ρ_{ij}/ρ. Hence,

$$(\gamma_{11}, \gamma_{12}, \gamma_{22}) \equiv (1.081, -0.163, 0.244) \text{ and}$$
$$\gamma_{11}\gamma_{22} - \gamma_{12}{}^2 = .237.$$

From the following data we can calculate the σ_{ij} terms:

$$V_d{}^0 = \text{dry dilatational velocity of rock}$$
$$= 3.67 \times 10^5 \text{ cm/sec},$$
$$V_s{}^0 = \text{dry shear velocity of rock}$$
$$= 2.71 \times 10^5 \text{ cm/sec},$$

bulk modulus of quartz

$$= 37.9 \times 10^{10} \text{ gm/cm/sec}^2,$$
$$\rho_R = \text{density of quartz} = 2.65 \text{ gm/cc},$$
$$V_F = \text{velocity of fluid} = 1.50 \times 10^5 \text{ cm/sec}.$$

We define:

$$V_q{}^2 = 37.9 \times 10^{10}/2.65 = 14.3 \times 10^{10} \text{ cm}^2/\text{sec}^2,$$

$$V^2 = (V_d{}^0)^2 - 4/3(V_s{}^0)^2 = 7.21 \times 10^{10} \text{ cm}^2/\text{sec}^2,$$
$$\alpha = 1 - V^2/V_q{}^2 = .485,$$
$$\beta = \alpha + (V_d{}^0)^2 V_q{}^2 = 1.449.$$

The elastic moduli, P, Q, H, R in Biot's theory are then related by:

$$Q/R = \alpha\frac{1 - \phi}{\phi} = 2.066,$$

$$P/R = \alpha\beta\left(\frac{1 - \phi}{\phi}\right)^2 + \frac{1 - \phi}{\phi}\frac{\rho_R(V_d{}^0)^2}{\rho_F V_F{}^2}$$
$$= 80.74,$$

$$H/R = (P + 2Q + R)/R = 85.87.$$

Finally, the σ_{ij} are given by:

$$\sigma_{11} = P/H = 0.94,$$
$$\sigma_{12} = R/H = 0.012,$$
$$\sigma_{22} = Q/H = 0.024,$$
$$\sigma_{11}\sigma_{22} - \sigma_{12}{}^2 = 0.0107.$$

The coefficient of Z in equation (7) may now be calculated:

$$\sigma_{22}\gamma_{11} + \sigma_{11}\gamma_{22} - 2\sigma_{12}\gamma_{12} = 0.2501.$$

The group of terms

$$\left| (Z_1 - 1)(Z_2 - 1) \right| (\sigma_{11}\sigma_{12} - \sigma_{12}{}^2),$$

which occurs in equation (6), is exactly equal to the value of the left-hand side of equation (7) when Z is put equal to one. Hence, the dilatational decrement is given by

$$\Delta_d = \pi(12.3)(.002) \frac{f}{f_c}$$

$$= .01 \text{ at } 20{,}000 \text{ cps.}$$

REFERENCES

Bardeen, Thomas, 1960, Amplitudes and attenuation of seismic reflections: Gulf Research & Development Co. Report.

Bayuk, E. I., 1960, The investigation of the elastic properties of rock samples taken from a deep borehole at high pressures: Bulletin, Academy of Sciences (USSR), English Translation, no. 12, 1960, pp. 1173–1177.

Biot, M. A., 1956, Theory of propagation of elastic waves in a fluid-saturated porous solid. 1. low-frequency range. 2. higher frequency range: J. of Acoust. Soc. Amer., v. 28, pp. 168–191.

———, 1955, Theory of elasticity and consolidation for a porous anisotropic solid: J. of Appl. Phys., v. 26, pp. 182–185.

Blizard, R. P., 1959, Private communication.

Birch, Francis, and Bancroft, Dennison, 1938, The effect of pressure on the rigidity of rocks: J. of Geology, v. 46, pp. 59–87, 113–141.

Born, W. T., 1941, The attenuation constant of earth materials: Geophysics, v. 6, pp. 132–148.

———, and Owen, J. E., 1935, Effect of moisture upon velocity of elastic waves in Amherst sandstone: Bull. Amer. Assoc. Petr. Geol. v., 19, pp. 9–18.

Gardner, G. H. F., 1962, Extensional waves in fluid-saturated porous cylinders: J. of Acoust. Soc. Amer. v. 34, pp. 36–40.

Geertsma, J., 1961, Velocity log interpretation: the effect of rock bulk compressibility, Trans. A.I.M.E., v. 222, pp. 235–246.

Hughes, D. S., and Cross, J. H., 1951, Elastic wave velocities in rocks at high pressures and temperatures: Geophysics, v. 16, pp. 577–593.

Kendall, J. M., 1941, The range of amplitudes in seismic reflection records: Geophysics, v. 6, pp. 149–157.

Knopoff, L., and Macdonald, G. J. F., 1958, Attenuation of small amplitude stress waves in solids: Reviews of Modern Physics, v. 30, pp. 1178–1192.

McDonal, F. J., Angona, F. A., Mills, R. L., Sengbush, R. L., Van Nostrand, R. G., and White, J. E., 1958, Attenuation of shear and compressional waves in Pierre shale: Geophysics, v. 23, pp. 421–439.

Niblett, D. H., and Wilks, J., 1960, Dislocation damping in metals: Advances in Physics, v. 9, pp. 1–8.

Peselnick, L., and Zietz, I., 1959, Internal friction of fine-grained limestones at ultrasonic frequencies: Geophysics, v. 24, pp. 285–296.

———, and Outerbridge, W. F., 1960, Internal friction and rigidity modulus of Solenhofen limestone over a wide frequency range: Geological Survey Research 1960—Short Papers in the Geological Sciences.

Sabbarao, K., and Ramchandra Rao, B., 1957, A simple method of determining ultrasonic velocity in rocks: Nature, v. 180, p. 978.

Schneider, W. C., and Burton, C. J., 1949, Determination of the elastic constants of solids by ultrasonic methods: J. of Appl. Physics, v. 20, pp. 48–58.

Tixier, M. P., Alger, R. P., and Doh, C. A., 1959, Sonic logging: Trans. AIME, v. 216, pp. 106–114.

Wyllie, M. R. J., Gregory, A. R., and Gardner, G. H. F., 1958, An experimental investigation of factors affecting elastic wave velocities in porous media: Geophysics, v. 23, pp. 459–493.

———, Gardner, G. H. F., and Gregory, A. R., 1961, Some phenomena pertinent to velocity logging: J. of Petr. Tech., v. 13, pp. 629–636.

Reprinted from Journal of Petroleum Technology, February 1964, p. 189–198.

Effects of Pressure and Fluid Saturation on the Attenuation of Elastic Waves in Sands

G. H. F. GARDNER
M. R. J. WYLLIE
MEMBERS AIME
D. M. DROSCHAK

GULF RESEARCH & DEVELOPMENT CO.
PITTSBURGH, PA.

ABSTRACT

The velocity and attenuation of elastic waves in sandstones were measured as a function of both pressure and fluid saturation. A large change occurs in these quantities if water is added and the rock is not compressed, but the change is small if the rock is subjected to a large overburden pressure. Measurements were made by vibrating cylindrical samples in both the extensional and torsional modes at frequencies up to 30,000 cycles/sec. Formulas were derived which enable the attenuation of dilatational waves in dry rocks to be deduced from the data.

Similar experimental methods were used to investigate the properties of unconsolidated sands. Velocities were found to vary with the 1/4 power of the overburden pressure and attenuations to decrease with the 1/6 power. The effects of grain size, amplitude and fluid saturation were studied. Formulas by which the effects produced by a jacket around the sample may be calculated were derived.

The practical application of these results to formation evaluation is discussed.

INTRODUCTION

The attenuation of elastic waves in the earth has been of interest to the seismologist and geophysicist for many years, but only recently to the petroleum engineer. Engineering interest has been brought about by the success of velocity logging devices, for it is possible by modification of these instruments to measure the attenuation of sound waves in addition to their velocity and, hence, deduce the mobility of formation fluids as well as the porosities of the rocks which contain them. The main problem is to decide whether field measurements can be made with sufficient accuracy to be of practical use. This problem can only be solved after we know the magnitude of the attenuations which are typical of the earth at various depths.

The logarithmic decrement of a fluid-saturated rock is the sum of a "sloshing" decrement and a "jostling" decrement, the former caused by the mobility of the fluid contained within the rock and the latter by the granular framework of the rock. Sloshing decrements can be calculated[1] using Biot's theory, but the jostling losses are less well understood. The present paper reports an experimental investigation of jostling losses in consolidated and uncon-

solidated sands, particularly with respect to the effect of overburden pressure and fluid saturation.

Born[2] showed that the decrement of a sandstone may increase dramatically when only a few per cent by weight of distilled water is added, and that the additional loss is proportional to the frequency of vibration. His measurements were made with no compressive stress on the framework of the rock. M. Gondouin[3] investigated similar phenomena for fluid-saturated plasters but also did not compress the samples. In the present paper it is shown that compression of the framework reduces this effect, so that at depth the jostling decrement of a sandstone may be expected to be almost independent of fluid saturation and frequency.

Decrements for many sedimentary rocks have been given by Volarovich,[4] but all for the state of zero overburden pressure.

Anomalously low velocities have been logged in shallow unconsolidated gas sands. Results of the present investigation confirm that these velocities are not caused by correspondingly high attenuations, because the jostling decrement in a packing of sand grains is small and much less than in a consolidated sandstone at the same depth. Velocities in sands have been measured by Tsareva[5] and by Hardin[6] as a function of pressure, but the corresponding decrements do not appear to have been measured previously.

The widely used "resonant bar method" of measuring velocities and decrements was employed. Comments on variations of this technique have recently been published by McSkimmin.[7] The main novelty of the present technique was the application of pressure to the samples. It was found possible to do this by placing the apparatus inside a pressure vessel, provided the conditions leading to large additional losses were avoided. These conditions are discussed below.

EXPERIMENTAL TECHNIQUE

Cylindrical samples were caused to vibrate in both the extensional and torsional mode of vibration and the amplitude of vibration was measured as a function of frequency in the neighborhood of a resonant frequency. The resonant frequency, f_r, is related to the corresponding elastic modulus by the formulas

$$E = \rho f_r^2 \lambda^2,$$
$$N = \rho f_r^2 \lambda^2 \qquad \ldots \ldots \ldots \ldots (1)$$

where E and N are Young's modulus and the modulus of rigidity, ρ is the density of the sample, and λ the wavelength of the vibration.

The bluntness of the resonant peak is proportional to the logarithmic decrement, Δ, as in the formula

$$\Delta = \frac{\pi \, \Delta f}{f_r}$$

where Δf is the difference between the two frequencies at which the amplitude is $1/\sqrt{2}$ times the peak amplitude at resonance. The amplitude of a plane progressive wave decays by a factor $exp(-\Delta x/\lambda)$ on traveling a distance x.

The experimental technique consists of a method of preparing the sample, supporting it, driving it, and detecting the vibration. The samples were supported by a triangular wire frame inscribed in a metal ring so that the three wires touched short portions of the central circumference of the cylindrical surface. The samples were driven by an electromagnet which acted on an 0.5 gm ferrite permanent magnet attached to the center of one plane end of the sample by a thin layer of beeswax. The electromagnet had two independent windings and four pole-pieces. One pair of pole-pieces was placed with a gap of about 2 mm from the poles of the permanent magnet; the other pair was at right angles and at the same distance. Thus the former pair tended to produce only longitudinal oscillations of the sample and the latter only torsional oscillations. The vibrations were detected by two piezo-electric bimorphs, each weighing about 0.1 gm and attached to the sample by a small piece of beeswax at the center of their length. The probe to detect longitudinal vibrations was placed at the center of the face at the end remote from the permanent magnet, and the probe to detect torsional oscillations was placed at the circumference of the same end as the first probe.

The voltage from the pickup was measured by a voltmeter,[*] and also fed to one pair of plates of an oscilloscope.[**] The output of a frequency generator[***] was connected to the other pair of plates and also to an electronic counter.[†] To measure the frequency of the signal from the pickup to within one-tenth of a cycle/sec the frequency from the generator was varied until a 10:1 standing wave pattern was obtained on the screen of the oscilloscope. The electronic counter then read 10 times the frequency of the signal of the pickup correct to 1 cycle/sec.

The voltage output from the pickup was also used to drive the pen of an X-Y recorder and the frequency to drive the X axis when an over-all picture of the frequency response of the sample was desired.

Fig. 1 gives a schematic drawing of the apparatus.

Consolidated samples were prepared by coring a cylinder from a large block of stone and squaring the ends. The rods were dried at 125C and their elastic properties measured by the resonant bar method. Then they were coated twice with an epoxy resin,[††] and an 0.5 cm length of brass tubing, 0.1 cm in diameter, fixed at the central circumference to give gas communication to the interior of the sample at this point only. When the apparatus was placed inside the pressure vessel, hypodermic tubing was used to connect the pore space of the sample with the exterior of the vessel. In this way any leakage of gas through the cement coating did not result in an increase of pressure in the pore space. This precaution was essential with the unconsolidated samples where some leakage always occurred, but was not essential with the consolidated rocks.

°Hewlett-Packard Model 400H
°°Tektronix Type 561, Type 63 Differential Amplifiers.
°°°Hewlett-Parkard Model 650A
†Hewlett-Packard Model 5512A
††Armstrong's Adhesive (A-1), made by Armstrong Products Co.

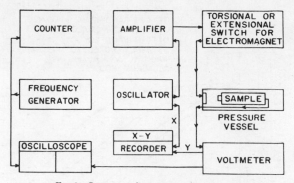

FIG. 1—SCHEMATIC DIAGRAM OF APPARATUS.

The unconsolidated samples were constructed with only slight modification of this method. Samples with either metal or rubber jackets were used. The former were made by first forming a cylindrical shell of foil, 0.002 in. thick, with one open end, then packing the grains inside through screens as described by Naar and Wygal,[°] and finally soldering on an end cap of foil. Pressure communication was established through a tube cemented with Eastman 910[†††] to the center of the cylinder. These samples were rigid enough to handle with equal pressure inside and outside the jacket. Samples were also constructed with jackets of Texin 480-A,[‡] 0.005 in. thick. A cylindrical bag with one closed end was first formed by heat-sealing the rubber, and the escape vent was cemented at the center of the curved surface. This bag was placed inside a Lucite tube and arranged so that a vacuum drew the rubber against the Lucite and established the cylindrical shape desired. The particles were packed through screens as before, and the open end sealed with a disc of rubber. The pressure was reduced inside the sample and because of the differential pressure the sample was then rigid enough to support its own weight and could be freely handled. All projecting rubber was sealed down; this is especially important to avoid large losses caused by the jacket.

Glass spheres and quartz sand grains were used in the packings. These particles were thoroughly washed with detergent and dried at 125C before being packed.

Helium was used throughout the experiments as the fluid to exert pressure on the samples. Preliminary experiments with nitrogen showed that standing waves could be excited inside the pressure vessel if the wavelength in the gas was comparable with the distance between the sample and the inner surfaces of the vessel. The greater velocity of sound in helium ensured that the wavelength in the gas was always too large for the formation of standing waves. Standing waves caused the resonance peaks to become jagged and made a measurement of the peak width impossible.

EXPERIMENTAL PROCEDURE

Both natural rocks and granular packings often are aeolotropic, and a check on this possibility was made. The higher harmonics of the extensional mode of vibration are not exact multiples of the first harmonic, but deviate by an amount dependent on the ratio of the diameter to the wavelength and also on Poisson's ratio.[°] The deviation is greatest at a diameter:wavelength ratio of about 0.6, and it is usually feasible to measure enough harmonics to

†††Made by Eastman Chemical Products, Inc.
‡Made by Mobay Chemical Co.

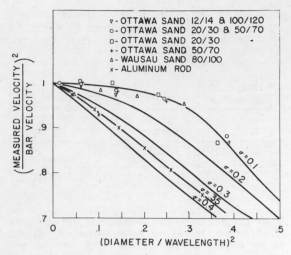

FIG. 2—DISPERSION DATA FOR EXTENSIONAL WAVES IN CIRCULAR CYLINDERS OF SAND AND ALUMINUM.

reach this range. Fig. 2 shows measurements of the extensional harmonic resonant frequencies of a cylinder of rolled aluminum and several cylinders of sand. If the samples are isotropic, the value of $E/2N-1$ as computed from the measured extensional and torsional fundamental frequencies should equal the value of σ as given by the dispersion curves. In the case of the aluminum rod the torsional and extensional resonant frequencies led to a Poisson's ratio of 0.31, while the dispersion of the extensional harmonics gave a value of 0.35, thus showing that the aluminum was not quite isotropic. Most of the packings of rounded sand grains gave a ratio of about 0.1 by both methods, showing that the packings were reasonably isotropic. An exception occurred with the angular Wausau sand. In general it was found that Naar and Wygal's technique of achieving low porosities tended to give aeolotropic packings with glass spheres, but isotropic packings with quartz grains.

The amplitude of vibration against the frequency was plotted on an X-Y plotter for each sample to enable the correct harmonics to be identified. It was found that flexural vibrations were often excited and sometimes a peak would overlap a peak of another mode causing a spuriously large decrement. Hence the flexural frequencies were calculated[10] from the measured first harmonic of the extensional mode and interferences with the other modes avoided.

EXTERNAL INFLUENCES ON THE SAMPLE

Several factors may have an important influence on the resonant frequency and bluntness of the resonance curve. They are: (1) radiation into the surrounding gas, (2) influence of the jacket, and (3) influence of the driver, pickups, and support.

The discussion here is aimed toward defining the conditions under which these influences are small rather than a quantitative correction. Some of these factors have been considered previously[11,12] and hence the relevant formulas are quoted without derivation.

EFFECT OF RADIATION

In torsional oscillation the outer surface of the cylinder does not alter its shape, and only shear waves are transmitted to the gas. These are damped out within a few

millimeters of the rod. The energy radiated into the gas increases the decrement of the sample by an amount Δ_1, where

$$\Delta_1 = \frac{A}{M}\left(\frac{\pi\mu_F\rho_F}{f}\right)^{1/2} \qquad (2)$$

and the resonant frequency f_r is reduced by the amount $\Delta_1/2\pi$. This factor is seldom limiting.

In the extensional oscillation of a cylinder, compressional waves are generated at both the curved surface and the plane ends. Radiation from the curved surface increases the decrement of the sample by an amount Δ_2, where

$$\Delta_2 = \frac{\pi^2 D^2 \sigma^2 \rho_F}{L^2 \rho_s} \qquad (3)$$

Radiation from the two plane ends increases the decrement by an amount Δ_3, where

$$\Delta_3 = \frac{\pi^2 D^2 \rho_F f}{L \rho_M C} \qquad (4)$$

For example, a cylinder with $D=5$ cm, $L=20$ cm, $\rho_M=2$ gm/cc, and $\sigma=0.2$, oscillating at a frequency of 1,000 cycles/sec in an atmosphere of helium at 1,000 psi for which $\rho_F=0.0024$ gm/cc, $\mu_F=2\times10^{-4}$ poise, Eqs. 2, 3 and 4 give

$$\Delta_1 = 7\times10^{-5},$$
$$\Delta_2 = 3\times10^{-5},$$
$$\Delta_3 = 15\times10^{-5}.$$

These losses are negligible compared with the losses in the samples. However, Eq. 4 shows that at high pressures and high frequencies radiation losses could be appreciable.

EFFECT OF JACKET

A thin elastic coating on the surface of a solid cylinder causes small changes in resonant frequency and decrement which are easily computed by use of Rayleigh's principle.[12] The resonant frequency for torsional oscillations of the coated sample is given by

$$f_r^2\lambda^2 = \frac{Nv + 2N_j v_{sj}}{\rho v + 2\rho_j v_{sj} + 2\rho_j v_{ej}}, \qquad (5)$$

and the corresponding decrement is given by

$$\Delta_{xc} = \frac{Nv\Delta_N + 2N_j v_j \Delta_{Nj}}{Nv + 2N_j v_j} \qquad (6)$$

The resonant frequency for extensional oscillations of the coated sample is given by

$$f_r^2\lambda^2 = \frac{Ev + B_1 v_{sj}}{\rho v + \rho_j v_{sj} + 2\rho_j v_{ej}}, \qquad (7)$$

and the corresponding decrement by

$$\Delta_{EC} = \frac{Ev\Delta_E + B_2 v_{sj}}{Ev + B_1 v_{sj}} \qquad (8)$$

where B_1, B_2 are defined by

$$B_1 = E_j(\sigma^2 - 2\sigma\sigma_j + 1)/(1-\sigma_j^2)$$
$$B_2(1-\sigma_j^2) = E_j[(\sigma^2 - 2\sigma\sigma_j + 1)\Delta_{Ej}$$
$$+ 2(\sigma_j - \sigma)(\sigma + 1)(\Delta_N - \Delta_E)$$
$$- 2(1-\sigma\sigma_j)(\sigma_j - \sigma)(\Delta_{Nj} - \Delta_{Ej})/(1-\sigma_j)]$$
$$\qquad (10)$$

In attempting to use these formulas for porous media it must be remembered that the jacket will be partly embedded in the sample and hence the elastic constants to be used in the formulas are some averages of those for the sample and the jacketing material.

EFFECT OF DRIVER, PICKUPS AND SUPPORT

These effects were examined experimentally. The distance between the electromagnet and the permanent magnet was varied without appreciable changes in the measurements. An additional permanent magnet was attached at the opposite end from the driver, in a similar position, and the increase in decrement was less than 2×10^{-4}. Various supports were tried, rubber pads, knife edges, etc., and it was concluded that the wire frame used introduced negligible loss and was the most convenient. Support losses have been discussed by Wachtman and Tefft.[13] The losses introduced by the presence of the extensional pickup were estimated to be less than 2×10^{-4} by measuring the increase in decrement when additional similar pickups were added. The losses introduced by the torsional pickup were found to be more variable, especially with unconsolidated samples with small total mass. Sometimes they were as large as 5×10^{-3} and sometimes smaller than 2×10^{-4}. To avoid large losses only samples with a mass in excess of 500 gm were used.

THEORY

The measurements lead to estimates of E and N by use of Eq. 1 and hence to the velocity of extensional waves and shear waves. The velocity of dilatational waves can be obtained from them by the formula

$$V_d^2 = V_s^2 \left[\frac{4V_s^2 - V_e^2}{3V_s^2 - V_e^2} \right] \quad . \quad . \quad . \quad . \quad . \quad (11)$$

This formula is applicable only to dry rocks. The corresponding formula for the relation between the decrements may be deduced from Eq. 11. Small losses can be satisfactorily accounted for by use of complex elastic constants. Thus we may replace E and N in the usual formula by the complex numbers $E(1 + i\Delta_E/\pi)$ and $N(1 + i\Delta_N/\pi)$. The imaginary parts of these equations then yield the corresponding equations between decrements. The equation corresponding to Eq. 11 may be written,

$$\Delta_D = a\Delta_E + (1-a)\Delta_N \quad . \quad . \quad . \quad . \quad . \quad (12)$$

where

$$a = (1+\sigma)/(1-\sigma)(1-2\sigma). \quad . \quad . \quad . \quad . \quad (13)$$

Since σ lies between 0 and 0.5, Δ_E lies between the values of Δ_N and Δ_D. It may also be noted that when any two of the decrements are equal, then all three are equal. The equation given by Born[2] is a special case of Eq. 12 and is obtained by assuming that the decrement for dilatation without shear vanishes. In general we have found that all the decrements for packings of particles are of equal magnitude.

The velocity of sound in packings of spheres has been the subject of several publications.[14,15] These have all been based on the Hertz[16] theory of contact between a pair of spheres. The assumption is usually made that the packing is sufficiently regular that the force of contact between spheres is the same at all points of contact. In this case the elastic moduli are proportional to the one-third power of the pressure applied to the packing. Up to pressures of 5,000 psi the moduli are more nearly proportional to the one-half power of the pressure, and this is in agreement with other experiments.[5] The half-power relationship may be explained qualitatively by an unequal distribution of pressures between spheres, but no simple quantitative theory has been found. At high pressures ($>5,000$ psi) the distribution of pressures will tend to become uniform and, hence, agreement with the simple theory might then be expected. An alternative hypothesis is that the Hertz theory fails because the spheres are not sufficiently smooth and that there are many points of contact where the spheres touch.

The decrement of packings has been calculated on the assumption that slip occurs in annular regions around each grain contact.[17] A conclusion of this theory is that the decrement is an increasing function of the amplitude of vibration, and some experimental support has been found at large amplitudes.[18] However, in all cases we have found that the decrement is independent of amplitude at small amplitudes. Hence, the theories referred to are not applicable, and we report at this time only the experimental measurements.

RESULTS FOR CONSOLIDATED SAMPLES

It is generally believed[19,20] that the decrements of dry rock are independent of frequency. However, it has been reported[21,22] that Solenhofen limestone at 10^6 cycles/sec has decrements several times greater than at a few cycles/sec. To examine this possible dependence on frequency, dry cylinders of Solenhofen limestone 1 and 2 in. in diameter were resonated at atmospheric pressure, and the torsional and extensional moduli and decrements were obtained for the lowest mode of vibration.

By shortening the length of each cylinder in ½-in. steps, the frequency range from 10 to 20,000 cycles/sec was covered for both modes of vibration. Within this range no tendency for the decrements to increase with frequency could be observed. The average decrements and moduli are given in Table 1 along with previously published results. It may be noted that the agreement between the moduli at all frequencies is satisfactory. The shear decrement at 10^4 cycles/sec is no greater than the shear decrement at 3.6 cycles/sec. However, the shear decrement at 10^6 cycles/sec is about four times larger. The shear and extensional decrements are almost equal in the resonant bar experiments.

It seems reasonable to conclude that at frequencies below 10^5 cycles/sec the decrements are independent of frequency; it is probable that the large decrement at 10^6 cycles is a consequence of a deficiency in the measuring technique used.

The effect of adding water to a rock was first examined by Born,[2] who concluded that a few per cent by weight of distilled water, when added to a sandstone, reduced its strength and added a component to the decrement which was proportional to the frequency. Figs. 3 and 4 show that overburden pressure substantially reduces this effect. The elastic moduli decrease when water is added at zero overburden pressure because of a weakening of the rock framework, but at high overburden pressure they remain almost constant. Similarly the large increase in decrement at zero overburden pressure is reduced to a slight increase, explainable by the mobility of the water, at high overburden pressure. The practical implication of this result is that at depth in the ground "jostling" losses have a simple nature. They are substantially independent of fluid content, independent of pressure, and independent of frequency.

Consequently the variation of the "sloshing" losses with fluid content and frequency may be used to separate the

TABLE 1—VELOCITIES AND DECREMENTS IN DRY SOLENHOFEN LIMESTONE MEASURED AT VARIOUS FREQUENCIES

Frequency cps	V_s	V_d	V_e	Δ_N	Δ_D	Δ_E	σ
3.59*	3.1	—	—	.005	—	—	—
10^{4**}	3.1	—	4.9	.0043	—	.0042	.27
10^{6***}	2.9	5.6	—	.0167	.0190	—	.31

*Peselnick and Outerbridge[22] **Present Work ***Peselnick and Zietz[21]

two types of losses. It is this fact which makes a permeability log theoretically feasible.

The same behavior as shown in Figs. 3 and 4 has also been observed with other sandstones. In general there is a hysteresis with respect to pressure. At a given pressure both the moduli and the decrements tend to be larger for decreasing pressure than for increasing pressure. The decrements in dry sandstones tend to decrease with a decrease in porosity or an increase in pressure; Fig. 5 shows a plot of the results. However, the number of data points is not large.

RESULTS FOR UNCONSOLIDATED PACKINGS

The large effects caused by distilled water, contrasted with oil,[1] suggests that a chemical or physical interaction between water and minerals, such as clay in sandstones, is responsible for the behavior observed by Born. It was considered that a better understanding of the mechanism by which the energy is dissipated might be reached if the loss in materials without cement was known. For this reason a study of dry packings of glass spheres and quartz sand grains was undertaken.

The conclusion drawn from this work is that the decrements for unconsolidated packings are extraordinarily small—so small that the method used to measure them is not suitable for high accuracy.* For example, it is believed that the loss in sand packs is less than 0.003 at compacting

pressures above 1,000 psi, that is, less than the loss measured in the dense Solenhofen limestone. At the same time the torsional rigidity is about 20 times less than for the limestone. This means that the fundamental torsional frequency for a cylinder 20 cm long is about 2,000 cycles/sec and that the width of the resonance peak is about 2 cycles. The error of measurement of this width is not less than 10 per cent.

It has been observed[24] that the damping capacity (that is, the modulus times the corresponding decrement) is fairly constant over a wide range of materials. Fig. 6 shows a plot of Young's modulus against the corresponding decrement on which the curves of constant damping capacity appear as straight lines. Unconsolidated sands have about the same damping capacity, $E\Delta_E$, as metals on the one ex-

*A method exists for measuring very small decrements with high accuracy.[23] In this method the time required for the amplitude to decline by some chosen factor is determined.

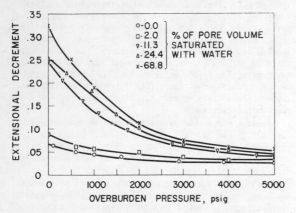

Fig. 3A—Variation of Extensional Decrements of a Berea Sandstone with Overburden Pressure at Fixed Water Saturations.

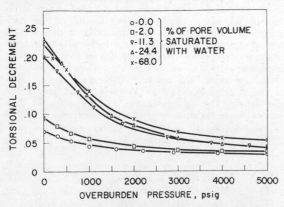

Fig. 3B—Variation of Torsional Decrements of a Berea Sandstone with Overburden Pressure at Fixed Water Saturations.

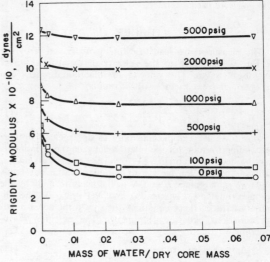

Fig. 4A—Variation of Rigidity Modulus of a Berea Sandstone with Water Saturation at Fixed Overburden Pressure.

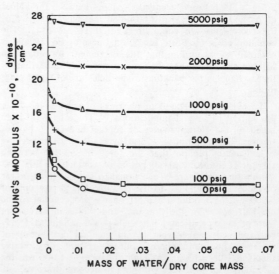

Fig. 4B—Variation of Young's Modulus of a Berea Sandstone with Water Saturation at Fixed Overburden Pressures.

treme of stiffness and elastomers on the other. Eqs. 6 and 8 show that the effect of a jacket on the decrement of a sample is proportional to its damping capacity. Thus the smallest effect is produced by a jacket with low damping capacity and small total volume. It is for this reason that metals and rubbers are suitable for jacketing unconsolidated sands. The damping capacity of consolidated sandstones, however, is comparatively large and is about the same as many polymers. For this reason a thin coating of Armstrong cement may be used to jacket consolidated samples with negligible effects.

It has been concluded from the present investigation that the moduli of unconsolidated sands increase with the square root of the pressure up to a pressure of 1,000 psi, and gradually approach proportionality to the one-third power above 5,000 psi. The decrements have been found to be inversely proportional to the sixth power of the pressure below 1,000 psi, and approach a constant value above 5,000 psi. Typical examples are shown in Figs. 7 and 10. The moduli are approximately linear functions of the porosity, at a fixed pressure, in the porosity range 20 per cent to 50 per cent, as shown in Fig. 8. In this range the decrements are equal and almost constant. The addition of a small volume of water increases the decrements by an amount proportional to the mass of fluid added. Both moduli and decrements are independent of particle size, frequency, and amplitude of vibration. A detailed account of these measurements is given below.

EFFECT OF THE JACKETS

The elastic parameters of the polyurethane rubber were measured in order to estimate the effect of the jacket on

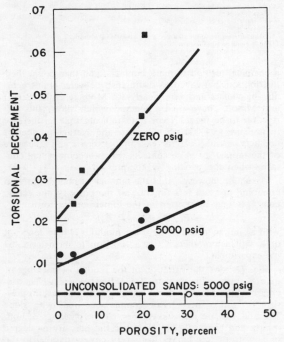

FIG. 5—DECREMENT VS POROSITY FOR SANDSTONES.

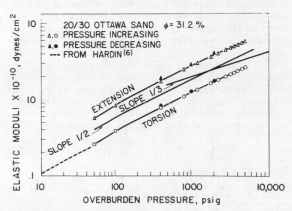

FIG. 7—CURVES TYPICAL OF THE VARIATION OF THE ELASTIC MODULI OF GRANULAR PACKINGS WITH PRESSURE.

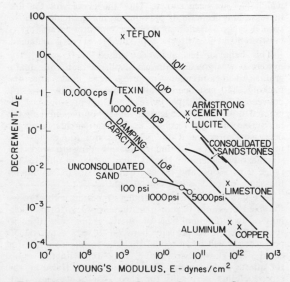

FIG. 6—COMPARISON OF THE DAMPING CAPACITY OF GRANULAR PACKINGS, CONSOLIDATED ROCKS AND VARIOUS SOLIDS.

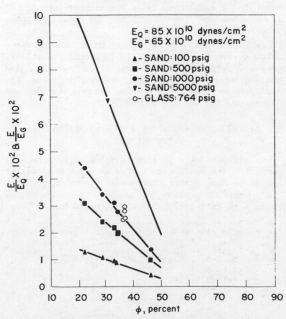

FIG. 8—VARIATION OF ELASTIC MODULI OF UNCONSOLIDATED PACKINGS WITH POROSITY AT FIXED PRESSURES.

the vibration of the samples. Young's modulus, the modulus of rigidity, and the two corresponding decrements were obtained in two ways: (1) progressive waves were transmitted along a strip of the rubber sheeting,[25] and (2) cylinders cast from granules of the material were resonated in the extensional and torsional modes. The agreement between the methods was satisfactory and the results obtained at several frequencies are given in Table 2.

Assuming that the numbers given in Table 2 are appropriate for use in Eqs. 5 to 8, it follows that the vibrations of a specimen 5 cm in diameter, for which the ratio of the volume of the packing to the volume of the jacket is about 100, are only slightly affected by the jacket except in extreme cases. From Fig. 6 it may be seen that the damping capacity of the rubber is less than 10 times that of an unconsolidated sand, provided the frequency of vibration is less than 5,000 cycles/sec. Hence for a volume ratio of 100 the increase in decrement caused by the jacket is less than 10 per cent. Most of the samples were constructed to be about 20 cm long and hence the fundamental frequencies were less than 3,000 cycles/sec at confining pressures up to 1000 psi. The odd harmonics, which have nodes at the supporting wires, occurred at a high enough frequency to be appreciably affected by the coating of rubber.

Data also shown in Fig. 6 show that the damping capacity of metals is about the same as rubber and, hence, for a volume ratio of 100, the losses caused by the jacket are again negligible. However, the measured decrements need to be corrected for the effect of the stiffness of the metal. Ignoring the term Δ_J in Eq. 6 we obtain

$$\Delta_N = \Delta_{NC}\left\{1 + \frac{2N_J v_J}{Nv}\right\} \quad \ldots \ldots \quad (14)$$

From Eq. 5 we obtain

$$1 + \frac{2N_J v_J}{Nv} = f_r^2 \lambda^2 (\rho v + 2\rho_s v_{sj} + 2\rho_j v_{ej})/Nv \quad \ldots \quad (15)$$

When N is known, Eq. 15 gives the term $1 + 2N_J v_J/Nv$ and Eq. 14 gives the corrected decrement, Δ_N. Measurements of the extensional decrement may be corrected in an analogous way. Table 3 illustrates the corrections made at several pressures.

Embedment of the grains in both the rubber and metal jackets occurred at high pressures, particularly with 20/30 mesh and larger grains, and created the possibility that the effective loss of the coating was not negligible. The actual loss in the packings must be less than that obtained by assuming the loss is negligible.

ELASTIC MODULI OF UNCONSOLIDATED PACKINGS

The elasticity of a packing of spherical particles is a function of the elastic constants of the grain material, the number of contacts per grain, and the confining pressure. The number of contacts per grain is fairly closely correlated with the porosity. Smith, Foote, and Busang[26] used

TABLE 2—ELASTIC PARAMETERS FOR TEXIN RUBBER (DENSITY = 1.2 GM/CC)

		1000	3000	5000	10,000	METHOD*
	Frequency of Vibration, cycles/sec					
E	{	$.055 \times 10^{10}$	$.066 \times 10^{10}$	$.071 \times 10^{10}$	$.076 \times 10^{10}$	1
	{	$.061 \times 10^{10}$				2
N	{	$.018 \times 10^{10}$	$.02 \times 10^{10}$	uncertain		1
	{	$.021 \times 10^{10}$				2
Δ_E	{	.6	.9	1	1.4	1
	{	.7				2
Δ_N	{	.6	.6	uncertain		1
	{	.7				2

*1—Progressive Waves. 2—Resonant Rod.

TABLE 3—ELASTIC MODULI OF PACKINGS OF GLASS SPHERES AND QUARTZ GRAINS AT EQUIVALENT PRESSURES OBTAINED WITH TEXIN JACKETS

Packing	Porosity %	N ×10⁻¹⁰ dynes/cm²	E ×10⁻¹⁰ dynes/cm²	E/2N	σ*	Bulk Density gm/cc
		at 1000 psi				
80-100 Mesh Wausau	46.3	0.60	1.16	0.97	.1	1.42
50/70 Mesh Wausau	34.6	1.05	2.33	1.11	.1	1.73
20/30 Mesh Ottawa	33.3	1.23	2.65	1.08	.1	1.78
20/30, 74% 50/70, 26%	29.0	1.30	2.89	1.11	.1	1.88
12/14, 75% 100/120, 25%	21.9	1.70	3.73	1.10	.1	2.07
		at 764 psi**				
400 Mesh Glass Spheres	36.8	.66	1.64	1.25	.24	1.50
20/30 Mesh Glass Spheres	34.7	.91	1.91	1.05	.25	1.60
60/70 Mesh Glass Spheres	36.6	.73	1.61	1.10	.30	1.59
60/70 Mesh Glass Spheres	36.9	.80	1.82	1.14	.30	1.58

*Estimated from the dispersion curve for extensional vibrations.
**764 is equal to 1000 E_G/E_Q.

a chemical method to mark contacts, and then count their distribution, but failed to distinguish between grains actually in contact and very nearly so. More recently, Bernal and Mason,[27] have used a method which allows this distinction to be made. Their data indicate that an increase in pressure to 5,000 psi will cause the average number of contacts per sphere to increase by between 1 and 2 because of the number of close contacts which become actual contacts when the packing is compressed.

Assuming, however, that the number of contacts at zero confining pressure is a function of the porosity, the elastic constants may be written in the dimensionless form

$$E/E_Q = f(\phi, \sigma_Q, P/E_Q),$$

with a similar form for the shear modulus, where for want of a satisfactory theory f is a function to be determined by experiment.

Fig. 7 shows the variation of the moduli of packings of quartz grains with pressure for a fixed porosity. The decrease of the porosity of the packing with increase in pressure is ignored here.[28] For comparison, lines with slope 1/2 and 1/3 are also shown. For all the packings of both quartz and glass spheres, whatever the porosity or degree of anisotropy, the curves were very closely parallel to the illustrated curves. The curves exhibited a definite hysteresis which has not been shown. Thus the curves for the first increase and decrease of pressure do not coincide with the second cyclic variation of pressure, but after several cycles between fixed pressures a reproducible curve is obtained.

This behavior has also been noted by Tsareva.[5] The hysteresis was greater with glass spheres than with quartz grain. All measurements reported were made after the packings had first been stabilized by cycling the pressure between 50 and 500 psi twice. Then the pressure was increased until the jacket leaked rapidly or the limit of the pressure vessel was reached, and measurements taken for decreasing pressure if the leak was not too rapid. The moduli for increasing and decreasing pressure were not significantly different. Hardin[6] has given results in the range 0 to 50 psi and his curve for the shear modulus at the same porosity is also shown in Fig. 7.

In Fig. 8 the variation of Young's modulus with porosity for fixed pressures is shown in dimensionless form. Packings of two different sizes of quartz grains were used to span the range of porosity between 20 per cent and 50 per cent. It will be observed that the results obtained for equal sized glass spheres agree fairly well with the results for quartz grains on this dimensionless plot. However, the reproducibility of results for glass spheres was less than for quartz grains.

The ratio of Young's modulus to the shear modulus for all the sand packings lies in the range 2 to 2.3. This includes packings which were not isotropic as shown by the dispersion of the extensional waves. Poisson's ratio for all the sand packings, obtained by the dispersion of the extensional waves, lies in the range 0.1 to 0.15. Most of the packings turned out to be almost isotropic. We have concluded from the results that Poisson's ratio for an isotropic random packing of quartz grains is 0.1, and for glass spheres is 0.25, and is independent of both pressure and porosity. Table 3 summarizes some of the data.

Fig. 9 shows the dependence of the dilatational wave velocity in sand on porosity and pressure, calculated from Eq. 11. The ratio of the dilatation velocity to the shear velocity is 1.5 for isotropic packings.

DECREMENTS OF UNCONSOLIDATED PACKINGS

Some decrements obtained for packings of glass spheres contained in both rubber and metal jackets are shown in Fig. 10 as a function of pressure. We have observed that the decrements for decreasing pressure always tend to be greater than for increasing pressure, at the same pressure, even though the corresponding moduli are almost equal. We refer here only to the decrements measured for increasing pressure.

It will be noted that the decrements for torsional and extensional oscillations are almost equal. This has been observed in almost all cases, whatever the porosity and pressure, and we conclude that this equality holds in general. Some of the data are listed in Table 4.

The decrease in decrement with pressure may be simply expressed by noting that the products $E\Delta_E p^{-1/3}$ and $N\Delta_N p^{-1/3}$ are almost constant in the range 50 to 5,000 psi. Hence if E and N become proportional to p raised to the one third power above 5,000 psi, then Δ_E and Δ_N become constant above 5,000 psi. Tables 4 and 5 give examples of this for packings of quartz grains and glass spheres.

Fig. 11 shows the effect of a partial water saturation on the shear decrement of two packings of sand. For the low permeability sand the frequency of vibration, $\sim$ 1,000 cycles/sec, is less than 0.15 times the critical frequency $\phi\mu_F/2\pi\rho_F K$, and hence lies in the low frequency range of

TABLE 4—MODULI AND DECREMENTS OF A PACKING OF GLASS SPHERES IN A TEXIN JACKET

P psi	$N\times10^{-10}$ dynes/cm²	$E\times10^{-10}$ dynes/cm²	Δ_N	Δ_E	$p^{-1/3}N\Delta_N \times10^{-5}$	$p^{-1/3}E\Delta_E \times10^{-5}$
50	.274	.563	.0123	.0124	91	189
100	.378	.782	.0115	.0119	94	201
200	522	1.089	.0103	.0088	92	164
300	.626	1.309	.0098	.0085	91	166
400	.710	1.490	.0094	.0085	90	172
500	.779	1.638	.0088	.0088	86	181
700	.891	1.887	.0098	.0084	98	178

Porosity 34.7%, Bulk Density 1.66 gm/cc, Diameter 5.1 cm, Length 20.36 cm, Mesh 20/30. No correction made for effect of jacket.

TABLE 5—MODULI AND DECREMENTS OF 20/30 MESH OTTAWA SAND COMPUTED FROM VIBRATIONS OF A SAMPLE WITH A JACKET OF PHOSPHOR BRONZE, .003 IN. THICK

Porosity 31.2%, Diameter 4.43 cm., Length 30.61 cm., Total mass of jacket 23.3 gm, Mass of both ends together 1.93 gm.

P psi	$N\times10^{-10}$ dynes/cm²	$\{\ \}*4L^2f^2\times10^{-10}$	$2N_j\nu_j\times10^{-10}$	$2N_j\nu_j/N\nu\times10^{-10}$	Δ_{Nj}	Δ_N	$p^{-1/3}N\Delta_N\times10^{-5}$
100	.387***	380	197	1.08	.0027	.0056	46
1200	1.345***	812	177	.28	.0026	.0034	42.5
3250	2.216	1223	177†	.17	.0023	.0027	40
5000	2.697	1450	177†	.14	.0023	.0026	41

P psi	$E\times10^{-10}$ dynes/cm²	$\{\ \}**4L^2f^2\times10^{-10}$	$B_j\nu_j\times10^{-10}$	$\dfrac{B_j\nu_j}{E\nu}\times10^{-10}$	Δ_{Ej}	Δ_E	$p^{-1/3}E\Delta_E\times10^{-5}$
100	.84***	729	331	.46	.0053	.0077	95
1200	3.00***	1646	230	.16	.0028	.0033	93
3250	4.85	2518	230†	.10	.0024	.0026	86
5000	5.85	2993	230†	.08	.0024	.0026	89

*$\{\ \} = \rho\nu + 2\rho_j\nu_{sj} + 2\rho_j\nu_{ej} = 905$ gm.

**$\{\ \} = \rho\nu + \rho_j\nu_{sj} + 2\rho_j\nu_{ej} = 885$ gm.

***These numbers were estimated from measurements made with Texin jackets and then $2N_j\nu_j$ and $B_j\nu_j$ computed by Eqs. 5 and 7.

†These numbers were assumed equal to the values at 1000 psi and then N and E computed by Eqs. 5 and 7.

Biot's theory.[1] The vibration of the high permeability sand, however, lies within the high frequency range of the theory.

The difference in shape of the two curves is striking. In the low frequency range the curve has approximately the shape of the relative permeability curve for water in unconsolidated sand;* in the high frequency range, however, structure parameters in addition to relative permeability must be introduced. A more detailed account of losses in fluid-filled unconsolidated sands will be given elsewhere.

CONCLUSIONS

The jostling decrement of damp rocks subjected to no confining pressure may consist of two components, as noted by Born—one independent of frequency and one proportional to frequency; but when the same samples are

*Exact agreement with the relative permeability curve cannot be expected because, at low water saturations for which the relative permeability is zero, a loss caused by the water can be detected.

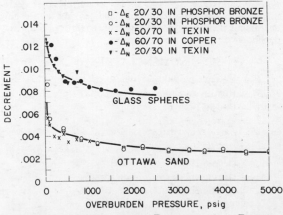

FIG. 10—UPPER BOUNDS FOR THE EXTENSIONAL AND TORSIONAL DECREMENTS IN PACKINGS OF QUARTZ AND GLASS AS A FUNCTION OF PRESSURE.

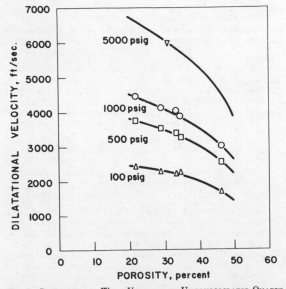

FIG. 9—DILATATIONAL WAVE VELOCITY IN UNCONSOLIDATED QUARTZ SANDS AS A FUNCTION OF POROSITY AT FIXED PRESSURES.

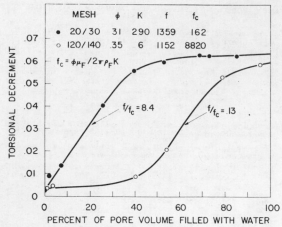

MESH	ϕ	K	f	f_c
● 20/30	31	290	1359	162
○ 120/140	.35	6	1152	8820

$f_c = \phi \mu_F / 2\pi \rho_F K$

$f/f_c = 8.4$ $f/f_c = .13$

FIG. 11—TORSIONAL DECREMENTS AT 700 PSIG VS PER CENT OF PORE VOLUME FILLED WITH WATER.

subjected to an overburden pressure of several thousand psi the jostling decrement of the damp rock is almost equal to the decrement of the dry rock. Thus at depth in the ground the jostling decrement should be independent of frequency and pressure.

Most sandstones under pressure have a jostling decrement less than 0.025. It decreases with decrease in porosity.

The jostling decrements for dilatational and shear waves are usually equal.

Unconsolidated sands have remarkably low decrements, which decrease only slightly with increase in pressure and are independent of both porosity and grain-size.

The decrements of unconsolidated sands increase when a small mass of water is added by an amount proportional to the mass added.

The decrements in all cases are independent of the amplitude of vibration at small amplitudes.

The increase in Young's modulus of sand is proportional to $p^{1/2}$ in the range 0 to 1000 psi, and is a linear function of porosity at a fixed pressure.

In principle a sloshing decrement can be measured and thus a permeability log made. This can be done by measuring total decrement at a single frequency and subtracting from it a jostling decrement of a magnitude fixed by the porosity and type of rock; or by measuring total decrements at two or more frequencies and deriving the sloshing decrement from the knowledge of its linear frequency dependence.

NOMENCLATURE

E = Young's modulus, dynes/sq cm,
N = Rigidity modulus, dynes/sq cm,
ρ = Density, gm/cc,
λ = Wavelength, cm,
f = Frequency, cycles/sec,
f_c = Critical frequency, cycles/sec = $\phi \mu_F / 2\pi \rho_F K$,
Δ_S = Decrement for shear wave,
Δ_E = Decrement for extensional wave,
Δ_D = Decrement for dilatational wave,
A = Total surface area of cylinder, sq cm,
M = Mass of cylinder, gm,
μ_F = Viscosity of fluid, poise,

ρ_F = Density of fluid, gm/cc,
D = Diameter of cylinder, cm,
L = Length of cylinder, cm,
ρ_s = Density of sample, gm/cc,
ϕ = Porosity of sample,
σ = Poisson's ratio of sample,
K = Permeability of sample, sq cm,
C = Velocity of sound in fluid, cm/sec,
N_j = Rigidity modulus of jacket, dynes/sq cm,
E_j = Young's modulus of jacket, dynes/sq cm,
σ_j = Poisson's ratio for jacket,
v_{sj} = Volume of jacket on curved surface of sample, cc,
v_{ej} = Volume of jacket on both ends of sample, cc,
ρ_j = Density of jacket, gm/cc,
Δ_{Nj} = Torsional decrement for jacket,
Δ_{Ej} = Extensional decrement for jacket,
Δ_{NC} = Decrement for torsional oscillation of coated rod,
Δ_{EC} = Decrement for extensional oscillation of coated rod,
V_d = Dilatational velocity, cm/sec,
V_s = Shear velocity, cm/sec,
V_e = Extensional velocity, cm/sec,
E_Q = Young's modulus of quartz $\simeq 85 \times 10^{10}$, dynes/sq cm,
E_G = Young's modulus of glass $\simeq 65 \times 10^{10}$, dynes/sq cm,
B_1, B_2 = Defined in Eq. 10.

ACKNOWLEDGMENT

We are indebted to W. P. Acheson for the derivation of Eq. 12 and the results in Table 2, and to J. H. Messmer and R. D. Wyckoff for assistance in constructing the apparatus.

REFERENCES

1. Wyllie, M. R. J., Gardner, G. H. F. and Gregory, A. R.: "Studies of Elastic Wave Attenuation in Porous Media", *Geophysics* (1962) **27**, 569.
2. Born, W. T.: "The Attenuation Constant of Earth Materials", *Geophysics* (1941) **6**, 132.
3. Gondouin, M.: "Analysis of a Method of Acoustic Logging", Thesis, Colorado School of Mines (1952).
4. Volarovich, M. P., Levykin, A. I. and Sizov, V. P.: "Investigation of the Attenuation of Elastic Waves in Rock Specimens", (Izv.) Acad. Sci. U.S.S.R. Geophys. Ser. (1960) 8, 793.
5. Tsareva, N. V.: "Propagation of Elastic Waves in Sands", (Izv.) Acad. Sci. U.S.S.R. Geophys. Ser. (1956) 9, 1044.
6. Hardin, B. O. and Richart, F. E.: "Elastic Wave Velocities in Granular Soils", *Jour. Soil Mech. and Foundation Div., Proc.,* ASCE (1963) **89**, 33.
7. McSkimmin, H. J.: "Notes and References for the Measurement of Elastic Moduli by Means of Ultrasonic Waves", *Jour. Acoust. Soc. Am.* (1961) **33**, 606.
8. Naar, J. and Wygal, R. J.: "Structures and Properties of Unconsolidated Aggregates", *Canadian Jour. of Physics* (1962) **40**, 818.
9. Bancroft, D.: "The Velocity of Longitudinal Waves in Cylindrical Bars", *Phys. Rev.* (1941) **59**, 588.
10. Spinner, S. and Tefft, W. E.: "A Method for Determining Mechanical Resonance Frequencies and for Calculating Elastic Moduli from These Frequencies", *Proc.,* ASTM (1961) **61**, 1221.
11. Browne, M. T. and Pattison, J. R.: "The Damping Effect of Surrounding Gases on a Cylinder in Longitudinal Oscillations", *Brit. Jour. Appl. Phys.* (1957) **8**, 452.
12. Gemant, A.: "The Measurement of Solid Friction in Plastics", *Physics* (1940) **11**, 647.
13. Wachtman, J. B. and Tefft, W. E.: "Effect of Suspension Position on Apparent Values of Internal Friction Determinations by Forster's Method", *Rev. Sci. Inst.* (1958) **29**, 517.

14. Gassmann, F.: "Elastic Waves Through a Packing of Spheres", *Geophysics* (1951) **16**, 673.

15. Brandt, H.: "A Study of the Speed of Sound in Porous Granular Media", *Trans.,* ASME (1955) **77**, 470.

16. Timoshenko, S. and Goodier, J. N.: *Theory of Elasticity*, McGraw-Hill Book Co., Inc.. N. Y., 2nd Ed. (1951) 372.

17. Mindlin, R. D.: "Mechanics of Granular Media", *Proc.,* 2nd Nat. Congr. Appl. Mech. (1955) 13.

18. Johnston, K. L.: "Surface Interaction Between Elastically Loaded Bodies Under Tangential Forces", *Proc.,* Roy. Soc. Series A (1955) **230**, 531.

19. Knopoff, L. and MacDonald, G. J. F.: "Attenuation of Small Amplitude Stress Waves in Solids", *Rev. Mod. Phys.* (1958) **30**, 1178.

20. Bruckshaw, J. McG. and Mahanta, P. C.: "The Variation of the Elastic Constants of Rocks with Frequency", *Petroleum* (1954) **17**, 14.

21. Peselnick, L. and Zietz, I.: "Internal Friction of Fine Grained Limestones at Ultrasonic Frequencies", *Geophysics* (1959) **24**, 258.

22. Peselnick, L. and Outerbridge, W. F.: "Internal Friction and Rigidity Modulus of Solenhofen Limestone over a Wide Frequency Range", *U. S. Geol. Survey*, Prof. Paper No. 400B (1960).

23. Pattison, J. R.: "An Apparatus for the Accurate Measurement of Internal Friction", *Rev. Sci. Instr.* (1954) **25**, 490.

24. Gemant, A.: *Frictional Phenomena,* Chem. Publishing Co. (1950) 327.

25. Kolsky, H.: *Stress Waves in Solids*, Oxford U. Press (1953) 143.

26. Smith, W. O., Foote, P. D. and Busang, P. F.: "Packing of Homogeneous Spheres", *Phys. Review* (1929) **34**, 1271.

27. Bernal, J. D. and Mason, J.: "Co-ordination of Randomly Packed Spheres", *Nature* (1960) **188**, 910.

28. Botset, H. G. and Reed, D. W.: "Experiment on Compressibility of Sand ", *Bull.,* AAPG (1935) **19**, 1053.   ★★★

Reprinted from Journal of Geophysical Research, v. 73, p. 3917–3935.

Velocity and Attenuation of Seismic Waves in Imperfectly Elastic Rock

R. B. Gordon and L. A. Davis

Kline Geology Laboratory, Yale University, New Haven, Connecticut 06520

The mechanical properties of crystalline rock have been studied in the laboratory as a function of temperature and pressure at the low frequencies and the strain amplitudes characteristic of seismic waves. The inelasticity observed arises at interfaces in the rock structure and is insensitive to temperature but highly pressure dependent. It cannot be described in terms of viscoelasticity but causes the rock to display static hysteresis. The internal friction ϕ of the rock is accompanied by a large modulus defect, $\delta M/M$; both ϕ and $\delta M/M$ are shown to be independent of frequency and strain amplitude throughout a very wide range. A model based on a network of cracks in an elastic medium accounts for these properties. The presence of a fluid phase in the rock does not in itself significantly increase the internal friction in the range of frequencies of seismic waves, but the presence of a small amount of intergranular fluid can make interface inelasticity persist in the presence of a large confining pressure. Interface inelasticity can occur at substantial depths in the earth, resulting in low seismic velocities and low Q. The low velocity zones in certain areas are interpreted in this way as arising from the presence of a fluid phase.

Introduction

If the material of the interior of the earth were everywhere perfectly elastic, the propagation velocity of a seismic signal at any point would be determined by the local value of the elastic constants and the density. From the seismically determined velocities (measured at frequencies ranging from 10 Hz downward), inferences about the constitution of the material traversed could be drawn on the basis of elastic constant and density data obtained in the laboratory (usually taken in the frequency range from 1 Hz upward, most often in the megahertz range). In fact, the materials of the earth's crust and mantle are imperfectly elastic, as is shown, for example, by the damping of seismic waves. In imperfectly elastic material the wave propagation velocity is, in general, dependent on the frequency and amplitude of the signal being observed. It may also be structure sensitive in ways not encountered in perfectly elastic materials. The occurrence of imperfect elasticity may arise from a multiplicity of causes [*Gordon and Nelson*, 1966]; this paper deals with measurement of the velocity changes and internal friction that occur in rock because of friction on interfaces within the rock structure. The experiments aim at determining the magnitude of the internal friction under conditions that would characterize a seismic wave traveling through the earth; measurements are made over a wide range of frequencies and strain amplitudes, and the influence of pressure and temperature is investigated. A summary of previous experimental work on the internal friction of rock, most of it confined to narrow ranges of experimental conditions, is given in the review article by *Knopoff* [1964].

Imperfect Elasticity

The mechanical properties of a perfectly elastic material are completely described by its elastic constants. When the elasticity is imperfect, it is generally assumed for the purpose of describing experimental results that the material is linearly viscoelastic, i.e., that stress σ and strain ϵ are related by the integral equation

$$\sigma(t) = \int_0^t G(t - \xi)\epsilon(\xi) \, d\xi \qquad (1)$$

and that the inelastic effects are small compared with the elastic ones. If a stress

$$\sigma = \sigma_1 e^{i\omega t} \qquad (2)$$

is applied to the viscoelastic material for a time sufficient to attain a steady state, the resultant strain may be resolved into in-phase and out-of-phase components ϵ_1 and ϵ_2:

$$\epsilon = (\epsilon_1 - i\epsilon_2)e^{i\omega t} \qquad (3)$$

(A parallel development starting from an impressed strain $\epsilon = \epsilon_1 e^{i\omega t}$ is possible.) The imperfectly elastic material is then characterized by an internal friction, ϕ, where

$$\phi = \tan^{-1}(\epsilon_2/\epsilon_1) \qquad (4)$$

and a dynamic modulus

$$M = \sigma_1/\epsilon_1 \qquad (5)$$

Another convenient assumption [*Nowick*, 1953] is to divide the total strain into an elastic strain ϵ' and a non-elastic strain ϵ'', such that

$$\epsilon' = \epsilon_1' e^{i\omega t} \qquad (6)$$

$$\epsilon'' = (\epsilon_1'' - i\epsilon_2'')e^{i\omega t} \qquad (7)$$

With periodic excitation, the elastic strain is always in phase with the driving stress, whereas the non-elastic strain contains in-phase and out-of-phase components ϵ_1'' and ϵ_2''. The internal friction is then

$$\phi \simeq \epsilon_2''/\epsilon_1' \qquad (8)$$

and the dynamic modulus is

$$M = M'[1 - (\epsilon_1''/\epsilon_1')] \qquad (9)$$

$M' = \sigma_1/\epsilon_1'$ is the elastic modulus of the material and

$$(M' - M)/M = \delta M/M = \epsilon_1''/\epsilon_1' \qquad (10)$$

is its modulus defect. The inelasticity is therefore characterized by the two quantities ϕ and $\delta M/M$. The separation of the total strain into elastic and non-elastic parts is in accord with the physical situation in a real material, where, for example, ϵ' may arise from the changes in interatomic distance under the applied stress and ϵ'' from the stress-induced motion of dislocations or point defects.

To analyze dynamic experiments performed on an inelastic material (this may include both laboratory and seismic experiments), a relation between σ and ϵ is needed. A linear viscoelastic material can be described by a complex modulus

$$\tilde{M} = M_1(\omega) + iM_2(\omega) \qquad (11)$$

for this purpose. In the case of a wave

$$e^{-\alpha(\omega)x} \sin \omega \left[t - \frac{x}{v(\omega)} \right]$$

traversing the viscoelastic material, the attenuation $\alpha(\omega)$ and the phase velocity $v(\omega)$ are then determined by $M_1(\omega)$ and $M_2(\omega)$:

$$\alpha(\omega) = f[M_1(\omega) + iM_2(\omega)]$$
$$v(\omega) = g[M_1(\omega) + iM_2(\omega)] \qquad (12)$$

There is thus a fixed relation between the magnitudes of α and v, which, however, cannot be given in explicit form except in special cases. An important special case is that of the standard linear solid of relaxation time τ [*Zener*, 1948], where

$$\phi = (\delta M/M)\omega\tau \qquad (13)$$

The inelasticity that results from the stress-induced ordering of solute atoms is found, for example, to be represented very accurately by the properties of a standard linear solid [*Nowick*, 1964].

The complex modulus concept is useful in the solution of the equations of motion that represent the behavior of various pieces of laboratory apparatus used to determine the dynamic mechanical properties of solids. By considering either the forced or free vibrations of a sample attached to an inertia member or the propagation of waves whose wavelength is small compared with the sample size, it may be shown that the various measures of internal friction are related as follows:

$$\phi = Q^{-1} = \Delta/\pi = \delta W/2\pi W = \alpha\lambda/\pi \qquad (14)$$

In these relations ϕ is the phase lag between the applied periodic stress and the resultant strain, Q^{-1} is determined by the half-width of a mechanical resonance curve, Δ is the logarithmic decrement for free vibration, δW is the amount of the strain energy W dissipated in a cycle, and α is the spatial attenuation of a traveling wave of wavelength λ. It is important to remember that these relations are exact only for the special case of linear viscoelasticity and small damping.

A number of different phenomena are expected to contribute to the inelasticity of a complex material such as a rock. Some inelastic processes, stress-induced ordering of solute atoms for example, make a crystalline solid behave as a viscoelastic body; the results of other processes cannot be so described. Among the other processes are two important sources of internal friction: inelasticity resulting from sliding on interfaces within the body and dislocation-breakaway damping [*Granato and Lücke*, 1956]. The inelasticity produced by these processes may be described as being due to static hysteresis. Under the excitation $\sigma = \sigma_1 \sin \omega t$, the steady-state strain of a viscoelastic body is

$$\epsilon = \epsilon_1 \sin \omega t - \epsilon_2 \cos \omega t$$

The stress-strain (σ-ϵ) curve is an ellipse that closes to a straight line at $\omega = 0$ and $\omega \to \infty$. In the case of static hysteresis, however, the σ-ϵ curve under periodic stress is a continuous loop that is independent of the rate of traverse. If a quasi-static experiment is done on a material displaying static hysteresis by stopping at successive stress values for strain measurements, the loop traversed is the same as in a dynamic experiment. Because of this frequency independence, the material cannot be described as viscoelastic, and equations 1, 11, and 12 are not applicable. It has been shown by E. T. Onat (private communication, 1967) that for the case of static hysteresis the σ-ϵ loop must have sharp corners (in contrast to a viscoelastic material, where the loop is an ellipse). For material that displays static hysteresis, the strain in the steady state must be represented by a series:

$$\epsilon = a_1 \sin \omega t + b_1 \cos \omega t + a_2 \sin 2\omega t$$
$$+ \, b_2 \cos 2\omega t + \cdots$$

Observation shows that in many cases the first two terms in this series are much more important than the rest. Then the σ-ϵ curve can be approximately represented as an ellipse. It is sometimes convenient to describe such a material as if it were viscoelastic, i.e., to characterize it by an internal friction, $\phi = \delta W / 2\pi W$, and a modulus defect, $\delta M/M$. The internal friction can be measured by any of the conventional laboratory methods, but the magnitude of $\delta M/M$ can only be determined when some way of suppressing the inelasticity can be found. For static hysteresis due to dislocation motion, appropriate irradiation may serve this purpose [*Gordon*, 1956]; for interface inelasticity, a large normal force on the sliding interfaces can be used.

Inelasticity in crystalline rock can arise from many sources. All these sources will have modulus defects, and hence sound velocity changes, associated with them. For the viscoelastic loss mechanisms, the magnitude of $\delta M/M$ is limited by equation 13 and by the fact that occurrence of $\phi > 0.1$ is rare. Large $\delta M/M$ can arise from static hysteresis, and the most important source of this in rock is interface inelasticity.

Experiments

Seismic waves are characterized by low strain amplitude and low frequency. The material that they traverse is for the most part at high temperature and under high pressure. These conditions cannot now be simultaneously obtained in laboratory experiments on wave propagation through solid materials. It is necessary to find the dependence of the inelasticity on each of the above variables in separate experiments. Observations of ϕ and M at low frequency (100 μHz to 10 Hz) and low strain amplitude ($\epsilon_0 \simeq 10^{-9}$) are wanted. In the laboratory, low frequency measurements of ϕ can best be made when $\epsilon_0 \gtrsim 10^{-6}$, owing to practical limitations on the direct measurement of strain. However, measurements can be made at strain amplitudes ranging down to 10^{-10} by ultrasonic resonance and pulse-echo methods. The procedure adopted for our experiments is to measure ϕ at high frequency over the amplitude range $10^{-10} < \epsilon_0 < 10^{-5}$ by the driven resonance method, and at low frequency in the range $10^{-5} < \epsilon_0 < 10^{-3}$ by direct determination of the stress-strain curve. Because the internal friction of rock is variable from sample to sample and is often sensitive to handling, measurements that are to be compared with one another must be made on a single sample; sample size and shape must therefore meet the requirements of both experimental methods.

In practice, the measurement by the driven resonance method is effected by use of the composite piezoelectric resonator illustrated in Figure 1. The quartz bar is made exactly 1 wavelength long for the frequency to be used. Driving voltage (ranging up to 1000 volts) is applied to the upper electrodes covering half the bar. The voltmeter connected to the narrow electrodes at the center of the lower half of the bar indicates the resultant vibration amplitude. The specimen to be studied is made

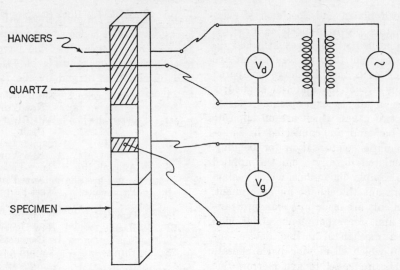

Fig. 1. Composite piezoelectric resonator used for measurement of internal friction and modulus at 90 kHz. Disposition of electrodes shown is for longitudinal vibration.

one-half wavelength long and may be cemented to either end of the quartz bar. Let subscripts c refer to the composite resonator, q to the quartz bar alone, and s to the sample alone. *Marx* [1951] has shown that the following relations hold for the internal friction, resonant frequency, and strain amplitude of the specimen:

$$\phi_s = (1/m_s)[m_c K_1(V_d/V_g) - m_q \phi_q] \quad (15)$$

$$f_s = (1/m_s)(m_c f_c - m_q f_q) \quad (16)$$

$$\epsilon_s = K_2 V_d \quad (17)$$

In these expressions m is mass, f is frequency of resonance, V_d and V_g are voltages measured as indicated in Figure 1, and K_1 and K_2 are constants for a particular resonator and are dependent on the circuit parameters and the properties of the quartz. By using the appropriate disposition of electrodes and orientation of the quartz, measurements may be made either in longitudinal or torsional vibration. For longitudinal vibration, the modulus M is Young's modulus, Y (correction is made for the finite thickness of the bar); for torsion, it is the shear modulus μ.

To make measurements at low frequency, a varying, cyclic displacement is applied to the sample and the resultant force is recorded. In general, the sample will be under some static bias force on which the varying force is super-

imposed. A schematic stress-strain (σ-ϵ) curve is shown in Figure 2. The internal friction is found by measuring the areas under the loading and unloading curves and by using equation 14. For the determination of ϕ by the σ-ϵ method to be successful, all sources of hysteresis in the force and strain measuring systems must be eliminated or allowed for through proper corrections. The system is tested by measurements on hardened steel blocks and samples of fused quartz that are free of measurable damping. In our experiments the sample is compressed by the motion of parallel steel platens. Strain may be measured by observing the platen displacement, provided that all lost

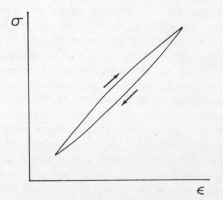

Fig. 2. Schematic stress-strain curve for a material displaying static hysteresis.

motion between platen and specimen is eliminated. Since this proves difficult in practice, we prefer to utilize strain gages attached directly to the sample. It is necessary that the active area of each gage be large compared with the grain size of the material studied, a condition met in all the experiments reported here. Use of two gages, which are on opposite sides of the bar and are connected in series, provides automatic compensation for bending. Stress is calculated directly from the applied force and the sample dimensions, which remain effectively constant throughout an experiment. The measurements are made in a standard testing machine and, for experiments under high pressure, in a mechanical testing device, the 'minitester,' in which all moving parts operate within the pressure vessel to avoid errors due to seal friction [*Gordon and Mike,* 1967]. True hydrostatic pressure is transmitted to the specimen by pentane containing a small amount of kerosene. When it is desired to isolate the sample from the pressure-transmitting fluid, a jacket of synthetic rubber is used. When the strain gages are separated from the rubber by a thin Teflon sheet, no measurable internal friction results from the presence of the jacket. Internal friction measurements can be made over strain amplitudes ranging from $\sim 10^{-5}$ upward to whatever maximum strain the sample will sustain during repeated loading. An upper limit of frequency is set by the response limit of the X-Y recorder system used; a lower limit, by the long-term stability of the amplifiers used to detect the strain gage and load cell signals. The practical frequency range of the apparatus used is $0.5 < f < 50$ mHz.

A variety of crystalline rocks was studied during the course of the experiments (see Table 1). The constraints on the selection of sample material are that the grain size be small compared with the specimen size and that the rock be sufficiently coherent to permit the cutting of specimen bars. The microstructure of each sample was observed by using both standard thin sections and polished sections examined with reflected light, the latter technique being more effective in revealing fine detail in the microstructure. It is found that the mineral constitution of the rock samples has little direct influence on their interface damping characteristics.

TABLE 1.   Rock Samples Used

Identifying Number	Rock Type	Locality
1	Anorthosite	Elizabethtown, New York
Q-2	Quartzite	Raft River Mountains, Utah
3	Granite	Tupper Lake, New York
5	Amphibolite	Carry Falls reservoir, New York
8	Pyroxenite	West Fort Ann, New York
15	Quartzite	Wyoming
25	Basalt	New Haven, Connecticut
28	Dunite	Tafjord area, Norway
29	Olivine basalt	New Mexico
33	Diopsidite	Northwest Adirondacks, New York

The experiments are made according to the following plan. First, the high frequency, longitudinal resonator is used to examine the internal friction of a large number of dry samples. This method is well suited for such a survey because of the speed with which data can be gathered. Measurements in each case are made over as wide a range of strain amplitudes as possible. The results show that the internal friction of most rock samples is highly structure sensitive but is almost amplitude independent. A number of measurements in the torsional mode of vibration are also made. Next, selected resonator samples are measured at low frequency using the σ-ϵ technique. These experiments show that the internal friction remains amplitude independent up to the highest attainable strains, and they demonstrate that the low frequency measurements yield a ϕ that would be characteristic of seismic conditions. Next, the influence of factors characteristic of the environment of rock in the earth is investigated. Because fluid phases are expected to be present in the earth, the effect of the presence of fluid on the internal friction is studied at both high and low frequency. Finally, the magnitude of the fluid pressure and the mechanical contact pressure on the sample is varied to reveal the pressure dependence of the internal friction. Under high contact pressure the modulus defect is eliminated, and the magnitude of the modulus defect associated with the internal friction may be found.

Internal friction of dry rock. Considerable care is required in the preparation of specimens to avoid mechanical damage or heating during cutting and grinding and to remove fluids introduced during the cutting process. The first measurements to be considered are on fully dry rock. The drying process found to be effective involves soaking the specimens that have been brought to their final dimensions first in pentane and next in alcohol; they are then held under vacuum. The process is repeated until it produces no further change in ϕ, at which point the rock is considered to be 'dry.' The samples really are dry in the sense that exposure to a damp atmosphere for a few minutes will, in many cases, result in a substantial increase in ϕ when measured at 90 kHz.

The results of measurements at 90 kHz in longitudinal vibration are shown in Figure 3. The striking characteristics of the results are that the internal friction of all the rocks is much greater than that of the single crystal quartz, that ϕ is highly structure sensitive (as illustrated by the fact that the highest and lowest ϕ's observed both occur in quartzite),

and that the ϕ is quite insensitive to the strain amplitude. 'Structure sensitive' means here that ϕ is sensitive to details of the microstructure of the particular sample studied. The magnitude of ϕ is not found to correlate with the mineral constitution of the rock; in fact, the contribution of the intrinsic ϕ of each mineral grain to the ϕ of the polycrystal is negligible. This is certainly true for quartz (where single crystal internal friction data are available) and almost certainly true for the feldspars, pyroxenes, and olivines which are the other principal mineral constituents of the various rocks studied. Single crystal internal friction data for these minerals are not available (because good single crystal specimens are difficult to find), but, from consideration of atomic and dislocation mobility in these materials, low ϕ at room temperature is expected. The observed ϕ must be due to the interfaces present in the rock. There is not, however, a correlation between the magnitude of ϕ and the grain size (true whether one considers both cracks and grain boundaries or grain boundaries alone as the active interfaces), indicating that interface properties are variable from sample to sample. Support for this hypothesis on the origin of the internal friction shown in the data of Figure 3 is found in an experiment in which a

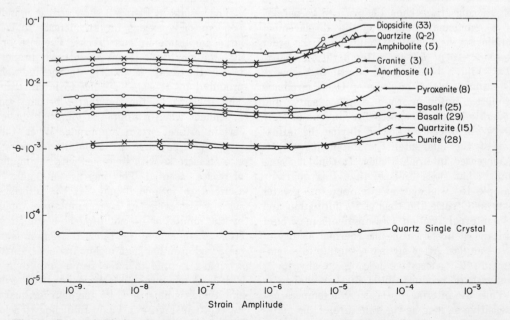

Fig. 3. Internal friction of dry rock samples measured in longitudinal oscillation at 90 kHz.

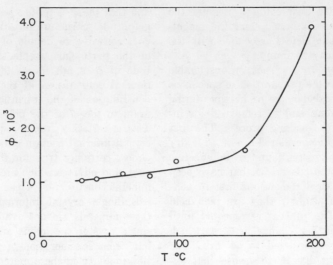

Fig. 4.   Internal friction at 90 kHz in longitudinal oscillation of quartzite (sample 15) measured at room temperature after heating to the temperature indicated.

sample of quartzite (sample 15) was heated to successively higher temperatures. Initially the ϕ of this material is very low. After heating to each temperature it is remeasured; the results are shown in Figure 4. Because of the anisotropy of the thermal expansion of quartz, localized strains develop at the interfaces between grains during heating. If the directions of greatest and least thermal expansion in neighboring grains should happen to be parallel on a particular boundary, the displacement parallel to the boundary would be of the order of $(5 \times 10^{-6})l \ \delta T$ for a temperature change $\delta T°C$ and a boundary of length l. A change of temperature of the order of 100°C produces significant grain-interface displacements, and moderate heating is expected to unbond or fracture grain interfaces bonded during the formation of the quartzite. The unbonding results in increased internal friction. It could be concluded that it is unlikely that this quartzite (sample 15) was ever at a temperature greater than 60°C since the time of its formation and that internal friction measurements may offer a means of investigating the thermal history of rock samples. After heating a sample to a temperature T_1, successive heating to any lower temperature produces little further change in ϕ. Thus the interface inelasticity is temperature insensitive so long as the structure of the material is unaltered by the temperature changes.

Further evidence that the observed ϕ has its origin at interfaces comes from observations on samples containing weathering products. Such material might, for example, cement together the interfaces in a rock, resulting in a reduction of ϕ. This has evidently occurred in dunite (sample 28); it contains a small amount of serpentinite running through its interfaces and is characterized by a low ϕ.

The results in Figure 3 show that ϕ is independent of ϵ_0 except for an increase in ϕ at large amplitude in some rocks. The rise in ϕ at large amplitude is not an intrinsic property but is the result of damage to the interfaces produced by repeated cycling to large strains. This is shown by the fact that ϕ becomes time dependent when a sample is vibrated above a certain critical strain amplitude. When there is an increase in ϕ with amplitude, there is also an increase with time, i.e., with the number of cycles executed. This increase of ϕ may be ascribed to fatigue damage at the interfaces, since a few cycles at large amplitude applied by the minitester do not produce a similar increase in ϕ when the internal friction is remeasured at either high or low frequency. After driving a sample to large strain amplitude at 90 kHz, ϕ remains high when remeasured at low amplitude, as illustrated by the data for quartzite (Q-2) in Figure 5. Upon returning to a low vibration amplitude, some of the fatigue damage

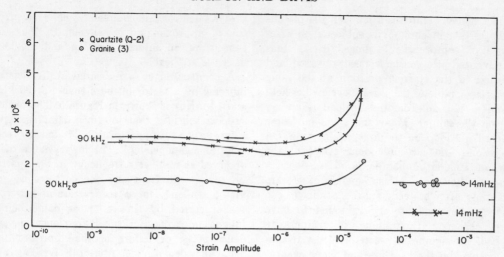

Fig. 5. Internal friction of granite and quartzite measured first at 90 kHz and then, by the σ-ε method, at 14 mHz.

is often observed to disappear spontaneously, as revealed by a tendency of ϕ to decrease with time. The initial value of ϕ is not, however, recovered even after prolonged resting of the specimen.

Measurements of ϕ by driven resonance may also be made in the torsional mode of vibration. Because ϕ is structure sensitive, a reliable comparison of the internal friction in torsional vibration ϕ_t, and longitudinal vibration ϕ_l, can be made only if both measurements are made on the same sample. The practical complication here is the difference in the length of sample required for 90-kHz resonance in the two cases. The ratio of resonant frequencies f_t/f_l varies from rock to rock, and it is not practical to prepare a quartz resonator for the torsional vibration of each of the 90-kHz longitudinal samples. The alternative is to match a sample for 90-kHz longitudinal vibration and then cut it down for 90-kHz torsional vibration. In samples with high internal friction, significant changes in ϕ occur during even the most careful cutting, as may be demonstrated by matching one sample to two different longitudinal resonators of different frequency. Consequently, the measurements in torsional resonance have been confined to the samples of relatively low ϕ, where experience shows that cuts in length can be made without greatly altering the internal friction. The results of these measurements are that the damping char-

acteristics in torsion are generally the same as in longitudinal vibration and that $\phi_t/\phi_l \approx 0.7$.

Because the resonator experiments show that the damping characteristics of the rock are insensitive to mineral constitution, most of the following, more detailed experiments on the frequency, amplitude, and pressure dependence of the interface damping are done on samples of the granite and quartzite alone, these being taken as representative rocks.

To find the internal friction of the rock samples at both high and low frequency, stress-strain and resonance measurements are done on the same sample, and that sample is so handled as not to change its internal friction. Strain gages are first attached to the specimen, after which ϕ_l is measured with the composite resonator. Mounting gages on a sample of fused quartz and measuring the apparent ϕ before and after mounting show that the increase in ϕ due to the presence of the gages is about 2×10^{-4} and thus is negligible compared with the magnitude of ϕ in the rock samples. High and low frequency measurements can then be made, leaving the specimen essentially undisturbed. This is demonstrated by an experiment on quartzite (Q-2), where ϕ_l is measured on the resonator, then in the minitester, again on the resonator, and finally again in the minitester; no changes in ϕ resulted during this sequence of tests.

After a rock sample is set up on the platens of the testing machine, it is subjected to a sequence of compressive loadings. The platens are advanced at a constant speed, reversed, and returned to the starting position at the same speed to complete the strain cycle. To hold alignment a small compressive load is always held on the sample. Many rocks will not show a closed σ-ϵ loop on the first strain cycle. On the second and all subsequent cycles of the same or smaller amplitude, a closed loop is obtained, however. Measurements of ϕ are made only after a steady state is attained. The σ-ϵ loops obtained are always found to have sharp corners. If straining is temporarily stopped, no significant creep or stress relaxation occur, and the loop size and shape eventually attained on completion of the displacement cycle are unaffected by the interruptions. Thus, the characteristics of static hysteresis are observed to occur in the rocks tested. Figure 6 shows a reproduction of a typical re-

corder trace for one complete cycle. This curve can be approximated by an ellipse; thus the concept of an internal friction and modulus defect is applicable.

A comparison of ϕ measured at frequencies differing by a factor of more than 10^6 (90 kHz and 14 mHz) is shown in Figure 5. The results support the idea that interface damping is, to a very good approximation, frequency and amplitude independent. The lower values of ϕ observed at millihertz frequency in the quartzite could arise from any one of three causes, viz., a tendency for ϕ to decrease at larger ϵ_0 (as suggested by the resonator data) before fatigue intervenes, a small frequency dependence of ϕ, or failure of the approximation made in calculating ϕ from $\delta W/W$. This approximation is that the observed sharp-cornered σ-ϵ curve can be represented as an ellipse. More detailed experiments would be required to distinguish between these possibilities.

A further test of the frequency independence of ϕ is made by varying the frequency used in the σ-ϵ curve measurements. Results of such a set of measurements are shown in Figure 7 for the same two types of rock for which data are shown in Figure 5. The observed frequency dependence of ϕ in the millicycle range is zero within experimental error. These results thus provide experimental support for the concept of a frequency-independent internal friction in rock developed by *Knopoff* [1964].

On the basis of the above observations showing the frequency and amplitude independence of ϕ, it is concluded that the laboratory data obtained in these experiments are representative of the inelastic response of the rock when excited by a wave having the low frequency and strain amplitude characteristic of seismic waves.

Internal friction of rock containing a fluid phase. Perfectly dry rock is not expected to occur very frequently in the earth because of the presence of ground water, of hydrothermal solutions, or, at greater depth, of partial melting. Experiments were therefore done to find the contribution to ϕ from the presence of a fluid phase in the rock. Results of two sets of measurements at high and low frequency on a sample of granite (sample 3) are shown in Figure 8. In one set the sample was dried as completely as possible; in the other it was saturated with water by soaking for several

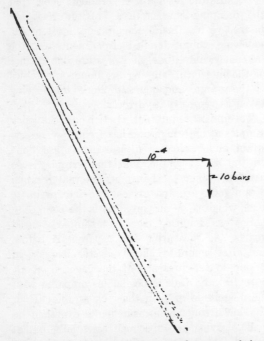

Fig. 6. Photograph of the recorder trace of the stress-strain curve of a bar of quartzite (Q-2) under cyclic loading at a frequency of 10 mHz. Compressive strain increases to the left along the horizontal axis; compressive stress increases upward. The two extreme ends of the loop are sharp corners.

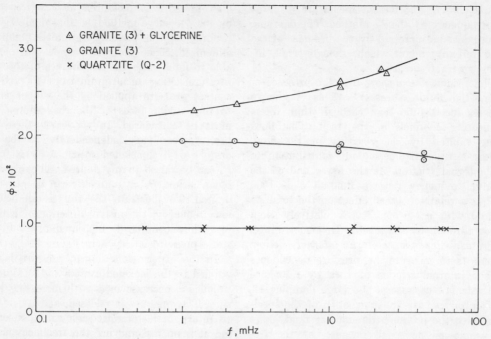

Fig. 7. Internal friction of dry granite and quartzite and of granite saturated with glycerine measured by the stress-strain curve method. Samples are not the same as those of Figure 5.

weeks. At 90 kHz the presence of water increases ϕ by almost an order of magnitude. At 14 mHz the ϕ of the wet and dry samples is almost the same. The fluid phase contribution to the total ϕ is therefore frequency dependent and much reduced at low frequency. The fatigue effect due to repeated cycling to high strain amplitude is still present in the wet sample at 90 kHz, but otherwise the internal friction is amplitude independent as in the case of the dry samples. Results of further experiments on the internal friction of rock containing a fluid phase are shown in Figure 7. In this case a saturating fluid of high viscosity, glycer-

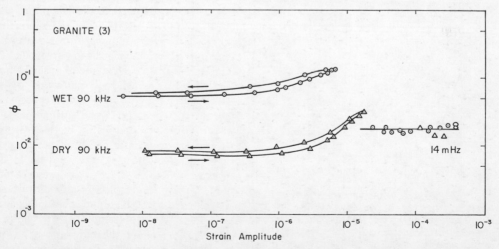

Fig. 8. Internal friction of granite (sample 3) measured at two frequencies when dry and when saturated with water.

ine, is chosen, and measurements are made at low frequency by the σ-ϵ method. The damping is found to be increased by the presence of the fluid by an amount which, though small, increases with frequency.

The results above show that the presence of interstitial fluid increases ϕ. This is due to stress-induced fluid flow through thin, interconnecting channels in the rock. This flow, being limited by the viscosity of the fluid, makes a frequency-dependent contribution to the internal friction. At the lower end of the seismic frequency range a fluid of $\eta \simeq 1500$ cp has a relatively small effect on the total ϕ; a fluid with $\eta \sim 1$ cp has a relatively small effect throughout the seismic frequency range.

Pressure dependence of the internal friction. Attempts to measure the pressure dependence of the internal friction of rock by resonance methods fail because of the large damping effects produced by the interaction of the specimen with the pressure-transmitting fluid. Pulse ultrasonic methods fail because, at the frequencies used for laboratory scale experiments, scattering losses dominate the observed attenuation. Therefore, the internal friction under

high pressure is measured at low frequency by the σ-ϵ curve method, so that damping due to drag of the pressure-transmitting fluid is minimized. Since the rock samples are permeable to fluid, both the magnitudes of the fluid pressure acting on all grain faces, P_f, and the contact pressure applied by the rubber jacket to the outside faces of the rock sample, P_o, must be considered. In the earth these two pressures can vary independently; in the laboratory with a jacketed specimen $P_f = 0$ and P_o can be raised to any desired value, or, without a jacket, $P_o = 0$ and P_f can be raised as desired. To illustrate the results obtained at low frequency, the internal friction of a block of granite (sample 3) measured under hydrostatic pressure is shown in Figure 9.

In the experiments shown, pressure is first applied to the jacketed sample and is found to produce a decrease in ϕ with increasing pressure. The pressure is released, and ϕ is found to return to its original high value. Next the jacket is opened and pressure again applied to make $P_o - P_f < 0$. This produces an increase in ϕ slightly greater than the small increase expected from the fluid damping alone. At a

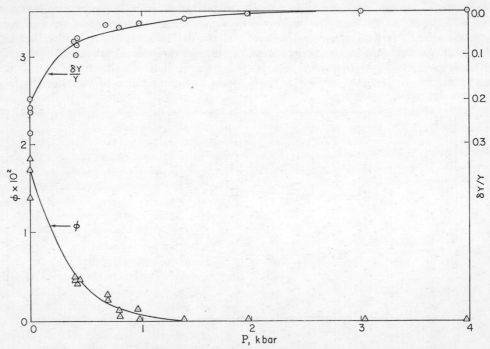

Fig. 9. Internal friction and modulus defect of granite (sample 3) measured at 10 mHz while jacketed and under hydrostatic pressure.

sufficiently high fluid pressure the interfaces in the rock would begin to lose contact, owing to compression of the solid phases. Such an extreme excess fluid pressure is not expected to occur in the earth, however, because of the permeability of layers surrounding a region containing fluid. Throughout the experiment the changes produced by the applied pressure are found to be reversible.

At each pressure used during the experiment, Young's modulus, Y, of the specimen is determined from the slope of the increasing-stress part of the σ-ϵ curve. As $P_o - P_f$ increases and the internal friction of the rock is reduced, the modulus is correspondingly increased. For $P_o - P_f \gtrsim 2$ kb, little further change in ϕ occurs. Under high excess contact pressure the absolute value of Y approaches the value expected to arise from the average of the intrinsic moduli of the constituents of the granite, i.e., about 0.8 Mb. The effect of a large contact pressure is to clamp the interfaces that are responsible for the internal friction; large contact pressure effectively stops the inelastic processes occurring at the interfaces of the sample and permits a determination of the modulus defect. The change of $\delta Y/Y$ with pressure is displayed in Figure 9. The experiment shows that a characteristic of interface inelasticity is that there is a large $\delta Y/Y$ associated with a large ϕ. This is in marked contrast to the behavior expected of a viscoelastic solid, where an upper limit to $\delta Y/Y$ is $\sim 10^{-2}$.

Results comparable with those obtained for the modulus defect by the σ-ϵ method are found

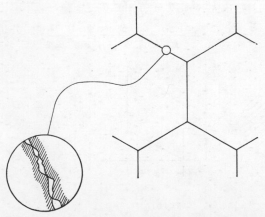

Fig. 10. Representation of the network of interfaces in a polycrystalline rock.

when a jacketed torsion resonator or pulse-echo specimen is subjected to hydrostatic pressure, as is, in fact, well known from the experiments that have been done to find the velocity of sound in rocks; a contact pressure of the order of 1–4 kb is required before a 'true' velocity is measured. The magnitude of $\delta M/M$ observed in the high frequency experiments is found to be variable among rocks samples in the same way as the ϕ of Figure 3. For example, *Christensen*'s [1966] pulse-echo data reveal a much reduced modulus defect in serpentinized peridotite as compared with unweathered material; the serpentinized dunite of Figure 3 likewise displays a low internal friction.

Discussion

Theory of the inelasticity. The inelasticity of the rock samples studied in the laboratory is characterized by an internal friction that is frequency and amplitude independent, insensitive to temperature and bulk mineral constitution of the rock, but sensitive to hydrostatic contact pressure and the mechanical history of the sample. Accompanying the internal friction is a large modulus defect. Since these are not the characteristics of inelasticity in homogeneous, crystalline materials, the non-elastic behavior must, as suggested many years ago by *Ide* [1937], arise at the interfaces within the rock. That relative motion can occur at the interfaces within an apparently solid rock sample is shown by the occurrence of microplasticity and of a volume compressibility in excess of the compressibilities of the constituent minerals [*Simmons and Brace*, 1965]. Microplasticity refers to the small, nonrecoverable deformation found in most rocks upon the first application of even a very small deviator stress. The interfaces responsible for inelasticity are identified as the grain boundaries, which we suppose to constitute a network of partially open channels running through the rock, and, if present, cracks penetrating the individual grains of the microstructure. A network array of interfaces is shown schematically in Figure 10. As indicated in the figure, the interfaces are not expected to be smooth. They are formed by the propagation of cracks during cool-down from an elevated temperature. The boundaries will not, in general, be along cleavage planes. Even if they were, the propagating crack would intersect many grown-in dislocations that thread through the crack plane with the resultant formation of steps on the interfaces.

In analyzing the inelasticity that will result from the crack network of Figure 10, it is important to establish the scale of the motion required on the interfaces relative to the network size. For a longitudinal inelastic strain ϵ, the relative displacement on one interface in a rock containing η_i boundaries per unit length must be of the order of $s = \sqrt{2}\ \epsilon/\eta_i$. For the range of strains used in our experiments, an upper limit to s ranges from 10^{-12} cm in the low strain amplitude experiments to a maximum of 10^{-5} cm at a strain corresponding to fracture, assuming that all the microscopically observed interfaces participate. That the strain is in fact distributed over all the interfaces is indicated by the following experiment carried out on a sample of granite (sample 3): One face of a rock bar is optically polished, and reference marks are made in many of the grains by means of a microhardness tester. The rock is then subjected to successively higher compression stresses until fracture occurs. Between successive stressings, the polished surface is examined, first with the reflected light microscope to see if the marked grains have undergone any relative displacement, and next with an interference microscope to detect any relative motion normal to the polished face. The sensitivity limits for these measurements are $\sim 4 \times 10^{-5}$ cm for motions in the polished plane and $\sim 2 \times 10^{-5}$ cm for motion normal to the plane. Up to the largest inelastic strain attained, where $s \sim 10^{-5}$ cm (when fracture occurred), no relative motion between the grains was detected. Therefore, nearly all the interfaces in the structure must have participated

in producing the total displacement. A remarkable characteristic of the inelasticity due to interfaces is that the resulting internal friction is amplitude independent for displacements ranging from 10^{-12} cm upward through seven orders of magnitude. The displacements are so small that the friction characteristics of the interfaces could be quite different from what would be observed in a macroscopic friction experiment.

It is not possible to observe and describe the detailed structure of an interface on a scale comparable to the displacements involved in the internal friction experiments; hence, it does not seem profitable to attempt a theory of the inelasticity in terms of the microstructure. Some model of the interface must be used instead. A start may be made by considering the interface friction machine shown in Figure 11a. It consists of a mass m sliding on a flat surface with an interface characterized by a coefficient of friction f. Force is applied at an angle θ to the interface, and the restoring force is supplied by the springs of constant k. The force mg represents a normal force across the interface which, in a rock, would be due to internal, static stresses; k is a measure of the looseness in the rock structure allowing relative motion to occur on the interfaces. The force-displacement curve is shown in Figure 11b; after the first cycle a closed curve results. This is a static hysteresis curve, and, hence, the energy dissipation is frequency independent. The 'internal' friction is

$$\frac{\delta w}{w} = \frac{k}{k'} \frac{2f \sin \theta}{\cos^2 \theta - f^2 \sin^2 \theta}$$
$$+ \frac{fmg}{k'} \frac{2 \cos \theta}{\cos^2 \theta - f^2 \sin^2 \theta} \frac{1}{s} \qquad (18)$$

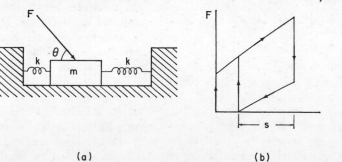

(a)   (b)

Fig. 11. (a) Interface friction machine having properties similar to an interface in rock. (b) Force-displacement curve of the machine.

where k' is the effective spring constant in series with F. A rock would be represented by an appropriately interconnected array of these machines with all possible values of θ. The resultant force-displacement curve would retain sharp corners at its extremities but otherwise would be rounded off.

The damping due to this interface machine is amplitude dependent, becoming large at small amplitudes. This results from the presence of the constant normal force mg; for $mg = 0$ an amplitude-independent damping results. Since the rocks studied show no increase in ϕ at low amplitude, it follows that the normal force on the interfaces in the rock specimens due to internal stresses must be negligibly small.

A limitation of the interface friction machine is that it does not allow for any motion perpendicular to the interface. As a result, the σ-ϵ curve arising from an array of these machines will always show either a constant or a decreasing slope as load is increased. Some rocks show a tendency for the slope of the σ-ϵ curve to increase at high uniaxial loads. This increase is due to a closing of interface openings under the applied stress. Since this upward curvature is small in the materials studied, it is neglected here.

A mechanical system that has properties similar to those of the interface friction machine and that may serve as a greatly simplified representation of an interface is a crack in an elastic medium (Figure 12). The internal friction that arises from a random array of cracks of equal size in an isotropic elastic solid has been calculated by *Walsh* [1966]. For a thin bar he finds that

$$\phi = \frac{1}{15} \frac{Y_0}{Y} c^3 n_v \left[2f - f^2 \frac{3 + 2f^2}{(1 + f^2)^{3/2}} \right]$$

$$= \frac{1}{15} \frac{Y_0}{Y} c^3 n_v h(f) \qquad (19)$$

where Y is Young's modulus, Y_0 is the effective modulus of the rock containing cracks, c is the crack length, and n_v is the number of cracks per unit volume. Walsh's assumption that only cracks that are just closed need be considered corresponds to our assumption above of neglecting motion and static forces normal to the interface. We want to consider the extension of this model to include the modulus defect. In

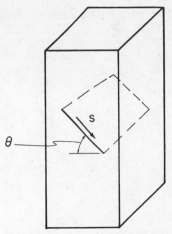

Fig. 12. Flat crack in an elastic medium used as an interface model.

the crack model the modulus defect arises from the relative motion on the crack surfaces; this motion gives a strain in addition to the elastic strain that would be present even if the cracks were absent. To calculate this additional strain we consider the case of a longitudinal wave in a slender bar and assume that the cracks are all in series, i.e., that their displacements add to give the total inelastic displacement in the bar. Let ϵ_e be the elastic strain that would occur in the bar without cracks and ϵ_c be the additional strain due to the cracks. The magnitude of the modulus defect is

$$\frac{\delta Y}{Y} = \frac{(\sigma/\epsilon_e) - \sigma/(\epsilon_e + \epsilon_c)}{\sigma/\epsilon_e} \qquad (20)$$

From Walsh's analysis we find that the mean displacement on a crack surface due to applied stress σ is

$$s = \frac{2\pi c}{Y} \cos \phi \, (\sin \phi - f \cos \phi)\sigma \qquad (21)$$

The total displacement resulting from motion on N cracks is then

$$S = \frac{\pi N \sigma c}{4Y} \int_{\tan^{-1} f}^{\pi/2} \cos \phi \sin^2 \phi$$

$$\cdot (\sin \phi - f \cos \phi) \, d\phi$$

$$= \frac{\pi N \sigma c}{4Y} \left[1 - \frac{\pi}{4} f - \frac{1}{2} \frac{f^2}{1 + f^2} + \frac{f}{2} \tan^{-1} f \right]$$

$$= \frac{\pi N \sigma c}{4Y} g(f) \qquad (22)$$

If the bar is of length L, $\epsilon_o = S/L$, so that

$$\frac{\delta Y}{Y} = \frac{1}{1 + (\pi n_l c/4)\, g(f)} \qquad (23)$$

where $n_l = N/L$ is the number of cracks per unit length. (A somewhat larger modulus defect is found from equations given by *Walsh* [1965] for the effective modulus of a rock containing cracks.) Since in this theory it is assumed that all the inelasticity is due to shearing motion on the cracks (motion normal to the crack surfaces is neglected), equation 19 should be written

$$\phi = \frac{1}{15}\left(1 - \frac{\delta Y}{Y}\right) c^3 n_l{}^3 h(f) \qquad (24)$$

where the relation $n_v = n_l{}^3$ has been used. The theory thus indicates a definite value for the ratio of ϕ to $\delta Y/Y$. For a continuous network of cracks such as that of Figure 10, $cn_l \simeq 1$. Taking $f = 0.5$, we find $\phi = 1.7 \times 10^{-2}$ and $\phi/(\delta Y/Y) = 5 \times 10^{-2}$. These are close to the observed values for the granite (sample 3) of Figure 9, where $\phi = 1.7 \times 10^{-2}$ and $\phi/(\delta Y/Y) = 6 \times 10^{-2}$. The crack theory expanded to include the modulus defect thus seems to be a reasonable approximation as a model to represent interface inelasticity in rock.

The crack model of inelasticity leads to the following conclusions: The internal friction is frequency and amplitude independent and is accompanied by a large modulus defect. There is a definite relation between ϕ and $\delta Y/Y$, which, for a continuous network of cracks, is fixed by the single parameter f. Correct magnitudes of ϕ and $\delta Y/Y$ are predicted for reasonable numerical values of f. Remaining to be explained are the structure sensitivity of ϕ and its dependence on hydrostatic pressure. Since different samples of quartzite show ϕ differing by an order of magnitude and since f is not expected to vary by so great an amount in a given material, the samples showing low ϕ must have an incomplete crack network; low ϕ could result from either or both n_l and c being small. Handling or heating changes ϕ by changing the number and size of the cracks. Hydrostatic pressure reduces ϕ and the modulus defect by reducing the number of active cracks in the rock. Since no amplitude dependence of ϕ is observed in rock samples under

pressure, the cracks must transform suddenly from active to inactive. Under sufficiently high pressure they all close, and this source of inelasticity is eliminated.

When a fluid phase is present there is an additional contribution to ϕ due to the stress-induced flow of the fluid. The experimental results indicate that at low frequencies this contribution to the inelasticity will be relatively small. The presence of fluid has, however, an important, indirect bearing on the occurrence of interface inelasticity, as is discussed below.

Applications. The experimental results show that the interface contributions to the inelasticity of crystalline rock may be characterized as static hysteresis. The most important practical consequence of this observation is that a large modulus defect and a drop in P- and S-wave velocities will accompany the presence of this type of internal friction. Inelasticity due to interfaces will be the dominant source of non-elastic behavior in rock near the surface of the earth. In dry rock the lithostatic pressure will eliminate this inelasticity when a depth of 6 to 8 km is reached. When a fluid phase is present, interface inelasticity may persist to much greater depths. Fluid in the open interface network of the rock will prevent the collapse and locking of these surfaces under the normal force resulting from the hydrostatic pressure. When the fluid can move freely through the interface network of the rock, the effective pressure suppressing interface damping is $P_o - P_f$, the contact pressure less the fluid pressure. That rock, even when under a high contact pressure, is permeable to fluid phase is shown by the following experiment. A bar of granite (sample 3) is driven in torsional resonance in a pressure vessel. The rock is jacketed. As shown in Figure 13, application of hydrostatic pressure reduces the modulus defect; when the jacket is punctured with the rock still under pressure, fluid penetrates the interface network and $\delta M/M$ returns after a period of a few minutes to its initial value, corresponding now to the condition $P_o = 0$, $P_f = 4$ kb. (A corresponding observation on the internal friction cannot be made because of the large contributions of intergranular fluid to the damping at 90 kHz.)

Most crustal rock is assumed to be water

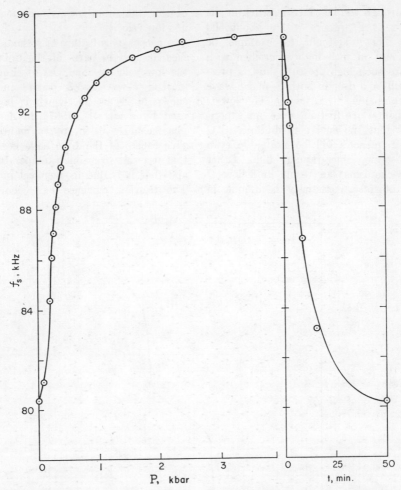

Fig. 13. Frequency of torsional resonance of a bar of granite (sample 3): first, as pressure is applied to the jacketed sample; second, as a function of time while under 4 kb with the jacket opened.

saturated. The lithostatic pressure is $P_o = \rho_r g d$, where ρ_r is the rock density and d is the depth. For a continuous open network of interstitial channels, the fluid pressure will be $P_f = \rho_f g d$; thus the effective pressure is $P_e = P_o - P_f = (\rho_r - \rho_f)gd$. Considering water to be the interstitial fluid and using the mean density of crustal rock, the depth to which interface inelasticity will occur will be increased by ~50%, owing to hydrostatic water pressure. Actual fluid pressure may be substantially greater than $\rho_f g d$ in some regions because of the compaction of sediments or tectonic activity. Where these processes are active in the crust, it is expected that both the P and

S seismic wave velocities will be lowered substantially as a result of interface inelasticity. Simultaneously, the sesmic Q will be low for both types of waves. Seismic P- and S-wave velocity and attenuation measurements made with low frequency signals along the same acoustic path in a tectonically active region could therefore be used to test for the presence or absence of a fluid phase.

The occurrence of interface inelasticity in rock need not necessarily be confined to the crust; under appropriate conditions it could also occur in the mantle. The necessary conditions are that a small volume of fluid be present in an otherwise solid rock, that the fluid penetrate through the interface network in the rock and be at a pressure within about 500 bars of P_e. The fluid could, for example, be a small amount of melt highly enriched with water. Where such high pressure fluid is present, there will be a drop in both P- and S-wave velocity accompanied by a low Q. The low Q results from interface friction, since our experiments show that the fluid contributions to Q at seismic frequencies will be negligible even for high viscosity intergranular fluid. Other damping mechanisms can result in a low Q zone, but interface inelasticity is unique in

causing the simultaneous occurrence of low Q and low velocity.

A conspicuous feature of seismic velocity profiles for certain parts of the upper mantle is the low velocity zone (LVZ). For example, a marked P-wave LVZ occurs in the profile shown in Figure 14, which is based on data used by *Anderson* [1967] in his analysis of phase changes in the upper mantle. Two possible causes of the LVZ have been advanced: that it results from high temperature gradients and that it is due to chemical inhomogeneity. The relevant evidence has been summarized

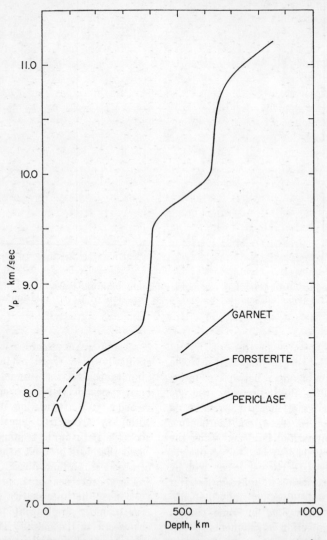

Fig. 14. *P*-wave velocity profile from *Anderson* [1967] showing a low velocity zone. Variation of velocity with depth for three mantle minerals is shown for comparison.

by *Clark and Ringwood* [1964]. It is very un-likely that temperature gradients alone could account for a *P*-wave LVZ, but they may be able to produce a *S*-wave LVZ. Take the curve of Figure 14 for example. From the high tem-perature elastic wave velocity data of *Soga* [1967] on garnet, it is found that a temperature peak of the order of 1000°C would be required to drop the *P*-wave velocity in this silicate mineral to the minimum of the observed ve-locity profile. Garnet is only one of the con-stituents of the pyrolite rock supposed to con-stitute the mantle at this depth, but it is unlikely that the olivines and pyroxenes, which are the other major constituents of this rock, would have markedly different elastic proper-ties. A change in composition at this depth may be hypothesized, but the fact that the LVZ occurs in a depth range characterized by a low seismic *Q* [*Anderson and Kovach*, 1964; *An-derson and Archambeau*, 1964] suggests an alternate hypothesis, namely, that the LVZ is due to a modulus defect in the upper mantle rock resulting from the presence of a small amount of fluid phase under high pressure. To test this hypothesis, suppose that the LVZ in Figure 14 is due entirely to the occurrence of a modulus defect. Then the velocity-depth rela-tion for the rock if no fluid were present would have to follow the dashed line shown in the fig-ure. This line has a reasonable slope for mate-rial that might be present in the upper mantle pyrolite, as may be seen by comparison with the straight line segments drawn for the elastic wave velocities of garnet, forsterite, and peri-clase. These line segments are based on a tem-perature gradient of 2.6°C/km [*Clark and Ringwood*, 1964] and velocity data of *Ander-son and Schreiber* [1965] for MgO, *Schreiber and Anderson* [1967] and *Soga and Anderson* [1967] for forsterite, and *Soga* [1967] for gar-net. Suppose that the results of Figure 9 char-acterize the inelasticity of the rock structure in the LVZ. This involves the assumption that the interface inelasticity is temperature insensi-tive, as is indicated by theory (no thermally activated processes participate in the inelastic-ity; the solid rock retains its strength to high temperature) and by experiments in the low temperature range. Then it is predicted that a damping peak of magnitude 7×10^{-3} will occur at a depth of 100 km. This is, in fact, approxi-mately the magnitude of damping observed at this depth. The seismic characteristics of the LVZ of Figure 14 can therefore be accounted for by the occurrence of interface inelasticity, and the hypothesis of chemical inhomogeneity is not necessary, though, of course, not ruled out. These hypotheses could be further tested if simultaneous *P*- and *S*-wave velocity pro-files could be obtained along the same acoustic path. (Because of regional variations in the velocity profiles, a comparison of data taken at different localities would be of dubious value.) If the LVZ is due principally to inter-face inelasticity, the minimum velocities for *P* and *S* waves should occur at the same depth, whereas in an all-solid upper mantle the possi-bility exists of the velocity minimum for *P* and *S* occurring at different depths. Since under some continental regions a small *S*-wave LVZ is observed with no minimum in *P*-wave ve-locities, it appears that the varying character of LVZ's may be useful as a means of dis-tinguishing localities where partial melting occurs in the upper mantle. The fact that the presence of a minute amount of fluid phase may substantially alter the seismic velocity through a rock indicates that great care would have to be taken in ascribing small changes in observed seismic velocities to changes in com-position. Comparison of seismic and laboratory velocities can be safely made only where it is known that the internal friction is very small. On the other hand, interface inelasticity is the only known source of a truly frequency-inde-pendent internal friction in the mantle. If it accounts for a substantial part of the observed *Q* of the mantle, then the analysis of the atten-uation of seismic waves will be much simplified.

According to the above hypothesis, the mate-rial of the LVZ of Figure 14 is solid, hetero-phase rock with a microstructure qualitatively similar to that of laboratory samples. The liquid phase in the interstitial spaces results from heating to the solidus temperature, which is lowered by the presence of small amounts (<1%) of amphiboles or other hydrous phases in the pyrolite. High pressure in the fluid phase results from sealing of the region of par-tial melt by the unmelted material above and below. The LVZ would not necessarily be a region of reduced mechanical strength so far as the usual processes of plastic deformation go, but it would be a favored site for the occurrence of earthquakes by *Frank's* [1965] dilatancy mechanism.

Acknowledgments. We would like to thank Professor E. T. Onat for helpful discussions on the mechanics of inelastic materials, Messrs. R. L. Stocker and T. J. J. Blanck for collecting and identifying rock samples, and Mr. Murray Ruggiero for assistance with the electronic equipment used.

This work was supported by the U. S. Office of Naval Research.

REFERENCES

Anderson, D. L., Phase changes in the upper mantle, *Science, 157,* 1165, 1967.

Anderson, D. L., and C. B. Archambeau, The anelasticity of the earth, *J. Geophys. Res., 69,* 2071, 1964.

Anderson, D. L., and R. L. Kovach, Attenuation in the mantle and rigidity of the core from multiply reflected core phases, *Proc. Natl. Acad. Sci. U. S., 51,* 168, 1964.

Anderson, O. L., and E. Schreiber, The pressure derivatives of the sound velocities of polycrystalline magnesia, *J. Geophys. Res., 70,* 5241, 1965.

Christensen, N. I., Elasticity of ultrabasic rocks, *J. Geophys. Res., 71,* 5921, 1966.

Clark, S. P., Jr., and A. E. Ringwood, Density distribution and constitution of the mantle, *Rev. Geophys., 2,* 35, 1964.

Frank, F. C., On dilatancy in relation to seismic sources, *Rev. Geophys., 3,* 485, 1965.

Granato, A., and K. Lücke, Theory of mechanical damping due to dislocations, *J. Appl. Phys., 27,* 583, 1956.

Gordon, R. B., The pinning of dislocations by X irradiation of alkali halide crystals, *Acta Met., 4,* 514, 1956.

Gordon, R. B., and L. F. Mike, Measurement of the mechanical properties of solids at high pressure, *Rev. Sci. Instr., 38,* 541, 1967.

Gordon, R. B., and C. W. Nelson, Anelastic properties of the earth, *Rev. Geophys., 4,* 457, 1966.

Ide, J. M., The velocity of sound in rocks and glasses as a function of temperature, *J. Geol., 45,* 689, 1937.

Knopoff, L., Q, *Rev. Geophys., 2,* 625, 1964.

Marx, J., Use of the piezoelectric gage for internal friction measurements, *Rev. Sci. Instr., 22,* 503, 1951.

Nowick, A. S., Internal friction in metals, in *Progress in Metal Physics,* vol. 4, edited by B. Chalmers, pp. 1–70, Pergamon, New York, 1953.

Nowick, A. S., Resonance and relaxation phenomena, in *Resonance and Relaxation in Metals,* pp. 1–43, Plenum, New York, 1964.

Schreiber, E., and O. L. Anderson, Pressure derivatives of the sound velocities of polycrystalline forsterite with 6% porosity, *J. Geophys. Res., 72,* 762, 1967.

Simmons, G., and W. F. Brace, Comparison of static and dynamic measurements of compressibility of rocks, *J. Geophys. Res., 70,* 5649, 1965.

Soga, N., Elastic constants of garnet under pressure and temperature, *J. Geophys. Res., 72,* 4227, 1967.

Soga, N., and O. L. Anderson, High temperature elasticity and expansivity of forsterite and steatite, *J. Am. Ceram. Soc., 50,* 239, 1967.

Walsh, J. B., The effect of cracks on the uniaxial elastic compression of rocks, *J. Geophys. Res., 70,* 399, 1965.

Walsh, J. B., Seismic attenuation in rock due to friction, *J. Geophys. Res., 71,* 2591, 1966.

Zener, C., *Elasticity and Anelasticity in Metals,* 170 pp., University of Chicago Press, Chicago, 1948.

(Received December 29, 1967; revised February 14, 1968.)

Reprinted from Journal of Geophysical Research, v. 73, p. 6051-6060.

The Effect of Temperature and Partial Melting on Velocity and Attenuation in a Simple Binary System[1]

Hartmut Spetzler and Don L. Anderson

California Institute of Technology, Seismological Laboratory
Pasadena, California 91105

A possible explanation of the low-velocity, low-Q zone in the upper mantle is partial melting, but laboratory data are not available to test this conjecture. As a first step in obtaining an idea of the role that partial melting plays in affecting seismic variables, we have measured the longitudinal and shear velocities and attenuations in a simple binary system that is completely solid at low temperatures and involves 17% melt at the highest experimental temperature. The system investigated was $NaCl \cdot H_2O$. At temperatures below the eutectic the material is a solid mixture of H_2O (ice) and $NaCl \cdot 2 H_2O$. At higher temperatures the system is a mixture of ice and NaCl brine. In the completely solid regime the velocities and Q change slowly with temperature. There is a marked drop in the velocities and Q at the onset of melting. For ice containing 1% NaCl, the longitudinal and shear velocities change discontinuously at this temperature by 9.5 and 13.5%, respectively. The corresponding Q's drop by 48 and 37%. The melt content of the mixture at temperatures on the warm side of the eutectic for this composition is about 3.3%. The abrupt drop in velocities at the onset of partial melting is about three times as much for the ice containing 2% NaCl; for this composition, the longitudinal and shear Q's drop at the eutectic temperature by 71 and 73%, respectively. If these results can be used as a guide in understanding the effect of melting on seismic properties in the mantle, we should expect sharp discontinuities in velocity and Q where the geotherm crosses the solidus. The phenomena associated with the onset of melting are more dramatic than those associated with further melting.

Introduction

Knowledge of the mechanical properties of multicomponent systems in the vicinity of their melting points is required in various geophysical problems. In particular, the behavior of seismic velocity and attenuation near the melting point is pertinent to discussions of the upper mantle low-velocity zone. In this study we have measured the velocity and attenuation of longitudinal and shear waves in the vicinity of the eutectic temperature in a simple dilute binary system. By varying the composition, we have been able to study the effects of temperature and partial melting. We chose dilute solutions so that we could investigate the region involving partial melting. In the system studied, $NaCl \cdot H_2O$, the amount of melt could be changed simply by varying either the temperature or the initial concentration of NaCl.

Previous studies of this sort have used pure materials or eutectic mixtures, both of which have sharply defined melting temperatures rather than a melting interval. *Mizutani and Kanamori* [1964] measured the compressional and shear velocities and the compressional wave Q for an alloy consisting of Pb, Bi, Sn, and Cd. The velocities varied approximately linearly with temperature until T/T_m was about 0.97, at which point they decreased rapidly. The P velocity dropped 20% upon melting, and the quality factor Q dropped by about an order of magnitude. The Q decreased very rapidly as the melting point was approached. Similar results were obtained by *Pokorny* [1965], who obtained a velocity drop of about 15% and a Q drop of an order of magnitude as melting progressed. Again the mechanical properties started to anticipate the melting point at a T/T_m of about 0.97.

Both of the preceding studies used an ultrasonic pulse method with frequencies in the high kilocycle or megacycle range, i.e., very short wavelengths. The actual amount of melt as a function of temperature and the configuration of the molten zones was not described.

Water and NaCl form a simple binary system

[1] Contribution 1528, Division of Geological Sciences, California Institute of Technology, Pasadena 91105.

that can easily be studied in the vicinity of the eutectic temperature. Both the phase diagram and the geometry of the components are well known. The melt phase (brine) occurs at the grain boundaries in cylindrical channels or thin layers, depending on the temperature. In polycrystalline specimens the melt occurs in irregularly shaped pockets between crystal and subcrystal boundaries. We used a resonance technique and large samples to assure that the wavelengths were always large compared with crystal or melt zone dimensions. Ice rods were frozen from dilute NaCl solutions, and the resonant frequencies were measured to obtain longitudinal and shear velocities. The quality factors, Q's, were obtained by measuring the width, and in some cases the decay, of the resonance peaks. The measurements were performed on pure H_2O ice and on NaCl-ice mixtures as a function of NaCl concentration and temperature for the fundamental mode and several overtones.

EXPERIMENTAL PROCEDURE

The experiments were performed in a So-Low Environmental Equipment Co. refrigerator. To assure temperature stability the refrigerator was packed with ice, and the experiment was placed inside a styrofoam insulating box.

A carefully measured amount of distilled water was heated close to its boiling point, and the appropriate quantity of NaCl was added. To remove entrapped air, the solution was placed in an airtight container and a vacuum (10 torr) was pumped until the solution was boiling slowly. The solution was pumped under vacuum into a Teflon tube 2.54 cm ID and 30.5 cm long. The ultrasonic transducers were supported with Teflon plugs at each end of the tube. To avoid separation of the ice crystals from the brine during the freezing process, the samples were frozen quickly at about $-30°C$ and were rotated at 1 rpm while in a horizontal position. The Teflon tubes were removed when the freezing was complete. After the resonance experiments were performed, the ice rods were melted and the salinity of the solution was remeasured. The salt content of the rod was approximately 10% less than the starting solution owing to concentration of salt at the surface of the rod during the early stages of freezing. Salinity was measured on one ice rod as a function of position within the rod. There was a

slight concentration of salt toward the center of the rod and away from the ends. The total variation, however, was less than $±0.15\%$ of salt concentration.

Photomicrographs of thin sections cut from the ice rods were taken in order to study the size and orientation of the ice platelets and the distribution of the brine. From the photomicrographs it was possible to determine the size and orientation of the ice platelets. Most platelets were oriented such that their c axes were in planes parallel to the surface of the rod. Figure 1 shows a radial cut close to one end of a 2%NaCl ice rod. The view is parallel to the c axis of the platelets. The temperature was $-5.3°C$. The hexagonal platelets have an average diameter of 0.5 mm, and their diameter-to-thickness ratio is about 8 to 1. Photographs of 1% ice NaCl show platelet diameters of approximately 1 mm.

A solenoid arrangement was used to excite longitudinal modes in the ice. A small bar magnet was frozen into each end of the ice rod, and the external transducer coils were enclosed in aluminum boxes that were covered with μ metal to avoid electromagnetic coupling between the driving and receiving transducers. The transducers at each end of the ice rod were identical and arranged symmetrically. The shear transducers consisted of small flat coils which were frozen into the ice at both ends. To reduce electromagnetic coupling between the coils, they were oriented 90° to each other. Permanent magnets were used at the driver and receiver end to complete the motor and dynamo action, respectively. The ice rods were supported by two narrow copper-band slings at the nodal points of the second harmonic. Various other supports were tried, including three slings and foam rubber pads, but the best reproducibility and the highest Q values were obtained with the two-sling arrangement, which was maintained throughout all measurements.

To monitor the temperature of the specimen, a separate ice rod was prepared under identical conditions to the one used for the velocity measurements. This control rod contained one thermocouple in the center and one on the outside. By connecting the constantan of the two iron-constantan couples, it was possible to record the absolute temperature and the difference temperature between the outside and the

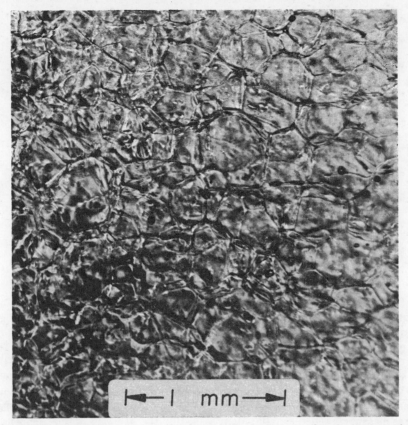

Fig. 1. A view parallel to the *c*-axis orientation of the ice platelets in ice containing 2% NaCl.

inside of the rod. The resistance between the top and the center and the center and the bottom of the rod was also measured. This measurement was performed to check for a possible brine drainage in the region warmer than the eutectic. No settling of brine was recorded on the time scale of the experiment. The absolute temperature was measured with the aid of a Leeds and Northrup potentiometer. Small temperature increments were read with a digital voltmeter. Periodic cross calibration between the voltmeter, the potentiometer, and various thermocouples suggests an absolute accuracy in temperature of ±0.15°C and a relative accuracy between measurements of ±0.05°C. The reference ice bath was aerated and carefully maintained by using distilled water and shaved ice.

INSTRUMENTATION

Figure 2 shows the circuitry used to measure and record the resonant frequencies and the Q's of the various modes of the ice rods. A Schomandle ND30M frequency synthesizer was driven by a synchronous motor to sweep through the appropriate frequency range. To excite the longitudinal modes, the output of the synthesizer was amplified and applied directly to the driving solenoid.

The shear transducer arrangement, shown in Figure 2, is somewhat more complicated. An arrangement of resistors between synthesizer and amplifier served to select amplitudes for various frequency ranges. This arrangement enabled the operator to perform all measurements without disturbing the output level of the synthesizer. The output of the receiving transducer was amplified in two stages and then detected. The detected signal was recorded on a strip chart recorder and on a digital voltmeter. Both the driving signal and the output signal were displayed on an oscilloscope.

A typical record of a frequency sweep is

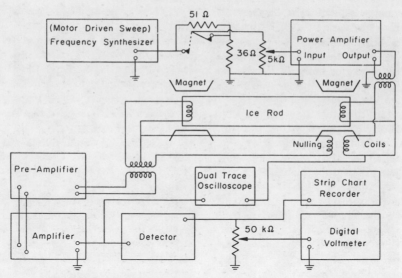

Fig. 2.   Schematic illustration of the shear resonance experiment.

shown in Figure 3. The frequency at the peak of the resonance was read from the synthesizer. The widths of the symmetric resonance peaks were measured by changing the frequency to the values where the digital voltmeter was 0.707 of the peak value, thus giving the half-energy points. In cases of high noise level the resonance width was measured from the graph. Many data points for Q were checked by measuring Q from the decay of the rod oscillations after the power was turned off. The results were compatible.

A few data points were obtained on rods that were frozen under identical conditions in order to check reproducibility. These Q and velocity data fell within the scatter of the data shown.

The perpendicular orientation of the shear-transducer coils did not completely eliminate the electromagnetic coupling between driver

and receiver. To cancel the coupling, a second driver and receiver coil were placed outside the refrigerator and adjusted so that the electromagnetic coupling was the same as that for the coils embedded in the ice. The two receiver coils were connected to a center trapped transformer. By tuning off the resonance and adjusting the coupling of the compensating coils, the output of the two receiver coils could be made to be 180° out of phase and thus cancel. The transformer served also to match the low impedance of the coils to the high input impedance of the amplifier. On the driver side, a transformer was used to match the impedance between the amplifier and the transducer coil.

EXPERIMENTAL RESULTS

Measurements of resonant frequency and peak width were made for the fundamental and

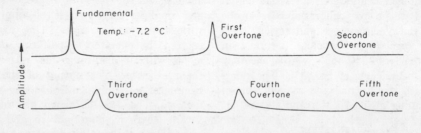

Fig. 3.   Tracing of strip chart recording of a typical frequency sweep.

several harmonics as a function of temperature. The temperature was varied slowly so that the difference temperature between the center of the rod and the outside never exceeded 0.1°C. This required a cooling or heating rate between 1 and 2 degrees per hour, slower in the eutectic region. A typical data run from −35° to −8°C and back through the eutectic took approximately 100 hours. The velocities were calculated from the fundamental resonance frequencies of the longitudinal and shear modes of the ice rods $V = 2lf/n$ where f is the fundamental frequency of the appropriate mode, n is the mode number, and l is the length of the ice rod. No correction was applied for the temperature dependence of l. According to estimates of *Weeks* [1961], the maximum length change in the temperature from −30° to −10°C for the 2% NaCl ice would be approximately 0.3%.

Velocity data from the fundamental mode of the 1 and 2% salt-ice-rods are presented in Figure 4 as a function of temperature. The eutectic temperature for the NaCl·H₂O system is −21.3°C. Figure 5 gives the same data as a function of brine content. At temperatures colder than the eutectic, the velocity decreases approximately linearly as the temperature increases and at a faster rate than pure ice. In these experiments the absolute temperature was

always greater than 0.94 of the eutectic temperature.

The normalized temperature derivatives of the velocities $-(1/V)\ dV/dT$ are 0.6×10^{-4}, 0.9×10^{-3}, and 1.4×10^{-3} °C^{-1} for longitudinal waves in pure ice and 1% NaCl and 2% NaCl ice, respectively, for temperatures colder than the eutectic. The corresponding shear velocity derivatives are 0.8×10^{-3}, 1.4×10^{-3}, and 2.8×10^{-3} °C^{-1}, significantly larger than the longitudinal derivatives. The drop in longitudinal velocity across the eutectic for the ice containing 1 and 2% NaCl was 9.5 and 28%, respectively. The corresponding drops for the shear velocities were 13.5 and 40%. These velocity decreases occur within 0.1°C but over a time span of several hours when the ice is warming up. When the ice is cooled through the eutectic point, the system is able to supercool by 1.5° to 2.5°C. In this case the eutectic region is spread over a larger temperature range but occurs in a shorter time interval. While the ice warms up, the velocities increase for several degrees immediately after the eutectic has been passed. When the temperature is reversed, the velocities do not show this dip.

Figure 5 gives the velocity as a function of brine content and salinity. To achieve a given brine content, the ice containing 1% NaCl

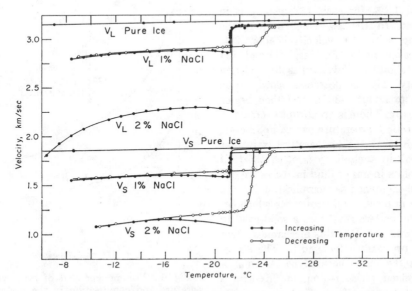

Fig. 4.   Longitudinal V_L and shear V_S velocities in pure ice, ice containing 1% NaCl, and ice containing 2% NaCl. Note the large drop in velocity at the eutectic temperature and the hysteresis between the warming and cooling cycles.

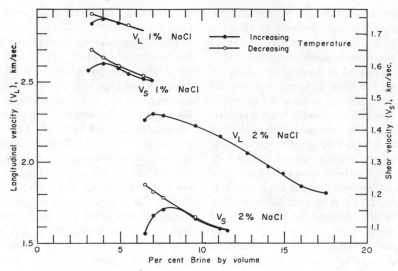

Fig. 5. V_L and V_S as a function of brine volume .The ice crystals in the 1% NaCl-ice system have linear dimensions approximately twice as large as in the 2% NaCl-ice system.

must be much warmer than the ice containing 2% NaCl. For example, a 7% brine content occurs at −8.4°C for the ice containing 1% NaCl and at −19°C for the ice containing 2% NaCl. Brine content was computed from the phase diagram in Figure 6 and *Weeks* [1961]. Contrary to our initial expectations, the velocity is not a unique function of the melt fraction. In the 1% NaCl system the velocities decrease by about 1.2 to 1.4% for each 1% increment in brine content. In the 2% NaCl system, the corresponding decrease in velocity is about 2%.

The Q measurements (Figures 7 through 12) were taken simultaneously with the velocity measurements. The Q decreases slowly with temperature to about −30°C and then begins to rise gradually. There is an abrupt decrease in Q at the eutectic temperature, which in all cases recovers slightly as the temperature is further raised. A peak in anelasticity is often observed at critical points in gas or fluid mixtures. In our case, this phenomenon is complicated by the large change in mechanical properties which occurs when the system goes from a solid-solid to a solid-fluid mixture. It is this mechanical effect that we are primarily interested in. The elastic wave upsets the local thermodynamic equilibrium in a mixed phase region, in this case a salt-water-ice system. This effect is superimposed on the grain boundary loosening associated with the onset of partial melting. When the temperature is reversed, i.e. when the sample is cooled, the ability to supercool (which is related to the difficulty of nucleation) permits nonequilibrium conditions to maintain during the passage of a stress wave, and there is no loss associated with thermodynamic relaxation

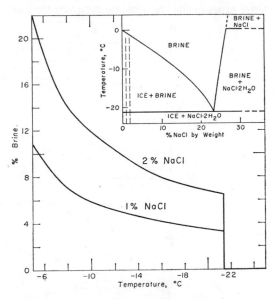

Fig. 6. Volume per cent of brine versus temperature and composition in the system studied. Insert shows the phase diagram of the NaCl system [*Weeks*, 1961]. Dashed lines are the compositions studied.

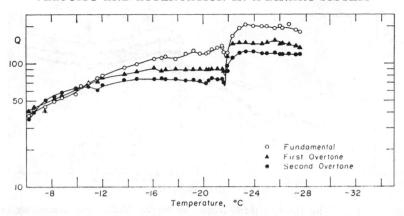

Fig. 7. Q for the first three longitudinal modes or the heating cycle for 1% NaCl.

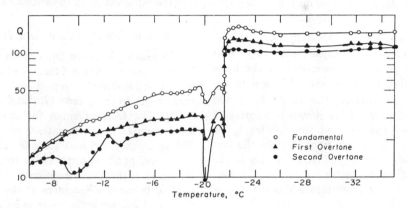

Fig. 8. Q for the first three shear or torsional modes or the heating cycle for 2% NaCl.

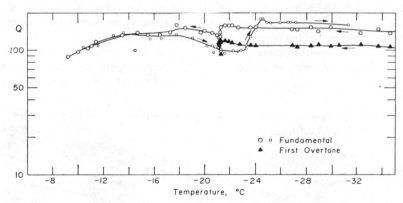

Fig. 9. Q for the first two harmonics, the shear or torsional modes for 1% NaCl. Arrows indicate whether the data were taken on the warming or cooling cycle.

effects. In this case the losses are associated with mechanical, presumably grain-boundary effects alone. On the warming cycle, the drop in Q at the eutectic point amounted to 48 and 71% for the fundamental frequency of the longitudi-

nal Q's for the 1 and 2% NaCl ice. The corresponding values for the shear Q's were 37 and 73%.

A further increase in temperature leads to a small increase in Q for the longitudinal modes.

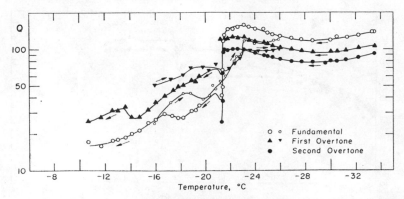

Fig. 10. Q in shear for the first three harmonics for 2% NaCl. Arrows pointing to the left indicate that the data were taken on the warming cycle. Arrows pointing to the right indicate that the data were taken on the cooling cycle; note the supercooling in the second case.

For the shear modes, especially for the 1% NaCl mixture, the increase was very pronounced. At warmer temperatures as the brine content further increases, the Q's begin to decrease. For comparison, the longitudinal and shear Q's for pure ice are shown in Figures 11 and 12. For pure ice, and for salt ice below $-35°C$, Q increases with frequency for the first few harmonics. For higher temperatures in the salt ice, Q decreases as the frequency increases. Above the eutectic temperature this generalization does not hold.

The shear Q data of pure ice show minima for the first three harmonics at $-28°$, $-16°$, and $-12°C$, respectively. Figure 12 shows the relation of these minima to temperature and relaxation time. The activation energy corresponding to this frequency response is approximately 7.5 kcal/mole.

SIZE OF ICE PLATELETS AND VELOCITY

The spacing between the centers of adjacent ice platelets is a linear function of salt concentration and is directly proportional to the square root of the freezing time [*Rohatgi and Adams*, 1967]. The freezing times for the 1 and 2% ice rods in this experiment are of the order of 100 to several hundred seconds. This freezing time corresponds approximately to that used in Figure 14 of the above mentioned reference. The spacing between the centers of the platelets of the 1 and 2% ice is therefore quite similar, and this is confirmed by the photographs described earlier. The other dimensions of the crystals are, however, approximately twice as large for the 1% ice as for the 2% ice.

It is clear from Figure 5 that the mechanical properties depend on more than just the brine or melt content. Since most of the brine is con-

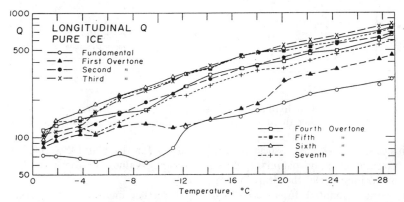

Fig. 11. Q as a function of temperature and frequency for pure ice; longitudinal modes.

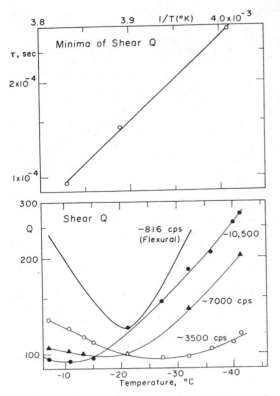

Fig. 12. Q as a function of temperature and harmonic for shear modes in pure ice. Also shown are the results in flexure obtained by *Kuroiwa* [1965].

tained in layers between subcrystal platelets, the function of the liquid (brine) seems to be one of decoupling at the grain boundaries. The number of grain boundaries per unit volume, an important parameter, is inversely proportional to the surface area of a grain, and the velocity is a function of the brine content times the surface area of the grains.

CONCLUSIONS

There are a number of generalizations that can be made from the data of this investigation. A small volume fraction of liquid has a large effect on the velocity and attenuation of shear and longitudinal waves. As expected, the effect on the shear velocity is considerably more than the effect on the longitudinal velocity. Because of the anisotropy of the ice rods, no direct comparison can be made between the velocities and Q's of the shear and the longitudinal modes.

The sharp dip in Q at the eutectic point is

of more than passing interest. This dip was observed on all the samples but was most pronounced in the shear data for ice containing 1% NaCl. A multicomponent, multiphase system that is in equilibrium will be disturbed by an acoustical signal. The degree to which the equilibrium is disturbed is a function of the frequency and amplitude of the signal. The rate at which the chemical equilibrium can follow temperature and pressure fluctuations (chemical kinetics) controls the frequency dependence of this part of the ultrasonic absorption. At a given frequency, the absorption is a function of the concentration gradients associated with the temperature and pressure fluctuations caused by the stress wave. Energy is absorbed from the acoustic signal and converted into chemical energy. Some of this chemical energy is released in the form of heat as the system attempts to return to equilibrium. The slopes of the curves in Figure 5 give a measure of the extent to which the equilibrium may be disturbed by a small change in temperature. At the eutectic where the slope is discontinuous, the absorption should be a maximum, as is observed. The above described absorption mechanism is well known and has been used to study reaction kinetics in liquids (see, for example, *Tabuchi* [1956, 1957]; *Yasunaga et al.* [1965]; *Tatsumoto* [1966]) in the megacycles-per-second range.

The preceding arguments suggest that a sharp dip in Q will be associated with the onset of partial melting in the mantle. Solid-solid phase changes may show similar dips. The low-velocity zones of the upper mantle has been interpreted in terms of (a) proximity to the melting point [*Press*, 1959], (b) high temperature gradients [*Gutenberg*, 1959; *Birch*, 1952; *Valle*, 1956], and (c) chemical inhomogeneity in the mantle [*Ringwood*, 1962a and b]. The data of the present investigation indicate that the seismic velocities and Q values will drop abruptly if the solidus crosses the geotherm. If due to partial melting, the boundaries of the low-velocity zone will be abrupt. This is consistent with recent tectonic and oceanic mantle models. The low-velocity zone will also be a zone of high attenuation and this also seems to be the case.

Acknowledgments. It is a pleasure to acknowledge the help of David F. Newbigging who assisted in all phases of the experiment. We are grateful for helpful discussions with Thomas

Ahrens, Charles Archambeau, and Samuel Epstein.

This research was partially supported by National Science Foundation grant GA 1003.

REFERENCES

Birch, F., Elasticity and constitution of the earth's interior, *J. Geophys. Res., 57,* 227–286, 1952.

Gutenberg, B., The asthenosphere low-velocity layer, *Annali di Geofisica, 12,* 439–460, 1959.

Kuroiwa, D., Internal friction of H_2O, D_2O, and natural glacier ice, *U. S. Army Material Command Res. Rept. 131,* Cold Regions Research and Engineering Laboratory, New Hampshire, 1965.

Mizutani, H., and H. Kanamori, Variation of elastic wave velocity and attenuative property near the melting temperature, *J. Phys. Earth, 12*(2), 43–49, 1964.

Porkorny, M., Variation of velocity and attenuation of longitudinal waves during the solid-liquid transition in Wood's alloy, *Studia Geophys. Geodaet. Ceskoslov. Akad. Ved, 9,* 1965.

Press, F., Some implications on mantle and crustal structure from *G* waves and Love waves, *J. Geophys. Res., 64,* 565–568, 1959.

Ringwood, A. E., A model for the upper mantle, *J. Geophys. Res., 67,* 857–867, 1962a.

Ringwood, A. E., A model for the upper mantle, 2, *J. Geophys. Res., 67,* 4473–4477, 1962b.

Rohatgi, P. K., and C. M. Adams, Jr., Ice-brine dendritic aggregate formed on freezing of aqueous solutions, *J. Glaciol., 6,* 47, 1967.

Tabuchi, D., Dispersion and absorption of sound in liquids in general chemical equilibrium and its application to chemical kinetics, *J. Chem. Phys., 26*(5), 993–1001, 1956.

Tabuchi, D., Dispersion and absorption of sound in ethyl formate and study of the rotational isomers, *J. Chem. Phys., 28*(6), 1014–1021, 1957.

Tatsumoto, N., Ultrasonic absorption in propionic acid, *J. Chem. Phys., 47*(11), 4561–4570, 1966.

Valle, P. E., On the temperature gradient necessary for the formation of a low velocity layer, *Ann. Geofis. Rome, 9,* 371–377, 1956.

Weeks, W. F., Studies of salt ice, 1, The tensile strength of NaCl ice, *U. S. Army Material Command Research Report 80,* U. S. Army Cold Regions Research and Engineering Laboratory, New Hampshire, 1961.

Yasunaga, T., N. Tatsumoto, and M. Miura, Ultrasonic absorption in sodium metaborate solution, *J. Chem. Phys., 43*(8), 2735–2738, 1965.

(Received April 8, 1968;
revised June 7, 1968.)

Reprinted from The Earth's Crust, AGU monograph 20, p. 197-213.

INTERNAL FRICTION MEASUREMENTS AND THEIR IMPLICATIONS IN SEISMIC
Q STRUCTURE MODELS OF THE CRUST

B. R. Tittmann

Science Center, Rockwell International
Thousand Oaks, California 91360

Abstract. Mitchell (1973) and more recently Herrmann and Mitchell
(1975) have presented the first detailed seismic Q profiles (quality
factor Q is proportional to reciprocal of internal friction) for a
region of the earth's crust. Their profiles show a dramatic increase
in seismic Q from values of a few hundred from the surface down to
15 km to as high as 2000 below 15 km. No ordinary rock types measured in laboratory air have shown such high values. Results of
laboratory internal friction measurements on strongly outgassed rocks
recently made as part of an effort to interpret the high seismic Q
values observed in the lunar crust show that moisture absorbed in
pores causes the low Q values. This suggests that the sharp increase
in Q might mark a boundary below which no appreciable moisture exists
in the crustal rocks. Our studies show that the Q of a low-porosity
olivine basalt without hydrated mineral phases ranges from about 100
measured dry in normal laboratory air to over 2000 outgassed at moderate temperatures in a high vacuum for about a week. These results
suggest that the Q of a dry sample of the same composition and mineralogy would be above 2000. Laboratory data on Q in rocks have
characteristics which suggest they are valid for rock in situ.

Introduction

Mitchell (1973) has recently presented the first detailed internal
friction profile for a region of the earth's crust. It is based on
data from the October 21, 1965, south central Missouri earthquake,
recorded at a number of stations located between the Rocky Mountains
and the Atlantic coast. The most conspicuous feature of this profile
is a sharp increase in Q (decrease in internal friction) from values
not exceeding a few hundred from the surface down to about 15 km to a
value at least as high as 2000 slightly below 15 km. No ordinary
rock types measured in laboratory air have shown such high Q values
(Knopoff, 1964). More recently, Herrmann and Mitchell (1975) and
Mitchell (this volume) have obtained similar but more detailed results
based on data from four earthquakes occurring in the New Madrid seismic
region, one earthquake in the northern Hudson Bay region, and three
underground nuclear explosions in the western United States. They
obtain somewhat lower Q values in the lower crust (Q ~ 1500) but

because of relatively large standard deviations do not rule out
$Q \approx 2000$ from a depth of 17 km to about 37 km. Lee and Solomon (1975)
have combined longer period data with the data of Mitchell (1973) to
infer the Q structure of the crust and upper mantle and draw conclusions about lithospheric thickness and properties of the asthenosphere.
These examples demonstrate the importance of Q measurements for gaining
insights into the natures of the lithosphere and asthenosphere. Comple-
mentary studies of material properties in the laboratory are necessary
in order to aid in the interpretation of seismic Q values in terms of
the nature of material in the upper and lower crust (Housley, et al.,
1974).

    This paper presents the results of recent laboratory studies of
internal friction and velocity made as part of an effort to interpret
the high seismic Q values observed near the surface of the moon
(Tittmann, et al., 1974). The results of these laboratory measurements
carried out in part on strongly outgassed rocks bear on the interpretation of the seismic Q-versus-depth profiles reported by Mitchell
(1973) and Herrmann and Mitchell (1975) in terms of the moisture content
of crustal rocks.

    In the following the technique of laboratory measurements is pre-
sented first, followed by the description of three key experiments.
The results of these experiments are then discussed and interpreted
on the context of the current seismic Q profiles for the crust.

Experimental Approach for Laboratory Internal Friction Measurements

    The experimental approach basically embraces the vibrating bar tech-
nique in which a uniform bar is excited in one of its natural modes of
vibration. The three modes selected by us for the measurements are
the flexural, torsional, and longitudinal modes. For excitation in
the torsional or longitudinal modes, our specimen is rigidly supported
at the mid-point, which becomes the mode for the fundamental free -
free mode. For excitation in the flexural mode, our specimen is
pivoted at the ends as described in detail in a later section. To
eliminate corrections due to dispersion, for example, the bar is given
a length-to-radius ratio of at least about 10 to 1 and is further given
a circular cross section for torsional resonance or a rectangular cross
section for flexural resonance (for longitudinal resonance the cross
sectional shape plays a minor role). The sharpness of resonance of a
mode of vibration has been observed as a measure of losses in metals
and other materials (Quimby, 1928), and this technique has been
applied to measuring losses in rock samples by for example, Birch and
Bancroft (1938). Many techniques, Norwick and Berry (1972), of
exciting the natural mode of the bar are available and Gordon and Davis
(1968), for example, used a piezoelectric element bonded to the end of
the specimen to form a composite piezoelectric resonator. In the
experiments described here, the excitation technique employs the magnetic drive and detection system used by Wegel and Walther (1935) in
which the specimen is excited by the use of an alternating magnetic
field acting on a ferrous electrode mounted on the specimen end. The
The vibrating bar technique was preferred over other techniques for
providing information helpful for the interpretation of seismic data
for the following reasons:

1) The technique enables measurements over a very large frequency range not accessible by other techniques. In particular with the flexural mode frequencies down to 1 kHz are typical for uniform bars of 3 cm length, but with end loading (Berry, 1955) the effective resonance can be lowered by another order of magnitude to bring the experiment very near to the high end of the seismic range of frequencies. The technique provides therefore, a valuable link between laboratory and seismic field determinations of Q.

2) The vibrating bar technique allows the determination of Q in at least two ways, by measuring the decay of the normal mode vibrations or by measuring the sharpness of the resonance. Since the Q values measured by the two methods should be the same, a valuable self-consistency check is obtained.

3) The vibrating bar technique provides a well-defined mode of vibration with a well-defined nodal point. Positioning of the support at the nodal point assures that essentially no energy is lost in the supports and that the attenuation of vibrational amplitude is accurately reflected in the width of the resonance or the decay of the vibrations.

4) The vibrational mode of excitation provides for a well-defined single frequency ω, which must be known accurately to determine the quality factor Q, i.e., for longitudinal resonance $Q_y = \omega/2\alpha_y c_y$ where c_y is the wave velocity in cm/sec and α_y is the internal friction (attenuation) in nepers/cm. Here, $c_y = (E/\rho)^{\frac{1}{2}}$ where E is Young's modulus in dynes/cm^2 and ρ is the density in gm/cm^3.

5) The method also minimizes errors introduced by the transfer of energy into unobserved vibrational modes which results in an apparent rather than real energy loss.

6) The use of this method allows measurements to be made at very small strain amplitudes to minimize nonlinear processes which are completely absent in terrestrial seismic waves whose strain amplitudes are typically of the order of 10^{-13} to 10^{-10}. Measurements of the strain with a commercial capacity microphone revealed strains in our measurement typically in the 10^{-8} range and as low as 10^{-9}.

Experiment 1: Internal Friction Q = 2400 Achieved in Igneous Rock

Sample Preparation

This experiment was carried out on samples of augite olivine basalt (Quaternary flow from Western Cascade volcanics designated W-8 from Weed, California, near Mt. Shasta). The open pore porosity was measured to be approximately 2.5%.

These rocks were attached to metal blocks with low melting point wax and sawed into bars measuring 8.0 x 0.4 x 0.5 cm with an oil-cooled diamond saw. Preliminary attempts to saw rocks into bars with a gas-cooled diamond saw always resulted in breaking the bars.

After sawing, the bulk of the wax and oil was removed from the samples by washing them in acetone at room temperature. After this procedure the sample typically had a Q of about 20. It was generally followed with a cleaning procedure in which the sample was washed first in boiling trichloroethylene for about an hour, then in boiling ethanol, and finally in boiling distilled water, since this sequence

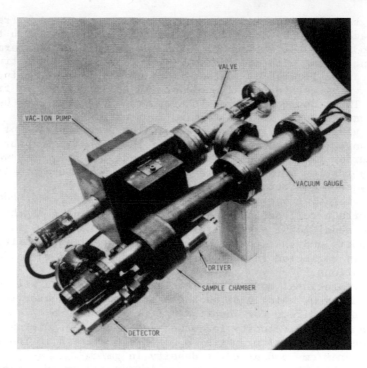

Fig. 1. Ultra-high vacuum apparatus, an all-metal system with
15 liter/sec Varian Vac-Ion pump.

is known from semiconductor technology to leave the least amount of
contaminants. Lately we have supplemented or partially bypassed the
cleaning procedure by boiling the samples in 30% H_2O_2 for up to
several hours, until excess bubbling stops, an indication that oxi-
dizable organics have been consumed. Although no systematic comparison has yet been made on the outgassing behavior of similar samples
cleaned by different procedure, our general impression is that the
H_2O_2 treatment is the more effective. With this procedure the measured
Q after cleaning is about 75 to 100, indicating a sample clean of the
oil and wax used in the cutting procedure. (Since the H_2O_2 is an
oxidizing agent, it does not further oxidize the rock minerals but
affects only the organic and biological material by transforming it
to CO_2 and H_2O)

The mode of vibration selected for this experiment was the longitudinal resonance mode and therefore the electrodes consisted of
small iron buttons cemented to the ends of the bars. Initially, we
used an epoxy to cement the buttons to the bars, but eventually, we
learned that at least for small samples, losses in the epoxy dominated
the other loss mechanisms above room temperature. We now use either
a fast-drying general-purpose household cement, which is satisfactory
at room temperature, or a high-temperature silicate cement. The
latter is cured by gradually warming the sample to about $150^{\circ}C$ in air
and holding it there for about a day. This procedure also drys the
sample and raises Q at room temperature to as high as ~ 400.

Apparatus

Initially (Tittmann, 1972), the sample, sample holder and associated hardware were placed in a bell jar and a conventional diffusion pump was used to free the rock from any remaining volatiles, such as adsorbed H_2O. More recently (Tittmann, et al., 1975), a second ultrahigh vacuum system was built to provide an even better and cleaner vacuum. This system is shown in Figure 1. Instead of a glass bell jar, diffusion pump, and mechanical forepump, this system is all stainless steel and employs a Vac-Ion pump and a molecular sieve sorption pump so that any backstreaming of oil vapor is eliminated. The system allows for the attachment of a residual gas analyzer to obtain quantitative information on the gases in the sample chamber. Perhaps even more important, the new system allows the sample chamber to be inserted in a liquid helium bath, so that vapor pressures in the chamber can be lowered sufficiently to freeze out all residual gases except He on the chamber walls. This technique should enable us to go to a substantially lower vacuum pressure than we can achieve now with the old system. Figure 1 is a photograph of the apparatus showing a U-shaped tubelike fixture terminated at one end by a 15 ℓ/s Varian Vac-Ion pump and at the other end by the magnetic driver, detector and sample chamber. Connecting these is a 2-in. (5.08 cm) pumping line with a gold seal valve and a Bayard-Alpert ionization gauge. The rock sample is held in the chamber at its mid point by a ring of four spring-loaded ball bearing tipped set screws. The sample is inserted through a port which is sealed with a Cu ring and a flange with a Varian-type stainless steel lip which can be tightened against the Cu seal by means of six bolts. The magnetic driver and detector are mounted on the outside of the chamber and interact with the sample ends through thin walls of nonmagnetic stainless steel. This has the advantage that all the electric circuitry and its associated contaminants (wire coating, solder joints, etc.) do not contribute to the gas load in the vacuum system. The iron buttons which complete the magnetic circuit of the driver and detector are attached to the sample with high-temperature quartz cement (American Fused Silica Corporation) which, after bakeout, does not appear to outgas a substantial gas load.

Results

A typical set of data obtained by the longitudinal resonance technique at room temperature after successive cleaning and outgassing treatments is shown in Table 1 at a frequency of 24 kHz. This frequency was chosen partly to provide a data point to determine the presence of any dispersion. After each outgassing step, Q was observed to increase, the highest value obtained being about 2400. A bandwidth plot at a high Q value is shown in Figure 2. The introduction of dry (< 10^{-8} torr partial pressure of H_2O) reactor grade nitrogen gas into the vacuum chamber after an outgassing treatment did not affect the Q. However, long term exposure to laboratory air at the end of a run brought the Q back down to about 100. These results strongly support the conclusion that igneous rocks display dramatically higher laboratory Q values after they are really freed from volatiles by strong outgassing techniques.

TABLE I.

Internal Friction Quality Factor Q Values in Igneous Rock W-8

Q	Cleaning and Outgassing Conditions
17–20	At room temperature in air at 1 atm after sawing with wet diamond saw, cleaning in acetone bath, and drying.
75–100	At room temperature in air at 1 atm after boiling in 30% H_2O_2 and drying.
400	At room temperature in air after gradual heating to $150°C$ in air and holding the temperature for about 1 day.
1500	At $\sim 1 \times 10^{-7}$ torr and room temperature after outgassing in 10^{-6} torr for 8 hours at $\sim 300°C$.
2375	At $\sim 1 \times 10^{-7}$ torr and room temperature after outgassing in 4×10^{-7} torr for 20 days at $\sim 400°C$.
100	After long term exposure to laboratory air at end of run.

Experiment 2.   Internal Friction Measurements at 50 Hz

In this experiment the mode of vibration selected was the flexural mode for which the resonant frequency is most readily shifted to lower frequencies by end-loading techniques. In particular, by end-loading our specimen 8.5 x 0.6 x 0.4 cm we lowered the resonance to 50 Hz, thus bringing the experiment into the high end of the seismic range of frequencies. (Seismic reflection studies on the upper and lower crust have been conducted at $\sim$ 100 Hz or slightly higher).

Very seldom has this technique been applied to brittle samples which are easily damaged by the conventional sample holder or to the measurement of internal friction on low loss materials because of the generally high "background" damping introduced by the apparatus itself. The present apparatus design was motivated by the desire to measure Q in fragile lunar return samples at zero static stress to shed light on the anomalously low attenuation of seismic waves on the moon. Furthermore, an additional requirement was that the sample be subjected to the low strain amplitudes and frequencies which are characteristic of seismic experiments.

Low Frequency Apparatus

Figure 3 shows a photographic overview of our measurement fixture. Here a rectangular bar-shaped sample of olivine basalt 8.5 x 0.6 x 0.4 cm is shown whose ends are held with a special clamping arrangement

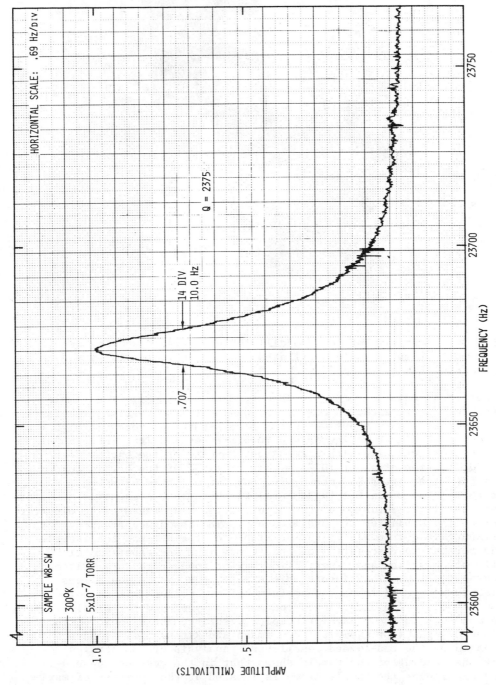

Fig. 2. Bandwidth data on a well-outgassed basalt sample showing a high Q value in the longitudinal mode of vibration.

Fig. 3.   Photograph of 50 Hz oscillator with insert showing detail
of sample clamp.   Sample oscillates around vertical axis formed by
supporting fibers at each end.

to a set of adjustable rotors in the shape of a barbell. The sample
and the rotors are anchored at the top and bottom by vertical fibers
to form vertical axes about which the rotors turn, as shown in Figure
3.   A magnetic driver acting on one rotor causes a slight rotation
about the vertical axis with a resultant bending of the bar of rock
with a deflection of about 10^{-5} cm.   The flexure of the bar in turn
causes the other rotor to rotate also but out of phase with the first
rotor for vibration in the fundamental flexure mode. A magnetic detector
senses the slight rotation of the second rotor and transmits an electric
signal to the amplifier and oscilloscope.   The resonance frequency of
the fixture is determined by the stiffness of the bar and the inertia
of the rotors.   A description of the theoretical and experimental
details of operation and construction may be found elsewhere (Tittmann
and Curnow, 1976).   The fixture is operable in a vacuum and allows
for sample heating (with a vacuum furnace) or cooling (with a hollow
cylindrical cold finger arranged concentrically around the sample).
     In contrast to the design principles described above, most of the
previous designs (for example, Sprungmann and Ritchie, 1971) have
been based on the end-loaded cantilever. Analysis of the strains
during the bending of the sample shows that in the case of the end-
loaded cantilever the sample clamp is located at the position of maximum bending strain, a situation inviting difficulties with concentrated
stresses.   On the other hand, in the present case the clamplike connections between the beam and the rotors are located at positions of mini-

mum bending strain. This result is most easily seen in the limit, where the sample's moment of inertia is very small. Then for the present design the strain distribution is uniform, whereas the cantilever has 50% greater strain at the sample clamp and zero strain at the load end. An additional advantage of the present design over the end-loaded cantilever is the absence of any static stress on the sample.

Fused Silica Sample

The Q value is typically determined by the decay method and Figure 4a shows the signal decay for a bar of fused silica (90 x 6.2 x 4.1 mm). Here the resonance frequency is 70.3 Hz, the time scale is 2 s/cm, and the measured Q = 3434 at 10^{-3} torr and at room temperature is in excellent agreement with Q = 3480 found previously (Sprungmann and Ritchie, 1971), on fused silica with a reed pendulum apparatus in the frequency range 1 to 30 Hz. The smooth monotonic decay of the signal demonstrates the absence of interference from other unwanted modes of vibration.

The maximum typical deflection of the sample at its center was measured by a capacitive microphone which gave an absolute reading for the deflection of $\delta\ell \approx 5 \times 10^{-5}$ cm, which was calculated to be equivalent to a strain of 10^{-7} for a sample with the above dimensions. This strain is considerably lower (factor of 10^2) than that encountered, for example, in the torsion pendulum and places the experiment near the upper end of the seismic range.

Olivine Basalt Sample

Having established the capability of measuring high Q (Q > 3400), we have begun to carry out Q measurements on W-8 samples at 56 Hz. For the rock as received, Q was equal to 51 at room temperature in laboratory air. When the rock was tested at 200 torr-vacuum pressure to remove the air drag on the rotors, the Q increased to 62. Then the rock was exposed to 10^{-6} torr, heated to 200°C for 48 hours, and allowed to cool in vacuum. A room temperature Q value of about 1100 was obtained, as is shown in Figure 4b. This result is analogous to the previous result at kilohertz frequencies in which similar increases in Q were obtained when strong outgassing procedures were carried out. Table II shows high temperature data obtained on the samples after outgassing at 460°C in 10^{-6} torr for about 20 hours. The data reveals high Q values to temperatures as high as 450°C, suggesting that high Q values are also concomitant with hot dry rock.

These results provide valuable evidence for supporting what the kilohertz measurements suggested, i.e., that the removal of volatiles in igneous rocks may be expected to produce dramatic increases in Q even at seismic frequencies. Thus the low-frequency measurements are an important link between the usual laboratory geophysical measurements and the seismic experiments that they are intended to simulate or clarify.

Experiment 3: Q Measurements on Rocks under Confining Pressure

Birch and Bancroft (1939) demonstrated the feasibility of resonating a rock sample under hydrostatic pressure by vibrating the fully encap-

f = 70.3 Hz   Q = 3434

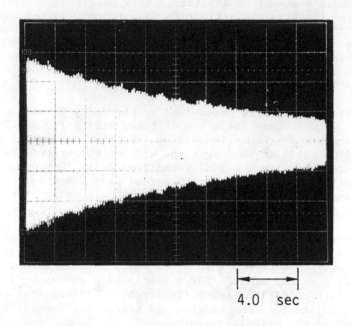

4.0 sec

f = 56.0 Hz   Q = 1100   1 sec/cm

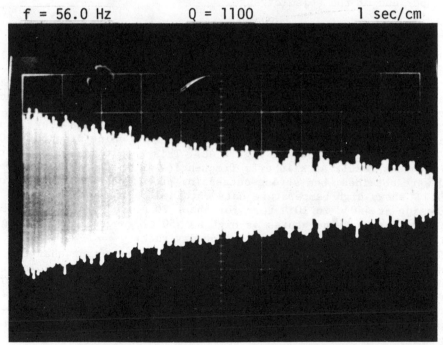

Fig. 4. Oscilloscope trace of signal decay pattern with:
a) quartz sample
b) strongly outgassed olivine basalt sample

TABLE II.

Temperature Dependence of Q at 56 Hz for Olivine Basalt Sample

T (°C)	Q	Conditions
450	900	
350	950	In vacuum of 10^{-7} torr after strong outgassing
250	900	
150	700	
50	1000	

sulated samples in the torsional mode. They used this approach to obtain some Q data and extensive data on the modulus of rigidity as a function of pressure for a wide variety of rock samples. We have begun to explore the feasibility of extending their approach to the measurement of Q as a function of hydrostatic pressure with the additional stipulation that the rock sample environment approach lunar conditions, i.e., a vacuum. Our initial experiments at moderate pressures have been sufficiently successful to make this approach seem promising and some of the results are described below.

Calibration of Background Damping

As was pointed out by Birch and Bancroft (1939), Q measurements are made very difficult by background damping due to such factors as direct pickup from the driving unit, stray induction from other sources, viscous damping due to the high-pressure confining gas, frictional effects of the capsule against the sample, and loss of energy to the supports. Considerable effort was spent in minimizing these factors with the aid of metallic high Q reference samples (polycrystalline Al) until a low background could be established. The progress to the present time is illustrated by high pressure Q values of about 1.2×10^4 which is an approximate measure of the Q of the apparatus (for the unclad Al, $Q \approx 4 \times 10^4$ in the atmosphere). Although this establishes a low loss measurement base, the effort of increasing the apparatus Q is being continued with the goal of making measurements to above 5 kbar, where viscous drag due to the confining gas is expected to be the main source of background damping.

Q of Outgassed Rock

In addition to studying rocks exposed to the atmosphere, our goal is to study rocks under thoroughly outgassed conditions. This is difficult because the rocks are porous and have to be encapsulated under vacuum in a thin wall Cu sheath while they are in a fully out-gassed state. We have accomplished this by sealing the capsule in a vacuum of 10^{-5} torr by electron beam welding. Figure 5 shows representative results on a fine grained basalt obtained from an exposed dike in the Santa Monica Mountains in southern California, and having a measured open pore porosity of about 1%. After initial saw cuts the samples were gently ground from square cross-section bars to polygonal bars with n sides (n increasing from 4 to 8) and finally to round cylinders about 1 cm in diameter and 15 cm long. These were then roughly cleaned in trichloroethylene, alcohol, and H_2O_2 and baked out for 3 days in 2×10^{-5} torr at $350°C$. The samples were then inserted into 4 mil wall seamless Cu tubing which was first heated so that when it cooled, it would shrink down tightly against the rock. The sheathed samples and their end caps were baked out again as had been done before and mounted under dry He gas in the welding chamber, where they were held at 1×10^{-5} torr for 12 hours. At this point a 4 kV electron beam was applied to seal the seam between the tube and the end cap. Repeated exposure of the samples, so encapsulated, to 0.5 kbar pressures showed no deleterious effects on the seam due to leaks in the sheath, so that the vacuum inside the capsule appears to have been maintained intact. Correspondingly, the results of Figure 5 show $Q \sim 1400$, which compares to $Q \approx 200$ for samples cleaned as was done before but sheathed while they were exposed to the atmosphere. Also shown in Figure 5 is the pressure dependence of the torsional frequency, which is seen to rise rapidly with pressure, up to 0.1 kbar as the Cu sheath establishes contact with the sample. At higher pressures the resonant frequency is seen to level out with a slope of $v^{-1}(\Delta v / \Delta p) \approx 0.01$ kbar^{-1} at a pressure $p \sim 0.25$ kbar, where v is the shear velocity. Figure 6 is a photo-graph showing the details of the sample holder, the Cu clad sample, and the magnetic drive. These results suggest that partial crack closure begins already at low pressures and supports the conclusion of Richter and Simmons (this volume) that most clean high-aspect ratio cracks are already closed at depths corresponding to 2-3 kbars and that these cracks heal over geological time. High pressure runs for the sample in the moist state carried out at 2.5 kbar (Tittmann, 1977) showed that the Q in the moist condition was still much smaller than the dry Q. This implies that the pressure of volatiles keeps the Q low, even when the cracks are partially closed corresponding to some depth below the surface, or alternatively that the presence of dry rock in the crust is concomitant with a high Q.

Discussion of Results

The three experiments discussed above show that the Q of a low-porosity (1-2%) olivine basalt without hydrated mineral phases ranges from about 100 measured dry in normal laboratory air to over 2000 outgassed at moderate temperatures in a high vacuum. Data so far obtained on the frequency dependence down to 50 Hz and limited data on the pressure dependence of the loss associated with small amounts of absorbed water lead us to believe that our laboratory Q values are probably applicable to seismic waves propagating in rocks subjected to high pressures under the earth's surface.

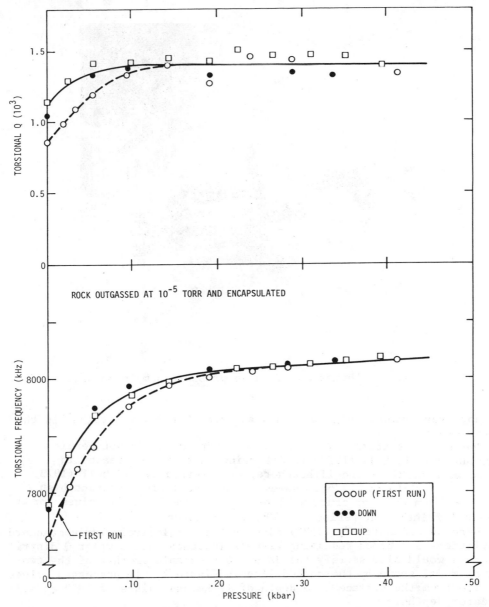

Fig. 5. Q and resonance frequency as a function of confining pressure for strongly outgassed basalt sample sheathed under vacuum of 10^{-5} torr.

Figure 7 shows the results of Hermann and Mitchell's (1975) calculation of Q-depth profile based on seismic field data. They find this profile similar to that of Mitchell (1973) in that it has relatively low values of Q (high Q^{-1}) in the upper crust and a rapid transition, at midcrustal depths, to high Q values (low Q^{-1}) in the lower crust. It differs from the model of Mitchell (1973) mainly by having a less

Fig. 6.   Photograph of high pressure sample holder.

abrupt transition to high Q values and somewhat lower Q values in the
lower crust.

The large standard deviations at greater depths do not permit
Hermann and Mitchell (1975) to determine whether Q decreases in
value with depth in the lithosphere, as inferred by Mitchell (1973)
and Lee and Solomon (1975). However, any model with constant high
Q values in the lower crust will predict extremely small values for the
internal friction at periods of 40 s and greater.

Hermann and Mitchell (1975) find that the relatively large standard
deviations and broad resolving kernels indicate that simpler Q distributions would also satisfy the data. One example is that of the two-
layer model shown by the dashed line in Figure 7. It possesses values
of 250 from the surface to a depth of 17 km and values of 2000 at all
greater depths.

The most conspicuous feature of the profiles of Hermann and
Mitchell (1975) and of Mitchell (1973) is the sharp increase in Q
(decrease in internal friction) from values not exceeding a few
hundred from the surface down to about 15 km to a value as high as
2000 below 15 km. No ordinary rock types measured in the laboratory
at room temperature and 1-atm pressure have shown such high Q values
(Knopoff, 1964).

As a result of the laboratory measurements on strongly outgassed
rocks discussed above we suggest that the sharp increase in Q probably
marks a boundary below which no free moisture exists in the crustal

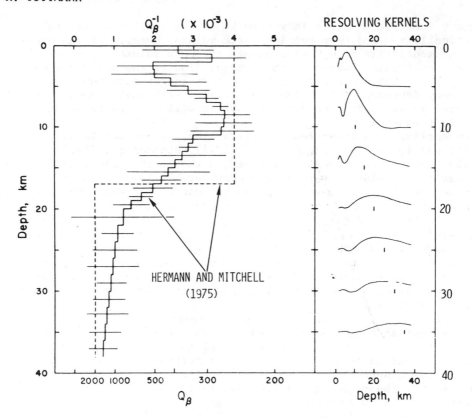

Fig. 7. The derived Q model (solid curve) determined by Hermann and Mitchell (1975). The bars are the standard deviations. The right hand side of the figure gives the resolving kernels at each depth. The simpler Q model (dashed curve) not obtained by the inversion process, provides an adequate, although less optimum fit to the data.

rocks. This boundary might, but need not necessarily, coincide with a transition between chemically distinct rock types. Our laboratory measurements also show that the observation of high Q values are not restricted to dry rocks at room temperature but extend also to temperatures in excess of 400°C, ie. to hot dry rock.

We have shown that Q of a low-porosity olivine basalt changes from about 100 measured dry in normal laboratory air to over 2000 outgassed at moderate temperatures in a high vacuum. The Q returns to roughly its original value after a few minutes reexposure of the sample to laboratory air, and this cycle can be repeated indefinitely. The difficulty which we experience in outgassing the samples well enough to achieve the maximum Q values indicates that just a few monolayers of adsorbed water on pore surfaces are very effective in drastically reducing the Q.

These results imply that the Q of a dry sample of the same composition and mineralogy would be above 2000. They also strongly

suggest that the Q values in wet rocks in the earth's crust are
drastically reduced from the intrinsic values, because of the presence
of even monolayers of H_2O even though the pore volume may be small
and the calculated losses due to viscous flow of the water may be
negligible. The limited data that we have so far obtained on the
frequency dependence of the loss associated with small amounts of
adsorbed water lead us to believe that it will be very important at
seismic frequencies.

In earlier work along the same direction, Gordon and Davis (1968),
Pandit and Tozer (1970), and Pandit (1971) obtained qualitatively
similar results on several rock types, although the fractional changes
and maximum Q values obtained were considerably lower, most probably
because they were not able to outgas the samples as thoroughly.

Some time ago, Birch and Bancroft (1939) showed that Q in the
kilohertz frequency range of some, but by no means all, igneous rocks
increased from values of a few hundred to values of about 2000 when
a pressure of 4000 kg/cm^2 was applied to the jacketed samples,
corresponding to a depth of roughly 10 km in the earth. In light of
our results we interpret the Birch and Bancroft results as implying
that those samples where Q did increase to the vicinity of 2000 were
free of moisture.

Based on our interpretation of the seismic data, it
appears that at depths below 15 km no free water exists. This conclusion could contribute to understanding the nature of earthquake
source mechanisms in the region. This interpretation also implies
that in the construction of theoretical electrical conductivity
profiles for the region it is appropriate to use values characteristic
of dry rocks below depths of about 15 km.

Acknowledgements. The author is grateful to L. Ahlberg, J. Curnow,
and H. Nadler for their help in the experiments; and to R. M. Housley,
G. A. Alers, and J. M. Richardson for many helpful discussions. The
work was partially supported by NASA contract NAS 9-13846.

References

Berry, G.W., Apparatus for the measurement of the internal friction of
    metals in transverse vibration, Rev. Sci. Instrum., 25, 884-886,
    1955.
Birch, F., and D. Bancroft, The effect of pressure on the rigidity of
    rocks, I, J. Geol., 46, 59-87, 1938.
Gordon, R.B. and L.A. Davis, Velocity and Attenuation of Seismic Waves
    in Imperfectly Elastic Rock, J. of Geophys. Res., 73, 3917-3935, 1968.
Hermann, R.B., and B.J. Mitchell, Statistical analysis and interpre-
    tation of surface-wave anelastic attenuation data for the stable
    interior of North America, Bull. Seismol. Soc. Amer., 65, 115-
    1128, 1975.
Housley, R.M., B.R. Tittmann, and E.H. Cirlin, Crustal porosity
    information from internal friction profile, Bull. Seismol. Soc.
    Amer., 64, 2003-2004, 1974.
Knopoff, L., Q, Rev. Geophys. Space Phys., 2, 625-660, 1964.
Lee, W.B., and S.C. Solomon, Inversion schemes for surface wave
    attenuation and Q in the crust and the mantle, Geophys. J., Roy.
    Astron. Soc., 43, 47-71, 1975.

Mitchell, B.J., Surface-wave attenuation and crustal anelasticity in
    central North America, Bull. Seismol. Soc. Amer., 63, 1057-1071,
    1973.

Pandit, B.I., Experimental studies on the mechanism of internal
    friction (Q^{-1}) of rocks, PH. D. Thesis, 117 pp., Univ. of Toronto,
    1971.

Pandit, B.I., and D.C. Tozer, Anomolous propagation of elastic energy
    within the moon, Nature, 226, 335, 1970.

Quimby, S.L., An experimental determination of the relation between
    viscosity and frequency in vibrating solids, Phys. Rev., 31, 1113,
    1928.

Tittmann, B.R., J.M. Curnow, and R.M. Housley, Internal friction
    quality factor Q > 3100 achieved in lunar rock 70215, 85, Proc.
    Lunar Sci. Conf. 6th, 3, 3217-3226, 1975.

Tittmann, B.R., and J.M. Curnow, Apparatus for measuring internal
    friction Q factors in brittle materials, Rev. Sci. Instrum., 47,
    1516-1518, 1976.

Sprungmann, K.W., and I.G. Ritchie, An improved reed pendulum apparatus
    and techniques for the study of internal friction of ceramic single
    crystals, U.S. At. Energy Comm. AECL-3794, 1971.

Tittmann, B.R., Rayleigh wave studies in lunar rocks, IEEE Trans.
    Sonics Ultrason. No. 72 CHO 708-8SU, 130-135, 1972.

Tittmann, B.R., R.M. Housley, G.A. Alers, and E.H. Cirlin, Internal
    friction in rocks and its relationship to volatiles on the moon,
    Proc. Lunar Sci. Conf. 5th, 3, 2913, 1974.

Tittmann, B. R., Unpublished.

Wegel, R., and H. Walther, Internal dissipation in solids for small
    cyclic strains, Physics, 6, 141-157, 1935.

Reprinted from Physics of the Earth and Planetary Interiors, v. 8, p. 246–250.

MECHANICAL HYSTERESIS IN ROCKS AT LOW STRAIN AMPLITUDES AND SEISMIC FREQUENCIES

B. McKAVANAGH and F.D. STACEY

Physics Department, University of Queensland, Brisbane (Australia)

Accepted for publication October 4, 1973

A sensitive capacitance displacement transducer has been used to record hysteresis loops in the stress–strain diagrams of laboratory samples of granite, basalt, sandstone and concrete subjected to cyclic axial strains with amplitudes of order 10^{-5} and periods of 10–300 sec. The ends of the loops are always cusped, whether the load cycle is sinusoidal or not, and at low strain amplitudes the loop shape becomes symmetrical and appears to be independent of amplitude. Thermal relaxation influences the observed loop shapes, so that the strain cycles represent a compromise between adiabatic and isothermal compressions. However, this does not affect the conclusion that stress–strain loops are always cusped. This observation does not appear to be consistent with linear theories of damping of acoustic and seismic waves, which indicate elliptical loops.

1. Introduction

Our understanding of the mechanism(s) of attenuation of acoustic and seismic waves in rocks, and in the earth, is still rudimentary. Representation in terms of the quality factor, Q (where $2\pi/Q = \Delta E/E$, the fractional energy loss per cycle) is most convenient, because empirical evidence (Knopoff, 1964; Gordon and Davis, 1968) indicates that Q is essentially independent of frequency, although the individual damping mechanisms which appear likely to be significant in the earth are all strongly frequency dependent (Jackson and Anderson, 1970). A common assumption is that, although the important processes are relaxation phenomena, the spectrum of relaxation times smears out the frequency dependence. But this is not theoretically satisfying, partly because it precludes reliable extrapolations outside the immediate ranges of experimental observations. However, in the absence of compelling evidence for frequency dependence in the earth, frequency independence is assumed as nearly as possible in empirical treatments of seismic data and deductions about Q-values in the interior of the earth.

A basic question to be examined is whether the attenuation mechanism is *linear* in the sense that

attenuation of a complex wave form is calculable from the attenuations of its Fourier components. The object of the experiment reported here was to examine this problem from the point of view of the shape of the stress–strain hysteresis loops, because linear theories of attenuation have indicated that hysteresis loops are elliptical (for sinusoidal stress cycles), whereas reported measurements (especially Gordon and Davis, 1968) have indicated loops with pointed ends. However, since Gordon and Davis did not use sinusoidal load cycles, but 'sawtooth' cycles, i.e. loads increasing and decreasing linearly between fixed limits, the problem required further experimental attention.

Another problem which arises is that stresses in seismic waves (except very close to the source) are very much smaller than those used in laboratory experiments which have produced observable hysteresis loops. For strain cycles of large amplitudes, Q becomes amplitude-dependent (and is thus certainly non-linear). According to data of Gordon and Davis (1968), amplitude independence in rocks is a reasonable approximation only for strain amplitudes smaller than about 10^{-5}. Measurements at greater amplitudes are therefore not unambiguously indicative of the seismic attenuation mechanism. Savage (1969) argued that at low strain amplitudes the attenuation mechanism must become linear.

We report here measurements with a new capacitance displacement transducer which has allowed us to observe hysteresis loops at strains down to 10^{-6}, cycled over a range of periods from 10 to 300 sec with both 'sawtooth', and sinusoidal waveforms. The object was to seek evidence of a transition to a linear mechanism at small strain amplitudes, as would be indicated by elliptical hysteresis loops.

2. Precise measurements of rock strain

The strain measurements have been made on blocks of rock of order 45 cm $\times$ 15 cm $\times$ 15 cm within which 4.3 cm diameter cores were cut to a depth of 35 cm and left secured to the blocks at their lower ends. Strains were applied by axially loading the cores whose contractions were measured relative to the surrounding blocks. To avoid end effects the top of each block was cut away, leaving 10 cm of core standing up and electrodes for the measurement of displacement were fixed at least 5 cm below the upper end of each core, so that the 'active' length of core was about 30 cm. Then measurement of strain smaller than 10^{-5} to the 0.1%-level required for the observation of details of loop shape involved the measurement of electrode displacement to better than 30 Å. Stress was inferred from the strain of a steel 'specimen', mechanically in series with the rocks, which had a high Q-value, i.e. more than an order of magnitude higher than the least dissipative of our rocks samples.

The principle of the displacement transducer system has been described by Stacey et al. (1969) and Stacey (1972). One annular electrode, with earthed guard rings on its inside and outside edges, is screwed to a steel ring shrunk onto the rock core at the appropriate point. This is the centre electrode of a three-electrode system, the other two being rigidly fixed together and mounted on the outer part of the block of rock, so that gaps between the inner electrode and each outer electrode are close to 0.5 mm. The pair of capacitances so formed is connected to a ratio transformer bridge, balance of which gives a seven-figure reading, being the ratio of one capacitance gap to the sum of the gaps (1 mm) and out-of-balance signal is proportional to displacement from

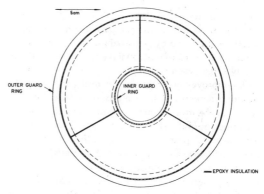

Fig. 1. Detail of electrode design, facilitating electrical checking of precise parallelism of displacement transducer electrodes. The active area of the central electrode is the area between the circles represented by broken lines. The pair of outer electrodes overlap the central electrode and each is cut into three sectors which may be used independently (for adjustment of parallelism) or together (for measurements of displacement).

the balance position. The electrical noise level in the system corresponds to an R.M.S. displacement noise of 0.2 Å, but most of the measurements have been made in hydraulic presses, in which mechanical vibration of the electrodes was at least ten times this level.

Two significant refinements of the transducer system as originally described (Stacey et al., 1969) have been applied in the present experiment. One is an electrical measurement of precise parallelism of the electrodes and the other concerns the time constant of the integrator on the synchronous detector used to measure the bridge output. Parallelism of the electrodes with electrode gaps nearly equal is obtained by mechanical adjustment of the angle of the outer pair of electrodes, which are made in three sectors, as in Fig. 1. Displacement of the central electrode with respect to one pair of outer electrode sectors is measured by balancing the bridge with those sectors connected to the bridge and the other sectors earthed. Repeated adjustment for 50-50 balance with each pair in turn ensures parallelism of the electrodes and when this is accomplished all three sectors of each outer electrode are connected together for displacement measurements. Apart from ensuring accuracy and linearity of the bridge calibration, the parallelism (at 50-50 balance) makes the bridge output insensitive to bending of the rock core, such as would be produced by non-axial loading. The records obtained are thus of axial strains and are independent of any other (shear) strains.

To improve noise rejection the bridge uses a synchronous detector with an integrator of time constant as long as the measurements permit. But a time lag in the bridge output, relative to a measured displacement would appear as a phase lag in the measured strain and hence a contribution to hysteresis, so that in the present measurements the time constant restriction is very stringent. Of course, it must still be substantially longer than period of the alternating current which drives the bridge. The frequencies of the oscillators used to drive the bridges in this experiment were therefore increased to 10–11 kHz instead of 2.5–3 kHz as generally favoured for ratio trans-

former bridges. Then an integrator time τ introduces a phase lag ϕ to a bridge output cycled with period T, where:

$$\phi \approx 2\pi \frac{\tau}{T} \text{ (for small } \phi)$$

If this is the phase lag in a strain relative to stress, it introduces a loop of width corresponding to an apparent Q of ϕ^{-1}. Since measured values of Q are of the order 100, it is evident that comparable spurious contributions are unacceptable and we require $\phi \ll 10^{-2}$, i.e. τ not much larger than 10^{-3} sec to record a stress–strain cycle of 10 sec period. It is this consideration which imposes the upper limit to the frequency of stress cycling.

Although care was taken to avoid a phase-lag problem, so that observed loop shapes could be of fundamental significance, two features of the experiment reduce the importance of the problem somewhat.

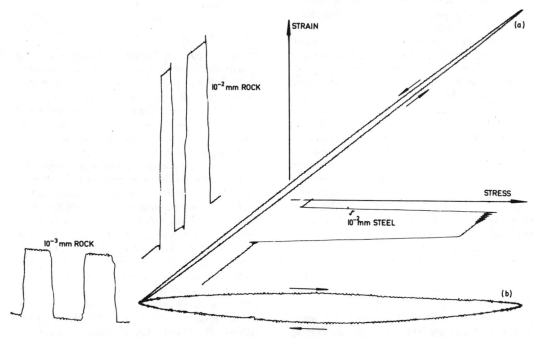

Fig. 2. X-Y recorder plot of the stress–strain loop of a sandstone sample (a) and the expanded form of the loop (b) with calibration pulses for stress, and for strain as plotted on both the original and expanded loops. The calibration pulses are applied simply by switching the appropriate decade of the ratio transformer bridge which is calibrated directly in terms of electrode displacement. This is a loop plotted with a strain amplitude of about $3.53 \cdot 10^{-5}$ about an ambient strain of $1.4 \cdot 10^{-4}$, using a servo-controlled press with a precise sinusoidal cycle of 60 sec period, but with obvious mechanical noise.

Firstly, identical circuits were used for the observation of strain and inferred stress (from the strain of a high Q steel), so that when we experimentally increased the integrator time constants substantially (which had the advantage of reducing the apparent noise level due to vibrations, by restricting the output bandwidths) the observed hysteresis loops were apparently unaffected having comparable lags on both stress and strain axes. Secondly, the phase lag effect could only produce an elliptical loop, or an elliptical (rounded end) contribution to a total loop, and since we were seeking evidence of departure from elliptical shape, the appearance of cusped ends on a loop could not have resulted from instrumental time constant effects.

The two bridge outputs were applied to an X-Y recorder for direct plotting of hysteresis loops. But to display the shapes of hysteresis loops in more detail than is apparent in a direct stress (ϵ)–strain (σ) plot the two signals were applied to a differential amplifier, so that we were able to plot instead $(\epsilon - \sigma/\bar{q})$ vs. σ, where $\bar{q}$ is the mean elastic modulus, i.e., the relative amplitudes of the input signals were adjusted to plot the loop about the stress (σ) axis. Then an increase in sensitivity expands the loop about this axis. An example is reproduced in Fig. 2.

3. Summary of results

We have experimented both with 'triangular' waves (stress varying linearly with time between fixed limits) and with sinusoidal waves having a 30:1 range of periods using hydraulic presses, and also with loads applied manually but less precisely controlled, always with the same general results – loops with cusped ends. Strain amplitudes smaller than about 3–$5 \cdot 10^{-6}$ could not be used very satisfactorily in the presses because of record noise resulting from mechanical vibration, but manually applied loads approximately sinusoids gave reasonably clean loops at strains smaller than 10^{-6}. Loop shapes for the 'sawtooth' and sinusoidal waves differed slightly, but we demonstrated conclusively by a sequence of experiments with 'square' waves that this effect arose from the thermal relaxation of the adiabatic heating and cooling associated with the load cycle. In fact our strains were neither isothermal nor adiabatic but

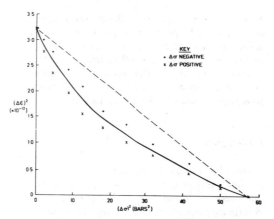

Fig. 3. Strain differences, $\Delta\epsilon$, between increasing-strain and decreasing-strain limbs of the cycle reproduced in Fig. 2, plotted as a function of the departure from ambient stress, $\Delta\sigma$. Both quantities are plotted as squares because on this scale an elliptical loop would give a linear relationship represented by the broken line. Asymmetry of the loop is indicated by the differences between values of $\Delta\epsilon$ for opposite signs of $\Delta\sigma$, the solid line representing the mean. This asymmetry disappears with decreasing strain amplitude, but the departure of the loops from elliptical form (represented by the difference between the solid and broken lines) remains.

some compromise and this necessarily restricts the possibilities for detailed interpretation of loop shape.

However, two essential observations are unaffected by the experimental limitations. Firstly, the hysteresis loops were always cusped at the ends, even at the lowest strain amplitudes and with sinusoidal waves. The thermal relaxation phenomenon cannot have contributed to this observation because if it acted alone it would give an elliptical loop and if superimposed upon another mechanism it would tend to round the loop ends. The cusped loops therefore represent an intrinsic effect in the attenuation mechanism for strain waves at least down to amplitudes of 10^{-6}. Secondly, at high strain amplitudes loops become asymmetrical, as in the example of Fig. 2. In Fig. 3 the 'width' of the loop in Fig. 2 is plotted as a function of stress (on a scale on which an ellipse would give a straight line). This emphasises the consistent departure of the loop shape from elliptical form. The loop shape is here clearly different for compressions and rarefactions, although cusped at both ends.

This asymmetry diminishes with decreasing strain amplitude and loops tend to an apparently constant shape, independent of strain amplitude as nearly as we have been able to discern. The general shape remains similar to Fig. 2. This encourages our supposition that extrapolation to indefinitely small strain amplitudes would still give cusped loops.

4. Discussion

Our observations clearly support the case for a non-linear attenuation mechanism in rocks under laboratory conditions. While we are not in a position to state categorically that linear mechanisms are excluded, at least the argument that attenuation is necessarily linear (e.g. Savage, 1969) should be examined for a basic flaw. The most important limitation on the usefulness of our results concerns the frequency dependence of loop shape. While our loops were very similar over a frequency range of 30:1, the problem of thermal relaxation imparted a subtle apparent rate-dependence which prevented a sufficiently accurate test for frequency dependence of the intrinsic attenuation mechanism. It is possible that this effect may be overcome by making measurements of torsional rather than compressional stress—strain loops.

Internal friction due to cracks or grain boundaries (as in the theories of Walsh, 1966 and White, 1966) is at least qualitatively consistent with our observations. Other mechanisms are probably more significant in the deep interior of the earth, but the fact is that all laboratory stress—strain type observations indicate cusped hysteresis loops. The contribution of the present work is to extend these measurements to much lower strain amplitudes than previous experiments, to overcome the objection that the cusps represented a non-linear effect of large strain amplitudes.

Our observations indicate that cusped loops extend to indefinitely small strain amplitudes.

It appears that the case for a non-linear attenuation mechanism merits further serious thought and conversely the necessity for a linear mechanism (Savage, 1969) should be critically re-examined.

Acknowledgements

This work is supported by the Australian Research Grants Committee. Some of our measurements were made using facilities of the University's department of Civil Engineering and all of the measurements with sinusoidal stress cycles were made using a servo-hydraulic press at the Aeronautical Research Laboratories of the Department of Supply, Melbourne. The interest and help of Dr. M.T. Gladwin, Dr. L.M. Hastie and Mr. J. Field are also greatly appreciated.

References

Gordon, R.B. and Davis, L.A., 1968. Velocity and attenuation of seismic waves in imperfectly elastic rock. J. Geophys. Res., 73: 3917–3935.

Jackson, D.D. and Anderson, D.L., 1970. Physical mechanisms of seismic wave attenuation. Rev. Geophys., 8: 1–63.

Knopoff, L., 1964. Q. Rev. Geophys., 2: 625–660.

Savage, J.C., 1969. Comments on paper by R.B. Gordon and L.A. Davis, 'Velocity and attenuation of seismic waves in imperfectly elastic rock'. J. Geophys. Res., 74: 726–728.

Stacey, F.D., 1972. Earth strain research at the University of Queensland. Aust. Physicist, 9: 5–13.

Stacey, F.D., Rynn, J.M.W., Little, E.C. and Croskell, C., 1969. Displacement and tilt transducers of 140 dB range. J. Phys. E (J. Sci. Instrum.), 2: 945–949.

Walsh, J.B., 1966. Seismic wave attenuation in rocks due to friction. J. Geophys. Res., 71: 2591–2599.

White, J.E., 1966. Static friction as a source of seismic attenuation. Geophysics, 31: 333–339.

Reprinted from Nature, v. 268, p. 220–222.

Frequency dependence of elasticity of rock—test of seismic velocity dispersion

LINEAR theories of attenuation of acoustic or seismic waves in media, such as rocks, which are characterised by constant or nearly constant Q factors (fractional energy loss per cycle equals $2\pi/Q$), imply velocity dispersion and, therefore, frequency dependence of elasticity. The effect is small, corresponding to a change of order 1 % over the period range of seismological interest (1 s to 1 h), and is consequently difficult to observe. However, it leads to an internal inconsistency in the development of earth models by inversion of free oscillation data and to discrepancies between these models and body wave data[1-3]. It is a matter of considerable geophysical interest to resolve the problem. Validity of the linearity assumption has been questioned[4] but we are now observing elliptical hysteresis loops in basalt and granite samples subjected to sinusoidal strains of order 10^{-6} and this is strong evidence that attenuation does become a linear phenomenon at low strain amplitudes. But whether or not perfect linearity applies, a direct demonstration of body wave dispersion is the most satisfying indication that earth model studies need revision.

The requirement of causality in linear attenuation theories (which we propose to review in a separate paper) leads to the relationship between phase velocity $v(\omega_0)$ at angular frequency ω_0 and an integral involving the attenuation coefficient, $\beta(\omega)$

$$\frac{1}{v(\omega_0)} = \frac{1}{c} - \frac{2}{\pi} P \int_0^\infty \frac{\beta(\omega)}{\omega_0^2 - \omega^2} \, d\omega \quad (1)$$

where P indicates the Cauchy principal value of the integral, c is a constant (phase velocity at $\omega \to \infty$) and a plane wave of any frequency ω decays with distance x as $\exp(-\beta x)$. For the present purpose, it suffices to assume that the behaviour of β at very high frequencies is such that the integral in equation (1) converges but that over a wide range of frequencies about ω_0, β is proportional to ω, which is very nearly equivalent to the constant Q assumption because

$$\beta = \omega/2Qv \quad (2)$$

the velocity dispersion being slight. Then it follows that, to first order in Q^{-1}, values of elastic modulus μ' (real part of the complex modulus), at frequencies ω_1 and ω_2, are related by

$$[\mu(\omega_2) - \mu(\omega_1)]/\mu(\omega_1) = (2/\pi Q)\ln(\omega_2/\omega_1) \quad (3)$$

The derivation of equation (3) demonstrates the equivalence for practical purposes of the alternative empirical linear theories[5-7], which differ only in the assumptions regarding the high frequency behaviour of β, as well as the mechanistic theory based on logarithmic creep[8-11]. The logarithmic dependence of elasticity or velocity on frequency is characteristic of frequency independent Q; other assumptions regarding $\beta(\omega)$ lead to quite different results, so that the direct observations which we report here, confirming the validity of this relationship, also support the constant Q assumption. The logarithmic dependence seems to have been quoted first by Kolsky[12], but we have been unable to find the earlier derivation. This dispersion law also satisfies the observations of pulse shape in acoustic experiments[12,13] and in particular the increase of pulse rise time τ with time of travel and Q for the path[13]

$$\Delta\tau \approx 0.5\int Q^{-1} \, dt \quad (4)$$

Our experiment is similar in principle to that of McKavanagh and Stacey[11], who measured the mechanical hysteresis of rock cores under cycled axial compressions. Our specimens are tubes of rock of length about 280 mm, diameter 44 mm and wall thickness 5 mm, and are subjected to torsion. The use of thin walled tubes ensures that the strain, which is pure shear, is approximately uniform through the material. The specimen under examination is connected to a standard tube, either of high Q steel or of fused quartz, subjected to the same torque and used to infer the stress. Rings, fixed about 20 mm from both ends of each specimen and standard, are used for mounting capacitance plates, which are centred about 200 mm from the specimens. Relative displacements of the plates record the twist of the specimen between its mounting rings. Balanced sets of plates are mounted on opposite sides of each specimen to minimise effects of asymmetry and thermal drift. Measurements of capacitance are made by means of ratio transformer bridges[14,15] with rapid response times (10^{-3} s) which avoid the appearance of any instrumental phase lag in the measured strains.

Small stresses are applied by means of loudspeaker coils mounted on a lever arm connected to one end of the standard specimen and driven by current from a low frequency sinusoidal oscillator. Strains in the rock specimen and standard are recorded on an X-Y recorder, giving, in effect, a stress–strain plot as the strain is cycled. With the granite and basalt specimens which we have used in this experiment, the hysteresis loop is barely apparent. However, by subtracting the two bridge outputs from one another (specimen minus standard) in a difference amplifier, with independently controlled gains on the two inputs, and further amplifying the difference signal, we can plot directly the departure of specimen strain from linearity with strain of the standard. This appears as a loop of adjustable width, whose area is a measure of the hysteretic loss in the specimen, relative to the standard. The slope of the axis of this loop is a very sensitive indicator of the elastic modulus. The fact that we are able to observe such fine details of loop shape at strain amplitudes as low as 10^{-6} (hopefully lower with improvements now planned) is due to the high sensitivity and wide linear range of the capacitance displacement transducer[14-17].

We have plotted hysteresis loops for specimens of basalt and granite and found that the elastic moduli vary with frequency precisely in the manner of equation (3) over the period range 2–1,000 s, which covers much of the seismic frequency band. Figure 1 gives plots of the data, represented as the fractional changes in rigidity modulus relative to the values at 0.1 Hz. The gradients of the two graphs do not agree perfectly with the theoretical values obtained from the measured Q values by equation (3). The discrepancies may arise because the standard specimens were not completely lossless or from frequency dependence of Q, although the measured Q values were independent of frequency within the accuracy of our measurements.

The hysteresis loops of both specimens were of elliptical form, not cusped at the ends, indicating that within the precision of these observations the attenuation mechanism is linear. This

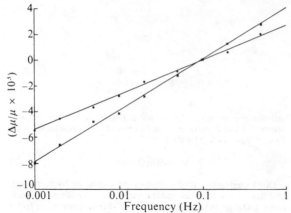

Fig. 1 Frequency dependences of rigidity moduli of basalt and granite specimens over the range 0.001–0.5 Hz relative to values of 0.1 Hz, which was used as a reference frequency in repeated comparisons. ●, Basalt, $Q = 590$, strain amplitude = 2.6×10^{-6}. ×, Granite, $Q = 320$, strain amplitude = 2.9×10^{-6}. By equation (3) the gradients correspond to Q values of 546 and 371 respectively.

was not the case with the sandstone specimen examined by McKavanagh and Stacey[14], even down to strains of 10^{-6}, or with granite and basalt specimens at strains of order 3×10^{-5}. It seems that the strain amplitudes used in the present experiment are at the upper limit of the linear range. Further instrumental improvements are required to examine with the same precision Q and modulus changes at strain amplitudes below 10^{-6}.

If we accept the validity of equation (3) then it should be possible to adjust earth free oscillation data by using measured

Q's of the identified modes to remove the effect of frequency dependence of elasticity. Comparison with body wave data also requires recognition of the fact that in the presence of dispersion, as represented by equation (3), the group velocity v_g of a body wave is related to phase velocity by

$$v_g = v(1 + 1/\pi Q) \tag{5}$$

to first order in Q^{-1}. This is the observed velocity in the case of a body wave, but free oscillations effectively measure phase velocity, which for the mantle is of order 0.1 % smaller.

This work is supported by the Australian Research Grants Committee. The mechanical part of the apparatus was constructed by R. Willoby.

B. J. Brennan
F. D. Stacey

Physics Department
University of Queensland
Brisbane, 4067, Australia

Received 17 May; accepted 2 June 1977.

1 Randall, M. J. *Phys. Earth planet. Interiors* **12**, 1–4 (1976).
2 Liu, H.-P., Anderson, D. L. & Kanamori, H. *Geophys. J. R. astron. Soc.* **47**, 41–58 (1976).
3 Kanamori, H. & Anderson, D. L. *Revs Geophys. Space Phys.* **15**, 105–112 (1977).
4 Stacey, F. D., Gladwin, M. T., McKavanagh, B., Linde, A. T. & Hastie, L. M. *Geophys. Surv.* **2**, 133–151 (1975).
5 Futterman, W. I. *J. geophys. Res.* **67**, 5279–5291 (1962).
6 Strick, E. *Geophys. J. R. astron. Soc.* **13**, 197–218 (1967); *Geophysics* **35**, 387–403 (1970); *Geophysics* **36**, 285–295 (1971).
7 Azimi, Sh. A., Kalinin, A. Y., Kalinin, V. B. & Pivovarov, B. L. *Izv. Earth Phys.* **(2)**, 88–93 (1968).
8 Lomnitz, C. *J. appl. Phys.* **28**, 201–205 (1957).
9 Pandit, B. I. & Savage, J. C. *J. geophys. Res.* **78**, 6097–6099 (1973).
10 Savage, J. C. & O'Neill, M. E. *J. geophys. Res.* **80**, 249–251 (1975).
11 Savage, J. C. *Phys. Earth planet. Interiors* **11**, 284–285 (1976).
12 Kolsky, H. *Phil. Mag.* **1**, 693–710 (1956).
13 Gladwin, M. T. & Stacey, F. D. *Phys. Earth planet. Interiors* **8**, 332–336 (1974).
14 McKavanagh, B. & Stacey, F. D. *Phys. Earth planet. Interiors* **8**, 246–250 (1974).
15 Stacey, F. D., Rynn, J. M. W., Little, E. C. & Croskell, C. *J. Phys. E* **2**, 945–949 (1969).
16 Gladwin, M. T. & Wolfe, J. *Rev. sci. Instrum.* **46**, 1099–1100 (1975).
17 Davis, P. M. & Stacey, F. D. *Geophys. J. R. astron. Soc.* **44**, 1–6 (1976).

Reprinted from Geophysics, v. 44, p. 681–690.

Attenuation of seismic waves in dry and saturated rocks: I. Laboratory measurements

M. N. Toksöz,* D. H. Johnston*, and A. Timur‡

The attenuation of compressional (P) and shear (S) waves in dry, saturated, and frozen rocks is measured in the laboratory at ultrasonic frequencies. A pulse transmission technique and spectral ratios are used to determine attenuation coefficients and quality factor (Q) values relative to a reference sample with very low attenuation. In the frequency range of about 0.1–1.0 MHz, the attenuation coefficient is linearly proportional to frequency (constant Q) both for P- and S-waves. In dry rocks, Q_p of compressional waves is slightly smaller than Q_s of shear waves. In brine and water-saturated rocks, Q_p is larger than Q_s. Attenuation decreases substantially (Q values increase) with increasing differential pressure for both P- and S-waves.

INTRODUCTION

The attenuation of compressional (P) and shear (S) waves in rocks strongly depends on the physical state and saturation conditions. Generally, attenuation varies much more than the seismic velocities as a result of changes in the physical state of materials. Thus, the anelastic properties of rocks supplement the elastic when inferring saturation conditions and pore fluids by seismic techniques. However, the experimental determination of attenuation is more difficult than the measurement of velocities.

Attenuation measurements have been carried out in the laboratory using different techniques over a fairly wide frequency range (Attewell and Ramana, 1966; Birch and Bancroft, 1938; Born, 1941; Bradley and Fort, 1966; Gardner et al, 1964; Hamilton, 1972; Jackson, 1969; Knopoff, 1964; Nur and Simmons, 1969; Peselnick and Zietz, 1959; Peselnick and Outerbridge, 1961; Spetzler and Anderson, 1968; Tittmann et al, 1974; Tullos and Reid, 1969; Warren et al, 1974; Watson and Wuenschel, 1973; Wyllie et al, 1962). These measurements indicate the attenuation coefficient is generally proportional to frequency (i.e., the quality factor Q is independent of frequency). Data on the dependence of attenuation on fluid saturation and pressure are relatively scarce. Available measurements indicate that (1) fluid saturation increases attenuation (Gardner et al, 1964; Wyllie et al, 1962;

Obert et al, 1946; Watson and Wuenschel, 1973), and (2) increasing pressure decreases attenuation (Gardner et al, 1964; Gordon and Davis, 1968; Levykin, 1965; Klima et al, 1964). However, more laboratory data are needed under controlled conditions to determine the effects of fluids and pressure on the attenuation in porous rocks.

In this paper, we present laboratory data on attenuation and its dependence on fluid saturation (methane, water, brine) and pressure. Part II of this paper (Johnston et al, 1979, this issue), discusses the mechanisms of attenuation and formulates theoretical models that fit the laboratory data.

ATTENUATION MEASUREMENTS

Accurate measurement of intrinsic attenuation is a difficult task and seriously limits the utilization of anelastic rock properties. Both in the laboratory and in the field, seismic wave amplitudes are strongly affected by geometric spreading, reflections, and scattering in addition to intrinsic damping. Thus, to obtain the true attenuation, it is necessary to correct for these other effects; this can be a formidable task.

In the laboratory, attenuation is generally measured by one of several techniques. These include the resonant bar method (Birch and Bancroft, 1938; Born, 1941; Gardner et al, 1964; Spetzler and Anderson,

Manuscript received by the Editor July 30, 1977; revised manuscript received February 27, 1978.
*Department of Earth and Planetary Sciences, M.I.T., Cambridge, MA 02139.
‡Chevron Oil Field Research Co., Box 446, La Habra, CA 90631.

1968); amplitude decay of multiple reflections (Peselnick and Zietz, 1959); slow stress strain cycling (Jackson, 1969); or a pulse transmission method where the amplitude decay of seismic signals traveling through samples is measured (Kuster and Toksöz, 1974; Tittmann et al, 1974; Watson and Wuenschel, 1973).

The pulse transmission technique is most suited for use in pressure vessels with jacketed and saturated samples, provided correction can be made for geometric factors such as beam spreading and reflections. We use the pulse transmission technique and measure attenuation relative to a reference sample which has very low attenuation. A description of the laboratory setup and high-pressure system is given by Timur (1977). The samples used in this Acoustic Measurements System (AMS) are cylindrical and generally 8.9 cm in diameter and 5.1 cm in length. Transmitter and receiver transducers (each 2.5 cm in diameter) are mounted at opposite ends of the samples. Only one-way transmission effects are measured. The sample to be studied and the reference sample have exactly the same shape and geometry. Essentially, two measurements are made using identical procedures, one with the rock sample of interest and the second with the reference sample.

The amplitudes of plane seismic waves for the reference and the sample can be expressed as

$$A_1(f) = G_1(x)e^{-\alpha_1(f)x}\,e^{i(2\pi ft - k_1 x)},$$

and

$$A_2(f) = G_2(x)e^{-\alpha_2(f)x}\,e^{i(2\pi ft - k_2 x)}, \qquad (1)$$

where A = amplitude, f = frequency, x = distance, $k = 2\pi f/v$ = wavenumber, v = velocity, $G(x)$ is a geometrical factor which includes spreading, reflections, etc., and $\alpha(f)$ is the frequency-dependent attenuation coefficient. Subscripts 1 and 2 refer to

the reference and sample, respectively. From available data it is reasonable to assume that over the frequency range of the measurements, 0.1–1.0 MHz, α is a linear function of frequency, although the method itself tests this assumption (Knopoff, 1964; Jackson and Anderson, 1970; McDonal et al, 1958). Thus one can write:

$$\alpha(f) = \gamma f, \qquad (2)$$

where γ is constant and related to the quality factor Q by

$$Q = \frac{\pi}{\gamma v}. \qquad (3)$$

When the same geometry is used for both the sample and standard (i.e., same sample dimensions, transducers, and arrangements), then G_1 and G_2 are frequency-independent scale factors. The ratio of the Fourier amplitudes is:

$$\frac{A_1}{A_2} = \frac{G_1}{G_2}e^{-(\gamma_1 - \gamma_2)fx}, \qquad (4)$$

or

$$\ln\left(\frac{A_1}{A_2}\right) = (\gamma_2 - \gamma_1)xf + \ln\left(\frac{G_1}{G_2}\right), \qquad (5)$$

where x is the sample length.

When G_1/G_2 is independent of frequency, $(\gamma_2 - \gamma_1)$ can be found from the slope of the line fitted to $\ln(A_1/A_2)$ versus frequency. If the Q of the standard reference is known, γ_2 of the sample can be determined. When the Q of the standard is very high (i.e., $Q_1 \cong \infty$), then $\gamma_1 = 0$, and γ_2 of the rock sample can be determined directly from the slope.

With the exception of the first set of measurements reported below, aluminum was used as the standard reference. Q for aluminum is about 150,000 (Zam-

Table 1. Measured porosities, velocities, and quality factors of three rock samples.

	Porosity (percent)	Velocities (km/sec)		Quality factors*	
		V_p	V_s	Q_p	Q_s
Navajo sandstone	12.5				
Room temperature		4.25	2.38	7.3	6.2
Frozen		5.59	3.63	—	—
Boise sandstone	25.0				
Room temperature		3.42	1.90	6.9	6.1
Frozen		4.92	2.90	—	—
Spergen limestone	14.8				
Room temperature		4.70	2.49	14.9	12.1
Frozen		5.83	3.12	—	—

*Assuming Q in frozen state is very high ($Q_{\text{frozen}} \cong \infty$)

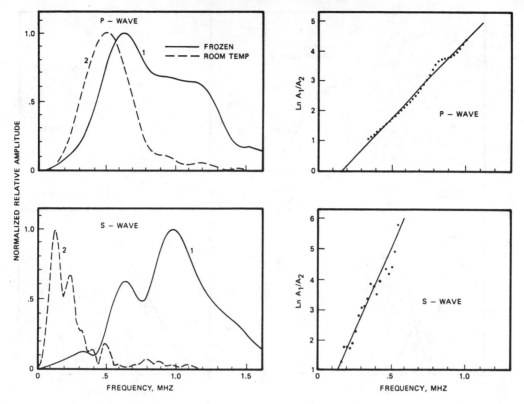

FIG. 1. Attenuation characteristics of the Navajo sandstone. Amplitude spectra and the natural logarithm of the spectral ratios versus frequency for frozen at $T = -40°C$ (curve 1) and water-saturated at room temperature (curve 2) samples. The top figure is compressional waves and the bottom figure shear waves.

anek and Rudnick, 1961), as opposed to $Q \leq 1000$ for rocks. Thus, taking $\gamma_1 = 0$ never introduces more than 1 percent error. For typical rocks where $Q = 10–100$, the error is less than 0.1 percent and is negligible. A more serious concern is the validity of the assumption that the geometric factors G_1 and G_2 have the same frequency dependence, and G_1/G_2 is independent of frequency. With polished rock surfaces and good coupling between the transducer holder and sample, one would not expect frequency-dependent reflection coefficients at the interface. Repeated measurements showed that pulse amplitudes, shapes, and spectra were duplicated. As a further test of the experimental method, Q values for Lucite were obtained using both the spectral ratios and a dynamic resonance technique. The results were in good agreement with $Q_p \approx Q_s \approx 50$. Finally, Q values using spectral ratios were found for Lucite and a Berea sandstone relative to aluminum for 1, 2, and 3 inch sample lengths. The measurements were done with uniaxial pressure. In each case, the Q values for P- and S-waves obtained with different length samples were consistent, supporting the assumption of a frequency independent G_1/G_2.

LABORATORY MEASUREMENTS

Two types of measurements were carried out in this study:

(1) Attenuation of P- and S-waves at one atm pressure in water-saturated and frozen sandstone and limestone samples; and

(2) attenuation of P- and S-waves as a function of pressure in a dry, gas, and brine-saturated Berea sandstone.

The purpose of the first set of measurements was to determine effects of lithology and porosity on attenuation in different rock types. The second set of measurements was important for understanding the mechanisms of attenuation and for extrapolating laboratory data to field applications.

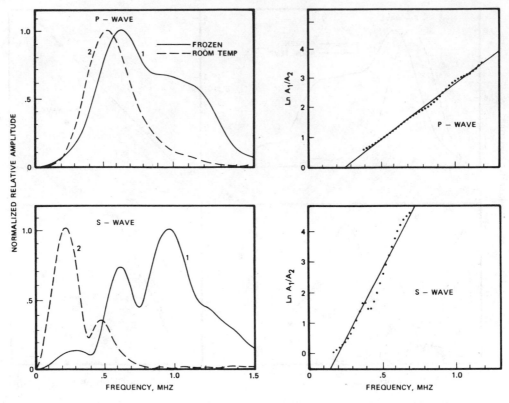

FIG. 2. Attenuation characteristics of the Spergen limestone. Curves are as described in Figure 1.

Attenuation measurements of water-saturated and frozen samples

Seismic velocities and Fourier spectra of P- and S-waves for three rocks were determined at 1 atm confining pressure saturated with water first at room temperature and then in the frozen state at $-40°C$. The three rocks, Navajo and Boise sandstones and Spergen limestone, have porosities and measured velocities listed in Table 1. The experimental technique and velocity measurements are discussed in an earlier paper (Toksöz et al, 1976). In this section we emphasize the attenuation data.

The samples were jacketed with a thin rubber sleeve and then saturated with distilled water using a vacuum technique. P- and S-wave velocities were measured and the waveforms were recorded with the sample at room temperature ($+20°C$). The samples were then frozen and maintained at $-40°C$ while similar measurements were made. After this, the samples were brought back to room temperature and measurements

were repeated. The repeated measurements of velocities at room temperature before and after freezing agreed to within 1 percent for P-waves and to within 2 percent for S-waves for all rocks, except for the Spergen limestone, for which velocities were lowered by about 5 percent as a result of the freezing cycle, indicating some matrix damage and crack formation as a result of freezing.

From the waveform spectra shown in Figures 1 and 2, it is clear that the attenuation is much greater in the room temperature case. In frozen rocks, attenuation is very small and Q is very high (Spetzler and Anderson, 1968). Thus, we use the frozen rock as a reference standard and determine Q in the water-saturated case relative to the frozen state with the spectral ratio method discussed in the previous section. Q in frozen rocks is greater than 100. For this lower limit, the relative Q values for water-saturated rocks may be about 10 percent higher than the absolute values. If Q in the frozen state is 500, relative Q is within 3 percent of the absolute value.

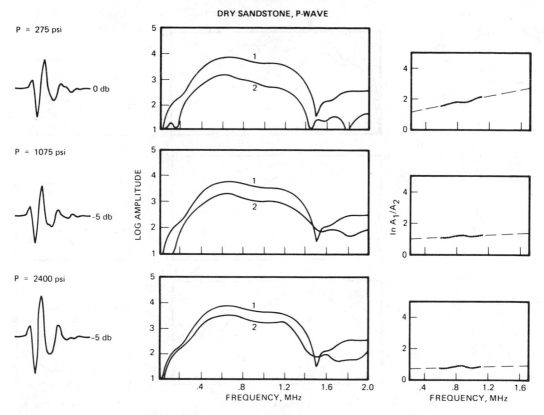

FIG. 3. Attenuation characteristics of P-waves in dry Berea sandstone at three different confining pressures: $P_c = 275$, 1075, 2400 psi ($P_f = 0$, thus $P_d = P_c$). Left: pulse waveforms in sandstone. Center: logarithm of Fourier amplitude as a function of frequency. Curve 1 is the aluminum reference and curve 2 is the rock. Right: natural logarithm of the aluminum to rock amplitude ratios as a function of frequency. The closely spaced points are actual ratios and the dashed line is the linear fit to the data. γ and Q are calculated from the slope of this line.

An examination of the Q values tabulated in Table 1 shows that, in water-saturated (unfrozen) rocks at 1 atm confining pressure for the samples studied; (1) Q of S-waves (Q_s) is slightly lower than Q of P-waves (Q_p) in all three samples; (2) in sandstones, Q decreases with increasing porosity but the decrease is small; and (3) for a given porosity, limestone has a higher Q than sandstone.

Attenuation measurements as a function of pressure and saturation

To determine the effects of pressure and saturating fluids on the attenuation properties of porous rocks, a detailed experiment was conducted with a Berea sandstone. The sample used has a porosity of 16 percent, permeability of 75 mD, and a matrix density of 2.61 g/cm. Berea is a relatively clean sandstone, and a

petrographic description of it is given by Timur (1968).

In the following experiments, jacketed samples were placed in a pressure vessel (Timur, 1977) where pore fluid pressure (P_f) and confining pressure (P_c) are controlled independently within a range of 0–3 kb. Measurements were made at discrete differential pressures ($P_d = P_c - P_f$) with sufficient time between measurements for the sample to reach equilibrium. Both increasing and decreasing pressure directions are used to test for repeatability, hysteresis, or any indication of sample damage.

Attenuation and velocity measurements were made as described before in the differential pressure range of 1 bar to 550 bars (8000 psi) in dry, methane, and brine-saturated samples. An aluminum sample with the same shape as the Berea sandstone sample was

DRY SANDSTONE, S-WAVE

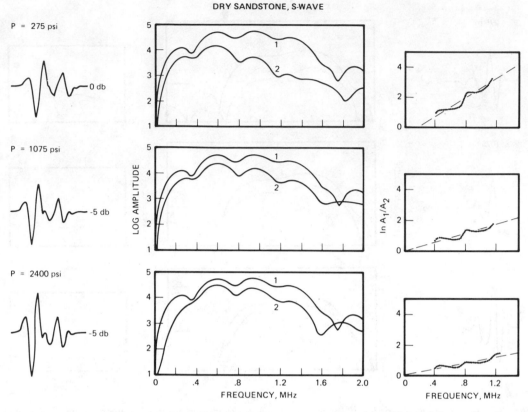

FIG. 4. Attenuation characteristics of S-waves in dry Berea sandstone. See Figure 3 for explanations.

used as a standard reference. Q values were calculated from spectral ratios of rock-aluminum pairs at each pressure.

In the first experiment, the sample was dried in a vacuum oven at 80°C, at a pressure of 50 μm of mercury while being periodically flushed with argon. P- and S-wave traveltimes and full waveforms were recorded in the confining pressure range of 1 bar (14.7 psi) to 550 bars (8000 psi) for both the rock sample and aluminum standard ($P_f = 0$; therefore, $P_d = P_c$). Examples of pulse waveforms, spectra, and spectral ratios at three confining pressures are shown in Figures 3 and 4 for P- and S-waves, respectively. The decrease of the attenuation coefficient (decreasing slope of the spectral ratio) with increasing pressure is obvious from these figures. Measured velocities as a function of pressure are shown in Figure 5 (Jones et al, 1977). Q values were calculated from the slopes of spectral ratios and velocities, and are shown in Figure 6. Both Q_p and

Q_s increase rapidly with confining pressure with a slight leveling off at higher pressures. Also, in this dry case, Q_s is slightly higher than Q_p.

In the second experiment, the sample was fully saturated with methane and the experiment repeated with $P_f = 0.465\ P_c$. This ratio of pore-fluid to confining pressures is a nominal value for saturated sedimentary rocks where the pore-fluid pressure is equal to the hydrostatic pressure of a water column. The behavior of the spectra and spectral ratios for methane-saturated rocks was similar to those shown in Figures 3 and 4. The resulting Q values are plotted as a function of differential pressure in Figure 7. Q_p increases rapidly at first with pressure but appears to level off at higher pressures. Q_s is again either equal to or slightly larger than Q_p. It exhibits a similar but slightly more gradual behavior. As would be expected, the differences between the dry rock and the methane-saturated rock are small.

Following the methane run, the sample was com-

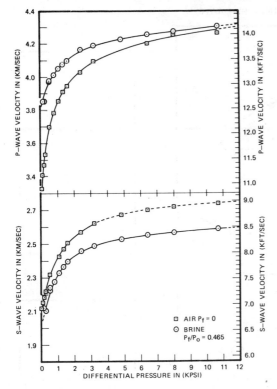

FIG. 5. Velocities of P- and S-waves as a function of differential pressure in dry and brine-saturated Berea sandstone. These velocities are used for the Q calculations.

pletely saturated with an NaCl brine of 67,191 ppm concentration. Again, P_f was maintained at 0.465 P_c throughout the experiment. Pulse waveforms, spectra, and spectral ratios are shown for three values of differential pressure in Figures 8 and 9 for P- and S-waves. An increase of the attenuation coefficient relative to the dry case (Figures 3 and 4) is obvious from the spectral ratios. Q values obtained are shown as a function of differential pressure in Figure 10. The same behavior as seen in Figures 6 and 7 is observed here. In the brine-saturated case, however, Q_s is lower than Q_p. This is in agreement with earlier results listed in Table 1, and it shows that this behavior for saturated rocks holds at high pressures as well.

A final brine-saturated (NaCl = 161,334 ppm) experiment was run in which P_c was fixed at 1035 bars (15,000 psi) and P_f was decreased from 1000 bars (14,500 psi). The results are shown in Figure 11 and are similar to those obtained in the previous experiment (Figure 10) at lower pressures ($P_d \leq$ 8000 psi). However, at a pressure of about $P_d =$ 13,000 psi, there is a definite increase in Q. This is most likely due to collapse of some of the thicker pore spaces and complete locking of some grain boundaries.

The interpretation of these results in terms of attenuation mechanisms is discussed in Part II. These data, in fact, provide strong constraints for determining the relative contributions of different mechanisms for attenuation of seismic waves in rock under different saturation conditions.

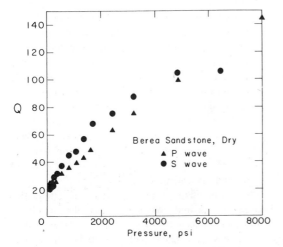

FIG. 6. Q values of P- and S-waves as a function of confining pressure in dry Berea sandstone ($P_f = 0$).

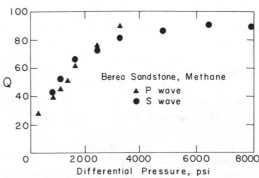

FIG. 7. Q-values of P- and S-waves as a function of differential pressure in methane saturated Berea sandstone ($P_f = 0.465\ P_c$).

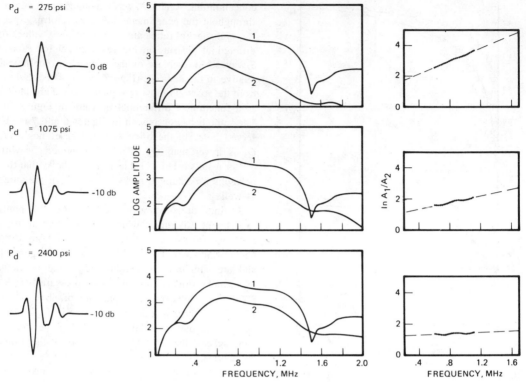

FIG. 8. Attenuation characteristics of P-waves in brine-saturated Berea sandstone for three differential pressures ($P_f = 0.465\ P_c$). Explanations of the different curves are the same as those given in Figure 3.

CONCLUSIONS

We described a laboratory method and presented data on the attenuation of seismic waves in rocks under pressure and in different saturation conditions. The laboratory measurements were made at ultrasonic frequencies (0.1–1.0 MHz). In this frequency range it is shown that:

1) Attenuation coefficients increase linearly with frequency (constant Q) for both P- and S-waves in both dry and saturated rocks. Although our measurement technique of spectral ratios requires that Q be constant, the verification comes from the fact that spectral ratios are linear with frequency over all frequency ranges where the signal-to-noise ratio is high.

2) Attenuation in brine- and water-saturated rocks is greater than in dry or methane-saturated rocks. Attenuation in frozen rocks is very much lower than in saturated rocks.

3) Attenuation decreases (Q increases) with increasing differential pressure both for P- and S-waves in all cases of saturation. The rate of increase is high at low pressures and levels off at higher pressures.

4) In water-saturated rocks, Q_p is higher (10 to 25 percent) than Q_s at both low and high pressures. In dry or methane-saturated rocks, Q_s is slightly higher than Q_p. Since the attenuation coefficient, $\alpha = \pi f / QV$, and the velocity of S-waves is lower than that of P-waves, the attenuation per unit distance of S-waves is higher than that of P-waves in both dry and water-saturated cases.

It is important to mention that the attenuation measurements and conclusions described apply to ultrasonic frequencies of 0.1 − 1.5 MHz and strain amplitudes associated with the pulse technique. Possible variations with frequency and strain amplitude are described in the following paper.

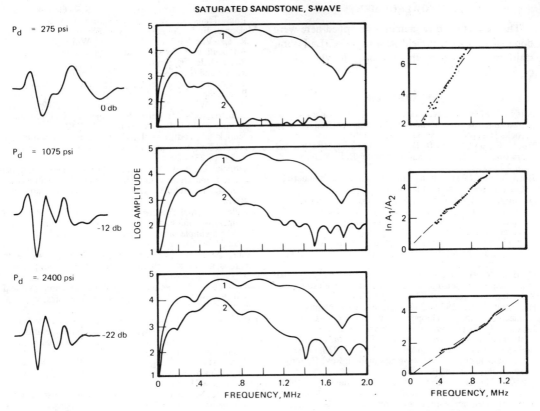

FIG. 9. Attenuation characteristics of S-waves in brine-saturated Berea sandstone. Explanations are the same as those in Figure 3.

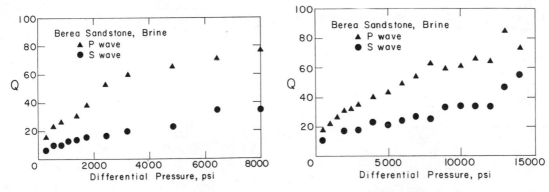

FIG. 10. Q values of P- and S-waves as a function of differential pressure $(P_f = 0.465\,P_c)$ in brine-saturated Berea sandstone. The NaCl concentration of the brine is 67,191 ppm.

FIG. 11. Q values of P- and S-waves as a function of differential pressure $(P_c = 15,000$ psi$)$ in brine-saturated Berea sandstone. The NaCl concentration of the brine is 161,334 ppm.

ACKNOWLEDGMENTS

The laboratory measurements described here were done at Chevron Oil Field Research Co., La Habra, California. We thank this company for the release of the data.

REFERENCES

Attewell, P. B., and Ramana, Y. V., 1966, Wave attenuation and internal friction as functions of frequency in rocks: Geophysics, v. 31, p. 1049–1056.

Birch, F., and Bancroft, D., 1938, Elasticity and internal friction in a long column of granite: Bull. SSA, v. 28, p. 243–254.

Born, W. T., 1941, Attenuation constant of earth materials: Geophysics, v. 6, p. 132–148.

Bradley, J. J., and Fort, A. N., Jr., 1966, Internal friction in rocks, in Handbook of physical constants: S. P. Clark, Jr., Ed., GSA Publ., p. 175–193.

Gardner, G. H. F., Wyllie, M. R. J., and Droschak, D. M., 1964; Effects of pressure and fluid saturation on the attenuation of elastic waves in sands: J. Petrol. Tech., v. 16, p. 189–198.

Gordon, R. B., and Davis, L. A., 1968, Velocity and attenuation of seismic waves in imperfectly elastic rock: J. Geophys. Res., v. 73, p. 3917–3935.

Hamilton, E. L., 1972, Compressional wave attenuation in marine sediments: Geophysics, v. 37, p. 620–646.

Jackson, D. D., 1969, Grain boundary relaxation and the attenuation of seismic waves: Ph.D. thesis, M.I.T.

Jackson, D. D., and Anderson, D. L., 1970, Physical mechanisms of seismic wave attenuation: Rev. Geophys. Space Phys., v. 8, p. 1–63.

Johnston, D. H., Toksöz, M. N., and Timur, A., 1979, Attenuation of seismic waves in dry and saturated rocks, Part II. Theoretical models and mechanisms: this issue, p. 691–711.

Jones, S. B., Thompson, D. D., and Timur, A., 1977, A unified investigation of elastic wave propagation in crustal rocks: submitted for publication.

Klima, K., Vanek, J., and Pros, Z., 1964, The attenuation of longitudinal waves in diabase and graywacke under pressures up to 4 kilobars: Studia Geoph. et Geod., v. 8, p. 247–254.

Knopoff, L., 1964, Q: Rev. Geophys., v. 2, p. 625–660.

Kuster, G. T., and Toksöz, M. N., 1974, Velocity and attenuation of seismic waves in two-phase media: Part II—Experimental results: Geophysics, v. 39, p. 607–618.

Levykin, A. I., 1965, Longitudinal and transverse wave absorption and velocity in rock specimens at multilateral pressures up to 4000 kg/cm²: U.S.S.R. Geophys. Ser., Engl. Transl., Phys. Solid Earth, no. 1, p. 94–98.

McDonal, F. J., Angona, F. A., Mills, R. L., Sengbush, R. L., Van Nostrand, R. G., and White, J. E., 1958, Attenuation of shear and compressional waves in Pierre shale: Geophysics, v. 23, p. 421–439.

Nur, A., and Simmons, G., 1969, The effect of viscosity of a fluid phase on velocity in low porosity rocks: Earth Planet. Sci. Lett., v. 7, p. 99–108.

Obert, L., Windes, S. L., and Duvall, W. I., 1946, Standardized tests for determining the physical properties of mine rock: U.S. Bur. Mines R.I. 3891.

Peselnick, L., and Zietz, I., 1959, Internal friction of fine grained limestones at ultrasonic frequencies: Geophysics, v. 24, p. 285–296.

Peselnick, L., and Outerbridge, W. F., 1961, Internal friction and rigidity modulus of Solenhofen limestone over a wide frequency range: USGS Prof. paper no. 400B.

Spetzler, H., and Anderson, D. L., 1968, The effect of temperature and partial melting on velocity and attenuation in a simple binary system: J. Geophys. Res., v. 73, p. 6051–6060.

Timur, A., 1968, Velocity of compressional waves in porous media at permafrost temperatures: Geophysics, v. 33, p. 584–596.

———— 1977, Temperature dependence of compressional and shear wave velocities in rocks: Geophysics, v. 42, p. 950–956.

Tittmann, B. R., Housely, R. M., Alers, G. A., and Cirlin, E. H., 1974, Internal friction in rocks and its relationship to volatiles on the moon: Geochim. et Cosmochim. Acta, Supp. 5, p. 2913–2918.

Toksöz, M. N., Cheng, C. H., and Timur, A., 1976, Velocities of seismic waves in porous rocks: Geophysics, v. 41, p. 621–645.

Tullos, F. N., and Reid, A. C., 1969, Seismic wave attenuation of Gulf Coast sediments: Geophysics, v. 34, p. 516–528.

Warren, N., Trice, R., and Stephens, J., 1974, Ultrasonic attenuation: Q measurements on 70215, 29: Geochim. Cosmochim. Acta. Supp. 5, p. 2927–2938.

Watson, T. H., and Wuenschel, P. C., 1973, An experimental study of attenuation in fluid saturated porous media, compressional waves and interfacial waves: Presented at the 43rd Annual International SEG Meeting, October 22 in Mexico City.

Wyllie, M. R. J., Gardner, G. H. F., and Gregory, A. R., 1962, Studies of elastic wave attenuation in porous media: Geophysics, v. 27, p. 569–589.

Zamanek, J., Jr., and Rudnick, J., 1961, Attenuation and dispersion of elastic waves in a cylindrical bar: J. Acoust. Soc. Am., v. 33, p. 1283–1288.

Reprinted from Nature, v. 277, p. 528–531.

Friction and seismic attenuation in rocks

Kenneth Winkler & Amos Nur
Department of Geophysics, Stanford University, Stanford, California 94305

Michael Gladwin
Department of Physics, University of Queensland, St Lucia, Brisbane, Australia 4067

Precise experimental results, combined with theoretical predictions, indicate that seismic energy loss caused by grain boundary friction is important only at low confining pressure and at strains greater than about 10^{-6}. Since these conditions are generally not encountered in seismology, frictional attenuation is not important in situ. *Other mechanisms such as fluid flow must dominate seismic attenuation in the upper crust.*

FAR more information about the constitution of the Earth is obtained from seismic studies than from all other geophysical methods combined. Of the two main aspects of wave propagation—velocity and attenuation—knowledge of velocities has provided most of our information about the Earth. Attenuation data is of much more limited use, partly because it is difficult to obtain, but mainly because it is difficult to interpret in terms of rock properties. This is due primarily to our lack of understanding of the physical processes involved in the attenuation of seismic waves. (Here we are not considering apparent attenuation caused by geometrical spreading, partial reflections and so on, but are only interested in those processes that convert seismic energy into heat.)

Many different techniques have been used to study the attenuation of acoustic waves propagating through rock. Although much data has been collected, no well-defined mechanism of energy loss has yet been firmly established on both experimental and theoretical grounds. One of the most intuitively appealing and widely discussed mechanisms proposed for seismic energy loss is based on simple Coulomb friction. In this mechanism the passing wave causes sliding at grain boundaries or across crack faces, thereby doing work against friction and converting seismic energy into heat. Since frictional heating is rate independent, it seems to account for the reported observation that the specific dissipation function[1], Q, is nearly independent of frequency. Friction also seems to explain why the introduction of cracks into a crystalline solid increases attenuation. The most severe criticism of a frictional mechanism is, however, that the mechanism is inherently nonlinear. This means that the velocity and attenuation of a sinusoidal wave will depend on how it is combined with other waves. It has been found that at large strain amplitudes ($>10^{-6}$) nonlinear effects such as cusped hysteresis loops[3,4] and amplitude-dependent velocity and attenuation[3,5,6] can be observed, but these effects disappear at strains more typical of seismic waves.

Using both theoretical arguments and experimental results, several authors[2,7-9] have expressed doubts about the validity of a frictional attenuation mechanism. For instance, Savage[7] has suggested that frictional effects should be important only at large strains where displacements across crack faces are sufficiently large for the concept of macroscopic friction to be valid. Despite these criticisms, frictional losses are often used to interpret experimental observations[10-12]. To clarify the role of friction in seismic wave attenuation, we carefully investigated the dependence of velocity and Q on strain amplitude. We were able to identify the range of conditions for which frictional losses may be significant and our results show that the frictional mechanism is generally not important for seismic waves in the Earth.

Theory of frictional losses

Two types of frictional attenuation models have been proposed. Walsh[13] developed a model based on thin elliptical cracks in which frictional sliding occurs between crack faces. Models based on spheres in contact were developed by Mindlin and Deresiewicz[14] and by White[15]. Mavko[16] has reviewed these theories and extended the contact model to more general geometries. The most significant feature of Mavko's model (which it shares with that of Mindlin and Deresiewicz) is that it predicts Q^{-1} to be proportional to strain amplitude. It has been found experimentally that frictional behaviour is often described by an equation of the form[17]

$$\sigma_f = S + \gamma\sigma_n$$

where σ_f is the frictional force, S is the cohesive shear strength, γ is the coefficient of friction and σ_n is the normal stress across the sliding surfaces. The contact models predict that attenuation will decrease as either γ or S increases. Attenuation will also decrease as σ_n increases and the sliding interfaces are effectively locked together.

In addition to the effects on attenuation, all frictional models predict that stress–strain loops will have discontinuities in slope at maximum strain. This is a static hysteresis effect. As stress reverses direction there is a simultaneous reversal of elastic strain and a delayed reversal of sliding strain.

Table 1 Sample descriptions

Sample	Dimensions (cm)	V_E/f_E (km s^{-1} Hz^{-1})	Porosity
Massilon sandstone	$100 \times 2.0 \times 1.9$	2.000	22%
Berea sandstone	$101.6 \times 2.54 \times 2.54$	2.032	20%
Sierra White granite	$101.6 \times 2.54 \times 2.54$	2.032	1%
Vycor 7930 glass	91.4×1.5 diam.	1.828	28%
Lucite (plastic)	98.7×1.51 diam.	1.974	0%
Aluminium	$100 \times 2.22 \times 2.22$	2.000	0%

Vycor 7930 porous glass is made by Corning Glass Works; the average pore size is 4 nm.

Past observations of frictional losses

Several of the predicted effects of frictional attenuation models have been observed. Energy loss has been found to increase with strain amplitude at large strains in rock[3] as well as between sliding glass and metal surfaces[19,20]. It has been shown that at low strains ($<10^{-6}$) attenuation is nearly constant[3] and that variation with strain amplitude occurs only at larger strain. Some attenuation in metals at large strain is associated with dislocation mechanisms, but this does not explain all of the observations.

Discontinuities in the slope of stress–strain loops have been observed in large strain quasi-static experiments[3,4]. These have been described as cusped loops. It has recently been shown[21] that as strain amplitude decreases, the cusped loops become elliptical at strains approaching 10^{-6}. It is interesting that two of the effects predicted by frictional attenuation models, amplitude-dependent Q and cusped hysteresis loops, both seem to disappear below the same strain amplitude.

It is also commonly observed that increasing confining pressure causes attenuation to decrease[22,23]. This is predicted by frictional models, but it is also predicted by many other possible mechanisms of energy loss and so by itself does not provide evidence for frictional loss.

New observations of frictional losses

We are using the well established bar resonance technique[22,23] to measure both attenuation and velocity in long thin bars of rock. (Experimental details will be presented elsewhere.) An important feature of our experiment is that pore pressure is controlled independently of confining pressure. This has not previously been achieved in a resonance experiment. The length of the samples (~1 m) puts the fundamental resonance frequencies at roughly 500–1,500 Hz. Sample dimensions are given in Table 1. Resonance frequencies are measured to 1 part in 10^3 and converted to velocities using the relation $v = 2Lf$ where L is the length of the bar. Geometric corrections given by Spinner and Tefft[24] are negligible for our samples in extensional resonance. Strain amplitudes are measured with crystal phonograph pick-ups calibrated against semiconductor strain gauges.

Table 2 Tests for nonlinear behaviour in various materials (negligible changes of Q_E and V_E are found for materials where frictional mechanisms are inoperative)

Material	Strain amplitude	Q_E	Velocity (m s^{-1})
Lucite	1.43×10^{-6}	23.4 ± 0.3	2,108
	3.04×10^{-8}	23.2 ± 0.4	2,108
Vycor	2.84×10^{-6}	437 ± 2	3,345
	1.58×10^{-8}	435 ± 4	3,345
Aluminium	2.41×10^{-6}	$34,800 \pm 300$	5,102
	2.46×10^{-8}	$35,200 \pm 400$	5,102
Granite	1.44×10^{-6}	185 ± 1	3,629
	4.15×10^{-8}	204 ± 1	3,637
Massilon sandstone	2.10×10^{-6}	107 ± 1	1,820
	1.55×10^{-8}	138 ± 2	1,834
Berea sandstone	2.10×10^{-6}	103 ± 3	1,937
	2.30×10^{-8}	140 ± 2	1,955

Attenuation measurements can be made either by plotting resonance peaks ($Q = f/\Delta f$) or by measuring the time constant of resonance decays ($\tau = Q/\pi f$). Before using either technique, resonance peak shapes were carefully observed to insure the purity of the resonances. For convenience, all attenuation data presented in this article was taken with the resonance decay technique (except as noted in Fig. 1). The precision of the attenuation measurements is 1% and the estimated absolute accuracy is 5–10%. Both torsional (shear modulus) and extensional (Young's modulus) resonance modes show similar nonlinear behaviour at large strains. Only extensional data (taken at the fundamental resonance) will be presented here.

Data for dry Massilon sandstone as a function of resonant strain amplitude are shown in Fig. 1. At strains below 5×10^{-7} both attenuation (Q^{-1}) and velocity are nearly independent of strain amplitude, whereas at relatively large strains ($>10^{-6}$) there is a clear increase in attenuation and decrease in velocity. Of the two measured quantities, Q^{-1} is more sensitive to strain amplitude than is velocity, varying by 19% and 0.7%, respectively. The similarity of the curve shapes suggests an underlying relationship between the amplitude dependence of Q^{-1} and velocity.

We have further investigated these nonlinear effects by varying effective stress—the difference between confining pressure and pore pressure. Figure 2 shows data for a jacketed sample of Berea sandstone, with helium used for both confining pressure and pore pressure. As confining pressure is increased from 10 to 50 bar (curves A–D), the variation of both Q^{-1} and velocity with strain amplitude gradually decreases. At a constant confining pressure of 50 bar (curves D and E) increasing pore pressure restores the dependence on strain amplitude. In Fig. 2a curves B

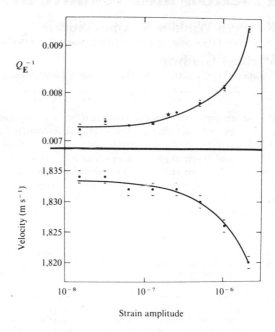

Fig. 1 Variation of attenuation (Q^{-1}) and velocity with strain amplitude for dry Massilon sandstone in extensional resonance. ★, The value found from the resonance peak half-width using the relation $Q^{-1} = \Delta f/f$.

and E show that Q^{-1} is almost a unique function of effective stress at a given strain amplitude.

The effects of changing water content on Q^{-1} in Berea sandstone are shown in Fig. 3. The data for partially saturated rock at very low saturation were obtained at ambient room conditions. Data for 'dry' rock were obtained after drying the sample in a vacuum oven, and the data for saturated rock after saturating the sample with distilled water. The most noticeable feature is that the addition of water greatly increases the attenuation. This effect, which is associated with a fluid flow mechanism, will be discussed more thoroughly elsewhere[29]. Of greater interest here is that the addition of a small amount of water greatly increases the variation of Q^{-1} with strain amplitude, whereas continued addition of water causes a smaller additional variation in Q^{-1}. This is indicated by the brackets in Fig. 3 which show the difference between Q^{-1} at a strain of 1.7×10^{-6} and the constant value of Q^{-1} at low strains.

To find out which structural features of a material are related to these nonlinear effects, we experimented with materials having physical compositions different from sandstone. Table 2 shows data for various materials measured at both large and small strain amplitudes. Lucite and aluminium show no evidence of any change in velocity or attenuation with strain (to within experimental uncertainty) although their respective Q values span more than three orders of magnitude and their internal structures are quite different. (The values of Q for aluminium are probably underestimates because they are at the limits of sensitivity of our system.) There is also no variation in the values for dry Vycor glass. Under the scanning electron microscope Vycor appears to be composed mainly of glass beads ~50 nm diameter. The contacts were beyond the resolution of the SEM used, (86,000×), but presumably the particles are welded together since they formed simultaneously from the melt. The flat cracks and partially cemented grain boundaries commonly found in rocks are probably scarce in Vycor. Table 2 also includes data for a room dry granite sample which shows the same nonlinear effects found in sandstone.

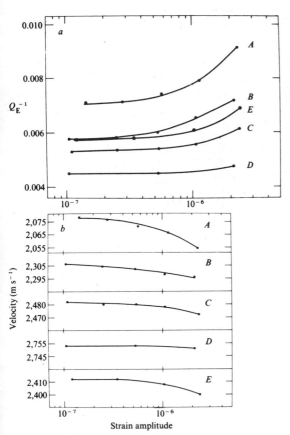

Fig. 2 *a*, Attenuation in dry Berea sandstone as a function of strain amplitude, confining pressure (P_c) and pore pressure (P_p). Helium was used as the pressure medium. (A) $P_c = 10$ bar; (B) $P_c = 20$ bar; (C) $P_c = 30$ bar; (D) $P_c = 50$ bar; (E) $P_c = 50$ bar, $P_p = 30$ bar. *b*, Extensional velocity of Berea sandstone. Data correspond to the attenuation data shown in (a).

Discussion

The data we have presented, together with the existing data we have discussed, provide strong support for the existence of a frictional attenuation mechanism in rock at large strain amplitudes and low effective stresses. Every predicted feature of the contact frictional loss mechanism has been observed experimentally. At large strains ($>10^{-6}$) Q^{-1} increases with increasing strain amplitude and the amount of the variation in Q^{-1} is strongly influenced by effective stress. A stress of 50 bar is enough to virtually eliminate the strain dependence at the strains and in the rocks which we have used.

The effect of water on the amplitude dependence of Q^{-1} is probably caused by water changing the frictional properties of the sliding interfaces. Figure 3 provides evidence for this. Adding a small amount of water to the sample presumably wets most of the internal surfaces, causing a slight increase in the low strain, linear attenuation (a fluid flow mechanism) and a large increase in the nonlinear, frictional component of attenuation. The addition of more water fills the remaining pore space, thus enhancing the flow losses but only slightly increasing the frictional energy loss.

The observations of cusped stress–strain loops[3,4] at strains greater than 10^{-6} also indicate that frictional sliding is occurring. It is significant that all of the evidence for a frictional attenuation mechanism disappears below strains of the order of 10^{-6}. The cusped loops become elliptical and Q^{-1} becomes independent of strain amplitude. The implication is that at low strains attenua-

tion is caused only by linear loss mechanisms. At larger strains, nonlinear frictional losses become observable. This leads to the question of why a strain of 10^{-6} or larger is needed before frictional attenuation is observed. We can assume that displacements across crack surfaces must be at least comparable with inter-atomic spacings ($\sim 10^{-8}$ cm) for friction to be present. A shear strain ε will produce a displacement d across the sides of a crack of length L given approximately by $d = \varepsilon L$. If $d = 10^{-8}$ cm and $\varepsilon = 10^{-6}$, then we find that $L = 10^{-2}$ cm. A crack length of 10^{-2} cm may well be a rough upper bound to the majority of crack lengths found in rock. Hadley[25] reported that in Westerly granite only 7% of the observed cracks had lengths greater than 10^{-2} cm and none were longer than 3×10^{-2} cm. Therefore, at strains below 10^{-6}, sliding displacements between crack faces are so small in the majority of cracks that friction does not adequately describe the interaction. As strain increases above 10^{-6}, frictional sliding begins to occur at more and more cracks and attenuation increases.

After eliminating friction as a loss mechanism at low strains, we still have to account for the observed energy loss at low strain amplitudes in 'dry' rock. We must also explain the change in attenuation that occurs as cracks are closed under confining pressure[3,22,23] or created through thermal cracking[3] or dilatancy[11]. Unfortunately, we do not know yet what mechanism is responsible for these losses. The data collected by Tittmann *et al.*[9] showing the strong effect that trace amounts of volatiles have on attenuation probably has a direct bearing on this mechanism. Another possibility is some form of grain boundary relaxation[26,27] or a dislocation mechanism similar to those discussed by Mason[8,28]. Whatever the cause may be, it is reasonably certain that the mechanism does not involve any macroscopic concept of friction.

Conclusion

On both experimental and theoretical grounds it seems clear that frictional energy losses occur in rock at room pressure only at strains greater than approximately 10^{-6}. This is far greater than strain amplitudes generally encountered in seismology except in regions near seismic sources. Even there the effect of confining pressure will virtually eliminate frictional energy losses in the Earth. Consequently, frictional losses are not important in seismology.

The mechanism by which cracks influence attenuation at low strains in 'dry' rock needs to be re-examined. The concept of frictional energy loss is so firmly established (on intuitive

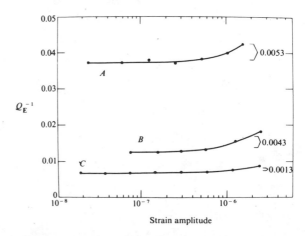

Fig. 3 Attenuation in Berea sandstone at various degrees of saturation as a function of strain amplitude. Brackets indicate frictional component of energy loss. *A*, saturated; *B*, partially saturated; *C*, dry.

grounds) that the term 'internal friction' is often interpreted literally to mean a frictional energy loss. Though the choice of words is unfortunate, internal friction is simply a general term for energy absorption and does not imply any particular mechanism. It must be kept in mind that actual frictional energy loss is a large-strain, low-pressure phenomenon that is often observed in the laboratory, but seldom observed in the Earth.

We thank Drs Gary Mavko and Einar Kjartansson for useful discussions, Peter Gordon for assistance in constructing equipment and Lori Dengler for SEM photographs. This research was supported in part by a grant from the NSF Division of Earth Sciences (EAR 76-22501).

Received 11 October; accepted 20 December 1978.

1. O'Connell, R. J. & Budiansky, B. *Geophys. Res. Lett.* **5**, 5–8 (1978).
2. Savage, J. C. & Hasegawa, H. S. *Geophysics* **22**, 1003–1014 (1967).
3. Gordon, R. B. & Davis, L. A. *J. geophys. Res.* **73**, 3917–3935 (1968).
4. McKavanagh, B. & Stacey, F. D. *Phys. Earth. Planet Int* **8**, 246–250 (1974).
5. Peselnick, L. & Outerbridge, W. F. *J. geophys. Res.* **66**, 581–588 (1961).
6. Gordon, R. B. & Rader, D. *Geophys. Monogr. Ser.* **14**, 235–242 (1971).
7. Savage, J. C. *J. geophys. Res.* **74**, 726–728 (1969).
8. Mason, W. P., Beshers, D. N. & Kuo, J. T. *J. appl. Phys.* **41**, 5206–5209 (1970).
9. Tittmann, B. R., Abdel-Gawad, M. & Housley, R. M. *Proc. Lunar Sci. Conf.* **3**, 2565–2575 (1972).
10. Birch, F. *J. geophys. Res.* **80**, 756–764 (1975).
11. Lockner, D. A., Walsh, J. B. & Byerlee, J. D. *J. geophys. Res.* **82**, 5374–5378 (1977).
12. Johnston, D. H., Toksoz, M. N. & Timur, A. *Geophysics* (in the press).
13. Walsh, J. B. *J. geophys. Res.* **71**, 2591–2599 (1966).
14. Mindlin, R. D. & Deresiewicz, H. *J. appl. Mech.* **20**, 327–344 (1953).
15. White, J. E. *Geophysics* **31**, 333–339 (1966).
16. Mavko, G. M. *J. geophys. Res.* (in the press).
17. Jaeger, J. C. & Cook, N. G. W. *Fundamentals of Rock Mechanics* (Halsted, New York, 1976).
18. Stacey, F. D., Gladwin, M. T., McKavanagh, B., Linde, A. T. & Hastie, L. M. *Geophys. Surveys* **2**, 133–151 (1975).
19. Mindlin, R. D., Mason, W. P., Osmer, T. F. & Deresiewicz, H. *Proc. 1st U.S. natn. Congr. appl. Mech.* 203–208 (1951).
20. Goodman, L. E. & Brown, C. B. *J. appl. Mech.* **29**, 17–22 (1962).
21. Brennan, B. J. & Stacey, F. D. *Nature* **268**, 220–222 (1977).
22. Birch, F. & Bancroft, D. *Bull. seis. Soc. Am.* **28**, 243–254 (1938).
23. Gardner, G. H. F., Wyllie, M. R. J. & Droschak, D. M. *J. Petrol. Tech.* **16**, 189–198 (1964).
24. Spinner, S. & Tefft, W. E. *Proc. ASTM.* **61**, 1221–1238 (1961).
25. Hadley, K. *J. geophys. Res.* **81**, 3484–3494 (1976).
26. Gordon, R. B. & Nelson, C. W. *Rev. Geophys.* **4**, 457–474 (1966).
27. Jackson, D. D. & Anderson, D. L. *Rev. Geophys. Space Phys.* **8**, 1–63 (1970).
28. Mason, W. P., Marfurt, K. J., Beshers, D. N. & Kuo, J. T. *J. acous. Soc. Am.* **63**, 1596–1603 (1978).
29. Winkler, K. & Nur, A. *Geophys. Res. Lett.* (in the press).

Reprinted from Geophysical Research Letters, v. 6, p. 1–4.

PORE FLUIDS AND SEISMIC ATTENUATION IN ROCKS

Kenneth Winkler and Amos Nur

Rock Physics Project
Geophysics Department,
Stanford University, Stanford, California 94305

Abstract. Seismic attenuation and velocities were measured in resonating bars of Massilon sandstone at various degrees of saturation. Whereas shear energy loss simply increases with degree of saturation, bulk compressional energy loss increases to ~95% saturation and then rapidly decreases as total saturation is achieved. This behavior is analogous to the behavior of shear and compressional velocities, but the effect on attenuation is larger by an order of magnitude. Our observations are in excellent agreement with the predictions of several models of energy loss involving partial or total saturation. Pore fluid attenuation mechanisms are expected to be dominant at least in the shallow crust.

Introduction

In recent years seismic attenuation has become of increasing interest to seismologists. This interest ranges from simultaneous inversion of velocity and attenuation data to obtain improved earth models [*Randall*, 1976], to the study of "bright spots" in hydrocarbon exploration [*Sheriff*, 1975]. To make full use of seismic data, it is necessary to interpret attenuation in terms of the physical properties of rocks. We need to know what mechanisms are responsible for loss of energy in seismic waves. Recently we have shown that the popular grain boundary friction mechanism is not important in the earth [*Winkler et al.*, 1979; *Mavko*, 1979]. In this paper we present evidence in support of another mechanism—fluid flow energy loss—which we believe to be dominant at least in the shallow crust.

The effect that pore fluids have on seismic velocities is well documented [*Nur and Simmons*, 1969; *Elliot and Wiley*, 1975; *Domenico*, 1976]. It is natural to expect that pore fluids will also influence seismic attenuation, but very little experimental work has been done in this area. There is, however, no shortage of theoretical models of fluid loss mechanisms. *Biot* [1956] has considered inertial effects and macroscopic flow, but the resulting energy losses are probably insignificant below ultrasonic frequencies [*White*, 1965]. Viscous shear relaxation has also been shown to be a high frequency (or low viscosity) mechanism [*Walsh*, 1969; *O'Connell and Budiansky*, 1977] and is not important below ultrasonic frequencies for water saturated rocks. Intercrack "squirting" flow [*Mavko and Nur*, 1975; *O'Connell and Budiansky*, 1977] may be important at low frequencies in fully saturated rock. *White* [1975] has presented a model for energy loss in macroscopically partially saturated rock that may be important in the earth, but is probably not important in the laboratory because of the small sample dimensions. A flow model based on partial saturation of individual cracks has been discussed by *Mavko and Nur* [1979], and a thermoelastic partial saturation mechanism has been presented by *Kjartansson and Denlinger* [1977]. Both models make similar predictions as to the relative size of shear and compressive energy losses, and either may explain the observations of *Born* [1941] and *Gardner et al.* [1964] that increasing fluid saturation causes increasing attenuation. *Johnston and Toksoz* [1977] have presented the first data on fully saturated rock, measured at ultrasonic frequencies. Although they concluded that fluid losses were negligible compared to frictional losses, other interpretations are possible.

Experimental Procedure

We have presented some experimental details in a previous paper [*Winkler et al.*, 1979] and will expand upon those here. However, a complete description of our experiments will be given in a forthcoming paper. We are using a bar resonance technique similar in principle to those used by *Born* [1941] and by *Gardner et al.* [1964]. Our samples are one meter long with a rectangular cross-section of 2.0 x 1.9 cm. Both torsional and extensional resonance modes are studied, with the fundamental resonance frequencies between 500–1700 hz, depending on satu-

Paper number 8L1301.
0094-8276/79/018L-1301$01.00

ration and stress conditions. Resonance frequencies are measured to 1 part in 10^3 and converted to velocities using geometric corrections given by *Spinner and Tefft* [1961]. For the samples used in this study the conversions are $V_E = 2f_E$ and $V_S = 2.18 f_T$ where V_E and V_S are the extensional velocity and shear velocity, respectively, in m/sec and f_E and f_T are the corresponding resonance frequencies in hz. P-wave velocity and Poisson's ratio are calculated from this data.

In this study the attenuation data was obtained by measuring the time constant of resonance decays using the relation $\tau = Q/\pi f$ where Q is the specific dissipation function. The same data could have been obtained from the half-width of resonance peaks, but we have used the peaks only to insure the purity of the resonance modes. Our system has been calibrated against an aluminum sample, and we estimate an absolute accuracy of 5–10% for our attenuation measurements. Relative precision is ~1%.

A unique feature of the experiments reported here is that pore pressure is controlled independently of confining pressure. The sample is jacketed with heat-shrink tubing with a light epoxy coating. Pore pressure is applied through capillary tubing connected to the sample at the support which is at the node of the fundamental resonance. The ability to control pore pressure is critical to the results presented here.

An additional feature involves our method of data analysis. Although we are measuring torsional (shear) and extensional (Young's modulus) attenuation, it is very useful to calculate P-wave attenuation (for use in seismology) and bulk compressional attenuation (for evaluation of mechanisms). To do this we have assumed that we can describe a solid using complex elastic moduli with small imaginary components and that the material is isotropic so that only two moduli are required. We use the correspondence principle [*Fung*, 1965] which lets us substitute complex moduli for their corresponding real moduli in the equations of linear elasticity. We also use the definition of attenuation (Q^{-1}) favored by *O'Connell and Budiansky* [1978], $Q^{-1} = \text{Im}(M)/\text{Re}(M)$, where M represents the complex modulus controlling a certain type of wave propagation. Straightforward algebra leads to the following results relating various measures of attenuation.

$$\frac{(1-\nu)(1-2\nu)}{Q_P} = \frac{1+\nu}{Q_E} - \frac{2\nu(2-\nu)}{Q_S} \qquad (1)$$

$$\frac{1-2\nu}{Q_K} = \frac{3}{Q_E} - \frac{2(\nu+1)}{Q_S} \qquad (2)$$

$$\frac{1+\nu}{Q_K} = \frac{3(1-\nu)}{Q_P} - \frac{2(1-2\nu)}{Q_S} \qquad (3)$$

Q_S, Q_E, Q_P, Q_K represent the Q's of shear waves, extensional waves, P-waves, and bulk compression, respectively. Poisson's ratio (ν) must be calculated from the velocities. It can also be shown that one of the following relations must be true.

$$Q_S > Q_E > Q_P > Q_K$$

or $$Q_S = Q_E = Q_P = Q_K$$

or $$Q_S < Q_E < Q_P < Q_K$$

The rock sample used in this study is Massilon sandstone. The porosity is 22% and the permeability is ~750 md. Anisotropy is less than 1% as measured by the velocities of ultrasonic P and S waves. The minimum Q measured in this study was ~20 and all data was taken at strains below those at which non-linear, frictional effects are observed [*Winkler et al.*, 1979]. Also, since we are relating Q's taken at slightly different frequencies ($f_E \approx 1.5 f_T$) we must assume that Q is independent of fre-

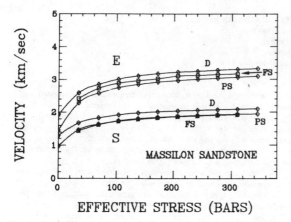

Fig. 1. Shear (S) and extensional (E) velocities in dry (D), partially (~95%) saturated (PS), and fully saturated (FS) Massilon sandstone. Pore fluid is water.

quency to use Eq. 1-3. We have verified this by measuring attenuation at several harmonics of the resonance frequency.

Observations

A series of three experiments was performed on a sample of Massilon sandstone. In the first experiment the sample was "room dry". After cycling confining pressure to eliminate hysteresis effects, attenuation and velocity data was taken vs. confining pressure. The rock was then evacuated to a pressure of 0.1 torr, then partially saturated with distilled water (pore pressure ~7 bars) and confining pressure was again varied. To achieve total saturation it was necessary to apply at least 15 bars pore pressure to the sample, so in the third experiment we held confining pressure constant at 345 bars and varied pore pressure. Because attenuation is a function of effective stress (in saturated rock), this procedure is equivalent to varying confining pressure in fully saturated rock. We do not have precise control over degree of saturation, but have estimated this from the velocities, as discussed below.

We will first discuss the velocity measurements because these help in understanding the attenuation data. Figure 1 shows the measured shear and extensional velocities and Figure 2 shows the shear velocities along with the computed P-wave velocities. Shear wave velocities decrease continuously as degree of saturation increases, whereas P-wave velocities decrease from dry to partial saturation, and sharply increase as total saturation is achieved. Similar results have previously been observed at

ultrasonic frequencies [*Elliot and Wiley*, 1975; *Domenico*, 1976], but this is the first such observation below 1 khz. Note in Figure 2 that shear velocities are almost identical in partially and fully saturated rock. From this evidence and the well documented effect of pore fluids on shear velocity we have estimated that the rock is ~95% saturated when partially saturated. Note also in Figure 2 that the P-velocity in fully saturated rock is much larger than in partially saturated rock. This increase in P-velocity is our main evidence that we have in fact achieved total saturation. The transition from partial to total saturation can be achieved by increasing pore pressure from ~7 bars to ~15 bars. Presumably this either dissolves any gas remaining in the rock or forces water into a few remaining undersaturated pores.

The attenuation data obtained in these experiments is shown in Figures 3–6. Figures 3 and 4 show the observed shear and extensional attenuation. Figs. 5 and 6 show the computed P-wave and bulk compressional attenuation. Error bars on shear and extensional data are ~1% and are not shown. They were computed, however, from the goodness of fit of the resonance decay to a decaying exponential, and these uncertainties were used in calculating the error bars in Figures 5 and 6. These relatively larger error bars result from uncertainties propagating through the calculations (Eq. 1 and 2), but they do not obscure the essential features of the data. Also, some of the calculated values of Q^{-1} are negative. Since negative values are physically impossible, these probably result from systematic errors in the data and should simply be interpreted as very small values of attenuation.

Figures 3–6 all show that attenuation decreases with increasing confining pressure. This feature is commonly observed [*Birch and Bancroft*, 1938; *Gardner et al.*, 1964] and provides little insight into physical loss mechanisms. Presumably crack closure is responsible for this behavior, and all proposed mechanisms will show some pressure dependence.

Considerable insight into pore fluid attenuation mechanisms is obtained by comparing shear and bulk compressional attenuation (Figures 3 and 6) at various degress of saturation. Shear attenuation is minimum in dry rock, is greater in partially saturated rock, and is maximum in fully saturated rock. However, while compressional loss is also minimum in dry rock and greater in partially saturated rock, the compressional loss is significantly reduced by total saturation. Also, in both dry and fully saturated rock, shear energy loss is greater than bulk energy loss. In partially saturated rock, however, shear energy loss is less than bulk loss. These results agree with those of *Wyllie et al.* [1962] and *Gardner et al.* [1964]. However, in neither of these studies was extensional attenuation measured in totally saturated rock and therefore they could not have observed the decrease in compressional energy loss. We are continuing these experiments with Berea sandstone, Sierra White granite, and porous Vycor glass. Although complete experiments have not yet been run, all data obtained thus far agrees with our observations on Massilon sandstone. In addition, qualitative attenuation measurements on P and S waves at ultrasonic frequencies [J. DeVilbiss, personal communication, 1978] show similar behavior as degree of saturation is varied.

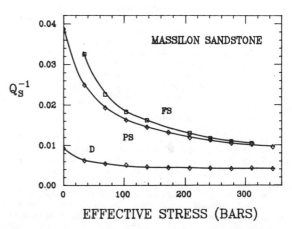

Fig. 2. Shear velocities (S) and computed P-wave velocities (P) in dry (D), partially saturated (PS), and fully saturated (FS) Massilon sandstone.

Fig. 3. Shear attenuation in dry (D), partially saturated (PS), and fully saturated (FS) Massilon sandstone.

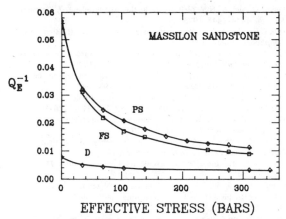

Fig. 4. Extensional attenuation in dry (D), partially saturated (PS), and fully saturated (FS) Massilon sandstone.

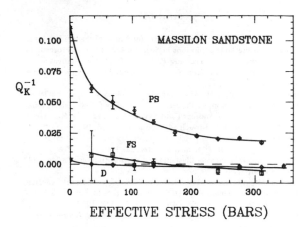

Fig. 6. Computed bulk compressional attenuation in dry (D), partially saturated (PS), and fully saturated (FS) Massilon sandstone.

Discussion

Our observations can be explained remarkably well with several existing theories of pore fluid attenuation mechanisms. *Mavko and Nur* [1979] have discussed a model in which liquid droplets in a partially saturated crack flow in response to crack compression or dilation. *Kjartansson and Denlinger* [1977] have presented a model in which compression of the gaseous phase of the pore fluid causes adiabatic heating of the gas followed by an irreversible flow of heat into the rock and pore water. Both mechanisms predict that attenuation should increase with degree of saturation and then rapidly decrease at total saturation. They also predict that bulk compressional energy loss in partially saturated rock should be approximately twice as large as shear energy loss, and this is very close to what we have observed. To explain the observations on fully saturated rock, we use a mechanism proposed by *Mavko and Nur* [1975] and developed by *O'Connell and Budiansky* [1977]. This involves "squirting" flow between cracks as cracks at different orientations to the passing wave (or cracks of different aspect ratio) undergo differential compression. *O'Connell and Budiansky* [1977] have shown that this mechanism may cause significant shear attenuation over a broad frequency range, and that shear attenuation should be much larger than compressional attenuation. Again, this agrees with our observations.

Conclusions

The contrasting behavior of shear and bulk compressional attenuation is very clear from the accurate new data we have presented. As water

is added to the pore space of rock, compressional energy loss is about twice as large as shear energy loss, and both increase with degree of saturation. The loss mechanism may be fluid flow [*Mavko and Nur*, 1979], thermoelastic [*Kjartansson and Denlinger*, 1977], or both. Shear attenuation continues to increase to 100% saturation. However, bulk attenuation reaches a maximum at approximately 95% saturation and then decreases with further saturation. At total saturation bulk loss is less than one third the shear loss. This minimization of bulk loss is predicted by both of the partial saturation mechanisms we have discussed as well as the total saturation mechanism developed by *O'Connell and Budiansky* [1977] involving flow between cracks.

The data also shows that in partially saturated rock (and for shear loss in fully saturated rock) pore fluid attenuation mechanisms clearly dominate over all mechanisms in dry rock. However, our samples were only "room dry" and so even here we may not have totally eliminated the fluid mechanisms. We do find, though, that in dry rock shear losses are greater than bulk losses, implying that the partial saturation mechanisms may not completely explain the behavior.

Although practical applications of these results must await further studies with different rocks and pore fluids, we may consider some potential uses. The results may enable us to better interpret the nature of gas related bright spots in reflection seismological surveys, as the attenuation due to the gas-liquid mix may be very significant. Attenuation in geothermal fields, particularly those with steam, may serve as a diagnostic tool due to the sensitivity to partial saturation.

There is, finally, an interesting speculation that can be made. The phenomenon of decreasing P-wave velocity with slight undersaturation of rock is the basis of the dilatancy-diffusion model of earthquake velocity precursors [*Nur*, 1972; *Anderson and Whitcomb*, 1975]. The great difficulty in detecting these anomalies [*Boore et al.*, 1975; *Allen and Helmberger*, 1973] may have thrown doubt on the applicability of the model, but it may also simply reflect the small magnitude of the effect being sought. The maximum velocity change possible is on the order of 10–15% [Figure 2; *Winkler and Nur*, 1977] and if the dilatant zone is a small fraction of the seismic ray path, the net change in travel-time will not be resolvable. However, our results suggest that P-wave attenuation may be a much more sensitive indicator of undersaturation caused by dilatancy. Whereas P-wave velocity varies by ~15% (Figure 2), P-wave attenuation varies by a factor of five or more (Figure 5). Although attenuation is more difficult to measure than is velocity, the large size of this effect may make it easier to detect attenuation anomalies as evidence of stress accumulation in the crust.

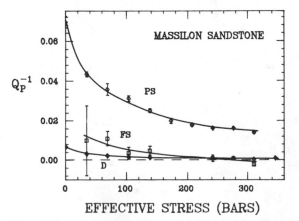

Fig. 5. Computed P-wave attenuation in dry (D), partially saturated (PS), and fully saturated (FS) Massilon sandstone.

Acknowledgments. We are grateful to J. Walls for providing permeability measurements and to T. Jones for ultrasonic velocity measurements. This research was supported by a grant from the Earth Science Division, U.S. National Science Foundation.

References

Allen, C.R., and D.V. Helmberger, Search for temporal changes in seismic velocities using large explosions in Southern California, in Proceedings of the Conference on Tectonic Problems of the San Andreas Fault System, *Stanford Univ. Publ. Geol. Sci. 13,* edited by R.L. Kovach and A. Nur, 436–445, 1973.

Anderson, D.L. and J.H. Whitcomb, The dilatancy diffusion model of earthquake prediction, in Proceedings of the Conference on Tectonic Problems of the San Andreas Fault System, *Stanford Univ. Publ. Geol. Sci. 13,* edited by R.L. Kovach and A. Nur, 417–426, 1973.

Biot, M.A., Theory of propagation of elastic waves in a fluid saturated, porous solid, I and II. *J. Acoust. Soc. Am., 28,* 168–191, 1956.

Birch, F. and D. Bancroft, Elasticity and internal friction in a long column of granite, *Bull. Seis. Soc. Am., 28,* 243–254, 1938.

Boore, D.M., A.G. Lindh, T.V. McEvilly, and W.W. Tolmachoff, A search for travel-time changes associated with the Parkfield, California earthquake of 1966, *Bull. Seis. Soc. Am., 65,* 1407–1418, 1975.

Born, W.T., The attenuation constant of earth materials, *Geophysics, 6,* 132–148, 1941.

Domenico, N.S., Effect of brine-gas mixture on velocity in an unconsolidated sand reservoir, *Geophysics, 41,* 5, 882–894, 1976.

Elliot, S.E. and B.F. Wiley, Compressional velocities of partially saturated unconsolidated sands, *Geophysics, 40,* 6, 949–954, 1975.

Fung, Y.C., Foundations of solid mechanics, Prentice-Hall, 1965.

Gardner, G.H.F., M.R.J. Wyllie, and D.M. Droschak, Effects of pressure and fluid saturation on the attenuation of elastic waves in sands, *J. Petrol. Tech., 16,* 189–198, 1964.

Johnston, D.H. and M.N. Toksoz, Attenuation of seismic waves in dry and saturated rocks, *Geophysics, 42,* 7, 1511, 1977.

Kjartansson, E., and R. Denlinger, Seismic wave attenuation due to thermal relaxation in porous media, *Geophysics, 42,* 7, 1516, 1977.

Mavko, G., Frictional attenuation: An inherent amplitude dependence, *J. Geophys. Res.,* in press, 1979.

Mavko, G. and A. Nur, Melt squirt in the asthenosphere, *J. Geophys. Res., 80,* 1444–1448, 1975.

Mavko, G. and A. Nur, Wave attenuation in partially saturated rocks, *Geophysics,* in press, 1979.

Nur, A., Dilatancy, pore fluids and premonitory variations in T_S/T_P travel times, *Bull. Seis. Soc. Am., 62,* 1217–1222, 1972.

Nur, A. and G. Simmons, The effect of saturation on velocity in low porosity rocks, *Earth. Plan. Sci. Lett., 7,* 183–193, 1969.

O'Connell, R.J. and B. Budiansky, Viscoelastic properties of fluid-saturated cracked solids, *J. Geophys. Res., 82,* 36, 5719–5736, 1977.

O'Connell, R.J. and B. Budiansky, Measures of dissipation in viscoelastic media, *Geophys. Res. Lett., 5,* 1, 5–8, 1978.

Randall, M.J., Attenuative dispersion and frequency shifts of the Earth's free oscillations, *Phys. Earth Planet. Int., 12,* 1–4, 1976.

Sheriff, R.E., Factors affecting seismic amplitudes, *Geophys. Pros., 23,* 125–138, 1975.

Spinner, S. and W.E. Tefft, A method for determining mechanical resonance frequencies and for calculating elastic moduli from these frequencies, *Proc. A.S.T.M., 61,* 1221–1238, 1961.

Walsh, J.B., New analysis of attenuation in partially melted rock, *J. Geophys. Res., 74,* 4333–4337, 1969.

White, J.E., Seismic waves: Radiation, Transmission and Attenuation, McGraw-Hill, 1965.

White, J.E., Computed seismic speeds and attenuation in rocks with partial gas saturation, *Geophysics, 40,* 2, 224–232, 1975.

Winkler, K. and A. Nur, Depth constraints on dilatancy induced velocity anomalies, *J. Phys. Earth, 25,* Suppl., 231–241, 1977.

Winkler, K., A. Nur, and M. Gladwin, Friction and seismic attenuation in rocks, *Nature,* in press, 1979.

Wyllie, M.R.J., G.H.F. Gardner, and A.R. Gregory, Studies of elastic wave propagation in porous media, *Geophysics, 27,* 569–589, 1962.

(Received October 24, 1978; accepted December 4, 1978.)

Attenuation: A state-of-the-art summary

David H. Johnston

Several new papers dealing with the laboratory measurement of attenuation have recently come to our attention. These new papers present data that space and time will only allow us to summarize here. In addition to this state-of-the-art summary, we will present a variety of conclusions and generalizations concerning attenuation behavior that we have drawn from the body of literature as it stood when we went to press.

While we wish to avoid presenting the reader with a compilation of Q values obtained from individual measurements, Table 1 presents some reference Q values for several nonporous materials. These data were taken primarily from Bradley and Fort (1966), who have compiled an extensive list of attenuation values, and an excellent review article on Q by Knopoff (1964).

The particular areas covered in this summary are the dependence of attenuation on frequency, pressure, wave strain amplitude, temperature, saturation, and general rock properties. Emphasis is placed on the qualitative behavior of attenuation in relation to these variables.

Frequency dependence

The common assumption that Q is independent of frequency is presently undergoing critical evaluation. Taken as a whole, the body of laboratory experience suggests that such a relationship exists (Attewell and Ramana, 1966). Some seismic data, however, definitely show attenuation as a frequency-dependent parameter. It is not yet clear how much of this is attributable to inherent frequency dependence in rocks and how much results from other factors. Most data suggest that, at least for dry rocks, Q is indeed independent of frequency (Birch and Bancroft, 1938; Born, 1941; Peselnick and Outerbridge, 1961; Pandit and Savage, 1973; Nur and Winkler, 1980; Tittmann et al, 1981). Several investigations, however, suggest that under some conditions for certain rocks, Q may, in fact, depend strongly on frequency.

In most cases, attenuation values obtained by ultrasonic methods are higher than those obtained by the lower frequency resonance techniques (Peselnick and Zietz, 1959; Johnston and Toksöz, 1980a). This difference in values may be, as suggested by Mason et al (1978), due to an inherent frequency dependence in dry rocks. Mason et al (1970) found Q^{-1} to increase with frequency (to 1 MHz) in Westerly granite. This increase was followed by a decrease at higher frequencies. However, data obtained from spectral ratios (*Toksöz et al, 1979;* Johnston and Toksöz, 1980a) show that Q is at least locally independent of frequency in the bandwidth of 0.2 to 0.8 MHz. Thus, differences between ultrasonic and resonance Q values for dry rocks may be attributable to some factor such as the strain amplitude of the exciting wave.

In fluid-saturated rocks, attenuation as a result of a flow-type mechanism is generally frequency dependent. The possibility of superposition of flow relaxation times, however, does not rule out an apparent constant Q over limited frequency bands. Frequency dependence has been observed in the laboratory for wet and saturated rocks. Born (1941) showed that for small amounts of water injected into an Amherst sandstone sample, Q^{-1} is linearly dependent upon frequency in the range of 1 to 4 kHz. Recently, Tittmann et al (1981) showed that attenuation in several sandstones is higher at 7 kHz than at 200 Hz and that the changes in Q^{-1} with pressure are less for the lower frequency data. Nur and Winkler (1980) observed an apparent peak in Q^{-1} for saturated Massilon sandstone at about 4 kHz, but they found a constant Q^{-1} for the dry case (Figure 1). The attenuation peak shifts to higher frequencies with increasing pres-

Table 1. Q and velocity for reference materials[*]

Material	Q	V(km/sec)[1]	Mode	Frequency range
Aluminum	200,000	5.00	Longitudinal resonance	1 to 200 kHz
	5,900	6.32	P-wave pulse	3.1 to 7.5 MHz
	7,630	6.32	P-wave pulse	5 to 15 MHz
	19,400	3.10	S-wave pulse	3.5 to 4.5 MHz
	17,200	3.10	S-wave pulse	3 to 6.8 MHz
Brass	655	3.48	Flexural resonance	—
Copper				
Unannealed	2,180	3.81	Longitudinal resonance	2.5 to 30 kHz
	4,380	2.32	Torsional resonance	3 to 30 kHz
	1,770	4.76	P-wave pulse	15 to 65 MHz
Annealed	5,830	5.01	P-wave pulse	25 to 75 MHz
Lead	36	1.21	Longitudinal resonance	1.6 to 15 kHz
	34	0.69	Torsional resonance	1 to 9 kHz
Magnesium	965	5.77	P-wave pulse	7 to 76 MHz
Nickel	980	4.90	Flexural resonance	12 to 33 Hz
Steel	1,850	5.20	Flexural resonance	2 to 8 Hz
Celluloid	7	2.81	Flexural resonance	0.5 to 18 Hz
Fused Quartz	44,500	3.76	S-wave pulse	5 to 19 Hz
Glass	490	5.36	Flexural resonance	12 to 27 Hz
Glass (Pyrex)	1,860	5.17	Longitudinal resonance	10 kHz
Glass (Soda lime)	1,450	4.54	Longitudinal resonance	5.6 to 6.1 kHz
	1,340	2.84	Torsional resonance	3.6 to 64 kHz
Lucite[2]	23	2.11	Longitudinal resonance	1 kHz
Plexiglas[3]	20	2.59	Longitudinal resonance	10 kHz
Polystyrene	240	2.24	Longitudinal resonance	20 to 60 kHz
Air				
Dry	562	0.343	Resonance	100 Hz
	3,485	0.343		10 kHz
100 percent humidity	4,139	0.345	Resonance	100 Hz
	1,434	0.345		10 kHz
Water				
Fresh (17°C)	210,000	1.48	Resonance	100 kHz
Salt (36 ppm)	63,000	1.52		150 kHz

[*] These data are taken primarily from the compilations of Bradley and Fort (1966), Knopoff (1964), and the Chemical Rubber Company Handbook of Chemistry and Physics.

[1] Velocities represent typical values for the mode of excitation listed.

[2] Winkler, K., 1979, Ph.D. thesis, Stanford Univ.

[3] Johnston, D.H., 1978, Ph.D. thesis, Massachusetts Institute of Technology.

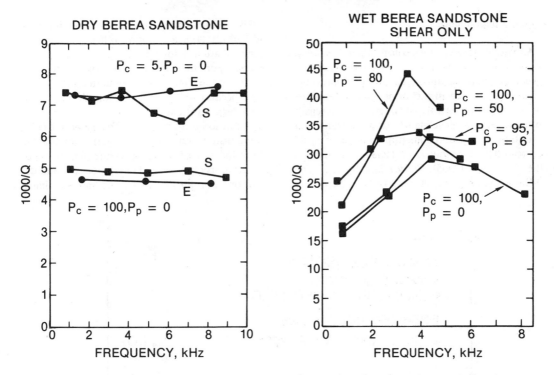

FIG. 1. Attenuation as a function of frequency for dry (Q_E^{-1} and Q_s^{-1}) and wet (Q_s^{-1} only) Berea sandstone. The data, shown for several effective pressures, are from Nur and Winkler (1980).

sure. A recent contribution by Spencer (1981) also suggests the presence of attenuation peaks in the low-frequency range. Attenuation was measured from the stress-strain phase angle for water-saturated Navajo sandstone (Figure 2), Spergen limestone, and Oklahoma granite. Data for the sandstone and granite exhibit an increase in attenuation with frequency, which Spencer models using a distribution of stress relaxation times. Calculated peaks in attenuation occur at about 500 Hz for the sandstone and 1 kHz for the granite. A peak is observed directly for the limestone at a frequency of 10 to 20 Hz. Data from Nur and Winkler (1980) and Spencer (1981) suggest a narrow range of relaxation times contributing to the overall attenuation in a particular rock. The data further suggest that extrapolation from laboratory to field data would require a detailed description of both the operative attenuation mechanisms and the geometry of the pore space.

Other experiments, using high-viscosity fluids as the saturant, have shown that a relaxation process can occur. Frequency dependence was implied by variations in attenuation observed with variations in the viscosity of the fluid obtained by changing the temperature. Gordon (1974) reported such data for extensional resonance at 50 kHz of glycerol-saturated granite. Nur and Simmons (1969) presented relative attenuation data at ultrasonic frequencies, also for glycerol-saturated granite. Data points for dry and water-saturated samples were also included. A distinct peak in attenuation was observed for shear waves. In both cases, however, the relaxations observed would probably not contribute to attenuation in water- or oil-saturated upper crustal rocks at seismic or even low ultrasonic frequencies.

Little work has been done on the frequency dependence of attenuation in partially saturated rocks. *Wyllie et al (1962)* presented data for alundum that show large and complex variations in attenuation in the range from about 5 to 12 kHz. Clearly, more work is required in this area.

Pressure dependence

As with velocity, attenuation depends strongly on pressure. Researchers are in near-unanimous agreement that for all frequencies and saturation conditions studied in the laboratory, attenuation decreases (Q increases) as a function of increasing applied pressure, leveling off to a constant value at high pressures. Experimental data verifying this are found in *Gardner et al* (*1964*), Klima et al (1964), Levykin (1965), *Gordon and Davis* (*1968*), Al-Sinawi (1968), Walsh et al (1970), *Toksöz et al*

(*1979*), *Winkler and Nur* (*1979*), and Johnston and Toksöz (1980a), among others. The rate of change of attenuation with pressure, though, does depend upon the physical state of the rock. In most cases the changes of attenuation are thought to be because of, or related to, the closure of microcracks in the rock.

At ultrasonic frequencies, *Toksöz et al* (*1979*) and Johnston and Toksöz (1980a) show that the rate of increase of Q with pressure depends upon saturation, rock type, and crack porosity and distribution. For air-dry rocks the rate of change

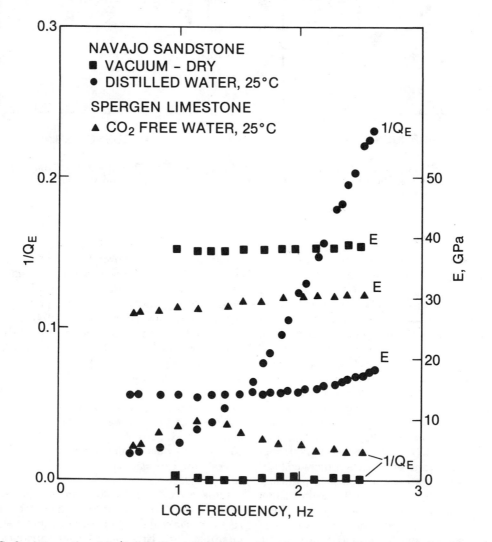

FIG. 2. Attenuation (Q_E^{-1}) and Young's modulus E as functions of frequency for vacuum-dry and water-saturated Navajo sandstone and water-saturated Spergen limestone. The data are from Spencer (1981).

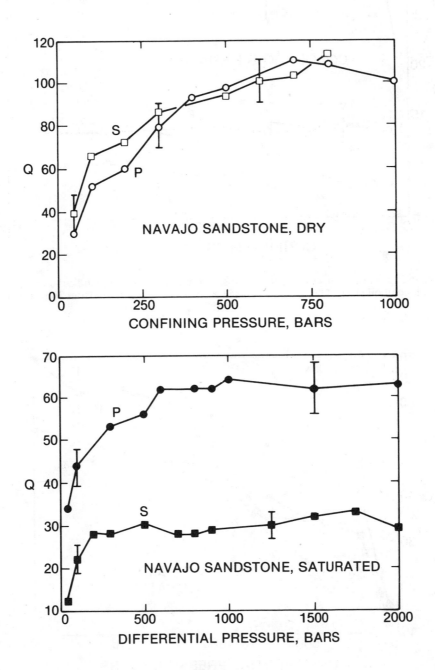

FIG. 3. Q_p and Q_s as functions of differential or effective pressure for air-dry and water-saturated Navajo sandstone at ultrasonic frequencies. The data are from Johnston and Toksöz (1980a).

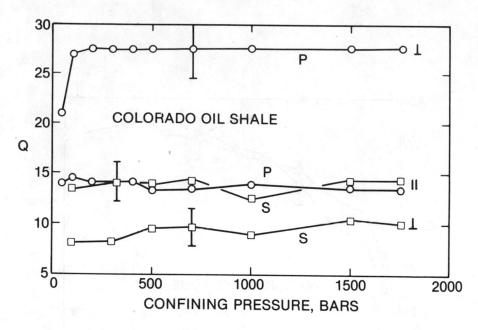

FIG. 4. Q_p and Q_s as functions of confining pressure for orientations perpendicular and parallel to the bedding plane in oil shale. The data, taken at ultrasonic frequencies, are from Johnston and Toksöz (1980a).

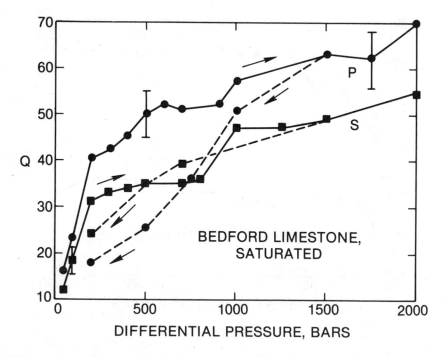

FIG. 5. Ultrasonic Q_p and Q_s as functions of differential or effective pressure for water-saturated Bedford limestone illustrating the effects of pore collapse. Data are shown for both loading (solid line) and unloading (dashed line) paths and are from Johnston and Toksöz (1980a).

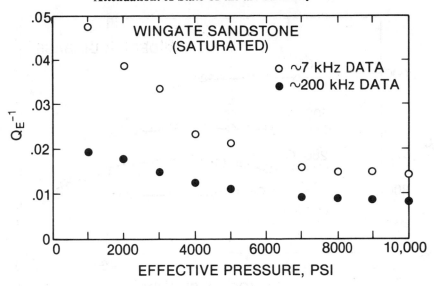

FIG. 6. Attenuation (Q_E^{-1}) as a function of effective pressure for water-saturated Wingate sandstone at 200 Hz and 7 kHz. The data are from Tittmann et al (1981).

with pressure is greater than for water-saturated rocks (Figure 3). Furthermore, Q_s for saturated samples, after a slight initial increase at low pressure, remains relatively constant at higher pressures. An exception is a Colorado oil shale (Johnston and Toksöz, 1980a) for which the mechanical properties do not depend on microcracks and thus Q is independent of pressure (Figure 4). As might be expected, attenuation exhibits strong anisotropy in the oil shale.

Another interesting case is the change in Q observed while a rock is undergoing pore collapse. As demonstrated by the Bedford limestone (Figure 5), Q is strongly correlated with cracks induced by pore collapse. These cracks open when the rock is unloaded. The relative changes in the anelastic properties with pressure track those of the elastic properties.

In the kilohertz frequency range, the most recent work is that of *Winkler and Nur* (*1979*) and Tittmann et al (1981). Their findings indicate that, again, attenuation decreases with increasing pressure. Also, for saturated rocks, limited data for sandstones offered by both investigators suggest that attenuation behaves according to an effective pressure law. That is, the effect of pressure on attenuation is determined from the difference between the confining or overburden pressure and

the pore fluid or reservoir pressure. Thus, for the cases studied so far, pore pressure and confining pressure have an equal effect on attenuation. While this behavior is commonly observed for velocity, it is not necessarily the case for flow properties such as permeability.

A major difference between the pressure dependence of attenuation measured at lower frequencies and that measured at ultrasonic frequencies in sandstones is that the change in attenuation for dry samples is smaller. Furthermore, as previously mentioned, Tittmann et al (1980) have observed that the rate of change of attenuation with pressure is smaller at 200 Hz than at 7 kHz (Figure 6) for several saturated sandstones.

Data for partially saturated samples are again limited, but both *Winkler and Nur* (*1979*) and Johnston and Toksöz (1980a) report a decrease in attenuation with increasing pressure in the kilohertz and megahertz regimes, respectively.

For an applied differential stress, attenuation appears to be anisotropic (Merkulova et al, 1972; Walsh et al, 1970). Lockner et al (1977) measured *P, SV,* and *SH* relative attenuation in a triaxially stressed rock and found a strong effect owing to crack orientation. Shear waves polarized normal to the axis of maximum compression suffered the least attenuation as a result of the closure of

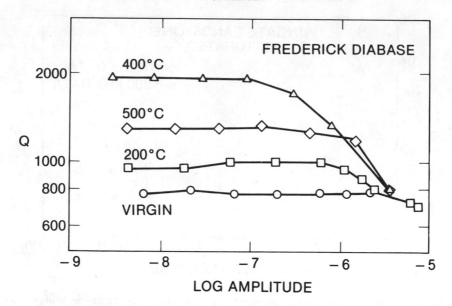

FIG. 7. Q_E (log scale) as a function of strain amplitude for thermally cycled Frederick diabase. The temperatures shown are the maximum achieved during slow cycling. Measurements were made at room temperature. The data are from Johnston and Toksöz (1980b).

cracks with faces in that plane. At high differential stresses, the onset of dilatancy, which causes cracks to open, results in increased attenuation.

Strain amplitude dependence

It is commonly assumed that attenuation is independent of strain amplitude. While this may be true for the low strains associated with seismic waves, laboratory data (*Peselnick and Outerbridge, 1961; Gordon and Davis, 1968;* Gordon and Rader, 1971; *McKavanagh and Stacey, 1974; Brennan and Stacey, 1977; Winkler et al, 1979;* Johnston and Toksöz, 1980b) show amplitude-dependent behavior at strains greater than about 10^{-6} that may be associated with a nonlinear attenuation mechanism such as frictional sliding. For example, cyclic loading tests run at low strain amplitudes have elliptical stress-strain loops, implying a linear mechanism. For strains greater than 10^{-6}, however, the loops become cusped.

Measurements of attenuation as a function of strain amplitude have recently been made in the 1 to 20 kHz range using resonance bar techniques (*Winkler et al, 1979;* Johnston and Toksöz, 1980b). Both investigators found that materials

containing no microcracks (Plexiglas, lucite, glass, aluminum) exhibited no amplitude dependence. For rocks, *Winkler et al (1979)* found that at 1 kHz, moderate effective pressures reduce or may eliminate amplitude dependence. In water-saturated rocks it was shown that amplitude dependence is enhanced.

Johnston and Toksöz (1980b) further examined the effect of microcracks on amplitude dependence, stimulating crack growth by thermal cycling of the rock sample. It was found that for low heating rates, where differential thermal expansion may tend to open preexisting cracks, the transition from constant Q to amplitude-dependent Q migrates to lower strain amplitudes as a function of the maximum temperature achieved during cycling (Figure 7). If the rock was heated above about 400°C, or if it was heated rapidly, new cracks were generated and the transition moved back to higher strains.

Temperature dependence

Very little work has been done on the temperature dependence of attenuation at temperatures found in the upper crust. Data from Volarovich and Gurevich (1957) and *Gordon and Davis (1968)* indicate that Q is generally temperature independent at temperatures less than 150°C in dry rocks. An increase in attenuation at temperatures greater than 150°C reported by *Gordon and Davis (1968)* may be a result of thermal cracking. However, resonant bar data on air-dry rock measured by Kissell (1972) from −200°C to 600°C show a pronounced peak in attenuation at room temperature. Kissell interpreted these results as evidence of varying moisture content in the rock.

Spencer (1981) reported a slight shift to higher frequencies in the peak attenuation as a function of temperature for water-saturated Navajo sandstone. If the observed attenuation is thermally activated, the data imply a range of activation energies from 16 to 22 kJ/mole.

Near the boiling temperatures of pore fluids, perhaps in geothermal areas, thermally activated attenuation mechanisms may be important, and the attenuation values will strongly depend upon temperature. Such effects are also noted in systems undergoing partial melting (*Spetzler and Anderson, 1968;* Stocker and Gordon, 1975).

Saturation dependence

Attenuation for fully or partially fluid-saturated rocks is higher than for dry rocks and depends upon the degree of saturation, the fluid type, and the frequency in complicated ways. In this section we shall deal first with the differences between vacuum-outgassed and air-dry rocks, air-dry and fully saturated samples, and fully and partially saturated rocks.

As an outgrowth of lunar studies it was found that vacuum outgassing of rock samples produced significant decreases in attenuation with respect to air-dry rocks (Pandit and Tozer, 1970; Warren et al, 1974). Perhaps the most exhaustive survey of the effects of outgassing and the subsequent readsorption of volatiles on attenuation has been carried out by Tittmann and his colleagues (Tittman et al, 1973, 1974, 1975, 1976; *Tittmann, 1977;* Clark et al, 1980). It was found, for example, that the high in-situ Q values (>3000) observed for the moon (Latham et al, 1974; Dainty et al, 1976) could be duplicated in the laboratory by repeated thermal cycling and application of hard vacuum (10^{-10} torr) for both lunar samples and their terrestrial analogs (Tittmann et al, 1975). Subsequent exposure of outgassed samples to saturated vapors of volatiles at ambient pressure generally resulted in a sharp increase in attenuation. Tittmann et al

(1976) found that the presence of several volatiles, free of water as an impurity, resulted in a significant increase in attenuation with small reductions in velocity. Water vapor, however, produced the largest changes. Interestingly enough, the viscosity of the volatile had little effect on the changes in attenuation. Volatiles with different dipole moments produced similar changes, but those with higher moments (e.g., water) proved more difficult to outgas.

The effect of volatile adsorption on attenuation in sedimentary rocks has recently been studied in great detail by Clark et al (1980). Increases much like those seen for lunar studies in Q upon evacuation were observed in several sandstones (400 to 700 percent) and limestones (100 to 300 percent). Q and velocity were then measured as a function of the partial pressure of water. The mass of the adsorbed water was determined and it was found that attenuation increased linearly with mass up to a 1 to 2 monolayer coverage of the pore space (Figure 8). Additional water had less effect until a high partial pressure was reached where, presumably, free water had condensed. The use of benzene vapor instead of water produced little change in attenuation, suggesting that the polar nature of water is a significant factor in the attenuation mechanism.

Similar experiments performed by Pandit and King (1979) on the Berea sandstone for both longitudinal and torsional resonances have given essentially the same results. It was found that after several monolayers of water had been adsorbed, a thickness was reached beyond which bulk water behavior was followed. A transition from frequency-independent to frequency-dependent Q was observed at this point.

In the preceding studies, "air-dry" conditions are those that exist after several monolayers of water have been adsorbed but before bulk water effects are observed (90 to 100 percent relative humidity). Clearly, most "dry" rock data reported in the literature are, in fact, air-dry data. The transition from air-dry behavior to bulk water behavior has not been well studied. Pandit and King estimate the thickness of the first bulk water layer as 110 Å for the Berea sandstone. This may serve as an upper bound on the thicknesses of the thinnest cracks that could contribute to bulk flow-losses. Bulk water effects undoubtedly involve flow that is not well understood for very low water saturations.

Numerous measurements comparing the air-dry state to the fully saturated state have been made. For ultrasonic frequencies (*Toksöz et al, 1979;* Johnston and Toksöz, 1980a), both Q_p and Q_s are higher in the air-dry case. The largest changes with complete saturation involve Q_s. In the air-dry case $Q_s \approx Q_p$, while for the saturated case $Q_s < Q_p$ (Figure 4). At resonance bar frequencies (*Winkler and Nur, 1979*) there are also large decreases in Q_s and Q_E going from the air-dry to the fully sat-

urated state. However, calculated Q_p values show little difference between the two conditions. Also, in the air-dry case $Q_s < Q_p$, a result not observed in ultrasonic data. Despite differences in detail, it is clear that the attenuation in a water-saturated rock is higher than that in an air-dry rock, although it may be possible at low frequencies that the two values converge (Tittmann et al, 1981) because of the frequency dependence of flow-type mechanisms.

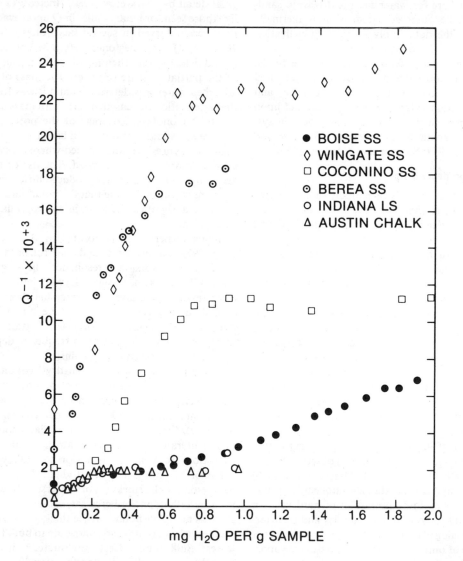

FIG. 8. Q_s^{-1} as a function of water vapor saturation in several rocks. The data are from Clark et al (1980).

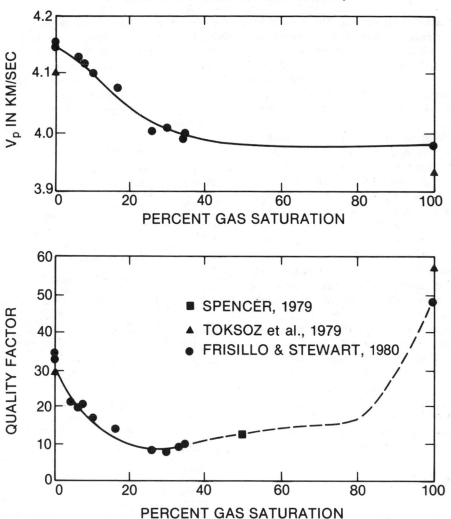

FIG. 9. Q_p and velocity as functions of nitrogen gas saturation in Berea sandstone at 1500 psi confining pressure. The figure is taken from Frisillo and Stewart (1980).

Work with saturants other than water is limited. *Wyllie et al* (*1962*) have, however, investigated Soltrol (chemically inert, deodorized kerosene) as a saturant. Attenuation data for Soltrol-saturated alundum suggest that it behaves qualitatively differently from water with respect to partial saturation. The absolute attenuation values for the two substances are similar. The effect of high-viscosity saturants (Nur and Simmons, 1969; Gordon, 1974) was discussed under frequency dependence.

Early experimental work concerning the effects of partial saturation on attenuation (*Wyllie et al, 1962; Gardner et al, 1964*) indicated little difference between the fully and partially saturated states. However, new, carefully controlled experiments show that Q_p for the partially saturated state can be lower than for the fully saturated state. *Winkler and Nur* (*1979*) showed for partially saturated Massilon sandstone and porous Vycor glass at 1 kHz that shear attenuation is less

134   Johnston

than compressional attenuation. The decomposition of extensional and shear losses into bulk and shear losses indicates that most of the attenuation is due to bulk processes. Spencer (1979), applying a similar decomposition to the full-saturation ultrasonic Berea data of *Toksöz et al* (*1979*) and new partial saturation data, reached a similar conclusion.

Frisillo and Stewart (1980) have determined Q_p in the Berea sandstone at ultrasonic frequencies as a continuous function of saturation and found a minimum at a gas saturation of about 35 percent (Figure 9). Q_s was not measured. Relative attenuation in several rock types was measured at ultrasonic frequencies during the water-steam transition at high temperature and moderately low pressure by DeVilbiss-Muñoz (1980). Again, a maximum in P-wave attenuation (minimum Q) was observed during the transition, which is assumed to be a partially saturated state. Little change was observed for shear waves.

Winkler and Nur (*1979*) point out that observations on the effect of partial saturation on attenuation may lead to an important diagnostic for gas reservoirs. For fully water-saturated zones, $Q_p/Q_s > 1$, while for partially saturated zones, $Q_p/Q_s < 1$. Thus, while absolute attenuation values may not be measured reliably in situ, the relative attenuation between compressional and shear waves may provide useful information.

Relationships between attenuation and rock properties

One of the goals of research regarding the nature of attenuation in crustal rocks is to correlate those data with meaningful and useful rock properties such as permeability, grain size, porosity, and ridigity. This correlation requires a complete characterization of the petrology and microfabric of the rock samples that has been lacking in work to date.

In a most general way, differences in attenuation between rock types under the same physical conditions are related to the seismic velocities through the shear modulus (Hamilton, 1972a) which is, in turn, related to the degree of consolidation and cementation. Rocks with higher velocity generally have higher Q. For unconsolidated sands, attenuation can be correlated with grain size and porosity as described by *Hamilton* (*1972b*) in chapter 4. In consolidated rocks, how-

ever, the microstructure, and in particular, the pore geometry, may be critical in defining relative differences in attenuation. Mechanisms involving intercrack flow, thought to be dominant under some conditions, are very sensitive to the microstructure. As a result, correlations between attenuation and bulk rock permeability may be difficult to establish.

REFERENCES

Al-Sinawi, S., 1968, An investigation of body wave velocities, attenuation on elastic parameters of rocks subjected to pressure at room temperature: Ph.D. thesis, St. Louis University.

Attewell, P.B., and Ramana, Y.V., 1966, Wave attenuation and internal friction as functions of frequency in rocks: Geophysics, v. 31, p. 1049–1056.

Birch, F., and Bancroft, D., 1938, Elasticity and internal friction in a long column of granite: SSA Bull., v. 28, p. 243–254.

Born, W.T., 1941, Attenuation constant of earth materials: Geophysics, v. 6, p. 132–148.

Bradley, J.J., and Fort, A.N., Jr., 1966, Internal friction in rocks, *in* Handbook of physical constants: S.P. Clark, Jr., Ed., GSA Publ., p. 175–193.

Brennan, B.J., and Stacey, F.D., 1977, Frequency dependence of elasticity of rock—Test of seismic velocity dispersion: Nature, v. 268, p. 220–222.

Clark, V.A., Spencer, T.W., Tittmann, B.R., Ahlberg, L.A., and Coombe, L.T., 1980, Effect of volatiles on attenuation (Q^{-1}) and velocity in sedimentary rocks: J. Geophys. Res., v. 85, p. 5190–5198.

Dainty, A.M., Goins, N.R., and Toksöz, M.N., 1976, Seismic investigation of the lunar interior, *in* Lunar science VII: Houston, Lunar Science Institute, p. 181–183.

DeVilbiss-Muñoz, J.W., 1980, Wave dispersion and absorption in partially saturated rocks: Ph.D. thesis, Stanford Univ.

Frisillo, A.L., and Stewart, T.J., 1980, Effect of partial gas/brine saturations on ultrasonic absorption in sandstone: J. Geophys. Res., v. 85, p. 5209–5211.

Gardner, G.H.F., Wyllie, M.R.J., and Droschak, D.M., 1964, Effects of pressure and fluid saturation on the attenuation of elastic waves in sands: J. Petr. Tech., p. 189–198.

Gordon, R.B., 1974, Mechanical relaxation spectrum of crystalline rock containing water: J. Geophys. Res., v. 79, p. 2129–2131.

Gordon, R.B., and Davis, L.A., 1968, Velocity and attenuation of seismic waves in imperfectly elastic rock: J. Geophys. Res., v. 73, p. 3917–3935.

Gordon, R.B., and Rader, D., 1971, Imperfect elasticity of rock: Its influence on the velocity of stress waves, *in* Geophysical monograph series, Vol. 14: J.G. Heacock, Ed., AGU, Washington.

Hamilton, E.L., 1972a, A correlation between Q_s of shear waves and rigidity: Mar. Geol., v. 13, p. M27–M30.

——— 1972b, Compressional wave attenuation in marine sediments: Geophysics, v. 37, p. 620–646.

Johnston, D.H., and Toksöz, M.N., 1980a, Ultrasonic P and S wave attenuation in dry and saturated rocks under pressure: J. Geophys. Res., v. 85, p. 925–936.

——— 1980b, Thermal cracking and amplitude dependent attenuation: J. Geophys. Res., v. 85, p. 937–942.

Kissell, F.N., 1972, Effect of temperature on internal friction in rocks: J. Geophys. Res., v. 77, p. 1420–1423.

Klima, K., Vanek, J., and Pros, Z., 1964, The attenuation of longitudinal waves in diabase and greywacke under pressures up to 4 kilobars: Studia Geoph. et Geod., v. 8, p. 247–254.

Knopoff, L., 1964, Q: Rev. Geophys., v. 2, p. 625–660.

Latham, G.V., Nakamura, Y., Lammlein, D., Dorman, J., and Duennebier, F., 1974, Structure and state of the lunar interior based upon seismic data (abst.), in Lunar science V: Houston, Lunar Sci. Inst., p. 434.

Levykin, A.I., 1965, Longitudinal and transverse wave absorption and velocity in rock specimens at multilateral pressures up to 4000 kg/cm^2: Izv., Physics of the Solid Earth, U.S.S.R. Acad. Sci., no. 1, p. 94–98.

Lockner, D., Walsh, J.B., and Byerlee, J., 1977, Changes in seismic velocity and attenuation during deformation of granite: J. Geophys. Res., v. 82, p. 5374–5378.

Mason, W.P., Beshers, D.N., and Kuo, J.T., 1970, Internal friction in Westerly granite: Relation to dislocation theory: J. Appl. Phys., v. 41, p. 5206–5209.

Mason, W.P., Marfurt, K.J., Beshers, D.N., and Kuo, J.T., 1978, Internal friction in rocks: J. Acoust. Soc. Am., v. 63, p. 1596–1603.

Merkulova, V.M., Pigulevskiy, E.D., and Tsaplev, V.M., 1972, Sound absorption measurements in uniaxially compressed rocks: Izv., Physics of the Solid Earth, U.S.S.R. Acad. Sci., no. 3, p. 166–167.

McKavanagh, B., and Stacey, F.D., 1974, Mechanical hysteresis in rocks at low strain amplitudes and seismic frequencies: Phys. Earth Planet. Int., v. 8, p. 246–250.

Nur, A., and Simmons, G., 1969, The effect of viscosity of a fluid phase on velocity in low porosity rocks: Earth Planet. Sci. Lett., v. 7, p. 99–108.

Nur, A., and Winkler, K., 1980, The role of friction and fluid flow in wave attenuation in rocks (abst.): Geophysics, v. 45, p. 591–592.

Pandit, B.I., and King, M.S., 1979, The variation of elastic wave velocities and quality factor of a sandstone with moisture content: Can. J. Earth Sci., v. 16, p. 2187–2195.

Pandit, B.I., and Savage, J.C., 1973, An experimental test of Lomnitz's theory of internal friction in rocks: J. Geophys. Res., v. 78, p. 6097–6099.

Pandit, B.I., and Tozer, D.C., 1970, Anomalous propagation of elastic energy within the moon: Nature, v. 226, p. 335.

Peselnick, L., and Outerbridge, W.F., 1961, Internal friction in shear and shear modulus of Solenhofen limestone over a frequency range of 10^7 cycles per second: J. Geophys. Res., v. 66, p. 581–588.

Peselnick, L., and Zietz, I., 1959, Internal friction of fine grained limestones at ultrasonic frequencies: Geophysics, v. 24, p. 285–296.

Spencer, J.W., Jr., 1979, Bulk and shear attenuation in Berea sandstone: The effects of pore fluids: J. Geophys. Res., v. 84, p. 7521–7523.

—— 1981, Stress relaxations at low frequencies in fluid saturated rocks: attenuation and modulus dispersion: J. Geophys. Res., v. 86, p. 1803–1812.

Spetzler, H., and Anderson, D.L., 1968, The effect of temperature and partial melting on velocity and attenuation in a simple binary system: J. Geophys. Res., v. 73, p. 6051–6060.

Stocker, R.L., and Gordon, R.B., 1975, Velocity and internal friction in partial melts: J. Geophys. Res., v. 80, p. 4828–4836.

Tittmann, B.R., 1977, Internal friction measurements and their implications in seismic Q structure models of the crust, in The Earth's crust: Geophys. monogr. 20, AGU, p. 197–213.

Tittmann, B.R., Ahlberg, L., and Curnow, J., 1976, Internal friction and velocity measurements, in Lunar Science Conf., 7th Proc., Geochim. et Cosmochim. Acta, suppl. 4, v. 3, p. 3123–3132.

Tittmann, B.R., Curnow, J.M., and Housley, R.M., 1975, Internal friction quality factor $Q \geq 3100$ achieved in lunar rock 70215,85, in Lunar Science Conf., 6th Proc., Geochim. et Cosmochim. Acta, suppl. 6, v. 3, p. 3217–3226.

Tittmann, B.R., Housley, R.M., Alers, G.A., and Cirlin, E.H., 1974, Internal friction in rocks and its relationship to volatiles on the moon, in Lunar Science Conf., 5th Proc., Geochim. et Cosmochim Acta, suppl. 5, v. 3, p. 2913–2918.

Tittmann, B.R., Housley, R.M., and Cirlin, E.H., 1973, Internal friction of rocks and volatiles on the moon, in Lunar Science Conf., 4th Proc., Geochim. et Cosmochim. Acta, suppl. 4, v. 3, p. 2631–2637.

Tittmann, B.R., Nadler, H., Clark, V.A., Ahlberg, L.A., and Spencer, T.W., 1981, Frequency dependence of seismic dissipation in saturated rocks: Geophys. Res. Lett., v. 8, p. 36–38.

Toksöz, M.N., Johnston, D.H., and Timur, A., 1979, Attenuation of seismic waves in dry and saturated rocks: I. Laboratory measurements: Geophysics, v. 44, p. 681–690.

Volarovich, M.P., and Gurevich, G.I., 1957, Investigation of dynamic moduli of elasticity for rocks in relation to temperature: Bull. Acad. Sci. USSR, Geophys., no. 4, p. 1–9.

Walsh, J.B., Brace, W.F., and Wawersik, W.R., 1970, Attenuation of stress waves in Cedar City quartz diorite: Air Force Weapons Lab. Tech. rep. No. AFWL-TR-70-8.

Warren, N., Trice, R., and Stephens, J., 1974, Ultrasonic attenuation, Q measurements on 70215,29: Geochim. et Cosmochim. Acta, suppl. 5, p. 2927–2938.

Winkler, K., and Nur, A., 1979, Pore fluids and seismic attenuation in rocks: Geophys. Res. Lett., v. 6, p. 1–4.

Winkler, K., Nur, A., and Gladwin, M., 1979, Friction and seismic attenuation in rocks: Nature, v. 227, p. 528–531.

Wyllie, M.R.J., Gardner, G.H.F., and Gregory, A.R., 1962, Studies of elastic wave attenuation in porous media: Geophysics, v. 27, p. 569–589.

Chapter 3
Attenuation mechanisms

The phenomenon of attenuation is complex. While elastic wave propagation is generally well understood, anelasticity is not. As observed from attenuation data, anelastic variations with changes in physical state are complicated and probably cannot be explained by a single model or mechanism. Therefore, in order to evaluate and interpret laboratory and field measurements reasonably, precise definitions of the possible attenuation mechanisms involved are needed.

The investigation of the behavior of Q in the earth has classically been approached along two lines. The first method is to explain the nature of attenuation in terms of a generalized equation of linear elasticity (Hooke's law) or by modified equations allowing certain nonlinearities. These phenomonological models (considered in chapter 5) have been well studied but yield little information concerning the microscopic properties of rock.

The second (mechanistic) approach uses the physical and mathematical description of possible attenuation mechanisms. Numerous mechanisms have been proposed and each may be considered to have a greater degree of importance to the overall attenuation under certain physical conditions. These mechanisms cover: matrix anelasticity, including frictional dissipation owing to relative motions at grain boundaries and across crack surfaces (*Walsh, 1966*); attenuation attributable to fluid flow, including relaxation owing to shear motions at pore-fluid boundaries (Walsh, 1968, 1969; Solomon, 1973); dissipation in a fully saturated rock because of relative motion of the frame with respect to fluid inclusions (Biot, 1956a, b; Stoll and Bryan, 1970); shearing "flow" of the fluid layer (Riesz, 1981) and "squirting" phenomena (Mavko and Nur, 1975; *O'Connell and Budiansky, 1977*); partial saturation effects such as gas-pocket squeezing (*White, 1975;* Dutta and Odé, 1979a, b; Dutta and Seriff, 1979); enhanced intracrack flow (*Mavko and Nur, 1979*); stress-induced diffusion of adsorbed volatiles (Tittmann et al, 1980); energy absorbed in systems undergoing phase changes (*Spetzler and Anderson, 1968*); and a large category of geometrical effects including scattering off small pores (Kuster and Toksöz, 1974) and large irregularities and selective reflection from thin beds (O'Doherty and Anstey, 1971; Spencer et al, 1977). The mechanistic approach is satisfying in that the physics of attenuation may be better understood. However, as will be seen, mathematical models based on these mechanisms suffer, many times, from excessive free parameters.

One of the earliest proposals for an attenuation mechanism in rocks, Coulomb frictional sliding between crack surfaces and grain boundary contacts, is still appealing for several reasons. First, it is a familiar physical process. Second, it provides a constant Q with frequency, observed for many dry rocks. Third, its dependence on contacts allows a simple understanding of the variation of attenuation with pressure. Fourth, an extension of dry friction to include boundary lubrication of cracks explains some (but not all) aspects of data observed in evacuated rocks exposed to small amounts of volatiles (Clark et al, 1980). In the first paper, *Walsh* (*1966*) analyzed this friction mechanism in terms of the contacts between the surfaces of very thin ellipsoidal cracks. Later, in the eighth paper, *Johnston et al* (*1979*) utilize the Walsh formulation to model pressure dependence for the attenuation of ultrasonic waves.

The validity of friction as a viable mechanism has recently come under question. Mavko's (1979) analysis of friction clearly demonstrates an inherent amplitude dependence that was only tacitly assumed by Walsh. Data concerning the amplitude dependence of attenuation discussed in the previous chapter (*Winkler et al, 1979*) imply that at the low strain amplitudes encountered in seismology or exploration, friction is not important. However, under certain laboratory conditions, namely high-amplitude ultrasonic experiments, and near seismic sources, it may be dominant. Finally, friction is a nonlinear mechanism. Observations of stress-strain hysteresis loops (chapter 2) are presently inconclusive as to the linearity of the dominant mechanism, but they may provide an important diagnostic for determining the relative importance of friction.

In the second paper, *Savage (1966)* proposes an alternative mechanism: thermoelastic attenuation in rocks containing cracks. Based on Zener's (1938) work, this model predicts a decrease in attenuation with increasing pressure. It also predicts an increase in Q for low frequencies. Recently, Armstrong (1980) has proposed a thermoelastic model for complex, randomly heterogeneous solids for which attenuation is essentially frequency independent. Kjartansson (1979) has considered the thermoelastic mechanism in terms of partially fluid-filled cracks in rock for which large changes in attenuation are predicted for small gas saturations. However, thermoelastic mechanisms produce a strong dependence on temperature. Unfortunately, experimental work in this regard is lacking.

As a further alternative to friction, *Mason (1969)*, in the third paper, discusses a nonlinear mechanism connected with the motion of dislocations. While the nonlinear mechanism produces a constant Q, later work (reviewed by Mason, 1971) suggests a peak in attenuation at about 1 MHz that is modeled by considering a combination of both linear and nonlinear dislocation mechanisms.

In a series of papers, Biot (1956a, b; 1962a, b) and Geertsma and Smit (1961) developed a theory for the dynamic response of a linear porous solid containing compressible fluid. While strictly a phenomenological model, the mechanism of inertial flow in rocks with attenuation resulting from motion of the pore fluid relative to the rock frame does fundamentally depend on bulk rock properties. Rather than including Biot's original work in this chapter, we have chosen a paper by *Stoll (1974)*. The model, first proposed by Stoll and Bryan (1970), is based on Biot's work but incorporates complex moduli to describe frame losses as well as fluid losses. Although it is generally concluded (White, 1965) that this type of mechanism produces negligible attenuation at low fre-

quencies in consolidated rocks, it may be important at ultrasonic frequencies (*Johnston et al, 1979*) or in permeable, unconsolidated sediments at intermediate frequencies.

In the fifth paper, *White (1975)* extends the bulk-flow model to include the effects of spherical gas pockets in an otherwise fluid-saturated porous rock. A subsequent paper (White et al, 1976) was also published. In this case, pressure differences at the fluid-gas interface may greatly enhance fluid flow, resulting in substantial attenuation even at seismic frequencies. Recently, a series of papers (Dutta and Odé, 1979a, b; Dutta and Seriff, 1979) in GEOPHYSICS developed a rigorous and exact theory of attenuation based on White's model and Biot's equations of motion. Analysis of the White model shows that the dissipation proposed is a result of the diffusion wave predicted by Biot and observed in the laboratory by Plona (1980). Also, the geometry of the gas pockets was extended to include alternating layers of gas- and liquid-filled rock and liquid-filled spheres surrounded by gas-filled rock.

While Biot-type mechanisms treat bulk fluid flow in porous rock, they ignore inter- and intra-crack flow (i.e., local flow), both of which may dissipate seismic energy. Walsh (1968, 1969) analyzed the effect of shear relaxation in isolated, thin penny-shaped cracks. For fluid viscosities encountered in exploration, such intracrack flow is extremely rapid and frequencies for which there is appreciable attenuation are far too high to be important. This mechanism has, however, been proposed for the earth's aesthenosphere.

Intercrack fluid flow, sometimes known as "squirt" flow, was first proposed as an attenuation mechanism by Mavko and Nur (1975); in the sixth paper, *O'Connell and Budiansky (1977)* systematically analyze the problem. The model, based on a fluid-filled solid containing a distribution of thin cracks, predicts four types of viscoelastic behavior including shear relaxation and squirting flow. Viscoelastic moduli are derived using a self-consistent approximation. Dissipation owing to intercrack flow is shown to be largest at frequencies ranging from 100 Hz to 1 MHz, depending upon the distribution of crack shapes assumed. In a recent paper, Budiansky and O'Connell (1980) present general results for heterogeneous materials based on the self-consistent method discussed in their 1977 paper. Mixing laws for attenuation are derived for polycrystals of anisotropic materials, composites of isotropic materials (e.g., dry porous solids), and fluid saturated porous solids with interconnected voids. It was found (for polycrystals and composites) that attenuation mechanisms that respond only to shear can result in bulk dissipation because of coupling between macroscopic volumetric strains and internal shear strains.

In the case of partial saturation, the presence of the gaseous phase allows fluid within a pore or crack to flow freely under pure compression. This type of intracrack flow is treated by *Mavko and Nur (1979)* in the seventh paper of this chapter. Attenuation for compressional waves is much greater than that for intercrack flow in the fully saturated case. Laboratory investigations (*Winkler and Nur, 1979;* Spencer, 1979; Frisillo and Stewart, 1980) confirm this behavior. Thus, it appears that attenuation may provide information independent of velocity with respect to the saturation condition of porous rocks.

Finally, in the last paper, *Johnston et al (1979)* model a data set of ultrasonic attenuation values in terms of several mechanisms including friction, Biot flow, shear relaxation, squirt flow, and scattering. The conclusions reached in that paper are strictly applicable only to the ultrasonic data; however, given the frequency dependencies of the mechanisms involved, the model may be extrapolated to lower frequencies.

Numerous other attenuation mechanisms have been proposed in the literature that are not included in this volume. One of the more recent is a stress-induced diffusion model based on an interaction between adsorbed layers of volatiles (notably water) and the solid surfaces of rock minerals in terms of thermally activated motions in the adsorbed film (Tittmann et al, 1980). At crack tips, opening and closing under compressional strains resulting from the passing seismic wave cause cyclic "hopping" (in and out of the crack tip) of the volatile molecules. At grain boundaries and asperities, shear strains cause a shearing of the film against interlayer forces. The activation energies for such processes correspond to bonding energies, depending upon the nature of the grain surfaces and the volatile coverage. Since it is presumed that a wide range of activation energies exist for the mechanism, Q will not depend appreciably on frequency. The mechanism is suggested to be dominant in air-dry rocks or in saturated rocks at frequencies low enough to eliminate intercrack or intracrack flow.

In partially saturated rocks, attenuation may be caused by gas bubble motions, losses being attributable to thermal, viscous, and radiation effects. A review of the physical properties of gas-bearing sediments may be found in Anderson (1974). Attenuation owing to gas bubbles may be important in ocean bottom sediments. Work is continuing in the fields of ocean acoustics and in biophysics, where gas in tissue can cause high attenuation of ultrasound.

Further work is being done on friction-type mechanisms and flow models as they relate to crack geometries. A crack contact model proposed by Walsh and Grosenbaugh (1979) that takes into account the actual physical dimensions of the pores and cracks holds a great deal of promise not only for understanding attenuation but also for a variety of other physical properties that depend upon crack connectivity. This model predicts the amplitude dependence of attenuation observed in laboratory measurements (Stewart et al, 1980).

In summary, the overall attenuation observed in upper crustal rocks can be caused by a number of mechanisms. The relative importance of each mechanism is a function of the physical conditions imposed upon the rock. Each mechanism depends on rock type, saturation state, pressure, frequency and amplitude of the acoustic wave, and other various rock properties. No one mechanism is responsible for the bulk of attenuation and, at present, no particular mechanism can be eliminated from consideration under all the conditions studied in the laboratory and field.

REFERENCES

Anderson, A.L., 1974, Acoustics of gas-bearing sediments: Appl. Res. Lab., Univ. of Texas at Austin, ARL-TR-74-19.

Armstrong, B.H., 1980, Frequency independent background internal friction in heterogeneous solids: Geophysics, v. 45, p. 1042–1054.

Biot, M.A., 1956a, Theory of propagation of elastic waves in a fluid-saturated porous solid. I. Low-frequency range: J. Acoust. Soc. Am., v. 28, p. 168–178.

———— 1956b, Theory of propagation of elastic waves in a fluid-saturated porous solid. II. High-frequency range: J. Acoust. Soc. Am., v. 28, p. 179–191.

———— 1962a, Mechanisms of deformation and acoustic propagation in porous media: J. Appl. Phys., v. 33, p. 1482–1498.

———— 1962b, Generalized theory of acoustic propagation in porous dissipative media, J. Acoust. Soc. Am., v. 34, p. 1254–1264.

Budiansky, B., and O'Connell, R.J., 1980, Bulk dissipation in heterogeneous media, *in* Solid earth geophysics and geotechnology: S. Nemat-Nasser, Ed., Appl. Mech. Div. Vol., ASME New York.

Clark, V.A., Spencer, T.W., Tittmann, B.R., Ahlberg, L.A., and Coombe, L.T., 1980, Effect of volatiles on attenuation (Q^{-1}) and velocity in sedimentary rocks: J. Geophys. Res., v. 85, p. 5190–5198.

Dutta, N.C., and Odé, H., 1979a, Attenuation and dispersion of compressional waves in fluid-filled rocks with partial gas saturation (White model)—Part I: Biot theory: Geophysics, v. 44, p. 1777–1788.

————— 1979b, Attenuation and dispersion of compressional waves in fluid-filled rocks with partial gas saturation (White model)—Part II: Results: Geophysics, v. 44, p. 1789–1805.

Dutta, N.C., and Seriff, A.J., 1979, On White's model of attenuation in rocks with partial gas saturation: Geophysics, v. 44, p. 1806–1812.

Frisillo, A.L., and Stewart, T.J., 1980, Effect of partial gas/brine saturations on ultrasonic absorption in sandstone: J. Geophys. Res., v. 85, p. 5209–5211.

Geertsma, J., and Smit, D.C., 1961, Some aspects of elastic wave propagation in fluid-saturated porous solids: Geophysics, v. 26, p. 169–181.

Johnston, D.H., Toksöz, M.N., and Timur, A., 1979, Attenuation of seismic waves in dry and saturated rocks: II: Mechanisms: Geophysics, v. 44, p. 691–711.

Kjartansson, E., 1979, Attenuation of seismic waves in rocks and applications in energy exploration: Ph.D. thesis, Stanford University.

Kuster, G.T., and Toksöz, M.N., 1974, Velocity and attenuation of seismic waves in two-phase media: Part I: Theoretical formulations: Geophysics, v. 39, p. 587–606.

Mason, W.P., 1969, Internal friction mechanism that produces an attenuation in the earth's crust proportional to frequency: J. Geophys. Res., v. 74, p. 4963–4966.

————— 1971, Internal friction at low frequencies due to dislocations: Applications to metals and rock mechanics, *in* Physical Acoustics, v. 8: W.P. Mason and R.N. Thurston, Eds., New York, Academic Press, p. 347–371.

Mavko, G.M., 1979, Frictional attenuation: An inherent amplitude dependence: J. Geophys. Res., v. 84, p. 4769–4776.

Mavko, G.M., and Nur, A., 1975, Melt squirt in the aesthenosphere: J. Geophys. Res., v. 80, p. 1444–1448.

————— 1979, Wave attenuation in partially saturated rocks: Geophysics, v. 44, p. 161–178.

O'Connell, R.J., and Budiansky, B., 1977, Viscoelastic properties of fluid saturated cracked solids: J. Geophys. Res., v. 82, p. 5719–5736.

O'Doherty, R.F., and Anstey, N.A., 1971, Reflections on amplitudes: Geophys. Prosp., v. 19, p. 430–458.

Plona, T.J., 1980, Observation of a second bulk compressional wave in a porous medium at ultrasonic frequencies: Appl. Phys. Lett., v. 36, p. 259–261.

Riesz, A.D., 1981, Seismic absorption due to induced fluid motions: submitted to J. Geophys. Res.

Savage, J., 1966, Thermoelastic attenuation of seismic waves by cracks: J. Geophys. Res., v. 71, p. 3929–3938.

Solomon, S.C., 1973, Shear wave attenuation and melting beneath the mid-Atlantic ridge: J. Geophys. Res., v. 78, p. 6044–6059.

Spencer, J.W., Jr., 1979, Bulk and shear attenuation in Berea sandstone: The effects of pore fluids: J. Geophys. Res., v. 84, p. 7521–7523.

Spencer, T.W., Edwards, C.M., and Sonnad, J.R., 1977, Seismic wave attenuation in non-resoluable cyclic stratification: Geophysics, v. 42, p. 939–949.

Spetzler, H., and Anderson, D.L., 1968, The effect of temperature and partial melting on velocity and attenuation in a simple binary system: J. Geophys. Res., v. 73, p. 6051–6060.

Stewart, R., Toksöz, M.N., and Timur, A., 1980, Strain dependent attenuation: Ultrasonic observations and mechanism analysis: Presented at the 50th Annual International SEG Meeting, November 18, in Houston.

Stoll, R.D., 1974, Acoustic waves in saturated sediments, *in* Physics of sound in marine sediments: L. Hampton, Ed., New York, Plenum Press.

Stoll, R.D., and Bryan, G.M., 1970, Wave attenuation in saturated sediments: J. Acous. Soc. Am., v. 47, p. 1440–1447.

Tittmann, B.R., Clark, V.A., Richardson, J., and Spencer, T.W., 1980, Possible mechanism for seismic attenuation in rocks containing small amounts of volatiles: J. Geophys. Res., v. 85, p. 5199–5208.

Walsh, J.B., 1966, Seismic wave attenuation in rock due to friction: J. Geophys. Res., v. 71, p. 2591–2599.

————— 1968, Attenuation in partially melted material: J. Geophys. Res., v. 73, p. 2209–2216.

————— 1969, New analysis of attenuation in partially melted rock: J. Geophys. Res., v. 74, p. 4333–4337.

Walsh, J.B., and Grosenbaugh, M.A., 1979, A new model for analyzing the effect of fractures on compressibility: J. Geophys. Res., v. 84, p. 3532–3536.

White, J.E., 1965, Seismic waves: Radiation, transmission, and attenuation: New York, McGraw-Hill Book Co., Inc.

————— 1975, Computed seismic speeds and attenuation in rocks with partial gas saturation: Geophysics, v. 40, p. 224–232.

White, J.E., Mikhaylova, N.G., and Lyakhovitskiy, F.M., 1976, Low-frequency seismic waves in fluid saturated layered rocks: Phys. Solid Earth, Trans. Izv, v. 11, p. 654–659.

Winkler, K., and Nur, A., 1979, Pore fluids and seismic attenuation in rocks: Geophys. Res. Lett., v. 6, p. 1–4.

Winkler, K., Nur, A., and Gladwin, M., 1979, Friction and seismic attenuation in rocks: Nature, v. 227, p. 528–531.

Zener, C., 1938, Internal friction in solids, 2, General theory of thermoelastic internal friction: Phys. Rev., v. 53, p. 90–99.

Reprinted from Journal of Geophysical Research, v. 71, p. 2591–2599.

Seismic Wave Attenuation in Rock Due to Friction

J. B. WALSH

Department of Geology and Geophysics
Massachusetts Institute of Technology, Cambridge
and Woods Hole Oceanographic Institution, Woods Hole, Massachusetts

The total attenuation in rock is probably due to a number of sources of dissipation. The mechanism proposed here as one source of attenuation is based on the frictional dissipation as crack surfaces in contact slide relative to one another during passage of a seismic wave. The attenuation due to friction at cracks is derived for dilatation and shear waves under the assumption that the cracks may be approximated as elliptical slits in plane strain. The results cannot be evaluated on an absolute basis; however, the ratio Q_a/Q_β of the quality factor for longitudinal waves to that for transverse waves predicted by the theory agrees reasonably well with published values (all of which equal approximately 0.5) found in laboratory experiments on granite and limestone.

INTRODUCTION

The attenuation of elastic waves in dry rock is found experimentally to be independent of frequency, over a large range of frequency (for a summary of data, see *Knopoff* [1964]). The process most likely responsible for this attenuation is friction [*Born*, 1941; *Förtsch*, 1957; *Knopoff and MacDonald*, 1960], for in the classical sense, frictional force, and therefore energy lost in friction, is independent of velocity. Another unique aspect of attenuation in geologic materials is the great difference in attenuation in rocks and in single crystals; for example, attenuation in two limestones is an order of magnitude greater than in calcite [*Peselnick and Zietz*, 1959]. Some feature of the grain structure apparently plays a vital role. One possibility is that grains or parts of grains shift relative to one another during passage of a sound wave. The energy lost through frictional forces developed during sliding could contribute to the attenuation.

Walsh [1965a] recently investigated a phenomenon closely related to attenuation. During a static loading-unloading cycle, the stress-strain curve of a rock in uniaxial compression often shows distinct hysteresis. The hysteresis was traced to friction at the surface of cracks, which are generally believed to be prevalent in crystalline rocks [*Birch and Bancroft*, 1938a; *Birch*, 1961]. To analyze the stress-strain behavior of such material, Walsh took as a model

an elastic solid containing cracks, which were idealized for analysis as elliptical slits. The effect of frictional sliding of one crack surface on another upon effective elastic properties was found in terms of various crack parameters, friction coefficient, and intrinsic elastic properties. In the present paper, this idea is carried one step further. Expressions for attenuation in a crack-filled solid are derived by analyzing the energy losses due to sliding at cracks.

In more detail, the proposed attenuating mechanism is as follows. Among the large number of cracks of all orientations and lengths in our model, some are open and others are closed at any given pressure [*Walsh*, 1965b]. As a compressional wave, for example, traverses the material, sliding of one crack face past the other occurs at cracks which have barely closed and which also have certain favorable orientations with respect to the direction of propagation. The motion is opposed by friction, and some of the elastic energy of the wave is dissipated. As the wave traverses the material, the normal stress between the crack faces increases, and thus the frictional shear stress increases. The relative motion between the crack faces continues, and the rate of dissipation increases until the maximum amplitude is reached. As the wave passes, the direction of the frictional shear stress is reversed, and again work must be done against friction as the crack returns to its equilibrium position.

It is interesting that this mechanism provides

the relationship between friction coefficient and displacement required by *Knopoff and MacDonald* [1960]. To give the proper frequency dependence for their attenuation model (a mass oscillating on a rough surface), they showed that the friction coefficient had to increase with distance from the equilibrium position. In the present model the same result is produced with a constant friction coefficient because, in effect, the normal force increases with displacement.

This mechanism cannot account for all the attenuation observed in dry rocks. Clearly, if the rock is put under sufficient confining pressure to close all cracks, an ordinary seismic wave will not have sufficient amplitude to cause relative motion between the crack faces, and no energy will be dissipated. This is not what is observed experimentally, however; *Birch and Bancroft* [1938a] found in experiments on granite at 4 kb confining pressure (which is high enough to close most cracks) that the attenuation, although sometimes lower by an order of magnitude than the value at atmospheric pressure, was not zero.

To accommodate this observation, we assume that attenuation has two parts: attenuation due to frictional losses at cracks and 'intrinsic' attenuation, i.e., the attenuation if the rock contained no cracks. For many rocks at low pressure, the intrinsic attenuation will probably be a rather small part of the total. The intrinsic attenuation is simply additive to that due to cracks and in the analysis will be taken as zero without loss of generality.

The purpose of the present analysis was primarily to explain the laboratory behavior of dry rock at frequencies below those at which scattering becomes important. The results should apply as well to very low frequency waves in shallow crustal rocks. Cracks in rock close completely under several kilobars confining pressure, so this mechanism is only possible in rocks to depths of a few tens of kilometers. Most crustal rocks also contain water, which, as the rock is deformed, is a source of attenuation due to viscous damping. The analysis should apply to long-period waves, however, where the shearing velocity in the water is small and the attenuation due to viscosity is insignificant.

ANALYSIS

For the purpose of analysis, rock is considered to be an isotropic, homogeneous medium which contains cracks of random orientation. The rock matrix is assumed to be perfectly elastic except that it is responsible for a small amount of attenuation, called the intrinsic attenuation. The cracks are very narrow, with width in the plane of the crack equal to length (as in Figure 2), and their behavior under stress is assumed for the moment to be given by the plane stress solution for cracks with elliptical cross section. The effect of using other analytical descriptions of crack behavior is discussed later. Cracks in actual rock have a spectrum of lengths, but here we shall assume that all cracks have the same length, called the average length. The relationship between this length and actual crack lengths could be defined rigorously, but this is of no immediate importance.

The wavelengths considered here are assumed to be much greater than the lengths of the cracks, and scattering can therefore be neglected. Finding the attenuation in such a rock model can be considered a problem in statics. The passage of a compressional pulse, for example, through a volume of material is equivalent to a compression test having been performed on that volume, and the attenuation depends upon the energy dissipated.

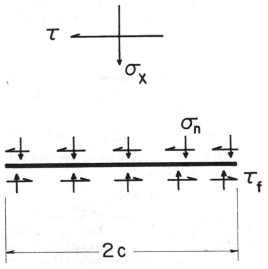

Fig. 1. Applied stress τ and σ_x and crack stress τ_f and σ_n for a closed crack.

Consider a crack of length $2c$ which has closed under some arbitrary stress state applied at infinity, as in Figure 1. In this analysis, compressive stresses will be assumed to be positive. The components of applied stress in which we are interested are the applied normal stress σ_z and the applied shear stress τ lying, respectively, perpendicular and parallel to the plane of the crack. In Figure 1, the normal stress σ_n and the frictional shear stress τ_f acting on the faces of the closed crack are uniform over the faces, as shown by *Berg* [1965]. When the applied stress at infinity is changed, relative motion between the faces of the cracks occurs only if the applied shear stress τ is greater than the frictional shear stress τ_f [see *Walsh*, 1965a]. If relative motion does occur, the energy dw dissipated by one crack due to an increase in applied stress is

$$dw = \tau_f \oint d\delta_f \, dA \qquad (1)$$

where $d\delta_f$, the displacement (relative to its original position) of a point on the crack surface, varies over the surface of the crack. Referring to analysis by *Walsh* [1965a], we find that the integral in (1) is given by

$$\oint d\delta_f \, dA = \frac{2\pi c^3}{E} d(\tau - \tau_f) \qquad (2)$$

where E is the intrinsic Young's modulus. The usual proportionality between frictional and normal forces in Coulomb's law is assumed to hold, so

$$\tau_f = \mu \sigma_n \qquad (3)$$

where μ is the friction coefficient. Combining (1), (2), and (3) results in an equation for the energy dissipated:

$$dw = (2\pi c^3/E)\mu \sigma_n \, d(\tau - \mu \sigma_n) \qquad (4)$$

provided that the crack is closed and that sliding occurs, i.e., that

$$|\tau| > |\mu \sigma_n| \qquad (5)$$

To evaluate (4), the value of σ_n must be related in some way to the applied stresses. First, we define σ_c as the value of σ_z which is required to close the crack. Once the crack is closed, any increase in the applied stress σ_z causes an equal increase in σ_n, the normal stress between the

crack surfaces; thus

$$\sigma_n = \sigma_z - \sigma_c \qquad (6)$$

The dissipation, from (4) and (6), is

$$dw = \frac{2\pi c^3}{E} \mu(\sigma_z - \sigma_c) \, d[\tau - \mu(\sigma_z - \sigma_c)] \qquad (7)$$

provided that (5) is valid, i.e., that

$$|\tau| > \mu |\sigma_z - \sigma_c| \qquad (8)$$

and provided that the crack is closed, i.e., if

$$\sigma_z > \sigma_c \qquad (9)$$

Otherwise, the dissipation is zero. The total dissipation is found by integrating (7) for each crack over the range of applied stress and summing the result over all the orientations and crack configurations which contribute to the dissipation. In practice, the calculations are very involved except for the simplest case, a longitudinal wave in which only a uniaxial stress is imposed on the sample.

Fortunately, a simplification of the problem is possible. At any initial state, some of the cracks are just closed. As the stress increases when the wave approaches, cracks close, and the total number of cracks contributing to the dissipation increases. In the limit, however, as the wave amplitude approaches zero, dissipation depends only upon the initial number. Thus, for low-amplitude seismic waves, we are concerned only with cracks for which the value of σ_c is zero, and (7), (8), and (9) become

$$dw = (2\pi c^3/E)\mu \sigma_z \, d(\tau - \mu \sigma_z) \qquad (10)$$

for

$$|\tau| > |\mu \sigma_z| \qquad (11)$$

and

$$\sigma_z > 0 \qquad (12)$$

The attenuation will be found from (10), (11), and (12) for seismic waves of several types: a longitudinal wave in a slender rod, a longitudinal wave in an infinite medium, and a transverse wave.

Longitudinal waves in slender rods. When a longitudinal wave passes down a slender rod, the material is subjected to a simple uniaxial compressive stress σ. The orientation of a crack is defined by the polar coordinates φ and θ of a normal to the plane of the crack, as in Figure 2. The stress σ is applied in direction 1. Stresses for this case are rotationally symmetric, so that only the φ coordinate is involved. Expressions for σ_x and τ for a crack of arbitrary orientation are found in a straightforward manner from Mohr's circle:

$$\tau = \sigma \sin \varphi \cos \varphi$$

$$\sigma_x = \sigma \cos^2 \varphi \qquad (13)$$

The dissipation due to a single crack, from (10) and (13), is

$$dw = (2\pi c^3/E)\mu \cos^2 \varphi$$
$$\cdot (\sin \varphi \cos \varphi - \mu \cos^2 \varphi)\sigma \, d\sigma \qquad (14)$$

if sliding takes place, i.e., from (11) and (13), if

$$\tan \varphi > \mu \qquad (15)$$

and if the crack is closed, i.e., from (12) and (13), if

$$\cos \varphi > 0 \qquad (16)$$

From (16) we see that cracks of all orientations tend to be closed by the applied stress but, from (15), that only cracks within a range of orientations contribute to the dissipation. Equa-

tion 14 is integrated to give the dissipation for a single crack.

$$w = (\pi\sigma^2 c^3/E)\mu \cos^3 \varphi(\sin \varphi - \mu \cos \varphi) \qquad (17)$$

The total dissipation W is found by summing (17) over the total number of cracks N.

$$W = (\pi\mu\sigma^2 c^3/E) \sum_N \cos^3 \varphi(\sin \varphi - \mu \cos \varphi) \qquad (18)$$

To evaluate (18), we think of the normals to the cracks as emanating from a common origin. The normals for a small number of cracks, ΔN, occupy the solid angle $\Delta\alpha$, where $\Delta\alpha/4\pi = \Delta N/N$. The angle of $\Delta\alpha$ is related to the polar coordinates by $\Delta\alpha = \sin\varphi\Delta\varphi\Delta\theta$. For very large N, the Δ quantities can be replaced by differentials and the summation in (18) by integration over all φ and θ for which sliding occurs:

$$W = 2\left(\frac{\pi\mu\sigma^2 c^3}{E}\right) \int_0^{2\pi} \int_{\tan^{-1}\mu}^{\pi/2} \cos^3 \varphi$$
$$\cdot \sin \varphi(\sin \varphi - \mu \cos \varphi) \frac{N}{4\pi} \, d\varphi \, d\theta$$

or

$$W = \left(\frac{\pi\sigma^2 N c^3}{15E}\right)\left[2\mu - \mu^2 \frac{3 + 2\mu^2}{(1 + \mu^2)^{3/2}}\right] \qquad (19)$$

The strain energy W' in a volume V at the maximum amplitude is $\sigma^2 V/2E_0$, where E_0 is the effective modulus of the rock. The specific attenuation factor Q (defined by $2\pi/Q = W/W'$) is given by

$$1/Q = (1/15)(E_0/E)(c^3/v)$$
$$\cdot \left[2\mu - \mu^2 \frac{3 + 2\mu^2}{(1 + \mu^2)^{3/2}}\right] \qquad (20)$$

where v is V/N, the average volume occupied by a single crack.

Longitudinal wave in unbounded medium. Lateral displacements must be zero for a plane longitudinal wave in an unbounded medium. The ratio of lateral strain to axial strain under uniaxial conditions is v_0, the effective Poisson's ratio. Therefore, in Figure 2, if the normal stress in direction 1 (the direction of propagation) is σ, lateral stresses of $v_0\sigma/(1 - v_0)$ must exist in directions 2 and 3. Again, from symmetry, only the φ coordinate of the crack orientation must be considered. From Mohr's circle,

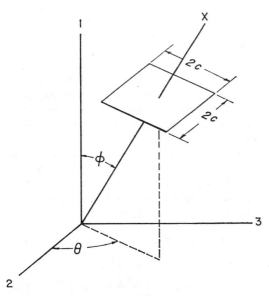

Fig. 2. Polar coordinates θ and φ for the normal x to the plane of the crack. The length and the width of the crack are $2c$.

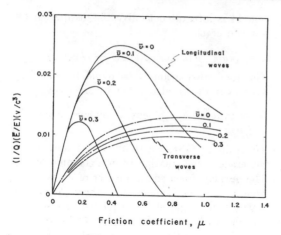

Fig. 3. Relationship between the specific attenuation factor Q, friction coefficient μ, and material constants for longitudinal and transverse waves. E_0/E is the ratio of effective and intrinsic Young's moduli, c^3/v depends upon the density of cracks just in contact, and $\bar{\nu}$, which corresponds to ν_0 in text, is the effective Poisson's ratio.

$$\tau = \sigma[(1 - 2\nu_0)/(1 - \nu_0)] \sin \varphi \cos \varphi \quad (21)$$
$$\sigma_x = \sigma[\cos^2 \varphi + \nu_0 \sin^2 \varphi/(1 - \nu_0)]$$

From (11), sliding takes place for cracks with orientation such that

$$(1 - 2\nu_0) \sin \varphi \cos \varphi/(1 - \nu_0)$$
$$> \mu[\cos^2 \varphi + \nu_0 \sin^2 \varphi/(1 - \nu_0)] \quad (22)$$

Solving (22) shows that dissipation occurs only for cracks with a limited range of orientations, the bounds of which are given by

$$\sin^2\varphi = [1/2(1 + \mu^2)]$$
$$\cdot \{1 + 2\mu^2(1 - \nu_0)/(1 - 2\nu_0)$$
$$\pm [1 - 4\mu^2\nu_0(1 - \nu_0)/(1 - 2\nu_0)^2]^{1/2}\} \quad (23)$$

where the plus or minus signs before the square root refer to upper and lower limits, respectively. Notice that the range of orientations over which dissipation takes place is zero when the expression under the square root is zero:

$$1 - 4\mu^2\nu_0(1 - \nu_0)/(1 - 2\nu_0)^2 > 0$$

or

$$\mu^2 < (1 - 2\nu_0)^2/4\nu_0(1 - \nu_0) \quad (24)$$

Thus, for any value of ν_0 greater than zero, dissipation is possible only for a limited range

of friction coefficients. As in the previous case, it is found that the restriction given by (12) does not further limit the range of orientations which contribute to the total dissipation. Equations 10 and 21 give

$$dw = (2\pi c^3/E)\mu\sigma \ d\sigma$$
$$\cdot [\cos^2 \varphi + \nu_0 \sin^2 \varphi/(1 - \nu_0)]$$
$$\cdot [(1 - 2\nu_0) \sin \varphi \cos \varphi/(1 - \nu_0)$$
$$- \mu(\cos^2 \varphi + \nu_0 \sin^2 \varphi/1 - \nu_0)] \quad (25)$$

Following the same procedure as before, we find that the total dissipation is

$$W = (\pi\sigma^2 Nc^3/E)[\mu(1 - 2\nu_0) \sin^3 \varphi/3(1 - \nu_0)$$
$$- \mu(1 - 2\nu_0)^2 \sin^5 \varphi/5(1 - \nu_0)^2$$
$$+ \mu^2\nu_0^2 \cos \varphi/(1 - \nu_0)^2$$
$$+ 2\mu^2\nu_0(1 - 2\nu_0) \cos^3 \varphi/3(1 - \nu_0)^2$$
$$+ \mu^2(1 - 2\nu_0)^2 \cos^5 \varphi/5(1 - \nu_0)^2] \quad (26)$$

Equation 26 is evaluated between the limits given in (23). The value of $1/Q$ is found as before, except here the maximum strain energy is given by $\sigma^2(1 - 2\nu_0^2/1 - \nu_0)V/2E_0$. The resulting expression for $1/Q$ is too long to be presented conveniently in closed form, and so it is given graphically in Figure 3 for several values of ν_0. Note that the previous case of a wave traveling down a slender rod is just a special case of the present solution for $\nu_0 = 0$.

Transverse wave. The transverse wave is more difficult to analyze because the stress system is no longer axially symmetric. In Figure 2, we let 3 be the direction of propagation; the shear stress applied as a transverse wave passes can be considered to be a compressive normal stress σ in the 1 direction and a tensile stress $-\sigma$ in the 2 direction. The expressions for σ_n and τ for a crack of arbitrary orientation depends upon both φ and θ and must be found by resolving components in three directions [see *Jaeger*, 1962, p. 18]:

$$\sigma_n = \sigma(\cos^2 \varphi - \cos^2 \theta \sin^2 \varphi) \quad (27)$$
$$\tau = \sigma[\cos^2 \varphi + \cos^2 \theta \sin^2 \varphi$$
$$- (\cos^2 \varphi - \cos^2 \theta \sin^2 \varphi)^2]^{1/2} \quad (28)$$

From (11), (27), and (28), sliding between crack faces is possible for orientations such that

$$\cos^2 \varphi + \cos^2 \theta \sin^2 \varphi$$
$$- (\cos^2 \varphi - \cos^2 \theta \sin^2 \varphi)^2$$
$$> \mu^2(\cos^2 \varphi - \cos^2 \theta \sin^2 \varphi)^2$$

or

$$(\cos^2 \varphi + \cos^2 \theta \sin^2 \varphi)/(\cos^2 \varphi$$
$$- \cos^2 \theta \sin^2 \varphi)^2 > 1 + \mu^2 \qquad (29)$$

For dissipation to take place, crack faces must be in contact or, from (12) and (27),

$$\cos^2 \varphi - \cos^2 \theta \sin^2 \varphi > 0$$

or

$$\operatorname{ctn} \varphi > \cos \theta \qquad (30)$$

The dissipation for a crack so oriented that (29) and (30) are valid is found from (10), (27), and (28). The total dissipation from cracks of all orientations is given by the double integral

$$W = (\sigma^2 N c^3/4E) \iint \mu \sigma_n(\varphi, \theta)$$
$$\cdot [\tau(\varphi, \theta) - \mu \sigma_n(\varphi, \theta)] \sin \varphi \, d\varphi \, d\theta \qquad (31)$$

where σ_n and τ are given by (27) and (28) and the bounds of integration are given by (29) and (30). A closed-form solution seems to be impossible because of the complicated nature of the integrand and the bounds; instead, the dissipation W was evaluated numerically by computer for a number of values of μ. Values of $1/Q$ for the transverse wave are plotted in Figure 3. For comparison with the other cases, the maximum strain energy for the transverse wave is expressed as $\sigma^2 V(1 + v_0)/E_0$, and consequently $1/Q$ is plotted in Figure 3 as a function of the effective Poisson's ratio v_0.

Discussion

In this analysis we have assumed that the attenuation in dry rock is composed of two parts. One part, the attenuation due to friction between crack faces sliding past one another as the rock deforms was analyzed in the preceding sections for dilatational and shear waves and is plotted in Figure 3. All other sources of dissipation are lumped together in the intrinsic attenuation, which might include losses due to thermoelastic effects, viscosity, and dislocation motion (for details, see *Mason* [1958]).

In Figure 3, the term c^3/v depends upon the density of cracks which are just in contact. Figure 3 cannot be used in computing attenuation for any particular case because we have no way of measuring this parameter. However, we can invert the problem to see if the density of cracks required for the attenuation is reasonable. The value of Q for rocks, to an order of magnitude, is 100 [*Knopoff*, 1964], and E_0/E is of the order of unity. Therefore, from Figure 3, the crack density necessary is of the order of unity or about one crack per grain. This value, which refers only to cracks that are barely closed, is much larger than seems reasonable intuitively.

A possible explanation is that rock is actually a much looser structure than the model indicates. Close correlation between electrical resistivity and crack porosity [*Brace et al.*, 1965] suggests that cracks in rock are in large part contiguous, whereas in the model they are not. In rock, therefore, the sliding displacement between contacting surfaces is greater than in the model, the dissipation per contacting surface is greater, and consequently the number of cracks does not need to be so large. If we had assumed in the derivations above that the cracks were penny-shaped or were elliptical slits in plane strain instead of plane stress, the resulting equations for $1/Q$ would be the same as those

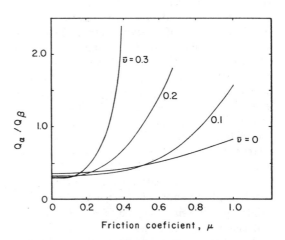

Fig. 4. The ratio Q_α/Q_β of the specific attenuation factor for longitudinal waves to that for transverse waves as a function of the friction coefficient μ. $\bar{v}$, which corresponds to v_0 in text, is the effective Poisson's ratio.

derived, except for a constant multiplier. If we could model the contiguous crack structure of rock analytically, the multiplier would presumably produce a more reasonable estimate of crack density.

Although the attenuation cannot be calculated on an absolute basis, the validity of the analysis can be checked by comparing the ratio of the specific attenuation factors for longitudinal and transverse waves with values available in the literature. For this purpose, the results in Figure 3 have been replotted in Figure 4, where Q_α/Q_β is the ratio of the specific attenuation factor of dilatational waves to that of shear waves. To compare the attenuation characteristics in Figure 4 with experiment, we need both the effective Poisson's ratio v_0 and the friction coefficient μ for the rock being considered. Values of Poisson's ratio for typical rocks are available, but it is somewhat more difficult to specify the friction coefficient. Should the value of μ for a marble, for example, be 0.1–0.2, such as is measured between smooth calcite single crystals [*Horn and Deere*, 1962], or a value near 0.7, such as *Jaeger* [1959] measured for sliding between fractured surfaces of marble? Some recent experiments (J. Byerlee, personal communication, 1965) suggest that the appropriate value for μ depends upon roughness and upon the displacements involved in the process. He found that the friction coefficient was about 0.1–0.2 for the first motion he could detect between highly polished surfaces of Westerly granite. The value of μ increased, evidently owing to the accumulation of wear particles, as the displacement increased during the course of the test. The initial value of μ was higher for surfaces ground with rougher grit; 0.5–0.6 was the highest value measured. The friction coefficient for sliding between smooth single crystals of quartz or feldspar, the two main constituents of granite, is found to be 0.1–0.2 [*Horn and Deere*, 1962]. The inference is thus that for small displacements the friction coefficient depends upon the single-crystal friction coefficients and the surface roughness.

The displacements involved in attenuation experiments on the laboratory scale are extremely small. Let us say the maximum strain encountered during passage of a stress wave in a typical experiment is 10^{-6}. The relative displacement between the faces of crack is of the order of the strain multiplied by the crack length; for a crack 1 mm long, for example, the displacement is only 10^{-7} cm. Clearly, in such experiments the friction coefficient is that associated with initial movement. In addition, the contacting surfaces probably should be considered smooth because only surface features which are much smaller than the displacements involved can be considered as roughness. Therefore, the value to be used is either that measured at the beginning of sliding between smooth rock surfaces or, equivalently, the value for sliding between smooth single crystals.

Granite is one rock type for which we have the necessary data in complete form. The effective Poisson's ratio for Quincy granite at low stress is about 0.1–0.14 [*Birch et al.*, 1942, p. 73]. The friction coefficient may be assumed to be 0.1–0.2, as discussed above; conceivably the value could be as high as 0.3, the value for biotite single crystals [*Horn and Deere*, 1962] if only cracks in the biotite flakes in the rock are involved. The value of Q_α/Q_β is found from Figure 4 to be between 0.3 and 0.5. *Birch and Bancroft* [1938a] found that Q_β for granite increased by an order of magnitude when the confining pressure was increased from 200 to 4000 kg/cm². For this case, then, the intrinsic attenuation is negligible in comparison with that due to cracks, and Q_α/Q_β in Figure 4 can be interpreted as the ratio of the total attenuation factors. *Birch and Bancroft* [1938b] measured Q_α and Q_β for Quincy granite, with an uncertainty which may have been as large as 20%. Their data give a value of about 0.5 for Q_α/Q_β. Thus agreement between theory and experiment is not unreasonable when we consider the approximate nature of the analysis and the uncertainty in measurements.

The ratio Q_α/Q_β from the attenuation measurements of Solenhofen and 'I-1' limestone [*Peselnick and Zietz*, 1959] is about 0.6 and 0.4, respectively. The value of μ for calcite sliding on calcite [*Horn and Deere*, 1962] is 0.1–0.2. Poisson's ratio for Solenhofen limestone from dynamic measurements is 0.25 [*Birch et al.*, 1942, p. 76]. Although v_0 is not known for the I-1 limestone, the result is insensitive to the value chosen, and, in Figure 4, we find that Q_a/Q_β is about 0.3 or 0.4. The experimental value is thus less than the experimental values but not unreasonably so, considering the over-all level of approximation.

As we have seen, in some experiments Q_a/Q_β is about 0.5. Let us consider the value of Q_a/Q_β which would be predicted by other mechanisms for attenuation which have been suggested (see *Mason* [1958]). Loss due to the Zener effect in polycrystalline materials is caused by the temperature increase which accompanies a decrease in volume. The change in volume is zero (or nearly zero) for shear waves but finite for dilatational waves. Thus the value of Q_a/Q_β predicted is zero (or at least very small). In models which involve dislocations, the energy loss should depend directly on the maximum shear stress imposed. The maximum applied shear stress for a transverse wave is larger than for a longitudinal wave. Therefore, the loss in shear is greater, and Q_a/Q_β should be larger than unity. A similar argument can be advanced to show that Q_a/Q_β should be greater than unity in viscous models also. *Anderson et al.* [1965] find for a linear model in which the imaginary part of the complex bulk modulus is zero that Q_a/Q_β is greater than 2. Thus the present theory seems to be unique in predicting a value of Q_a/Q_β which is near the measured value for the particular experiments cited.

The proposed attenuation model does not permit use of the complex modulus representation sometimes found in the literature (see, for example, *Peselnick and Zietz* [1959]). For instance, the magnitude of the imaginary part of the complex bulk modulus is a measure of the dissipation when the applied stress is pure hydrostatic pressure. In the present model, this case can be treated by considering a dilatational wave in a material with v_0 equal to 0.5. From (24) we see that the dissipation is zero, and therefore the imaginary part of the complex bulk modulus should be zero. On the other hand, the imaginary part of the complex bulk modulus calculated from the dilatational and shear components given in Figure 3 will be zero only for selected values of v_0 and μ. The apparent contradiction arises because the relationships between complex moduli generally derived are valid only for isotropic materials, whereas the model proposed here behaves anisotropically under stress.

It is of interest to consider the effect of confining pressure on Q_a/Q_β at pressures less than that required to close all the cracks. As cracks in the rock close owing to the pressure, Poisson's ratio increases [*Walsh*, 1965c]. The values of v_0 for granite, for example, increase from about 0.1 at atmospheric pressure to about 0.25 at a confining pressure of 4 kb [*Birch*, 1961]. On the other hand, the friction coefficient does not vary with confining pressure because attenuation is affected by only the cracks that are barely closed. In Figure 4 we see that if μ is about 0.1, Q_a/Q_β will be virtually independent of pressure, but if μ is greater than 0.3 or so, Q_a/Q_β should increase appreciably with increasing v_0. Values of μ greater than 0.3 for sliding between mineral single crystals are not uncommon (values for serpentine greater than 0.6 were measured by *Horn and Deere* [1962]); also μ for joints in the earth, where displacements are probably larger than those in laboratory experiments, may be higher than the single-crystal values used here. Therefore, the possibility of values of Q_a/Q_β greater than 0.5 cannot be ruled out.

Finally, let us investigate the effect of finite wave amplitude upon attenuation. As a wave of finite amplitude passes, cracks that were initially slightly open will close, and some of these will contribute to the dissipation. Thus the model predicts that the attenuation should increase with increasing wave amplitude. A crack which participates in the dissipation only over part of the cycle clearly is not so effective as one which contributes over the whole cycle. Therefore, the dissipation does not increase directly with number of cracks involved. The effect of finite amplitude upon the dissipation was derived for a dilatational wave with $v_0 = 0$. The derivation even for this simple case was rather lengthy and will not be presented. The result was that the dissipation was found to increase by about 20% for a wave with sufficient amplitude to double the number of cracks in contact. Thus a rather slight increase in attenuation with amplitude is predicted; *Peselnick and Outerbridge* [1961] found such behavior in their experiments on Solenhofen limestone.

Acknowledgments. W. F. Brace, Gene Simmons, J. C. Savage and M. Nafi Toksöz suggested changes in the manuscript which were incorporated in the final version. James D. Byerlee kindly allowed the use of unpublished experimental results. T. R. Madden and the Computation Center at the Massachusetts Institute of Technology, Cambridge, Massachusetts, helped with the numerical integration.

The research reported in this document was sponsored by The Air Force Cambridge Research Laboratories, Office of Aerospace Research, United States Air Force, Bedford, Massachusetts, under contract AF 19(628)-3298.

REFERENCES

Anderson, D. L., A. Ben-Menahem, and C. B. Archambeau, Attenuation of seismic energy in the upper mantle, *J. Geophys. Res., 70*(6), 1441–1448, 1965.

Berg, C. A., Deformation of fine cracks under high pressure and shear, *J. Geophys. Res., 70*(14), 3447–3452, 1965.

Birch, F., The velocity of compressional waves in rocks to 10 kilobars, 2, *J. Geophys. Res., 66*(7), 2199–2224, 1961.

Birch, F., and D. Bancroft, The effect of pressure on the rigidity of rocks, 1, *J. Geol. 46*, 59–87, 1938a.

Birch, F., and D. Bancroft, Elasticity and internal friction in a long column of granite, *Bull. Seismol. Soc. Am., 28*, 243–254, 1938b.

Birch, F., J. F. Schairer, and H. C. Spicer, *Handbook of Physical Constants*, Geol. Soc. Am. Spec. Paper 36, 325 pp., 1942.

Born, W. T., The attenuation constant of earth materials, *Geophysics, 6*, 132–148, 1941.

Brace, W. F., A. S. Orange, and T. M. Madden, The effect of pressure on the electrical resistivity of water-saturated crystalline rocks, *J. Geophys. Res., 70*(22), 5669–5678, 1965.

Förtsch, O., Die Ursachen der Absorption elastischer Wellen, *Ann. Geofis., 9*, 469–524, 1965.

Horn, H. M., and D. U. Deere, Frictional characteristics of minerals, *Geotechnique, 12*, 319–335, 1962.

Jaeger, J. C., The frictional properties of joints in rocks, *Geofis. Pura Appl., 43*, 148–158, 1959.

Jaeger, J. C., *Elasticity, Fracture and Flow*, John Wiley & Sons, New York, 1962.

Knopoff, L., Q, *Rev. Geophys., 2*(4), 625–660, 1964.

Knopoff, L., and G. J. F. MacDonald, Models for acoustic loss in solids, *J. Geophys. Res., 65*(7), 2191–2197, 1960.

Mason, W. P., *Physical Acoustics and the Properties of Solids*, D. Van Nostrand Company, Princeton, N. J., 1958.

McDonal, F. J., F. A. Angona, R. L. Mills, R. L. Sengbush, R. G. Van Nostrand, and J. E. White, Attenuation of shear and compressional waves in Pierre shale, *Geophysics, 23*, 421–439, 1958.

Peselnick, L., and W. F. Outerbridge, Internal friction in shear and shear modulus of Solenhofen limestone over a frequency range of 10^7 cycles per second, *J. Geophys. Res., 66*(2), 581–588, 1961.

Peselnick, L., and I. Zietz, Internal friction of fine-grained limestones at ultrasonic frequencies, *Geophysics, 24*(2), 285–296, 1959.

Walsh, J. B., The effect of cracks on the uniaxial elastic compression of rocks, *J. Geophys. Res., 70*(2), 399–411, 1965a.

Walsh, J. B., The effect of cracks on the compressibility of rock, *J. Geophys. Res., 70*(2), 381–389, 1965b.

Walsh, J. B., The effect of cracks in rock on Poisson's ratio, *J. Geophys. Res., 70*(20), 1965c.

(Manuscript received December 27, 1965.)

Reprinted from Journal of Geophysical Research, v. 71, p. 3929–3938.

Thermoelastic Attenuation of Elastic Waves by Cracks

J. C. SAVAGE

Geophysics Laboratory, University of Toronto
Toronto, Ontario, Canada

Zener's theory of thermoelastic attenuation by inhomogeneities has been applied to a medium containing cracks. The cracks have been approximated by elliptic cylinders, and energy losses per wave cycle have been calculated for both two-dimensional hydrostatic pressure and shear. Absorption spectrums vary less than an order of magnitude over a frequency range greater than 10^4 hz. The values of Q for both longitudinal waves (Q_a) and transverse waves (Q_β) for a distribution of such cracks have been derived in terms of parameters measured in static testing. For granite the theoretical values ($Q_a = 200$, $Q_\beta = 350$) are about twice the experimental values; however, the difference is not significant in view of the rather large uncertainties in some of the parameters. The theory also predicts that Q_a/Q_β should be near ½ for most rocks; laboratory observations yield about the same value. Finally, the theory predicts a temperature dependence for Q which has not been observed.

INTRODUCTION

There appears to be no general agreement on the mechanism which accounts for the attenuation of elastic waves in solids at frequencies below 1 Mhz. Several reviews of the pertinent data and possible theories are available (e.g., *Knopoff* [1964]). The experimental observation which has attracted the greatest attention appears to be the general result that the internal friction $1/Q$ (defined as $\Delta E/2\pi E$, where ΔE is the energy loss per unit volume per wave cycle and E is the energy density) remains the same order of magnitude over a frequency range as large as 1 to 10^6 hz. This observation has suggested to *Knopoff and MacDonald* [1958] that the attenuation mechanism is nonlinear. *Savage* [1965] noted that measurements of the attenuation of seismic pulses agreed with the attenuation measured in sinusoidal waves. This implied that the principle of superposition was applicable to the attenuation mechanism, and, as a consequence, the attenuation mechanism must be linear. More convincing evidence has been produced by *Wuenschel* [1965], who showed that the dispersion which should accompany attenuation in a linear theory was indeed present in both Pierre shale and Plexiglas. Thus there appears to be solid experimental evidence that the attenuation mechanism is linear.

Zener [1938] showed that the thermoelastic effects in solids could account for measurable losses, and he demonstrated convincingly that

at least some of the attenuation in metals was produced by thermoelastic losses associated with elastic anistopy of individual crystallites. Zener also discussed the thermoelastic losses associated with cavities, imperfections, and inhomogenieties in the medium. In the present paper, only the effect of cavities is considered, and particular attention is directed toward long flat cavities which can represent cracks in the medium. Except for a consideration of attenuation by spherical pores, the calculations presented here are two-dimensional, i.e., cracks are represented by elliptic-cylinder cavities and the states of stress restricted to those producing plane strain.

The attenuation is calculated for the two basic stress systems, dilatation and shear. Let Q for pure dilatation be denoted by Q_κ and for pure shear by Q_μ. Since the elastic energy may be resolved unambiguously into a dilatational part $\frac{1}{2}\kappa e^2$ (where e is the dilatation and κ the bulk modulus) and a shear part $\mu e'_{ij} e'_{ij}$ (where e'_{ij} is the strain deviator and μ the rigidity), we can calculate Q for a process involving both dilatation and shear from

$$\frac{1}{Q} = \frac{Q_\kappa^{-1}\kappa e^2/2 + Q_\mu^{-1}\mu e'_{ij}e'_{ij}}{\kappa e^2/2 + \mu e'_{ij}e'_{ij}} \quad (1)$$

An equivalent procedure would be to employ the concept of complex elastic moduli as *Zener* [1948] has done in his treatment of thermoelastic attenuation. Then the formalism employed by *Press and Healy* [1957] can be used.

149

The two approaches are, of course, equivalent. The complex moduli theory is preferable because it also includes the attendant dispersion.

Finally, it should be noted that in all calculations the cracks are assumed to be separated so far from one another that neither elastic nor thermal interaction occurs. It is quite probable that for rocks this assumption is not fulfilled, and for this reason it seems likely that this theory may underestimate the thermoelastic attenuation.

THEORY

We consider an elastic wave of angular frequency ω propagating through an elastic body which contains a small cavity. The wavelength of the disturbance is assumed to be much greater than the dimensions of the cavity. Thus, not only may scattering of the wave be neglected, but also a quasi-static approximation may be employed to represent the elastic field near the cavity; i.e., the strain field may be represented by the static solution e_{ij} multiplied by $\exp(-i\omega t)$. As is the usual custom, the time-dependent factor $\exp(-i\omega t)$ will not be written explicitly in what follows.

Let $e_{ij}{}^A$ be the strain which would be induced in a uniform elastic body subject to the specified stress system. Then $e_{ij}{}^A + e_{ij}{}^c$ will be the strain in the body containing the cavity when subjected to the same stress. The strain perturbation associated with the cavity is obviously $e_{ij}{}^c$. (The notation employed here is that used by *Eshelby* [1957] in studying a more general problem. We shall have occasion to use Eshelby's results frequently.) It will be assumed that the volume of the over-all sample is sufficiently large compared with the cavity for the variable part of $e_{ij}{}^c$ to approach zero near the outer boundary of the elastic medium.

Owing to the thermoelastic effect, e^c, the dilatational part of the perturbation strain field, will induce a temperature fluctuation ΔT in the medium. This fluctuation will be governed by the inhomogeneous diffusion equation

$$\partial(\Delta T)/\partial t = D\nabla^2(\Delta T) - \Gamma T_0(\partial e^C/\partial t) \qquad (2)$$

where D is the thermal diffusivity, Γ is Grüneisen's ratio ($\Gamma = \alpha\kappa/\rho c$, where α is the coefficient of thermal expansion, ρ the density, c the specific heat, and κ the bulk modulus), and T_0 the mean local temperature of the medium. There may be an additional thermal fluctuation associated with the dilatational part of $e_{ij}{}^A$; that fluctuation will be essentially uniform over large distances, and thus it will not significantly contribute to the thermoelastic losses. Since e^c is a harmonic function (i.e., $\nabla^2 e^c = 0$) a particular solution of (2) is $\Delta T = -\Gamma T_0 e^c$. Thus the general solution is

$$\Delta T = (\Delta T)_h - \Gamma T_0 e^c$$

where $(\Delta T)_h$ is a solution of the homogeneous part of (2). The sum ΔT must satisfy the condition of no heat flow across the cavity boundary; i.e.

$$\partial(\Delta T)_h/\partial n = \Gamma T_0(\partial e^c/\partial n) \qquad (3)$$

(where n is the normal to the cavity boundary) at the boundary, and also the variable part of ΔT must vanish at large distances from the cavity. Thus finding $(\Delta T)_h$ is no more than a straightforward exercise involving the diffusion equation.

The entropy gain can be calculated by the method of *Zener* [1938]. Let ΔS be the net entropy created per unit volume per wave period $(2\pi/\omega)$ by the conduction of heat from regions of strain concentration. Then

$$\Delta S = \rho c \int_0^{2\pi/\omega} (D/T)\nabla^2 T \, dt$$

We can replace $D\nabla^2 T$ by its equivalent from (2) and write $T = T_0 + \Delta T$. It is then easily shown that for $\Delta T \ll T_0$

$$T_0\Delta S = -\rho c\Gamma \int_0^{2\pi/\omega} \Delta T(\partial e^c/\partial t) \, dt$$

since both first-order integrals and the other second-order integral all vanish when integrated over a complete wave period. Since the quantities involved depend upon t as $\exp(-i\omega t)$, we can write

$$T_0\Delta S = -(\pi\rho c\Gamma/\omega) \, \mathrm{Re} \, \{(\Delta T)_h(\partial e^c/\partial t)^*\}$$

where Re indicates the real part of the following quantity and the asterisk indicates the complex conjugate. It follows that the mechanical energy which becomes unavailable in the entire volume in one wave cycle is

$$\Delta E = -\frac{\pi\rho c\Gamma}{\omega} \, \mathrm{Re} \int (\Delta T)_h \left(\frac{\partial e^c}{\partial t}\right)^* \, dv$$

If we note that the integrand may just as well be written $(\partial(\Delta T)_h/\partial t)$ $(-e^c)^*$ and replace the time derivative by its equivalent $D\nabla^2(\Delta T)_h$, we have

$$\Delta E = \frac{\pi\rho c\Gamma D}{\omega} \operatorname{Re} \int \nabla^2(\Delta T)_h (e^c)^* \, dv$$

We can subtract $(\Delta T_h)\nabla^2(e_c)^*$ from the integrand because this term is identically zero (e^c is a harmonic function). The volume integral may then be reduced to an integral over the surface of the cavity

$$\Delta E = \frac{\pi\rho c\Gamma D}{\omega} \operatorname{Re} \int \Delta T \left(\frac{\partial e^c}{\partial n}\right)^* dS \qquad (4)$$

where n is the normal outward from the cavity and dS is an element of the cavity surface.

Spherical Cavity

An elementary example of the theory concerns the thermoelastic losses associated with a spherical cavity. This problem has already been treated by *Zener* [1938], but the relations found here differ somewhat from his results. It may be shown either from the equations given by *Eshelby* [1957] or *Love* [1944] that an applied strain $e_{12}{}^A$ produces a dilatation about a spherical cavity of radius a given by

$$e^c = -15\frac{1 - 2\sigma}{7 - 5\sigma} e_{12}{}^A \left(\frac{a}{r}\right)^3 \sin^2\theta \sin 2\phi \qquad (5)$$

where σ is Poisson's constant for the elastic medium and r, θ, and ϕ are the spherical coordinates with origin at the center of the cavity. The polar angle θ is measured from the z axis and the azimuthal angle ϕ from the x axis (Figure 1). Notice that (5) differs by a factor of $6a/r$ from the expression used by *Zener* [1938]; it appears that Zener's expression is in error. The appropriate solution of the homogeneous diffusion equation is

$$(\Delta T)_h = Bh_2{}^{(1)}(\beta r/a) \sin^2\theta \sin 2\phi \qquad (6)$$

where

$$\beta^2 = i\omega a^2/D \qquad (7)$$

and $h_2{}^{(1)}(\beta)$ is the spherical Hankel function of the first kind, order 2, argument β. The constant B may be evaluated from the boundary condi-

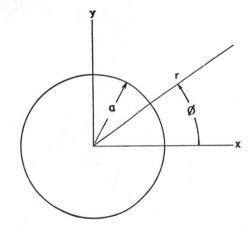

Fig. 1. Position of coordinate system relative to cavity for spherical or circular-cylinder cavities.

tion (3). Then from (4) it is found that

$$\Delta E = (20a\rho c\, DT_0/\omega)$$
$$\cdot [6\pi\Gamma e_{12}{}^A(1 - 2\sigma)/(7 - 5\sigma)]^2$$
$$\cdot \operatorname{Re}\{3h_2{}^{(1)}(\beta)/(\beta h_2{}^{(1)}(\beta)) + 1\} \qquad (8)$$

Let us now suppose that the number of spherical cavities per unit volume is N. Then the energy dissipated per unit volume is $N\Delta E$. The energy density for the wave is $E = 2\mu'(e_{12}{}^A)^2$, where μ' is the effective rigidity of an elastic medium with N spherical pores of radius a per unit volume. From *Eshelby* [1957, p. 390]

$$\mu' = \mu/[1 + 15(1 - \sigma)\nu/(7 - 5\sigma)] \qquad (9)$$

where $\nu = 4\pi a^3 N/3$ is the porosity of medium. Thus Q for the medium will be given by

$$Q_\mu{}^{-1} = (16/3)(\mu/\mu')p(\sigma)\nu\alpha\Gamma T_0$$
$$\cdot (1 - 2\sigma)(1 + \sigma)F_s(\omega) \qquad (10)$$

where

$$p(\sigma) = (135/4)(7 - 5\sigma)^{-2}$$

$$F_s(\omega) = \chi^2(2\chi^2 + 5\chi + 4)/[(2\chi^3 - 9\chi - 9)^2$$
$$+ \chi^2(2\chi^2 + 8\chi + 9)^2] \qquad (11)$$

$$\chi^2 = \omega a^2/2D$$

Notice that for the usual values of σ, $p(\sigma)$ is about 1 (e.g., for $\sigma = 0.25$ and 0.3, $p(\sigma) = 1.02$ and 1.18, respectively). A graph of $F_s(\omega)$ as a function of $\omega a^2/D$ is given in Figure 2. It

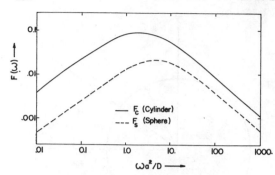

Fig. 2. Thermoelastic absorption spectrums for spherical and circular-cylinder cavities.

is worth noting that the porosity ν might be obtained from a measurement of κ', the effective bulk modulus of the porous medium. The relation is [*Eshelby*, 1957, p. 390]

$$\kappa' = \kappa/[1 + 3(1 - \sigma)\nu/(2 - 4\sigma)] \qquad (12)$$

A more direct measurement is, of course, feasible for spherical pores; however, we shall find this method of estimating ν quite useful in the case of flat cracks.

We now turn to a calculation of the energy loss in pure dilatation. From the equations of *Eshelby* [1957] or *Love* [1944] we can deduce that an applied dilatation e^A produces a constant perturbation dilatation

$$e^C = -e^A\nu \qquad (13)$$

Thus $\partial e^C/\partial r = 0$, and, from (4), $\Delta E = 0$; hence $Q_\kappa^{-1} = 0$. This result was noted by *Zener* [1938, p. 96]. It follows from (1) that the ratio of Q for longitudinal waves to that for transverse waves is

$$Q_\alpha/Q_\beta = 3(1 - \sigma')/2(1 - 2\sigma')$$

where σ' is the effective Poisson ratio for the porous medium. This ratio is always greater than 3/2.

CIRCULAR-CYLINDER CAVITY

It may be shown from the equations given by *Eshelby* [1957] or *Timoshenko and Goodier* [1951] that an applied strain e_{12}^A produces a dilatation about a circular cylinder cavity (radius a) which is given by

$$e^C = -4(1 - 2\sigma)e_{12}^A(a/r)^2 \sin 2\phi \qquad 14)$$

where r, ϕ, and z are the usual circular-cylinder coordinates. The z axis coincides with the axis of the cylinder, and ϕ is measured from the x axis (Figure 1). We have, of course, assumed that plane strain obtains. The appropriate solution of the homogeneous diffusion equation is

$$(\Delta T)_h = BH_2^{(1)}(\beta r/a) \sin 2\phi \qquad (15)$$

where $H_2^{(1)}(\beta)$ is a cylindrical Hankel function. The constant B can then be evaluated from the boundary condition (3). Then from (4) it is found that

$$\Delta E = (32pc\,DT_0/\omega)[\pi\Gamma(1 - 2\sigma)e_{12}^A]^2$$
$$\cdot \mathrm{Re}\,\{H_1^{(1)}(\beta)/H_2^{(1)'}(\beta)\} \qquad (16)$$

$H_2^{(1)'}(\beta)$ is the derivative of $H_2^{(1)}(\beta)$ with respect to β (β was defined in (7)).

For an elastic medium with a density of N such cylindrical holes per unit area, the energy dissipated per unit volume is $N\Delta E$. The energy density of the wave is $2\mu'(e_{12}^A)^2$, where $\mu' = \mu/[1 - 4\nu(1 - \sigma)]$ is the effective rigidity of the porous medium. Thus Q for the medium is given by

$$Q_\mu^{-1} = (16/3)(\mu/\mu')\nu\alpha\Gamma T_0$$
$$\cdot (1 - 2\sigma)(1 + \sigma)F_c(\omega) \qquad (17)$$

where

$$F_c(\omega) = |\beta|^{-2}\,\mathrm{Re}\,\{H_1^{(1)}(\beta)/H_2^{(1)'}(\beta)\} \qquad (18)$$

A plot of $F_c(\omega)$ as a function of $\omega a^2/D$ is given in Figure 2.

Let us now turn to the case of two-dimensional hydrostatic strain (i.e., $e_{11} = e_{22}$, $e_{33} = 0$, $e_{ij} = 0$ if $i \neq j$). From *Timoshenko and Goodier* [1951] it is found that the perturbation dilatation is a constant,

$$e^C = 2(1 + \sigma)e^A\nu/3$$

Thus once again $\Delta E = 0$, and hence $Q_\kappa^{-1} = 0$. It follows from (1) that the ratio $Q_\alpha/Q_\beta = 3(1 - \sigma')/2(1 - 2\sigma')$, a ratio which is always greater than 3/2.

We notice that (10) and (17) are of the same form. The extra factor $p(\sigma)$ which occurs in (10) is very closely equal to 1 for reasonable values of σ. The difference between the two internal frictions consists of the difference between $F_s(\omega)$ and $F_c(\omega)$ (Figure 2). The cylinder is a more efficient absorber by a factor of about 4. However, the two calculations are not strictly comparable because the cylinder calculation is intrinsically two-dimensional. All of the cylindrical cavities are oriented with their axes perpendicular to the plane of strain. In a three-dimensional body with cylindrical cavities of random orientation, only about ⅓ (i.e., those having their axes roughly perpendicular to the plane of strain) of the cylinders would produce attenuation as described in the calculation. Thus the effective Q would be multiplied by 3, and the values of Q for bodies having either cylindrical or spherical pores would be quite comparable.

ELLIPTIC CYLINDER CAVITY

Let us consider an applied shear $e_{12}{}^A$ and a state of plane strain in a medium containing an elliptic cylinder cavity in which the cylinder axis coincides with the z axis of the coordinate system and the major axis of the elliptical cross section is parallel to the x axis of the coordinate system (Figure 3). The dilatation in elliptic cylinder coordinates may be found from the equations given by *Timoshenko and Goodier* [1951]:

$$e^C = -\frac{2(1 - 2\sigma)e_{12}{}^A e^{2\xi_0} \sin 2\eta}{\cosh 2\xi - \cos 2\eta} \quad (19)$$

where $\xi = \xi_0$ is the equation of the surface of the ellipse and the elliptic cylinder coordinates ξ, η are related to the x, y coordinates by

$$x = h \cosh \xi \cos \eta \quad (20)$$
$$y = h \sinh \xi \sin \eta$$

Here $2h$ is the interfocal distance for the elliptic cross section. The appropriate solution of the

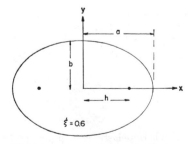

Fig. 3. Position of coordinate system relative to cavity for an elliptic-cylinder cavity.

homogeneous diffusion equation is

$$(\Delta T)_h = \Gamma T_0 \sum_{n=1}^{\infty} C_{2n}$$
$$\cdot \mathrm{N} e_{2n}{}^{(1)} (\xi, q)\, se_{2n} (\eta, q) \quad (21)$$

where

$$C_{2n} = \frac{8(1 - 2\sigma)e_{12}{}^A e^{2\xi_0} E_{2n}}{\mathrm{N} e_{2n}{}^{(1)\prime}(\xi_{0,q})} \quad (22)$$

$$E_{2n} = \sum_{m=1}^{\infty} m B_{2m}{}^{(2n)} e^{-2m\xi_0} \quad (23)$$

$$q = i\omega h^2/4D$$

$se_{2m} (\eta, q)$ is the odd Mathieu function of the first kind, order $2n$, $\mathrm{N} e_{2n}{}^{(1)} (\xi, q)$ is the odd modification Mathieu function of the third kind, first type, order $2n$, and the prime indicates differentiation with respect to the argument ξ_0. The coefficients $B_{2m}{}^{(2n)}(q)$ are the odd Mathieu coefficients of order $2n$ normalized so that

$$\sum_{m=1}^{\infty} [B_{2m}{}^{(2n)}]^2 = 1$$

All the notation for Mathieu functions and elliptic cylinder coordinates in this paper is the same as that given by *McLachlan* [1947].

The energy loss per wave cycle is then found from (4) to be

$$\Delta E = 8\rho c T_0 [\pi \Gamma (1 - 2\sigma)e_{12}{}^A h]^2 F_e{}^{(\mu)}(\omega) \quad (24)$$

where

$$F_e{}^{(\mu)}(\omega) = |q|^{-1}$$
$$\cdot \sum_{n=1}^{\infty} \mathrm{Re} \left\{ \frac{2\mathrm{N} e_{2n}{}^{(1)}(\xi_0, q) E_{2n}{}^2}{\mathrm{N} e_{2n}{}^{(1)\prime}(\xi_0, q)} + E_{2n} K_{2n} \right\} \quad (25)$$

$$K_{2n} = \sum_{m=1}^{\infty} B_{2m}{}^{(2n)} e^{-2(m-1)\xi_0} \quad (26)$$

A plot of $F_e{}^{(\mu)} (\omega)$ as a function of $\omega a^2/D$ is given in Figure 4 for several ratios of b/a (here a and b are the semimajor and semiminor axes of the elliptical cross section of the cavity). The last term in (25) may be summed exactly:

$$\sum_{n=1}^{\infty} E_{2n} K_{2n} = (1 - e^{-4\xi_0})^{-2} \quad (27)$$

This relation has been valuable in checking the reliability of the coefficients $B_{2m}{}^{(2n)}$ found numerically, but the convergence of the series in

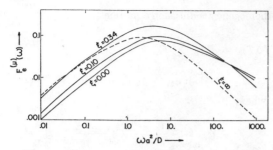

Fig. 4. Thermoelastic absorption spectrums for elliptic-cylinder cavities subjected to a shear $e_{12}{}^A$. The values of ξ_0 determine the ellipticity of the cylinder cross section.

(25) is improved by retaining the term $E_{2n}K_{2n}$ within the sum.

For a porous medium having N such cylindrical cavities per unit volume,

$$\frac{1}{Q_\mu} = \frac{8\mu\nu\alpha\Gamma T_0(1 - 2\sigma)(1 + \sigma)F_e{}^{(\mu)}(\omega)}{3\mu' \sinh 2\xi_0} \quad (28)$$

where the porosity is now

$$\nu = N\pi h^2 \sinh \xi_0 \cosh \xi_0 = N\pi ab \quad (29)$$

and the effective rigidity μ' may be found from the equations given by *Eshelby* [1957] to be

$$\mu' = \mu/[1 + \nu(1 - \sigma)(a + b)^2/ab] \quad (30)$$

it can be verified that (28) does indeed reduce to (17) as the cross-section of the cavity approaches a circle (i.e., $h \to 0$ and $\xi_0 \to \infty$, so that $he^{\xi_0} \to a$).

We now turn to a calculation of the energy loss in pure dilatation. From the equations given by *Timoshenko and Goodier* [1951], the perturbation dilation associated with a dilatation e^A is found to be

$$e^C = -\frac{2(1 + \sigma)e^A \sinh 2\xi}{3(\cosh 2\xi - \cos 2\eta)} \quad (31)$$

The appropriate solution of the homogeneous diffusion equation is

$$(\Delta T)_h = \Gamma T_0 \sum_{n=1}^{\infty} D_{2n}$$

$$\cdot \mathrm{Me}_{2n}{}^{(1)} (\xi, q) \, \mathrm{ce}_{2n} (\eta, q) \quad (32)$$

where

$$D_{2n} = \frac{8\Gamma T_0(1 + \sigma)e^A G_{2n}}{3 \, \mathrm{Me}_{2n}{}^{(1)'} (\xi_0, q)} \quad (33)$$

$$G_{2n} = \sum_{m=1}^{\infty} m A_{2m}{}^{(2n)} e^{-2m\xi_0} \quad (34)$$

$\mathrm{ce}_{2n} (\eta, q)$ is the even Mathieu function of the first kind, order $2n$, $\mathrm{Me}_{2n}{}^{(1)} (\xi_0, q)$ is the even modified Mathieu function of the third kind, first type, order $2n$, and the prime indicates differentiation with respect to the argument ξ_0. The coefficients $A_{2m}{}^{(2n)}$ are the even Mathieu coefficients of order $2n$ normalized so that

$$2(A_0{}^{(2n)})^2 + \sum_{m=1}^{\infty} (A_{2m}{}^{(2n)})^2 = 1$$

The energy loss per wave cycle is then found from (4) to be

$$\Delta E = 8\rho c T_0[\pi\Gamma(1 + \sigma)e^A h/3]^2 F_e{}^{(\kappa)}(\omega) \quad (35)$$

where

$$F_e{}^{(\kappa)}(\omega) = \frac{1}{|q|}$$

$$\cdot \sum_{n=0}^{\infty} \mathrm{Re}\left\{ \frac{2 \, \mathrm{Me}_{2n}{}^{(1)} (\xi_0, q)G_{2n}{}^2}{\mathrm{Me}_{2n}{}^{(1)'} (\xi_0, q)} + G_{2n}L_{2n} \right\} \quad (36)$$

$$L_{2n} = \sum_{m=0}^{\infty} A_{2m} e^{-2m\xi_0} \quad (37)$$

A plot of $F_e{}^{(\kappa)} (\omega)$ as a function of $\omega a^2/D$ is given in Figure 5 for several ratios b/a. The last term of (36) may be summed exactly:

$$\sum_{n=0}^{\infty} G_{2n}L_{2n} = (e^{2\xi_0} - e^{-2\xi_0})^-$$

Although this equation is useful for checking the values of $A_{2m}{}^{(2n)}$, the convergence of the series in (36) is improved by retaining the last term within the summation.

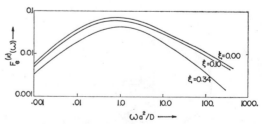

Fig. 5. Thermoelastic absorption spectrums for elliptic-cylinder cavities subjected to two-dimensional hydrostatic pressure.

For a porous medium having N such cylindrical cavities per unit volume (porosity ν),

$$\frac{1}{Q_\kappa} = \frac{16\kappa(1+\sigma)^2\nu\alpha\Gamma T_0(1-2\sigma)F_e^{(\kappa)}(\omega)}{9\kappa'(1-2\sigma)\sinh 2\xi_0} \quad (38)$$

In (38) the obvious combinations and cancellations have not been made in order to keep the equation as nearly comparable to (28) as possible. The effective bulk modulus κ' may be found from the equations given by *Eshelby* [1957].

$$\kappa' = \kappa\left[\frac{1+\nu(a^2+b^2)(1-\sigma)}{ab(1-2\sigma)}\right]^{-1} \quad (39)$$

So far we have considered only the case in which the plane of the crack (i.e., the plane containing the axis of the cylinder and the semimajor axis of the elliptical cross section) coincides with a plane of maximum shear. It is easy to generalize this to the case in which the angle between the two planes is γ. From the equations of *Timoshenko and Goodier* [1951] we find for that case

$$e^C = -2(1-2\sigma)e^{2\xi_0}e_{12}{}^A[1 + (\sinh 2\xi \sin 2\gamma$$
$$- \cos 2\gamma \sin 2\eta)/(\cosh 2\xi - \cos 2\eta)] \quad (40)$$

The first term in the brackets, being constant, may be neglected, and the other two terms have been treated individually in the preceding sections. The cross terms in the quadratic expressions for the energy loss make no net contribution. Thus it is readily shown that

$$\Delta E = 8\rho c T_0[\pi\Gamma h(1-2\sigma)e_{12}{}^A]^2$$
$$\cdot[(e^{2\xi_0}\sin 2\gamma)^2F_e^{(\kappa)}(\omega) + \cos^2 2\gamma F_e^{(\mu)}(\omega)] \quad (41)$$

and

$$Q_\mu^{-1} = (16/3)(\mu/\mu')\nu\alpha\Gamma T_0(1-2\sigma)$$
$$\cdot(1+\sigma)(e^{2\xi_0}\sin 2\gamma)^2F_e^{(\kappa)}(\omega)$$
$$+ \cos^2 2\gamma F_e^{(\mu)}(\omega)]/(2\sinh 2\xi_0) \quad (42)$$

If we suppose that cracks of all orientations are equally likely, the effective ΔE is found by averaging over-all values of γ. Thus

$$\frac{1}{Q_\mu} = \frac{8\mu\nu\alpha\Gamma T_0(1-2\sigma)(1+\sigma)F_e^{(s)}(\omega)}{3\mu'\sinh 2\xi_0} \quad (43)$$

where

$$F_e^{(s)}(\omega) = \tfrac{1}{2}[e^{4\xi_0}F_e^{(\kappa)}(\omega) + F_e^{(\mu)}(\omega)] \quad (44)$$

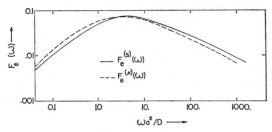

Fig. 6. Thermoelastic absorption spectrums for a distribution of flat elliptic-cylinder cavities of arbitrary orientation subjected to pure shear ($F_e^{(s)}$) and two-dimensional hydrostatic pressure ($F_e^{(\kappa)}$).

Both $F_e^{(s)}(\omega)$ and $F_e^{(\kappa)}(\omega)$ are shown in Figure 6 as functions of $\omega a^2/D$ for small ξ_0 (i.e., for flat cracks). The differences between the two functions is relatively small, and in what follows we shall consider them equal. Then from (38) and (43) we have

$$\frac{Q_\mu}{Q_\kappa} = \frac{2(1+\sigma)^2(1-2\sigma')}{3(1-2\sigma)^2(1+\sigma')} \quad (45)$$

where σ' is the effective Poisson ratio of the porous medium and use has been made of the identity $\kappa/\mu = 2(1+\sigma)/3(1-2\sigma)$.

Equation 1 relates the above values of Q_μ and Q_κ to the values associated with the attenuation of longitudinal waves (Q_α) and transverse waves (Q_β). A transverse wave has no dilatational energy, so $Q_\beta = Q_\mu$. A longitudinal wave propagating in the x direction has $\kappa'e_{11}{}^2/2$ dilatational energy and $2\mu'e_{11}{}^2/3$ shear energy. Thus

$$\frac{1}{Q_\alpha} = \frac{(1+\sigma')/Q_\kappa + 2(1-2\sigma')/Q_\mu}{3(1-\sigma')} \quad (46)$$

where σ' is the effective Poisson ratio. From (45) and (46) we find

$$\frac{Q_\beta}{Q_\alpha} = \frac{2(1-2\sigma')}{3(1-\sigma')}\left[1 + \frac{(1+\sigma)^2}{3(1-2\sigma)^2}\right] \quad (47)$$

The above equation is valid only for small ξ_0 (i.e., flat cracks).

Finally, it should be pointed out that (38) and (43) are in terms of quantities which may be observed in experiments not related directly to attenuation measurements. Measurement of the parameters ν and ξ_0 would, in general, be difficult, but all that is required for flat cracks (small ξ_0) is the ratio ν/ξ_0; that ratio can be calculated directly from (39):

$$\nu/\xi_0 = (\kappa/\kappa' - 1)(1 - 2\sigma)/(1 - \sigma) \qquad (48)$$

Thus we have the data on hand to estimate Q for various rocks.

DISCUSSION

Although we have admitted a random distribution in the orientation of the cracks, we have not yet taken account of the distribution in crack sizes. It is seen from Figures 4 and 5 that once ξ_0 is less than, say, 0.1 it ceases to be an important parameter in the absorption spectrum. Also from (48) the ratio ν/ξ_0 ceases to depend upon ξ_0 if ξ_0 is small. Thus the expressions (38 and 43) for Q do not depend upon ξ_0 if ξ_0 is small, and the only significant crack size parameter is then the half-length a. The equations in the preceding section will then give the partial internal friction associated with each crack size; the total internal friction is, of course, the sum of the partial internal frictions. Without knowing the crack length distribution function we cannot calculate the total internal friction. However, we can make a good estimate with very little trouble. The effect of adding up many partial internal frictions is to smear out the absorption spectrum (Figure 6) so that it has a smaller maximum value but is more nearly flat over the frequency range. Let us smear out the absorption spectrum so that it is relatively flat over the range $0.1 < \omega a^2/D < 1000$ (this range is suggested by the observation that Q is essentially constant over a range of about four decades for many materials). This, of course, means we must replace $F_e(\omega)$ by its average value over that interval. From Figure 6 this average appears to be $\langle F_e(\omega) \rangle = 0.03$. It seems quite unlikely that this estimate of $\langle F_e(\omega) \rangle$ will be in error by more than a factor of 2 anywhere in the frequency range of interest.

We are now able to use the theory to estimate the values of Q_α and Q_β for certain rocks. Granite has been chosen as an example because the required parameters have been experimentally determined. From (43) and (48) we find that

$$\frac{1}{Q_\beta} = \frac{4\kappa}{3\kappa'}\left(\frac{\kappa}{\kappa'} - 1\right)\frac{(1 - 2\sigma)^3(1 + \sigma')}{(1 - \sigma)(1 - 2\sigma')}$$
$$\cdot \alpha \Gamma T_0 \langle F_e(\omega) \rangle \qquad (49)$$

We recall that the primed quantities refer to the bulk properties of the rock, whereas the unprimed quantities refer to the constituent minerals. For the parameters α, Γ, and D, which refer to a mineral constituent of granite, we shall use the physical properties of quartz [Birch et al., 1942]: $\alpha = 5.6 \times 10^{-5}$ deg^{-1}, $D = 5 \times 10^{-2}$ cm^2/sec, and $\Gamma = 1.1$. Brace [1965] has found that κ/κ' is about 4 for Westerly granite and 7 for Stone Mountain granite, we choose $\kappa/\kappa' = 5$ as a typical value. Typical values for σ and σ' [Birch et al., 1942] seem to be 0.25 and 0.1, respectively. Thus, from (49) we estimate $Q_\beta = 365$ and from (47) $Q_\alpha = 225$; these values compare with values of $Q_\beta = 150$ to 200 and $Q_\alpha = 100$ given by Birch et al. [1942]. Had we used the value $\kappa/\kappa' = 7$ found by Brace [1965] for Stone Mountain granite, we would have duplicated the observed values of Q very nicely. However, it should be recalled that the theory is two-dimensional, with all cracks oriented so that the cylindrical axis is perpendicular to the plane of strain. In a three-dimensional body with randomly oriented cracks, only about $\frac{1}{3}$ of the cracks would contribute to the attenuation as calculated here. Thus, if the restriction to plane strain is relaxed, it is likely that the above theoretical values of Q should be multiplied by 3. On the other hand, the theory assumes that neither the elastic nor thermal fields of any two cracks interact. This is not likely to be true for rocks, and the effect of such interaction would be to increase the dissipation. Thus all that can be claimed is that the theory is capable of giving the correct order of magnitude for Q.

The values of Q calculated above should obtain in the frequency range $0.1 < \omega a^2/D < 1000$, where $2a$ is now the mean crack length. Observation indicates that Q does have approximately the above value in the frequency range 20 to 2×10^5 hz. Thus, to fit observations, the mean crack length must be 0.08 mm, i.e., about 1/10 of the grain size of a typical granite. All that can be said here is that this crack length seems reasonable.

Other properties of thermoelastic attenuation have already been discussed by Savage [1965] on the basis of an order of magnitude estimate. It was shown there that thermoelastic attenuation by cracks provides an obvious explanation of the dependence of Q upon pressure and also an explanation of what appears to be an increase in Q as frequency decreases at frequencies below about 10 hz.

A quantity that could not be estimated in the previous order of magnitude calculations [*Savage*, 1965] is the ratio Q_α/Q_β. In the present theory that ratio is given by (47), which indicates that $Q_\alpha/Q_\beta = (1-\sigma')/2(1-2\sigma')$ for $\sigma = \frac{1}{4}$ and about half that value for $\sigma = \frac{1}{3}$. As σ' ranges from 0.05 to 0.25, Q_α/Q_β ranges from 0.52 to 0.75 if $\sigma = \frac{1}{4}$ and from 0.26 to 0.38 if $\sigma = \frac{1}{3}$. Thus we should expect Q_α/Q_β to be somewhere in the range 0.25 to 0.75 for ordinary materials containing cracks. The ratio has been measured for two rocks: for Quincy granite [*Birch and Bancroft*, 1938] $Q_\alpha/Q_\beta = 0.5$ to 0.7 and for limestone [*Peselnick and Zietz*, 1959] 0.4 to 0.6.

The theory is subject to criticism on at least two points. One of these concerns the dependence of Q_β^{-1} upon $(\kappa/\kappa' - 1)$ (see (49)). *Brace* [1965] lists κ and κ' for nine different rocks. Of these, five (two granites, a quartzite, a gneiss, and a marble) have values of the ratio κ/κ' in the range 3 to 7, one (a dolomite) has $\kappa/\kappa' = 19$, and three (a limestone, a dolomite, and a diabase) have ratios κ/κ' very close to 1. Equation 49 implies that the last group of rocks would have negligible internal friction from the thermoelastic effects discussed here. Thus internal friction exhibited by such rocks must be attributed to other loss mechanisms. This secondary loss mechanism would presumably be a property of other rocks as well, and it is probable that it becomes the primary mechanism at pressures above that required to close up the cracks in a rock. Typical values of Q under such pressures are of the order of a few thousand. Thus we should expect rocks for which κ/κ' is close to 1 to have Q's in excess of 1000. I do not believe such values of Q have been observed in rocks at ordinary pressures. However, it is quite possible that Q has not been determined for a rock sample for which κ/κ' is close to 1. The matter can be settled only by further measurements.

Probably a more serious criticism concerns the dependence of Q upon temperature. This dependence may be more complicated than the explicit variation of Q as $1/T_0$ displayed in (43), since presumably there is also an implicit dependence through μ'/μ. Nevertheless, as a rough estimate we should expect Q to vary about as $1/T_0$. Unfortunately, the experimental determination of the variation of Q with temperature in rocks is quite difficult because it must be measured on jacketed specimens under confining pressure (unconfined rocks break up readily under heating). The only measurements of this type of which I am aware are those reported by *Birch and Bancroft* [1938]. They found that Q did not change perceptibly in a temperature range of 20°C to 100°C, a range in which (43) suggests that a change of about 25% should occur. A strong dependence of Q upon temperature has been found for most other substances; attention is particularly directed to glass [*Birch et al.*, 1942], a material in which the internal friction is very likely associated with the small cracks which are known to occur.

Volarovich and Gurvich [1957] have determined for several rocks the variations in both Q and Young's modulus as a function of temperature at atmospheric pressure. Because new cracks in the rock are formed in the heating process, the decrease of the effective Young's modulus with increasing pressure is particularly rapid. A similar decrease should be expected in the ratio κ'/κ. The thermoelastic theory presented here would then imply a rapid decrease in Q as temperature increases (see (49)). However, such an increase was observed only at temperatures above about 300°C; below that temperature Q was generally an increasing function of temperature. Thus the thermoelastic theory discussed here is not consistent with the observations of Volarovich and Gurvich at temperatures in the range 20°C to about 300°C.

If, as seems possible, the attenuation of seismic waves in rocks measured in the laboratory can be attributed to thermoelastic attenuation by cracks, such laboratory experiments are not relevant to the attenuation of seismic waves in the earth. This follows from the obvious fact that open cracks could not exist at any appreciable depth within the earth. It may be that thermoelastic attenuation associated with fluid-filled cracks occurs at depth, but, if so, neither the theory presented here nor the laboratory experiments on dry rocks are pertinent. It is worth remarking that thermoelastic attenuation by fluid-filled cracks is an attractive hypothesis because it might account for the observation

[*Anderson et al.*, 1965] that in the mantle Q_{κ} is very large compared with Q_{μ}. This could follow from the fact that the discontinuity in compressibility at the cavity would be quite small, but that in rigidity quite large. The attenuation might, however, be complicated by viscous effects [*Knopoff*, 1964].

Only one form of thermoelastic attenuation has been discussed here. Other forms may be equally important. For example, in a rock composed of two or more minerals either shear or compression will induce different temperature fluctuations in the constituent minerals. Even for monomineralic rocks, anisotropy effects will contribute to thermoelastic attenuation.

Acknowledgment. Preparation and publication of this paper was supported by grants from the National Research Council of Canada.

REFERENCES

Anderson, D. L., A. Ben-Menahem, and C. B. Archambeau, Attenuation of seismic energy in the upper mantle, *J. Geophys. Res., 70*(6), 1441–1448, 1965.

Birch, F., and D. Bancroft, The effect of pressure on the rigidity of rocks, *J. Geol., 46*, 59–87 and 113–141, 1938.

Birch, F., J. F. Schairer, and H. C. Spicer, Handbook of physical constants, *Geol. Soc. Am. Spec. Paper 36*, 1942.

Brace, W. F., Some new measurements of linear compressibility of rocks, *J. Geophys. Res., 70*(2), 391–398, 1965.

Eshelby, J. D., The determination of the elastic field of an ellipsoidal inclusion, and related problems, *Proc. Roy. Soc. London, A, 241*, 376–396, 1957.

Knopoff, L., and G. J. F. MacDonald, Attenuation of small amplitude stress waves in solids, *Rev. Mod. Phys., 30*(4), 1178–1192, 1958.

Knopoff, L., Q, *Rev. Geophys., 2*(4), 625–660, 1964.

Love, A. E. H., *Mathematical Theory of Elasticity*, 4th ed., Dover Publications, New York, 1944.

McLachlan, N. W., *Theory and Application of Mathieu Functions*, Oxford University Press, London, 1947.

Peselnick, L., and I. Zietz, Internal friction of fine-grained limestones at ultrasonic frequencies, *Geophysics, 24*, 285–296, 1959.

Press, F., and J. Healy, Absorption of Rayleigh waves in low-loss media, *J. Appl. Phys., 28*, 1323–1325, 1957.

Savage, J. C., Attenuation of elastic waves in granular mediums, *J. Geophys. Res., 70*(16), 3935–3942, 1965.

Timoshenko, S., and J. N. Goodier, *Theory of Elasticity*, McGraw-Hill Book Company, New York, 1951.

Volarovich, M. P., and A. S. Gurvich, Investigation of dynamic moduli of elasticity for rocks in relation to temperature, *Bull. Acad. Sci. USSR, Geophys. Ser., English Transl.*, no. 4, 1–9, 1957.

Wuenschel, P. E., Dispersive body waves—An experimental study, *Geophysics, 30*(4), 539–551, 1965.

Zener, C., Internal friction in solids, 2, General theory of thermoelastic internal friction, *Phys. Rev., 53*, 90–99, 1938.

Zener, C., *Elasticity and Anelasticity of Metals*, University of Chicago Press, 1948.

(Manuscript received February 28, 1966.)

JOURNAL OF GEOPHYSICAL RESEARCH   VOL. 72, No. 24   DECEMBER 15, 1967

Correction to Paper by J. C. Savage, 'Thermoelastic Attenuation of Elastic Waves by Cracks'

J. C. SAVAGE

Geophysics Laboratory, University of Toronto
Toronto, Ontario, Canada

On page 3934 of the August 15, 1966, issue of the Journal, the first sentence preceding equation 31 should refer to 'two-dimensional dilatation' not 'pure dilatation.' For this reason the right-hand sides of both equations 38 and 45 should be multiplied by $2(1 + \sigma')/3$. The energy partition (see the sentence preceding equation 46 on page 3935) should be $3\kappa'e_{11}^2/[4(1 + \sigma')]$ dilatational energy and $\mu'e_{11}^2/2$ shear energy. Thus, equation 46 should be

$$\frac{1}{Q_\alpha} = \frac{1/Q_\kappa + (1 - 2\sigma')/Q_\mu}{2(1 - \sigma')}$$

The final result is to change the first term in the square brackets in equation 47 from 1.0 to 0.75. Fortunately, these changes do not affect the numerical values significantly; therefore, the conclusions are unchanged.

Finally, a typographical error occurs in equation 39, page 3935. That equation should read

$$\kappa' = \kappa\left[1 + \frac{\nu(a^2 + b^2)(1 - \sigma)}{ab(1 - 2\sigma)}\right]^{-1}$$

(Received August 18, 1967.)

159

Reprinted from Journal of Geophysical Research, v. 74, p. 4963-4966.

Internal Friction Mechanism That Produces an Attenuation in the Earth's Crust Proportional to the Frequency

WARREN P. MASON

Department of Engineering and Applied Science, Columbia University
New York, New York 10027

To explain an attenuation proportional to the frequency (internal friction Q^{-1} independent of the frequency) present for waves in the earth's crust, a nonlinear mechanism connected with the motion of dislocations is proposed. With the kink model for dislocation motion, it is shown that an energy loss occurs when the kink goes over the kink barrier and becomes unstable. This loss is proportional to the number of kink displacements, and for a given strain level the number of kink displacements per cycle, i.e. the Q^{-1} value, will be independent of the frequency. The energy loss measured is consistent with theoretical calculations of the dissipation stress, and the loop length l_A and the number of dislocations found are consistent with the results for polycrystalline metals and rocks.

INTRODUCTION

It has long been established that waves in the earth's crust are attenuated by amounts proportional to the frequency. This corresponds to an internal friction Q^{-1}, which is independent of the frequency. This proportionality holds for frequencies from 3×10^{-4} Hz (longest period of the earth) to frequencies as high as 500 kHz for fine-grained material. The strains are in the range from 10^{-11} to 10^{-5}.

Knopoff [1965] has discussed the various assumptions that have been considered to explain this effect and concludes that a nonlinear elastic mechanism is the most likely source. This note suggests a nonlinear elastic mechanism, connected with the motion of dislocations, that appears to explain this effect. This method can be applied to any type of crystal structure and is valid for any temperature range or amplitude range and for frequencies from very low values up to a frequency for which the drag coefficient, which introduces an internal friction proportional to the frequency, becomes dominant. For the impure type of polycrystalline rocks present in the earth's crust this is probably in the megahertz range. Furthermore, the observed internal friction values are consistent with reasonable values of dislocation densities in polycrystalline metals.

DISLOCATION MODEL

The dislocation lies in the glide plane, which is the (110) plane for cubic crystals or the (0001) plane for hexagonal crystals. (See Figure 1 for dislocation model.) The kink model assumes that parts of the dislocation lie in the Peierls troughs and that these sections are connected by kinks crossing the Peierls barriers. When a shearing stress τ is applied in the glide plane, a force τab is applied to the kink, which tends to move it into the position shown by the dashed line in Figure 1. Here a is the height of the kink and b is the Burger's distance. To move under the application of a stress alone, however, the kink must cross the kink barrier W shown on the energy diagram. To make this crossing requires a kink stress σ_k and a kink strain S_k, which is of the order of the lowest yield strain. This value may be of the order of 10^{-4}; hence, for all strains of interest for seismic waves the stress alone is not sufficient and thermal energy is required to cause the kink to cross the barrier.

The effect of a stress lowers one well by an amount $\tau ab^2/2$ and raises the other by the same amount. Since $\tau ab^2/2kT$ can be shown to be much smaller than 1, reaction rate theory can be used to show that the difference between the number of jumps per second of a kink in the forward direction and the number in the backward direction is

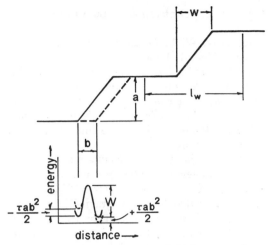

Fig. 1. Dislocation kinks and kink energy barriers [*Mason*, 1969].

$$(\alpha_{12} - \alpha_{21}) = \omega_0(\tau ab^2/kT)e^{-W/kT} \qquad (1)$$

where ω_0 is the angular attempt frequency, which is taken to be $2\pi \times 10^{13}$, since the resonant frequency of a kink in its potential well is usually considered to be of the order of 10^{13} Hz. W is found to be of the order of 1.2×10^{-4} ev, and hence for any temperatures above $5°K$ the last term can be taken as unity.

For strains in the range 10^{-11} to 10^{-5} (with the shear stiffness μ taken to be 4×10^{11} dynes/cm^2) the jump frequency will be of the order of

$$(\alpha_{12} - \alpha_{21}) = 1.1 \times 10^7 \quad \text{to} \quad 1.1 \times 10^{13} \qquad (2)$$

If it takes 2π kink jumps to complete a cycle in the time of the applied frequency, the upper limit of the frequency range over which the thermal mechanism will be operative will be from 1.7×10^6 to 1.7×10^{12} Hz, which is amply fast for any seismic effect.

The feature of the kink model that results in an internal friction independent of the frequency (attenuation proportional to the frequency) can be described as follows: when the dislocation crosses the Peierls barrier, it becomes unstable and generates lattice vibrations that carry off energy. The energy loss is proportional to the number of kink displacements rather than the kink velocity, and hence for a given strain level the number of kink displacements per cycle will be independent of the frequency. Since the internal friction is propor-

tional to the ratio of the energy dissipated per cycle divided by the maximum energy stored, this mechanism gives an internal friction independent of the frequency.

The value of the energy lost on crossing the barrier has been investigated by *Weiner* [1964] and *Atkinson and Cabrera* [1965]. Using a one-dimensional Frenkel-Kontorowa model, they have shown that, once the dislocation has surmounted the barrier, it takes a much smaller stress (called the dynamic Peierls stress τ_{dp}) to keep the dislocation moving. This stress is required to replace the energy lost to the lattice vibrations. Its value has been evaluated to be from 0.01 to 0.1 of the kink stress σ_k. In this model the energy returned from the nondissipative part of the negative slope of the barrier is not applied to sending the dislocation up the next barrier, but rather is applied to pushing kinks closer together. This represents a storage of energy, which can be shown to equal to $\frac{1}{2}n\sigma_k ab^2$, where n is the number of kink displacements. The energy dissipated is

$$\tfrac{1}{2}n\tau_{dp}ab^2 = \tfrac{1}{2}n\beta\sigma_k ab^2 \qquad (3)$$

where β is the ratio of the dynamic Peierls stress to the kink stress. Since the definition of the internal friction Q^{-1} is the ratio of the energy dissipated to the energy stored, this result is

$$Q^{-1} = \beta \qquad (4)$$

for the dislocation system alone. To determine the internal friction for the complete crystal, we must multiply the value of β by $\Delta\mu/\mu$, where $\Delta\mu$ is the part of the modulus due to dislocations and μ is the shearing modulus of the crystal. Hence, independent of the number of dislocations or the loop length distribution, the internal friction value is

$$Q^{-1} = \beta(\Delta\mu/\mu) \qquad (5)$$

For sinusoidal motions it is simpler to consider a string model whose area of displacement $\bar{N}\bar{x}lb$ is equal to nab, where $\bar{N}$ is the number of dislocations of loop length l and $\bar{x}$ is the average displacement. The force applied by the thermal effect plus the force due to the stretching of the dislocation can be written as

$$\tau^*b = \tau(1 - j\beta)b \qquad (6)$$

This relation takes account not only of the conservative force applied to the kink by the negative slope plus that due to stretching, but also of the dissipative force due to lattice vibrations. If we apply this force to the string model, neglecting the mass since the dislocation is overdamped, we obtain the result [*Mason*, 1969]

$$\Delta\mu/\mu = \frac{\bar{N}l^2/6}{1 + (\omega Bl^2/6\mu b^2)^2}$$

$$Q^{-1} = (\Delta\mu/\mu)[\beta + \omega Bl^2/6\mu b^2] \qquad (7)$$

where $\bar{N}$ is the number of dislocations of loop length l. For another type of stress, such as a longitudinal or shear stress, (7) can be generalized to

$$\Delta E/E = \frac{\bar{N}Rl^2/6}{1 + (\omega Bl^2/6\mu b^2)^2}$$

$$Q^{-1} = (\Delta E/E)[\beta + \omega Bl^2/6\mu b^2] \qquad (8)$$

where $\Delta E/E$ is the modulus defect caused by the type of wave propagated and R is the orientation factor that relates the average shearing stress in the glide planes to the applied stress. R is related to the crystal structure type since different glide planes are involved. It is generally found that the loop lengths follow an exponential type distribution rather than the single loop length of (8). The effect of this change can be evaluated by employing the calculations of *Oen et al.* [1960], as shown in Figure 2. Curves 1 and 2 show the ratio of the modulus change $\Delta E/E$ and the internal friction Q^{-1} to the value $\bar{N}Rl_A^2$, where l_A is the average loop length, plotted against ω/ω_0 (ω_0 equals $\mu b^2/Bl_A^2$). Here B is the drag coefficient which produces an energy loss proportional to the velocity of dislocation motion. These curves are valid under the assumption that $\beta = 0$. As shown by (8), however, the internal friction for the kink displacement model consists of two parts: one equal to β times $\Delta E/E$ and the other given by curve 2 of Figure 2. Hence, the total value will be given by curve 2 plus curve 1 times the value β. Some idea of the value of β can be obtained for copper from the work of *Thompson and Holmes* [1956], who measured both the internal friction and the modulus change that can be removed by neutron irradiation. The ratio varied from 0.04 to 0.09, the lower value being the most probable since

the drag coefficient can contribute to the higher value.

Recently, W. P. Mason and J. Wehr (paper to be published, 1969) found the value for an alloy of titanium to be 0.03. Hence, from the two curves the internal friction will be constant until ω/ω_0 reaches a value of 10^{-2}. Above this point the drag coefficient exceeds the constant value and the remainder of the curve will follow curve 2 of Figure 2.

To see if this theory accounts quantitatively for the values measured, we use the curve given by *Knopoff and Porter* [1963]. Their curve shows that for a fine-grained granite the attenuation is proportional to the frequency up to 1000 kHz with a Q of about 100. Since deviations from linearity occur above a ratio of

$$\frac{\omega}{\omega_0} = 10^{-2} \quad \text{or} \quad f = \frac{10^{-2}\,\mu b^2}{2\pi Bl_A^2} \qquad (9)$$

using the values $f = 1,000,000$ Hz, $\mu = 4 \times 10^{11}$ dynes/cm^2, $b = 3 \times 10^{-8}$ cm, $B \doteq 10^{-3}$ dyne (sec/cm^2), we have

$$l_A = 2.4 \times 10^{-5} \text{ cm} \qquad (10)$$

which is a reasonable value for an impure metal or rock. With this value, a Q^{-1} value of 0.01 is obtained for a dislocation density of

$$Q^{-1}/\bar{N}Rl_A^2 = 0.04 \qquad (11a)$$

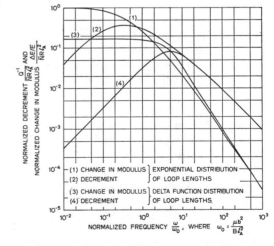

Fig. 2. Internal friction and modulus change for an exponential dislocation distribution assuming a velocity drag coefficient damping [after *Oen et al.*, 1960].

$$\bar{N} = \frac{0.01}{0.25 \times 5.75 \times 10^{-10} \times 0.04} \qquad (11b)$$

if the orientation factor R is assumed to be 0.25, which is a usual value. This gives a dislocation density of

$$\bar{N} = 1.7 \times 10^9 \text{ dislocations/cm}^2 \qquad (12)$$

which is a reasonable value for a strained and impure sample.

Acknowledgment. This work was supported by the Institute of Fatigue and Reliability.

References

Atkinson, W., and N. Cabrera, Motion of a Frenkel–Kontorowa dislocation in a one-dimensional crystal, *Phys. Rev., 138,* A763–A766, 1965.

Knopoff, L., chapter 7, in *Physical Acoustics,* vol. IIIB, edited by W. P. Mason, pp. 287–324, Academic Press, New York, 1965.

Knopoff, L., and L. D. Porter, Attenuation of Rayleigh waves in fine-grained Westerly granite, *J. Geophys. Res., 68,* 6317–6321, 1963.

Mason, W. P., A source of dissipation that produces an internal friction independent of the frequency, Conference on Fundamental Aspects of Dislocation Theory, National Bureau of Standards, April 21–25, 1969.

Oen, O. S., D. K. Holmes, and M. T. Robinson, A computor calculation of the internal friction and modulus defect of a dislocation string model, *U. S. At. Energy Comm. Rept. ORNL-3017-3,* 1960.

Thompson, D. O., and D. K. Holmes, Effect of neutron irradiation upon the Young's modulus and internal friction of copper single crystals, *J. Appl. Phys., 27,* 713–723, 1956.

Weiner, J. H., Dislocation velocities of a linear chain, *Phys. Rev., 136,* A863–A868, 1964.

(Received May 14, 1969.)

Reprinted from Physics of Sound in Marine Sediments, Plenum Press, p. 19–39.

ACOUSTIC WAVES IN SATURATED SEDIMENTS

ROBERT D. STOLL

School of Engineering and Applied Science

Columbia University, New York, New York 10027

ABSTRACT

This paper discusses a phenomenological model that describes the propagation of sound waves in saturated sediments. The compressibility and shearing stiffness of the skeletal frame, the compressibility of the fluid, and two major sources of attenuation are included in the model. Attenuation is attributed to two fundamentally different types of energy loss, one resulting from inelasticity of the skeletal frame and the other due to motion of the pore fluid relative to the frame, with each significant in a different frequency range.

Attenuation, dispersion, and wave velocities are found to be in favorable agreement with experimental results for both sands and fine-grained sediments over a wide range of frequencies.

INTRODUCTION

In order to accurately describe the propagation of mechanical waves in ocean sediments, it is necessary to take into account the basic mechanisms by which energy is dissipated in such a medium. Over the frequency range of interest in geophysical research, from less than 1 Hz to over 100 kHz, there are undoubtedly a number of discrete physical processes to which energy dissipation can be attributed. However, most of the available data seems to indicate that the overall response can be described, in a phenomenological sense, by grouping the losses into two fundamentally different categories--one which accounts for the effects of inelasticity of the skeletal frame in a water environment and a second which accounts for the viscosity of the water moving relative to the frame. In granular materials, such as sands and silts, the inelasticity of the frame can be traced primarily to friction losses occurring at the grain-to-grain contacts. In finer materials such as clays and silty clays, the losses that occur during small distortions of the frame have been attributed to a variety of rate-dependent processes associated with the electrochemical bonds between particles. In composite materials containing both sand and clay as well as in cemented or indurated material, it is probable that both rate-dependent processes and friction combine to produce the observed response. Fortunately (for those who attempt to formulate mathematical models for this behavior) experimental results seem to indicate that a relatively simple model is capable of reproducing the combined effects of all of these mechanisms. This model, proposed by Stoll and Bryan (1970), is based on the comprehensive theory for the behavior of saturated porous media developed by Maurice Biot. Using the generalized formulation given by Biot (1962a), including his modifications for higher frequencies, and the notion of a constant complex modulus to describe the response of the skeletal frame, most of the observed behavior can be reproduced over a wide frequency range. The theory predicts that losses in the skeletal frame dominate at low frequencies, while the viscous losses due to motion of the interstitial water become predominant at higher frequencies. The terms high and low frequency are used in a relative sense with their actual values depending on the physical properties of the particular sediment being modeled. For example in very fine clayey materials losses attributable to the skeletal frame dominate over most of the frequency range of interest while in sands the viscous losses become important over a significant portion of this range. As a result, the variation of attenuation with frequency may be quite complex, making it virtually impossible to extrapolate outside the range of experimental values without the help of a comprehensive physical theory such as the one to be described.

THEORY

Starting with a paper on consolidation of porous, elastic material in 1941 (Biot, 1941), Biot has developed a comprehensive theory for the static and dynamic response of linear, porous materials containing compressible fluid. He has considered both low and high frequency behavior (Biot, 1956a, 1956b) and has included the possibility of viscoelastic or viscodynamic response in various components of his model (Biot, 1962a, 1962b).

In the course of developing and generalizing the theory, Biot introduced several changes of notation and a number of generalizations so that some effort is necessary to extract the form most suitable for a particular application. For this reason an abbreviated derivation leading to one form of his equations is outlined below. This derivation is included to help in identifying the variables that are used and in visualizing how the response of the sediment is modeled in a mathematical way. For more rigorous and complete derivations the reader is referred to Biot's original papers, particularly (Biot, 1962a), and to a paper by Geertsma and Smit (1961).

Biot's theory predicts that in the absence of boundaries, three kinds of body waves, two dilatational and one shear, may exist in a fluid saturated porous medium. One of the dilatational waves, which is called the first kind, and the shear wave are similar to waves found in ordinary elastic media. The second kind of compressional wave is highly attenuated in the nature of a diffusion process. Compressional waves of the second kind become very important in acoustical problems involving very compressible pore fluids such as air, whereas for geophysical work in water-saturated sediments waves of the first kind are of principal interest. One exception to this may be the case of very gassy sediments where the effective compressibility of the pore fluid has been greatly reduced by the presence of dissolved or free gas.

To obtain equations governing the propagation of dilatational waves, we will first consider the case of a plane wave in a porous elastic medium filled with fluid. The model will then be generalized to include the inelasticity of the frame and the frequency dependence of viscous losses to yield a realistic model of natural sediments.

If $\overline{u}$ is a vector function giving the displacement of points in the skeletal frame and $\overline{U}$ a vector function giving the displacement of the fluid, then the volume of fluid that has flowed in or out of an element of volume attached to the frame or the "increment of fluid content" is

$$\zeta = \beta \text{ div } (\overline{u} - \overline{U}) \qquad (1)$$

where β is the ratio of the volume of pores to the volume of solids (porosity) of the medium. For small strains the dilatation or volumetric strain of the element attached to the frame may be written as

$$e = e_x + e_y + e_z = \text{div } \bar{u} \tag{2}$$

where e_x, e_y, and e_z are the components of small compressional or extensional strain in a cartesian coordinate system. If the porous frame and the pore fluid are elastic, the strain energy, W, of the system depends on the strain components and the increment of fluid content

$$W = W(e_x, e_y, e_z, \gamma_x, \gamma_y, \gamma_z, \zeta) \tag{3}$$

where γ_x, γ_y, and γ_z are the components of shear strain. For an isotropic, linear material, W is a quadratic function of the invariants of strain, I_1 and I_2, and the increment of fluid content, ζ

$$W = C_1 I_1^2 + C_2 I_2 + C_3 I_1 \zeta + C_4 \zeta^2 \tag{4}$$

where

$$I_1 = e_x + e_y + e_z = e,$$

and

$$I_2 = e_x e_y + e_y e_z + e_z e_x - \frac{1}{4}\left(\gamma_x^2 + \gamma_y^2 + \gamma_z^2\right).$$

The constants C_1, C_2, C_3, and C_4 may be identified with one set used by Biot (H, C, M and μ) by writing Eq. (4) as

$$W = \frac{H}{2} e^2 - 2\mu I_2 - Ce\zeta + \frac{M}{2} \zeta^2 . \tag{5}$$

Considering total stresses on the element of volume attached to the frame, τ_{ij}, and the pressure in the pore fluid, p_f, a set of stress-strain relationships may be obtained from the strain energy, Eq. (5), by differentiation (i.e., $\tau_{xx} = \partial W/\partial e_x$, $\tau_{xy} = \partial W/\partial \gamma_z$, $p_f = \partial W/\partial \zeta$, etc). The stress-strain relations are

$$\tau_{xx} = He - 2\mu(e_y + e_z) - C\zeta$$

$$\tau_{yy} = He - 2\mu(e_z + e_x) - C\zeta$$

$$\tau_{zz} = He - 2\mu(e_x + e_y) - C\zeta \tag{6}$$

$$\tau_{xy} = \mu \gamma_z$$

$$\tau_{yz} = \mu \gamma_x$$

$$\tau_{zx} = \mu \gamma_y$$

$$p_f = M\zeta - Ce$$

From these equations it is clear that μ is the shear modulus. However, in order to understand the significance of the constants H, C, and M it is helpful to visualize two idealized quasistatic tests, involving isotropic loading. In one kind of test, termed a "jacketed" test (Biot and Willis, 1957) the saturated porous medium (shown as a granular material in Fig. 1) is placed in an impervious, flexible bag and loaded by an external pressure. The interstitial fluid in the sample is free to flow out of the bag via a tube so that the fluid pressure remains unchanged during slow loading. In the other test, called an "unjacketed" test, an uncased sample is completely immersed in fluid which is subsequently pressurized from an external source. If p' is the externally applied isotropic pressure in both cases, then

$$\tau_{xx} = \tau_{yy} = \tau_{zz} = -p'$$

$$\tau_{xy} = \tau_{yz} = \tau_{zx} = 0$$

and adding the first, second, and third of Eqs. (6),

$$-p' = (H - 4\mu/3)\, e - C\, \zeta. \tag{7}$$

For the "jacketed" test, p_f remains unchanged so that the bulk modulus of the free draining porous frame, K_b, is

$$K_b = -p'/e = H - 4\mu/3 - C^2/M \tag{8}$$

from the last of Eqs. (6) and Eq. (7). In a practical test the dilatation of the sample, e, could be obtained by measuring the quantity of fluid expelled from the sample provided the effect of

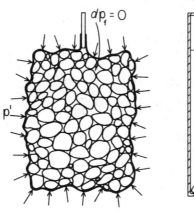
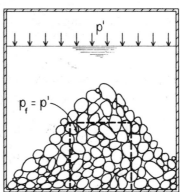

"Jacketed" Test   "Unjacketed" Test

FIGURE 1

ISOTROPIC TESTS TO DETERMINE MODULI

membrane penetration on the outer surface of the specimen was negligible (which is often not so). This type of drained, isotropic loading is utilized in several standard tests used to determine the engineering properties of soil (Bishop and Henkel, 1957).

In the unjacketed test, the pressure in the pore fluid equals the applied isotropic pressure, and we may derive two measures of compliance, δ, the "unjacketed compressibility,"

$$\delta = - e/p' = \frac{1 - C/M}{H - 4\mu/3 - C^2/M} \tag{9}$$

and, γ, the "coefficient of fluid content,"

$$\gamma = \zeta/p' = \frac{H - 4\mu/3 - C}{(H - 4\mu/3 - C^2/M)M} \tag{10}$$

utilizing the last of Eqs. (6) and Eq. (7) in both cases. If the ratio of pore volume to solid volume remains constant (i.e., constant porosity) during unjacketed loading, δ equals the compressibility of the solid material composing the porous frame and γ may be expressed in terms of δ, β, and the compressibility of the pore fluid. In terms of the reciprocals of compressibility (bulk moduli)

$$\gamma = \beta(1/K_f - 1/K_r) \tag{11}$$

and

$$\delta = 1/K_r \tag{12}$$

where K_f is the bulk modulus of the fluid and K_r is the bulk modulus of the solid material composing the porous frame (the bulk modulus of the individual particles in the case of granular media). While Eqs. (11) and (12) are strictly true for materials where the porous frame is isotropic, homogeneous, and linear, they are also reasonable for cases where the frame is heterogeneous but behaves approximately like a homogeneous mass in that it undergoes the same volumetric strain as the pores.

Finally, using Eqs. (8) through (12) we may solve for Biot's coefficients in terms of the bulk moduli of the porous frame, the pore fluid, and the solid material or discrete particles composing the following frame.

$$H = \frac{(K_r - K_b)^2}{D - K_b} + K_b + 4\mu/3$$

$$C = \frac{K_r(K_r - K_b)}{D - K_b}$$ (13)

$$M = \frac{K_r^2}{D - K_b}$$

where

$$D = K_r(1 + \beta(K_r/K_f - 1)) \ .$$

Having established a set of constitutive equations and the meanings of the various parameters, equations can now be written for the motion of an element of volume attached to the skeletal frame and for fluid moving into or out of the element. To simplify the derivation we consider one-dimensional motion in the x direction. The stress-equation of motion for the volume attached to the frame is

$$\frac{\partial \tau_{xx}}{\partial x} = \frac{\partial^2}{\partial t^2} \left[\beta\rho_f U_x + (1 - \beta)\rho_r u_x \right]$$

$$= \frac{\partial^2}{\partial t^2} \left[\beta\rho_f u_x + (1 - \beta)\rho_r u_x - \beta\rho_f(u_x - U_x) \right]$$ (14)

where ρ_f is the density of the pore fluid and ρ_r is the density of the solid material composing the frame (density of individual grains for granular material). Differentiating with respect to x, and substituting for τ_{xx} from Eq. (6) we obtain the one-dimensional form of one of the equations given by Biot (Eq. (6) of Biot, 1962a). His equation is

$$\nabla^2 (He - C\zeta) = \frac{\partial^2}{\partial t^2} (\rho e - \rho_f \zeta)$$ (15)

where ρ is the total density of the saturated medium.

The second equation, which describes the motion of the fluid relative to the frame, is

$$\beta \frac{\partial p_f}{\partial x} = \frac{\partial^2}{\partial t^2} \left[\beta\rho_f U_x \right] + \frac{\beta\eta}{k} \frac{\partial}{\partial t} \left[\beta(U_x - u_x) \right]$$

or

$$\frac{\partial p_f}{\partial x} = \frac{\partial^2}{\partial t^2} \left[\rho_f u_x - \frac{\rho_f}{\beta} \left(\beta(u_x - U_x) \right) \right] - \frac{\beta\eta}{k} \frac{\partial}{\partial t} \left[\beta(u_x - U_x) \right] \ .$$ (16)

The last term on the right side of this equation gives the viscous
resistance to flow which depends on η, the viscosity of the pore
fluid, and k, the coefficient of permeability. By differentiating
Eq. (16) with respect to x and substituting for p_f from Eq. (6) we
obtain the one-dimensional form of the second equation given by
Biot. His equation is

$$\nabla^2 (Ce - M\zeta) = \frac{\partial^2}{\partial t^2} (\rho_f e - m\zeta) - \frac{\eta}{k} \frac{\partial \zeta}{\partial t} . \qquad (17)$$

In Eq. (17), a parameter m, greater than ρ_f/β, has been substituted
for ρ_f/β in the part of the inertial term corresponding to the
increment of fluid flow. This has been done to account for the
fact that not all of the pore fluid moves in the direction of the
macroscopic pressure gradient because of the tortuous, multi-
directional nature of the pores. As a result less fluid flows in
or out of an element for a given acceleration than if all the pores
were uniform and parallel to the gradient. The parameter m may be
written as

$$m = \alpha \rho_f/\beta, \quad \alpha \geq 1 . \qquad (18)$$

For uniform pores with axes parallel to the gradient, α would
equal 1, while for a random system of uniform pores with all possible
orientations the theoretical value of α is 3. In real granular
materials it is impossible to calculate α from theory so that it
must be considered one of the variables to be determined from
experiments.

Equations (15) and (17) are a pair of coupled differential
equations that determine the dilatational motion of a saturated
porous medium with a linear elastic frame and a constant ratio of
fluid flow to pressure gradient (Poiseuille flow). Solving these
equations leads to a relationship between attenuation and frequency
such as shown by the broken curve of Fig. 2. It is clear from this
figure that the model is not adequate to predict the behavior of
real sediment at this stage of development. In fact, two major
modifications are required to accomplish this. First, the viscous
resistance to fluid flow must be made frequency dependent to correct
for the deviation from Poiseuille flow that occurs at all but very
low frequencies and, second, the inelastic nature of the frame must
be taken into account.

In order to incorporate the frequency dependence of viscous
resistance, Biot derived a complex correction factor to be applied
to the fluid viscosity by considering the actual microvelocity
field that exists within the pore channels. This problem of
oscillatory motion in a closed channel is quite well known, having
been solved as early as 1868 by Kirchoff. Biot's solution is

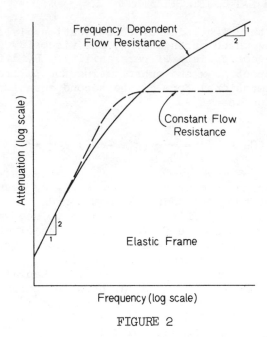

FIGURE 2

ATTENUATION VERSUS FREQUENCY FOR A LINEAR ELASTIC FRAME

written in such a way that the factor $\beta^2 F\eta/k$ gives the ratio of the friction force exerted by the fluid on the frame to the average relative velocity for oscillatory motion. Hence, η/k may be replaced by $F\eta/k$ in the frequency domain, where

$$F(\kappa) = F_r(\kappa) + i\, F_i(\kappa) = \frac{1}{4}\frac{\kappa T(\kappa)}{1-2\,T(\kappa)/i\kappa}\quad,$$

$$T(\kappa) = \frac{ber'(\kappa) + i\, bei'(\kappa)}{ber(\kappa) + i\, bei(\kappa)}\quad,$$

and

$$\kappa = a(\omega\rho_f/\eta)^{1/2}$$

The functions $ber(\kappa)$ and $bei(\kappa)$ are the real and imaginary parts of the Kelvin function, ω is angular frequency, and a is a parameter with the dimension of length that depends on both the size and shape of the pores. $F(\kappa)$ approaches unity for very low frequencies, thus resulting in the same equation as when Poiseuille flow is assumed. Like α, the parameter a cannot be derived theoretically for real sediments so that it must be obtained from experiments.

The solution on which Biot's complex correction factor is based is valid only for frequencies where the wavelength is large

compared to the pore size. For sands, this puts the upper limit on
frequencies at about 10^5 to 10^6 Hz, which is high enough to cover
the frequency range of interest. When the complex correction factor
is incorporated into Eq. (17), by replacing η by ηF, a relationship
between frequency and attenuation such as shown by the solid curve
in Fig. 2 is obtained.

Finally, to complete a physically realistic model of saturated
sediment, the inelastic nature of the frame must be considered. In
order to study the effects of inelasticity separate from those of
viscous losses in the fluid, one may work at frequencies low enough
that the inelastic effects dominate. Since low frequencies require
long wavelengths, this approach often results in complications both
in laboratory and in situ studies. Alternatively, tests may be
performed on dry or partly saturated specimens with corrections
made for the effects of wetting and full saturation. Still another
possibility is to examine losses occurring during volume-constant
shearing deformation and then to interpret the results in terms of
dilatational motion. All three of these approaches have been used
in various experimental programs. In addition many theoretical
models incorporating friction losses, relaxation at grain boundaries,
and other forms of dissipation have been studied. To describe all
this work in detail is beyond the scope of this paper. However, we
will make use of one major observation that is common to much of
the work that has been reported, that is, that the logarithmic
decrement is independent of frequency or amplitude for oscillations
of very small amplitude. This is equivalent to saying that the
material has a constant Q, or, for propagating waves, that the
attenuation constant varies linearly with frequency. It is
difficult to reproduce this kind of behavior with a linear viscoelastic model without resorting to an integral formulation that
leads to a very cumbersome complex modulus (see Stoll and Bryan,
1970). On the other hand certain "slightly" nonlinear models are
very appealing since they assume loss mechanisms that are very close
to physical reality. Included in these are several models that are
based on frictional losses at the grain-to-grain contacts of the
particle. These models may be used to approximate the response of
the skeletal frame and still result in a linear system of equations
by replacing the slightly nonlinear terms by their linear equivalents
according to the procedure of Kryloff and Bogoliuboff (1947). This
technique, which is often used in electrical circuit analysis, leads
to a set of linear equations in the frequency domain. Moreover, it
bypasses the need to restrict the results to a particular physical
model because many of the interesting models lead to the same form
of frequency equation. In particular, several of the models for
which the specific damping is constant (i.e., constant Q) may be
shown to result in a constant complex modulus with small imaginary
part.

In order to incorporate the inelasticity of the frame into the present model, H,C,M, and μ in Eqs. (15) and (17) are regarded as operators that may be linear viscoelastic or "slightly" nonlinear. For the conditions in most sediments it is reasonable to consider K_r and K_f, the bulk moduli of the individual particles and the fluid, respectively, as elastic constants and concentrate the inelastic effects in the operators K_b and μ which describe the response of the skeletal frame in a water environment.

To obtain a frequency equation, solutions of Eqs. (15) and (17) of the form

$$e = A_1 \exp(i(\omega t - \ell x))$$

and

$$\zeta = A_2 \exp(i(\omega t - \ell x))$$

are considered, with $\ell = \ell_r + i\,\ell_i$, the constants H,C,M and μ replaced by the appropriate operators, and η replaced by $F(\kappa)\eta$. Upon transformation to the frequency domain the following equation results

$$
\begin{vmatrix}
\overline{H}\,\ell^2 - \rho\omega^2 & \rho_f\omega^2 - \overline{C}\,\ell^2 \\[2ex]
\overline{C}\,\ell^2 - \rho_f\omega^2 & m\omega^2 - \overline{M}\,\ell^2 - \dfrac{i\omega F\eta}{k}
\end{vmatrix} = 0 \qquad (19)
$$

where, in general, $\overline{H},\overline{C},\overline{M}$ and $\overline{\mu}$ are complex functions of frequency derivable by Laplace transformation in the case of linear viscoelastic operators or by the method of Kryloff and Bogoliuboff for the case of slightly nonlinear equations. Following the latter procedure, and restricting to physical models that result in constant complex moduli for the reasons mentioned above,

$$\overline{H} = \frac{(K_r - \overline{K}_b)^2}{D - \overline{K}_b} + \overline{K}_b + 4\overline{\mu}/3$$

$$\overline{C} = \frac{K_r(K_r - \overline{K}_b)}{D - \overline{K}_b}$$

$$\overline{M} = \frac{K_r^2}{D - \overline{K}_b}$$

where $\overline{K}_b = K_b + i\,K_b'$ and $\overline{\mu} = \mu + i\,\mu'$. The roots of Eq. (19) give the attenuation ℓ_i and phase velocity ω/ℓ_r as functions of frequency for the first and second kinds of dilatational wave. A similar procedure may be followed to obtain equations describing shear waves.

APPLICATIONS

When Eq. (19) is solved, utilizing physical properties typical
of ocean sediments, curves such as the ones shown in Fig. 3 are
obtained. For clays and other materials of very low permeability,
the relationship between attenuation and frequency is linear over
almost the entire range of interest. Apparently for these materials
the observed dissipation is largely due to losses in the skeletal
frame with viscous losses in the fluid producing a perturbing
effect only at very high frequencies. On the other hand in coarser
materials such as sand, the losses due to fluid motion relative to
the frame appear to have a strong influence over much of the
frequency range of interest. Only at very low and very high
frequencies does the curve indicate a tendency for a linear relationship between frequency and attenuation. In fact the slope of the
log-log curve is greater than 2 at some points and less than 1/2 at
others. The position of the two curves at the low frequency end of
the scale depends primarily on the damping characteristics of the
skeletal frame so that either curve may be shifted up or down
depending on the nature of the particular material. Also the point
where deviation from a linear relationship occurs will depend on

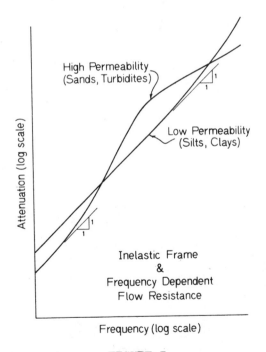

FIGURE 3

TYPICAL CURVES OF ATTENUATION VERSUS FREQUENCY
FOR A REALISTIC MODEL OF OCEAN SEDIMENT

such characteristics as porosity, permeability, damping in the
frame, and the density of the fluid and solid particles.

In order to study the applicability of the proposed model to
real sediments, Eq. (19) has been evaluated for several sets of
data from laboratory and in situ studies. Figure 4 shows one set
of curves that correspond to physical properties of sandy sediments.
Figure 4 also shows most of the experimental data on sands that are
available in the open literature. Some additional data from other
sources may be found in papers by Hamilton (1972) and Hampton and
Anderson (1973). The parameter values used to obtain the curves of

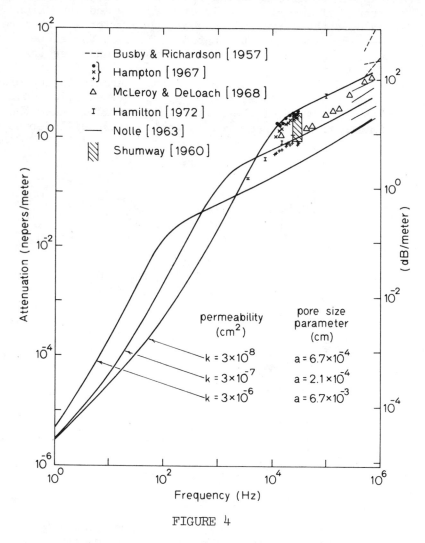

FIGURE 4

ATTENUATION VERSUS FREQUENCY FOR SANDS

Fig. 4 are given in Table 1. The complex moduli, $\overline{K}_b$ and $\overline{\mu}$, were evaluated on the basis of experiments with vibrating columns of dry and saturated sand (see Hall and Richart, 1963, and Hardin, 1965). The logarithmic decrement and the wave velocity measured in these tests were used to evaluate the required complex moduli. For example, the complex shear modulus is related to the logarithmic decrement in torsional vibrations, δ_s, by the relationship

$$\delta_s = \pi\mu'/\mu \ ,$$

and for small damping the real part of the modulus can be obtained from the shear wave velocity, v_s, by the relationship

$$v_s = (\mu/\rho)^{1/2} \ .$$

A complex modulus for longitudinal vibrations of a long column of sediment confined in a flexible membrane may be found using similar equations. Finally, the bulk modulus of the frame may be obtained from the moduli based on shear and longitudinal vibrations by using

MATERIAL PROPERTY	Values for Fig. 4	Values for Fig. 5	Values for Fig. 6
Bulk Modulus of Grains, $K_r(\text{dyn/cm}^2)$	3.6×10^{11}	3.6×10^{11}	3.6×10^{11}
Bulk Modulus of Fluid, $K_f(\text{dyn/cm}^2)$	2.0×10^{10}	2.0×10^{10}	2.0×10^{10}
Mass Density of Grains, $\rho_r(\text{g/cm}^3)$	2.65	2.3	2.65
Mass Density of Fluid, $\rho_f(\text{g/cm}^2)$	1.0	1.0	1.0
Absolute Viscosity of Fluid, η (dyn-sec/cm^2)	1.0×10^{-2}	1.0×10^{-2}	1.0×10^{-2}
Coefficient of Permeability, $k(\text{cm}^2)$	3.0×10^{-6} to 3.0×10^{-8}	2.6×10^{-11}	3.0×10^{-6} to 3.0×10^{-10}
Pore Size Parameter, $a(\text{cm})$	6.7×10^{-3} to 6.7×10^{-4}	1.3×10^{-5}	6.7×10^{-3} to 6.7×10^{-5}
Rod Velocity of Frame, $v_E(\text{m/sec})$	3.0×10^2	3.0×10^2	3.0×10^2
Shear Velocity of Frame, $v_s(\text{m/sec})$	2.1×10^2	2.1×10^2	2.1×10^2
Structure Constant, α	1.25	3.0	1.25
Porosity, β	0.40	0.76	0.40 to 0.70
Log decrement for longitudinal vibrations of frame, δ_E	0.15	0.45	0.15
Log decrement for shear vibrations of frame, δ_s	0.20	0.60	0.20

TABLE 1

MATERIAL PROPERTIES USED TO OBTAIN FIGS. 4, 5 and 6

Eq. (17) of Stoll and Bryan (1970). At this point it is worth noting that there is no low frequency data available for sands other than those used in the calculations described above. Were this not the case, one could use direct field measurement of velocity (shear and dilatational) and attenuation to obtain the appropriate moduli.

The range of permeability coefficients used to obtain Fig. 4 is typical for naturally occurring sands (see Lambe and Whitman, 1969) and the constants a and α were chosen so that the theoretical curves would fit the bulk of the available experimental data. In choosing the pore size parameter a, the usual assumption that permeability is proportional to the square of pore size was made with the same proportionality constant used for all three curves. The reader is referred to Stoll and Bryan (1970) for curves showing the effects of varying each of the other parameters individually.

A second example, for parameters typical of a very fine-grained material, is shown in Fig. 5. In this case the complex moduli that determine the response of the frame and the other parameters of the model were chosen to match the data of Wood and Weston (1964). Other experimental data for similar materials may be found in Hamilton (1972) and Hampton and Anderson (1973) as well as in many other papers.

As a third example of results to be expected from the theory, the family of curves on the left side of Fig. 6 was obtained for a range of permeabilities from 3×10^{-6} to 3×10^{-10} cm^2. As before, the pore size parameter was assumed proportional to the square root of the permeability. For a fixed frequency, such as 30 kHz, a bell-shaped curve relating attenuation to pore size parameter may be derived as shown on the right side of Fig. 6. This curve is similar in all respects to experimental curves (also for 30 kHz) presented by Shumway (1960), McCann and McCann (1969), and Hamilton (1972). If the peak of the theoretical curve is brought into coincidence with the peak of the experimental curves, the pore size parameter and the mean grain diameter may be related resulting in the upper scale shown in Fig. 6. In this case the pore size parameter appears to be between 1/6 and 1/7 of the mean grain diameter. For each curve shown in Fig. 6 the porosity was chosen to match the experimental data given by Shumway (1960) and Hamilton (1972) (e.g., see Fig. 8 of Hamilton's paper). However, no attempt was made to vary the complex moduli of the frame since insufficient data is available at the required low frequencies, particularly for sands.

On the basis of the small amount of data that is available, it appears that the theory will reproduce the experimental results even better when enough data becomes available to include the variation of frame losses with size, porosity, etc.

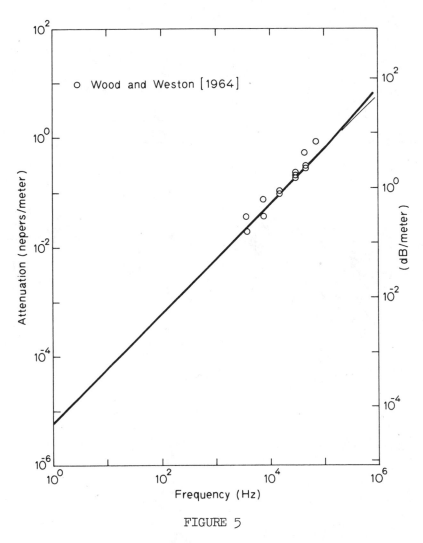

FIGURE 5

ATTENUATION VERSUS FREQUENCY FOR FINE SEDIMENT WITH LOW PERMEABILITY

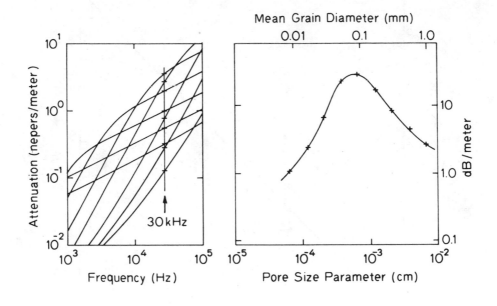

FIGURE 6

RELATIONSHIP BETWEEN ATTENUATION AND GRAIN SIZE AT FIXED FREQUENCY

SUMMARY AND CONCLUSIONS

On the basis of the foregoing discussion it appears that a relatively simple model based on Biot's theory for porous media is capable of reproducing most of the behavior that has been observed in saturated sediments. However, since there is virtually no experimental data available for sands at frequencies less than 10 kHz, it is necessary to obtain much additional information before the full range of usefulness can be evaluated. Nevertheless, several conclusions can be drawn at this time.

First, because of the complex variation of attenuation with frequency that is evident from Figs. 3, 4, 5 and 6, it is clear that extreme care should be exercised in extrapolating experimental curves out of the range where data was taken. Such extrapolation cannot be based strictly on the trend within the data interval nor should a simple first power law be used for all cases. The latter procedure has been suggested by some authors on the basis of the general trend indicated by plotting data for all different kinds of sediment on one diagram. The figures mentioned above show that large errors may result from this procedure for certain materials such as sands and other sediments with high permeability.

One of the strongest arguments for a unifying theory, such as the one discussed herein, is that it can serve as a guide in the acquisition and interpretation of meaningful experimental data. Furthermore it can be used as a starting point in the development of approximate formulas for practical use because it indicates the significant variables and those that may be neglected in certain cases. As an example, consider the problem of low frequency waves. Both the Navy and geophysical researchers in general have recently become more interested in acoustic waves at lower frequencies, so that it is probable that more work will be done in this area in the near future. For this reason it is important to examine which parameters are significant in field and laboratory experiments at low frequencies. In the higher frequency ranges, porosity and grain size have been used quite successfully as correlating parameters. On the basis of the proposed theory this result is to be expected, at least for coarser materials, because permeability is known to be primarily dependent on these two variables. On the other hand at lower frequencies, where frame losses become important, other factors such as intergranular stress and cementation become very significant. On the sea floor intergranular or effective stress depends on the overburden pressure so that it increases rapidly with depth. Thus, field studies involving long waves that pass through regions where there is a significant variation of effective stress may be difficult to interpret. In the laboratory the effective stress depends on the size of the specimen and on the amount of static confining pressure that is applied, as well as other factors. Comparisons of data from different kinds of tests can be made only after these factors have been taken into account.

Finally, it should be noted that the influence of effective stress can easily be incorporated into the proposed theory. This may be done by making the complex moduli representing the frame response dependent on this variable, when more experimental data are obtained.

REFERENCES

Biot, M. A., "General Theory of Three-Dimensional Consolidation," J. Appl. Phys., 12, 155-164, 1941.

Biot, M. A., "Theory of Elastic Waves in a Fluid-Saturated Porous Solid. I. Low-Frequency Range," J. Acoust. Soc. Am., 28, 168-178, 1956a.

Biot, M. A., "Theory of Elastic Waves in a Fluid-Saturated Porous Solid. II. Higher Frequency Range," J. Acoust. Soc. Am., 28, 179-191, 1956b.

Biot, M. A., "Mechanics of Deformation and Acoustic Propagation in Porous Media," J. Appl. Phys., 33, 1482-1498, 1962a.

Biot, M. A., "Generalized Theory of Acoustic Propagation in Porous Dissipative Media," J. Acoust. Soc. Am., 34, 1254-1264, 1962b.

Biot, M. A., and D. G. Willis, "The Elastic Coefficients of the Theory of Consolidation," J. Appl. Mech., 24, 594-601, 1957.

Bishop, A. W., and D. J. Hankel, The Measurement of Soil Properties in the Triaxial Test, Edward Arnold, Ltd., London, 1957.

Busby, J., and E. G. Richardson, "The Absorption of Sound in Sediments," Geophysics, 22, 821-828, 1957.

Geertsma, J., and D. C. Smit, "Some Aspects of Elastic Wave Propagation in Fluid-Saturated Porous Solids," Geophysics, 26, 169-181, 1961.

Hall, J. R., Jr., and F. E. Richard, Jr., "Dissipation of Elastic Wave Energy in Granular Solid," J. Soil Mech. Found. Div., A.S.C.E., 89(SM6), 27-56, 1963.

Hamilton, E. L., "Compressional-Wave Attenuation in Marine Sediments," Geophysics, 37, 620-646, 1972.

Hampton, L. D., "Acoustic Properties of Sediments," J. Acoust. Soc. Am., 42, 882-890, 1967.

Hampton, L. D., and A. L. Anderson, "Acoustics and Gas in Sediments - Applied Research Laboratories Experience," Proc. ONR Conf. Natural Gases in Marine Sediments and Their Mode of Distribution, Lake Arrowhead, Calif., 1973.

Hardin, B. O., "The Nature of Damping in Sands," J. Soil. Mech. Found. Div., A.S.C.E., 91 (SM1), 1965.

Kryloff, N., and N. Bogoliuboff, Introduction to Non-Linear Mechanics, Princeton University Press, Princeton, N. J., 1947, pp. 55-63.

Lambe, T. S., and R. V. Whitman, Soil Mechanics, John Wiley and Sons, Inc., New York, 1969, pp. 281-294.

McCann, C., and D. M. McCann, "The Attenuation of Compressional Waves in Marine Sediments," Geophysics, 34, 882-892, 1969.

McLeroy, E. C., and A. DeLoach, "Sound Speed and Attenuation, from 15 to 1500 kHz, Measured in Natural Sea-Floor Sediments," J. Acoust. Soc. Am., 44, 1148-1150, 1968.

Nolle, A. W., W. A. Hoger, J. F. Mifsud, W. R. Runyan, and M. B. Ward, "Acoustical Properties of Water-Filled Sands," J. Acoust. Soc. Am., 35, 1394-1408, 1963.

Shumway, G., "Sound Speed and Absorption Studies of Marine Sediments by a Resonance Method," Geophysics, 25, 451-467, 659-682, 1960.

Stoll, R. D., and G. M. Bryan, "Wave Attenuation in Saturated Sediments," J. Acoust. Soc. Am., 47, 1440-1447, 1970.

Wood, A. B., and D. E. Weston, "The Propagation of Sound in Mud," Acustica, 14, 156-162, 1964.

Reprinted from Journal of Geophysical Research, v. 82, p. 5719-5735.

Viscoelastic Properties of Fluid-Saturated Cracked Solids

RICHARD J. O'CONNELL

Department of Geological Sciences and Center for Earth and Planetary Science
Harvard University, Cambridge, Massachusetts 02138

BERNARD BUDIANSKY

Division of Applied Sciences and Center for Earth and Planetary Science
Harvard University, Cambridge, Massachusetts 02138

The effective elastic moduli of a fluid-saturated solid containing thin cracks depend on the degree of interconnection between the cracks. Three separate regimes may be identified: (1) dry (drained), in which fluid in cracks can flow out of bulk regions of compression, (2) saturated isobaric, in which fluid may flow from one crack to another but no bulk flow takes place, and (3) saturated isolated, in which there is no communication of fluid between cracks. Transitions between these cases involve fluid flow, resulting in dissipation of energy. Relaxation of shear stresses in viscous fluid inclusions also results in dissipation. Viscoelastic moduli are derived, by using a self-consistent approximation, that describe the complete range of behavior. There are two characteristic frequencies near which dissipation is largest and the moduli change rapidly with frequency. The first corresponds to fluid flow between cracks, and its value can be estimated from the crack geometry or permeability. The second corresponds to the relaxation of shear stress in an isolated viscous fluid inclusion; its value may also be estimated. Variations of crack geometry result in a distribution of characteristic frequencies and cause Q to be relatively constant over many decades of frequency. Fluid flow between cracks accounts for attenuation of seismic waves in water-saturated rocks and attenuation observed in laboratory measurements on water-saturated rocks and partially molten aggregates. Attenuation in a partially molten upper mantle is probably due to fluid flow between cracks, although grain boundary relaxation in an unmelted upper mantle could also account for the seismic low-velocity zone. Grain boundary relaxation in the mantle may cause the long-term shear modulus to be around 20% less than that measured from seismic observations.

INTRODUCTION

The presence of cracks can have a large effect on the effective elastic properties of a solid even though the fractional volume occupied by the cracks is extremely small. The state of fluid saturation of the cracks has a correspondingly large effect. Because the fluid can flow in response to pressure changes, the elastic response of a saturated cracked solid to an applied load may be time dependent as the pore fluid flows in response to pressure changes. The internal flow of fluid in the pore space of the solid will result in the dissipation of energy, and the bulk solid will exhibit anelastic or viscoelastic behavior. An elastic wave in such a medium will be attenuated owing to such dissipative mechanisms, and its velocity may depend on its frequency.

Such effects may be important in the earth wherever a fluid phase exists either in cracks in rocks or as thin films, along grain boundaries, for instance. Variations of seismic wave velocities preceding earthquakes have been interpreted in terms of changes in the state of saturation of cracks in crustal rocks [*Nur*, 1972; *Scholz et al.*, 1973; *Whitcomb et al.*, 1973; *O'Connell and Budiansky*, 1974]. Such changes in the state of saturation will cause corresponding changes in the elastic properties of the affected region which may in turn have a bearing on the development of the strain field associated with impending earthquakes. In fact, such processes are implicit in the dilatancy models of earthquakes that have been proposed to account for the velocity changes.

The low-velocity seismic region in the upper mantle has been interpreted as one in which thin inclusions of fluid exist owing to partial melting [*Anderson and Sammis*, 1970]; thus this region may also exhibit a time or frequency dependent elastic response. In this case the elastic properties of the region

inferred from seismic observations may be substantially different from those appropriate for longer-term deformations. In addition, delayed elastic response due to fluid flow may have an influence on the history of deformation before and after large lithospheric earthquakes [*Mavko and Nur*, 1975].

In this paper we analyze the time or frequency dependent elastic properties of a solid permeated with fluid-filled cracks. We find that such a solid exhibits linear viscoelastic behavior and that there are three important relaxation mechanisms. The first is due to the relaxation of a viscous fluid in a single crack in shear, which has previously been analyzed by *Walsh* [1968, 1969]. The second involves fluid flow between cracks with different orientations. Although this mechanism has been recognized before [*Mavko and Nur*, 1975], it has up to now not been systematically analyzed. The last mechanism involves the bulk flow of fluid out of regions of compression and is essentially that of consolidation [cf. *Terzaghi*, 1943].

The characteristic times associated with these relaxation mechanisms can be related to the properties of the fluid, the geometrical properties of the cracks, and the effective permeability of the cracked solid. Thus the analysis results in a continuum model with characteristic parameters that can be estimated from measurements and considerations independent of the model itself.

The analysis in this paper is an extension of that of *O'Connell and Budiansky* [1974] and *Budiansky and O'Connell* [1976], hereafter referred to as OB and BO, respectively. They calculated the static elastic moduli for a solid permeated with cracks on the basis of the consideration of the energy change associated with the introduction of a set of cracks into an initially uncracked homogeneous, isotropic elastic body in a state of uniform stress. The energy change associated with the cracks was estimated from the energy change due to a single isolated crack in an infinite medium having the effective elastic

Paper number 7B0604.

properties of the cracked medium. This 'self-consistent' approximation attempts to take into account interactions between cracks and gives results that may be useful for nondilute concentrations of cracks. The accuracy of the self-consistent approximation at arbitrary crack concentrations is not known, since the correct results must depend on the details of higher-order geometrical crack statistics, which are unspecified. Details of the analysis are given by BO, and comparison of the theoretical results with laboratory data is given by OB. The review by *Watt et al.* [1976] discusses the self-consistent approximation in the general context of the elastic properties of heterogeneous bodies.

STATIC ELASTIC MODULI

Consider a solid permeated with N cracks per unit volume, which are approximated as ellipsoidal cavities with semiaxes $a > b \gg c$, each filled with a fluid of bulk modulus $\check{K}$ and viscosity η. The concentration of the cracks is specified by the crack density

$$\epsilon = (2N/\pi)(A^2/P) \tag{1}$$

where A is the area and P the perimeter of the cracks. For circular cracks of radius a this reduces to

$$\epsilon = N\langle a^3 \rangle \tag{2}$$

It has been shown (OB and BO) that the static elastic moduli of a fluid-saturated cracked solid are nearly independent of the shape of the cracks so long as the crack density is defined as it is in (1). We shall therefore continue to consider only circular cracks, with the understanding that the results may be used for other crack shapes with the crack density defined by (1).

We consider initially several limiting cases for which the static elastic moduli may be immediately determined from the results of *Budiansky and O'Connell* [1976]. We restrict our attention to the case for which the bulk modulus of the fluid satisfies $K \geq \check{K} \gg Kc/a$ (K being the bulk modulus of the uncracked solid), which is appropriate for water in cracks with aspect ratios $c/a \sim 10^{-3}$, for example. (The restriction $K \geq \check{K}$ is to eliminate the case of very stiff disclike inclusions, which do not concern us here.)

The first, trivial, case is that in which the viscosity of the fluid is sufficiently great that the shear stresses in the fluid do not relax and the crack faces are essentially glued together over the time of interest. This will be the case for elastic waves of sufficiently high frequency, and the moduli of the solid will be completely unaffected by the presence of the cracks.

In the second case the shear stresses in the fluid relax completely, but no fluid is able to flow out of any of the cracks. Application of stress to the saturated body will induce changes in the pore fluid pressure which will be superposed on the ambient pore pressures.

These changes of fluid pressure may not be the same in every crack and will be given by $p_f = -\tilde{\sigma} = -\sigma_{ij}n_in_j$, where σ_{ij} is the stress tensor in the solid matrix and $\mathbf{n}$ is the unit normal to the crack. This case will be called saturated isolated, for which the effective bulk modulus $\check{K}$, shear modulus $\check{G}$, and Poisson ratio $\bar{\nu}$ have been derived by BO and OB:

$$\check{K}/K = 1 \qquad \frac{\check{G}}{G} = 1 - \frac{32}{15}\left(\frac{1-\bar{\nu}}{2-\bar{\nu}}\right)\epsilon \tag{3}$$

$$\bar{\nu} = \nu + \frac{32}{45}\frac{(1-\bar{\nu}^2)(1-2\nu)}{(2-\bar{\nu})}\epsilon \tag{4}$$

where K, G, and ν are the moduli of the uncracked solid. Note

that the root of (4) must be used to evaluate the moduli (3).

The third case involves the flow of fluid out of (or into) individual cracks in response to pressure gradients resulting from changes in fluid pressure in the cracks. If the fluid is in communication with the outside of the sample, held at constant fluid pressure, then the fluid pressure in the cracks will ultimately reequilibrate to its former value after a change caused by deformation of the solid. The solid will thus ultimately respond as if the fluid were not there, and the effective elastic moduli are those appropriate for dry or empty cracks (BO):

$$\frac{\check{K}}{K} = 1 - \frac{16}{9}\left(\frac{1-\bar{\nu}^2}{1-2\bar{\nu}}\right)\epsilon$$

$$\frac{\check{G}}{G} = 1 - \frac{32}{45}\frac{(1-\bar{\nu})(5-\bar{\nu})}{(2-\bar{\nu})}\epsilon \tag{5}$$

$$\bar{\nu} = \nu - \frac{16}{45}\frac{(1-\bar{\nu}^2)(10\nu - 3\nu\bar{\nu} - \bar{\nu})}{(2-\bar{\nu})} \tag{6}$$

This case is called drained in the soil mechanics literature [e.g., *Rice and Cleary*, 1976].

The last case is intermediate between the second (saturated isolated) and the third (dry). All the cracks are in communication with each other, but no fluid flow is permitted out of bulk regions of the sample. The change in fluid pressure will be the same in all the cracks and will be $p_f = -\tilde{\sigma} = -\sigma_{ii}/3$, but there will be no bulk flow of fluid. This case will be called saturated isobaric; it is the same as that called undrained in the soil mechanics literature [e.g., *Rice and Cleary*, 1976].

For an externally applied stress that is purely isotropic ($\sigma_{ij} = \sigma\delta_{ij}$, where δ_{ij} is the Kronecker delta) the fluid pressure will be the same in all cracks regardless of whether they are in communication or not. Thus the effective bulk moduli will be the same for the saturated isobaric case ($p_f = -\sigma_{ii}/3 = -\sigma$) as it is for the saturated isolated case ($p_f = -\sigma_{ij}n_in_j = -\sigma n_in_i = -\sigma$) given by (3). If a shear stress is applied, there will be no net volume change of the sample and no net change in the total crack volume. Although some cracks will be compressed, other cracks, at different orientations, will be equally dilated. If the cracks are in communication, fluid will flow from the compressed to the dilated cracks, and there will be no change in fluid pressure ($p_f = \sigma_{ii}/3 = 0$). Thus the cracked solid will respond in shear as if the cracks were empty, and the effective shear modulus will be the same as it is for the case of dry cracks (5).

Consequently, the effective bulk and shear moduli for the saturated isobaric (undrained) case are

$$\check{K}/K = 1 \qquad \frac{\check{G}}{G} = 1 - \frac{32}{45}\frac{(1-\nu')(5-\nu')}{(2-\nu')}\epsilon \tag{7}$$

where the value of ν' to be used in the expression for $\check{G}$ is the root of

$$\nu' = \nu - \frac{16(1-\nu'^2)(10\nu - 3\nu\nu' - \nu')}{45\quad(2-\nu')} \tag{8}$$

The effective Poisson ratio for the bulk sample will not be given by (8), however. Instead it can be calculated from standard relations among elastic moduli and is

$$\bar{\nu} = \frac{(1+\nu)\check{K}/K - (1-2\nu)\check{G}/G}{2(1+\nu)\check{K}/K + (1-2\nu)\check{G}/G} \tag{9}$$

with $\check{K}/K$ and $\check{G}/G$ given by (7). We note that for the saturated

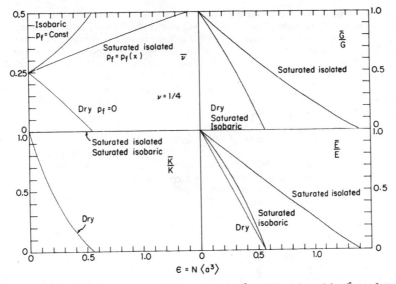

Fig. 1.   Effective Poisson ratio $\bar{\nu}$, bulk modulus $\bar{K}$, shear modulus $\bar{G}$, and Young's modulus $\bar{E}$ as a function of crack density ϵ for a solid permeated with dry or fluid-saturated cracks. In the saturated isobaric case the fluid in each crack is in communication with other cracks, and the fluid pressure is the same in all cracks. In the saturated isolated case the fluid in each crack is not in communication with other cracks, and the fluid pressure depends on the orientation of the crack. The shear modulus is the same for the dry and saturated isobaric cases. Poisson's ratio of the uncracked solid is $\frac{1}{4}$.

isolated and dry cases the expressions (4) and (6) for Poisson's ratio are equivalent to (9).

The elastic moduli $\bar{K}$, $\bar{G}$, $\bar{\nu}$, and $\bar{E}$ (Young's modulus) as a function of crack density ϵ are shown in Figure 1 for the three cases: dry, saturated isobaric, and saturated isolated. Poisson's ratio of the uncracked solid is $\frac{1}{4}$. For the saturated isobaric case the shear modulus decreases more rapidly with an increase in crack density than it does for the saturated isolated case, and it vanishes at $\epsilon = \frac{9}{16}$, at which point, Poisson's ratio reaches a limiting value of $\frac{1}{2}$. Thus saturated cracks have a larger effect on the elastic moduli of a solid under isobaric conditions than under isolated conditions.

Figure 2 shows the compressional wave velocity $\bar{V}_P$, the shear wave velocity $\bar{V}_S$, and the ratio $\bar{V}_P/\bar{V}_S$ for the same cases as appear in Figure 1. The velocities are calculated from the moduli directly:

$$\bar{V}_S/V_S = (\bar{G}/G)^{1/2}$$

$$\bar{V}_P/V_P = [(1 - \bar{\nu})(1 + \nu)\bar{K}/(1 + \bar{\nu})(1 - \nu)K]^{1/2} \qquad (10)$$

$$\frac{\bar{V}_P/\bar{V}_S}{V_P/V_S} = [(1 - \bar{\nu})(1 - 2\nu)/(1 - 2\bar{\nu})(1 - \nu)]^{1/2}$$

The ratio $\bar{V}_P/\bar{V}_S$ increases much more rapidly with increasing

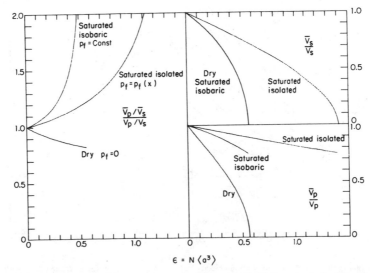

Fig. 2.   Compressional, $\bar{V}_P$, and shear, $\bar{V}_S$, wave velocities and their ratio $\bar{V}_P/\bar{V}_S$ for a cracked solid with dry cracks and fluid-saturated cracks which are in communication (saturated isobaric) or not connected (saturated isolated).

crack density for isobaric conditions than it does for isolated cracks; furthermore, a transition from isolated to isobaric conditions at a constant crack density could drastically reduce the shear wave velocity (and shear modulus).

For the interpretation of measured elastic properties in terms of theoretical results, OB proposed a plot of $\bar{V}_P/\bar{V}_S$ against $\bar{V}_S$. Figure 3 is such a plot, when a fraction of the cracks is saturated under isobaric conditions and the remaining fraction is dry. This figure is analogous to Figure 8 of OB, which is for a mixture of saturated isolated and dry cracks. We note that the interpretation of the laboratory data of OB would not be substantially changed if it were based on the saturated isobaric case rather than the saturated isolated case. In particular, the curves for complete saturation ($\xi = 1.0$) are the same for the two cases, except for the values of crack density along the curve. This is so because the changes in $\bar{V}_S$ and $\bar{V}_P/\bar{V}_S$ are due only to changes in the shear modulus, the bulk modulus being unaffected by cracks for either case.

Which of the four limiting cases is best applied to any given situation depends on the time history of the loading (or frequency of elastic waves) and on the characteristic times for the transitions between the cases. These will be considered in the next section, where time or frequency dependent results are presented that are applicable to all limiting cases as well as intermediate cases.

VISCOELASTIC BEHAVIOR

The transitions between the four cases described in the last section involve the relaxation of shear stresses in a viscous fluid and the flow of a viscous fluid in a permeable medium. These effects, coupled to the elastic deformation of the solid, lend themselves to a description in terms of linear viscoelasticity. The analysis for the viscoelastic case is carried out in the appendix. The results can be represented as frequency dependent effective elastic moduli, which are appropriate for sinusoidal variation of stress and strain. The moduli are complex quantities, and the imaginary parts give rise to dissipation of energy or attenuation of elastic waves. The variations of the moduli with frequency depend on parameters that represent characteristic frequencies for the transition from one type of limiting behavior to another. The values of the parameters are determined by the geometric properties of the cracks and the nature of interconnections between cracks, and they are estimated in the appendix.

The frequency dependent complex moduli $\bar{K}$ and $\bar{G}$ are defined by the constitutive equation

$$\sigma_{ij} = 2\bar{G}e_{ij} + (\bar{K} - 2\bar{G}/3)e_{kk}\,\delta_{ij} \qquad (11)$$

with the understanding that the time-varying strain Re $[e_{ij}\exp(i\omega t)]$ produces the stress Re $[\sigma_{ij}\exp(i\omega t)]$. For fluid-saturated cracks (i.e., excluding the 'dry' case) the effective moduli are derived in the appendix as

$$\bar{K}/K = 1$$

$$\frac{\bar{G}}{G} = 1 - \frac{32}{45}\left(\frac{1-\nu'}{2-\nu'}\right)[(2-\nu')D + 3C]\epsilon \qquad (12)$$

where ν' is the root of

$$\nu' = \nu + \frac{16}{45}\left(\frac{1-\nu'^2}{2-\nu'}\right)[2(1-2\nu)C$$
$$- (2-\nu')(1+3\nu)D]\epsilon \qquad (13)$$

where

$$D = \frac{-id\omega_1/\omega}{1 - id\omega_1/\omega} \qquad (14)$$

with

$$d = \frac{9}{16}\left(\frac{1-2\nu}{1-\nu}\right)\frac{1}{1-\nu'}\frac{\bar{G}}{G} \qquad (15)$$

and

$$C = \frac{-ie\omega_2/\omega}{1 - ie\omega_2/\omega} \qquad (16)$$

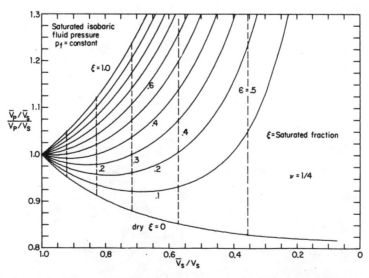

Fig. 3. A plot of the ratio of P wave to S wave velocity versus S wave velocity for a cracked solid with both dry and communicating (saturated isobaric) fluid-filled cracks. The fraction of saturated cracks is ξ, and contours of constant crack density ϵ are shown as dashed lines; these are vertical because the shear velocity is independent of saturation for this case.

with

$$e = \frac{\pi}{4} \left(\frac{2 - \nu'}{1 - \nu'} \right) \frac{\bar{G}}{G} \qquad (17)$$

The value of Poisson's ratio for the bulk sample is not given by (13); instead it may be calculated from the bulk and shear moduli, as given by (9).

In (14) and (16), $i = (-1)^{1/2}$, ω is the frequency, and ω_1 and ω_2 are characteristic frequencies. The characteristic frequency for the flow of fluid between cracks, corresponding to the transition from saturated isolated to saturated isobaric behavior, is ω_1. The value of this parameter can be estimated from several simple models, which are discussed in the appendix. One estimate, which is appropriate for closely spaced cracks having many intersections, is

$$\omega_1 \approx (K/\eta)(c/a)^3 \qquad (18)$$

where η is the fluid viscosity and c/a is the aspect (thickness to diameter) ratio of an oblate spheroidal crack. Other estimates that depend on the permeability of the cracked solid are discussed in the appendix.

The characteristic frequency for the relaxation of shear stress in the viscous fluid in the crack, which corresponds to the transition from the glued case to the saturated isolated case, is

$$\omega_2 \approx (G/\eta)(c/a) \qquad (19)$$

Equations (12)–(17) must be solved simultaneously for $\bar{G}/G$. The bulk Poisson ratio $\bar{\nu}$ can be calculated from (9), and the elastic wave speeds from (10). The resulting moduli describe the cracked solid for the complete range of behaviors from the saturated isobaric case ($\omega \ll \omega_1$, $D = 1$; $\omega \ll \omega_2$, $C = 1$) through the saturated isolated case ($\omega \gg \omega_1$, $D = 0$; $\omega \ll \omega_2$, $C = 1$) to the glued case ($\omega \gg \omega_1$, $D = 0$; $\omega \gg \omega_2$, $C = 0$). The moduli will be complex in general, but for the limiting cases the imaginary parts will be vanishingly small, and the moduli agree with those calculated for these cases from static considerations in the previous section.

The imaginary parts of the moduli give rise to dissipation of energy, and their relation to various measures of dissipation or attenuation are discussed by O'Connell and Budiansky [1977].

Wave Velocities and Attenuation

The velocities of shear (S) and longitudinal (P) waves can be calculated from the complex moduli by use of (10). The velocities so calculated are complex quantities, and the phase velocity and attenuation for each type of wave are related to the real and imaginary parts. The displacements associated with either deviatoric (S) or longitudinal (P) waves in a viscoelastic medium can be written

$$u = B \exp(-\alpha x) \exp[i\omega(t - x/c)]$$

where the spatial attenuation is

$$\alpha = \omega v_i/(v_r^2 + v_i^2) \qquad (20)$$

and the phase velocity is

$$c = (v_r^2 + v_i^2)/v_r \qquad (21)$$

where the complex velocity is $v = v_r + iv_i$.

A common measure of dissipation is the quality factor Q. A definition of Q in terms of stored elastic and dissipated energy [O'Connell and Budiansky, 1977] yields

$$Q^{-1} = M_i/M_r \qquad (22)$$

where M_r and M_i are the real and imaginary parts of the appropriate complex viscoelastic modulus. Another nonequivalent expression has been deduced by *Futterman* [1962] for traveling waves, which we will call the 'seismic' Q:

$$Q_{\text{seismic}}^{-1} = [1 - \exp(-4\pi v_i/v_r)]/2\pi \qquad (23)$$

Note that for large dissipation (or small Q), (22) gives a limiting value of zero, while (23) gives a limiting value of 2π.

The phase velocities for P and S waves as a function of frequency for different crack densities ϵ are shown in Figure 4. The frequency range is centered on the characteristic frequency for viscous relaxation of the fluid in the cracks, ω_2 (equation (19)), and covers the transition from the saturated isolated to the glued case. (The effect of flow between cracks has been omitted by setting $D = 0$ in (12) and (13).) Also shown is the seismic Q for each wave. It can be seen that the dissipation can be quite large, Q being as low as 10, which is very near the lower limit of 2π. The velocity dispersion is also large and is spread out over two decades in frequency around $\omega = \omega_2$.

Figure 5 shows the velocities and seismic Q for the transition from saturated isobaric to saturated isolated behavior in the frequency range near the characteristic frequency for flow between cracks ω_1 (equation (18)). The effect of viscous relaxation has been removed by setting $C = 1$ in (12) and (13). The dissipation and velocity dispersion are greater for this effect than for viscous relaxation and are spread over a slightly larger range of frequencies. The shift of the peak in Q^{-1} to lower frequency for larger crack densities is a consequence of the self-consistent approximation and the change of $\bar{G}$ and ν' in (15). The flat peaks of the Q curves are due to the definition of the seismic Q with a lower limit of 2π.

Figure 6 shows the effects of both dissipative mechanisms. The characteristic frequencies ω_1 and ω_2 differ by a factor of 10^4, which is appropriate for cracks with an aspect ratio $c/a \sim 10^{-2}$ (cf. (18) and (19)), and viscous relaxation occurs at a considerably higher frequency than flow between cracks. There are consequently two well-separated peaks in dissipation with associated velocity dispersion, with 'plateaus' of nearly constant velocity at both sides of the peaks. At low frequencies, dissipation is small, and the velocities correspond to the saturated isobaric case; thus (7)–(9) give the relaxed moduli. At high frequencies also, dissipation becomes small, and the intrinsic moduli for the uncracked solid (the glued case) are the unrelaxed moduli. The region between the two peaks corresponds to saturated isolated behavior.

Small Crack Density

For sufficiently small crack densities the self-consistent approximation can be dropped, and (12)–(17) evaluated with ν' set equal to the intrinsic value of Poisson's ratio ν. In this case the behavior is that of a 'standard linear solid' [*Zener*, 1948], and the effective shear modulus $\bar{G} = \bar{G}_r + i\bar{G}_i$ is given by

$$\bar{G}_r = G - \delta G_1/[1 + (\omega\tau_1)^2] - \delta G_2/[1 + (\omega\tau_2)^2] \qquad (24)$$

$$\bar{G}_i = \delta G_1 \omega\tau_1/[1 + (\omega\tau_1)^2] + \delta G_2 \omega\tau_2/[1 + (\omega\tau_2)^2]$$

where the relaxation of the modulus due to each relaxation mechanism is

$$\delta G_1 = 32(1 - \nu)G\epsilon/45$$

$$\delta G_2 = 32(1 - \nu)G\epsilon/15(2 - \nu) \qquad (25)$$

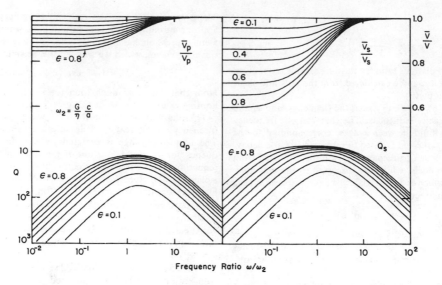

Fig. 4.   Elastic wave velocities and seismic Q for P and S waves for frequencies near the characteristic frequency ω_2 for relaxation of shear stress in the viscous fluid filling the cracks. At high frequencies the cracks are glued, and the wave velocities are unaffected by the presence of cracks. At low frequencies the velocities are those for the saturated isolated case. Curves are shown for values of the crack density ϵ from 0.1 to 0.8.

and the relaxation times (at constant stress [cf. *Nowick and Berry*, 1972, p. 491]) are

$$\tau_1 = 1/d\omega_1 \qquad \tau_2 = 1/e\omega_2 \qquad (26)$$

A value of Poisson's ratio $\nu = \frac{1}{4}$ results in $d = \frac{1}{2}$, $e = 7\pi/12$, $\delta G_1/G = 8\epsilon/15$, and $\delta G_2/G = 32\epsilon/35$. Thus at small crack densities, viscous relaxation is nearly twice as effective a relaxation mechanism as is flow between cracks. Inspection of Figure 6 shows that the opposite holds at large crack densities.

For the case with viscous relaxation only and no flow be-

tween cracks, (24)–(26) are equivalent to *Walsh*'s [1969] results for small aspect ratio c/a and small crack density ϵ, which is implicit in his results.

The attenuation of seismic waves for small crack density is

$$Q^{-1} \approx \bar{G}_i(\omega)/G$$

for shear waves and

$$Q^{-1} \approx 4\bar{G}_i(\omega)/9G$$

for P waves, with $\nu = \frac{1}{4}$.

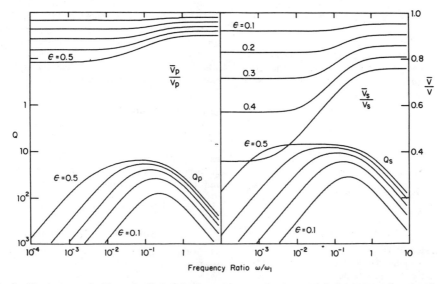

Fig. 5.   Elastic wave velocities and seismic Q for frequencies near the characteristic frequency ω_1 for relaxation due to fluid flow between cracks, showing the transition from saturated isobaric behavior at low frequency to saturated isolated at high frequency. Note the large velocity dispersion and very low values of Q that are possible.

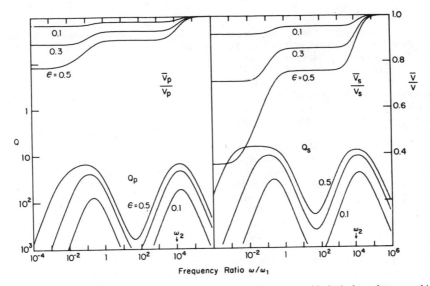

Fig. 6. Elastic wave velocities and seismic Q showing the transition from saturated isobaric through saturated isolated behavior to the glued case. The ratio of the characteristic frequency for viscous relaxation (ω_2) to that for fluid flow between cracks (ω_1) is 10^4, which is appropriate for cracks with aspect ratios of $\sim 10^{-2}$.

Distributed Characteristic Frequencies

Discrete characteristic frequencies for viscous relaxation and flow between cracks would not be expected in real materials owing to variability in the crack shapes and widths and the varied interconnections between cracks. The results obtained so far are easily generalized to treat continuous distributions of characteristic frequencies. Each crack of size $s = 2A^2/\pi P$ (cf. (1)) is assumed to have characteristic frequencies ω_1 and ω_2 for stress relaxation, and the number of cracks with parameters in the range s to $s + ds$, ω_1 to $\omega_1 + d\omega_1$, and ω_2 to $\omega_2 + d\omega$ is

$$dN = n'(s, \omega_1, \omega_2) \, ds \, d\omega_1 \, d\omega_2$$

If we assume for simplicity that the characteristic frequencies are not correlated with size, the increment in crack density may be written

$$d\epsilon = \epsilon n_1(\omega_1) n_2(\omega_2) \, d\omega_1 \, d\omega_2$$

where we have assumed that the joint distribution $n(\omega_1, \omega_2)$ is separable and

$$\int_0^\infty n_i(\omega_i) \, d\omega_i = 1$$

Owing to the additive contribution of the energy of each crack (cf. the appendix) the effective moduli for a distribution of characteristic frequencies are given by (12) and (13) with

$$
D = \int_0^\infty \frac{-i\omega_1 d/\omega}{1 - i\omega_1 d/\omega} n_1(\omega_1) \, d\omega_1
$$

$$
C = \int_0^\infty \frac{-ie\omega_2/\omega}{1 - ie\omega_2/\omega} n_2(\omega_2) \, d\omega_2 \tag{27}
$$

For a uniform distribution of ω_1 between Ω_1 and Ω_2 and ω_2 between Ω_3 and Ω_4 this yields

$$
D = 1 + \left[\log\left(\frac{1 - id\Omega_2/\omega}{1 - id\Omega_1/\omega} \right) \right] \Big/ [id(\Omega_2/\omega - \Omega_1/\omega)]
$$

$$ \tag{28} $$

$$
C = 1 + \left[\log\left(\frac{1 - ie\Omega_4/\omega}{1 - ie\Omega_3/\omega} \right) \right] \Big/ [ie(\Omega_4/\omega - \Omega_3/\omega)]
$$

If the distribution is uniform with respect to $\log \omega_1$ and $\log \omega_2$, then

$$
D = \log\left[(1 - id\Omega_2/\omega)/(1 - id\Omega_1/\omega) \right]/\log(\Omega_2/\Omega_1)
$$

$$ \tag{29} $$

$$
C = \log\left[(1 - ie\Omega_4/\omega)/(1 - ie\Omega_3/\omega) \right]/\log(\Omega_4/\Omega_3)
$$

In particular we note that if the aspect ratio c/a of the cracks is uniformly distributed with respect to $\log(c/a)$ between α_1 and α_2, then the characteristic frequencies ω_1 and ω_2 will have a similar distribution owing to their dependence on c/a in (18) and (19).

Similar results can be obtained for other distributions, either assumed or measured (where possible). A log normal distribution of aspect ratios or characteristic frequencies may be appropriate for many cases [cf. *Nowick and Berry*, 1972, p. 94], but this does not allow the integrals (27) to be expressed in terms of standard functions.

For a static case the effective moduli depend on the crack density ϵ, which is independent of the thickness of the cracks and hence the porosity of the cracked solid. For a dynamic case the thickness of the cracks enters through the characteristic frequencies, which depend on the aspect ratio. Thus the porosity, given by

$$
\varphi = \frac{4\pi\epsilon}{3} \int_0^\infty \alpha n(\alpha) \, d\alpha \tag{30}
$$

where $n(\alpha)$ is the distribution of crack aspect ratios, is related to the frequency dependence of the effective moduli.

DISCUSSION

The model that we have analyzed is directly applicable to solids containing a fluid phase distributed as thin films. The most obvious examples are water-saturated rocks and partially molten polycrystalline aggregates in which the melt is distributed as thin films along grain boundaries. The model also lends itself to a treatment of grain boundary relaxation [*Zener*, 1948] or any relaxation mechanism that relieves shear stress on a plane, as well as to pressure-induced freezing or thawing of a melt.

In spite of the idealization of the model we expect it to be applicable to real materials for two primary reasons. The first is that the results of the analysis are essentially independent of the shape of the cracks, so long as the crack density is specified by (1). The crack density so defined can in turn be directly related to measurements of crack traces in a plane (OB and BO), and it is thus susceptible to measurement or rational estimation. The second is that the crack thickness (and degree of interconnection) enters only into the characteristic frequencies for relaxation, and these will be expected to be distributed over some range corresponding to the spectrum of crack shapes and interconnections present. In this case a plausible distribution of relaxation frequencies may well cover many cases. In addition, the frequencies can be related (through simple models such as those in the appendix) to the crack aspect ratio and the permeability of the solid, both of which can be estimated or measured.

Although we have restricted our attention to the effects of thin cracks, other dissipative mechanisms may be included. Any intrinsic dissipation in the solid can be taken into account by letting the solid moduli G and ν be complex in (12)–(17). The effect of a compressible fluid phase as well as mixtures of dry and saturated cracks can be included as was done by OB and BO, and the effects of different mechanisms (such as grain boundary sliding between some phases and partial melt between others) can be combined by a suitable specification of the relaxation frequency distribution.

Fluid-Saturated Rocks

Figure 7 shows the wave velocities and associated seismic Q values for a water-saturated rock ($\eta = 0.01$ P $= 0.001$ N s/m²) with a distribution of crack aspect ratios that is uniform in log (c/a) between 10^{-4} and 10^{-2}. The characteristic frequencies were calculated from (18) and (19), and (29) was used to account for the distribution of aspect ratios. The aspect ratio distribution is consistent with that measured for Westerly granite by *Hadley* [1976], but it should be regarded only as a plausible representative distribution, and details in the curves of velocity and Q versus frequency should be interpreted accordingly. The crack density of Hadley's sample was 0.2–0.3, which seems representative (cf. OB). For frequencies from 1 to 10^7 Hz the dissipation is due to fluid flow between cracks, and over this range, Q is roughly independent of frequency. Thus a frequency independent Q can result from a superposition of linear relaxation mechanisms [cf. *Liu et al.*, 1976], and it is not necessary to invoke nonlinear mechanisms [*Knopoff*, 1964] (although they may well be present). Laboratory measurements using resonance (10^4–10^5 Hz) or ultrasonic (10^6 Hz) techniques are in the range in which fluid flow between cracks dominates dissipation. This is therefore consistent with the suggestion of *Gordon and Davis* [1968] that intercrack flow was responsible for the increased attenuation induced by the saturation of their granite specimen. For frequencies in and below the seismic range the velocities approach those for the saturated isobaric case, for which the shear velocity is independent of the degree of saturation, although the P wave velocity is not. This is in accord with *Gordon and Davis*'s [1968] measurements at 0.014 Hz. Viscous relaxation in water-saturated cracks becomes important only for frequencies greater than around 10^7 Hz. It therefore appears that for water-saturated rocks with thin cracks, seismic and laboratory wave velocities will correspond to a state ranging between saturated isobaric and saturated isolated and that fluid flow between cracks will be the dominant dissipative mechanism. In addition, the marked velocity dispersion between 1 and 10^6 Hz indicates

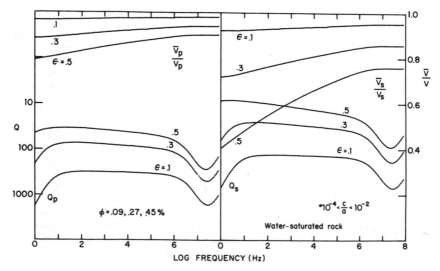

Fig. 7. Wave velocities and seismic Q for a water-saturated rock with aspect ratios distributed (uniformly in log c/a) from 10^{-4} to 10^{-2}. Note that Q can be relatively independent of frequency over a wide range of frequencies. From 1 to 10^6 Hz, dissipation is due to fluid flow between cracks; this range includes most laboratory measurements and seismic measurements. The velocity dispersion between laboratory and seismic frequencies can be appreciable. The crack porosity ϕ is shown for the three crack densities considered.

that ultrasonically measured velocities may be substantially higher than those at seismic frequencies. Note the correspondence between low Q and rapid change of velocity with frequency. This is in accord with the approximate relation

$$Q^{-1} = \frac{M_i}{M_r} \approx \frac{\pi}{2} \frac{d(\log M_r)}{d(\log \omega)}$$

[Nowick and Berry, 1972, p. 90; Pipkin, 1972, p. 46], which is appropriate when Q varies slowly with frequency. This relation may permit laboratory results to be extrapolated to seismic frequencies if bounds on Q can be assigned over the frequency interval.

Nur and Simmons [1969] measured velocities and obtained a rough measure of relative attenuation in a sample of Barre granite saturated with glycerin. By varying the temperature the viscosity of the glycerin and the velocities and attenuation were varied; this gave an effect similar to varying the frequency with constant viscosity. They noted a peak in attenuation for a viscosity of 0.1 P, which they interpreted as being due to viscous relaxation of the fluid in individual cracks, after Walsh's [1969] analysis. This required an aspect ratio of 2×10^{-7} (9.5×10^{-7} including an omitted factor of 2π), which seems implausibly small, since it implies a thickness of a few angstroms for 1-mm cracks. On the other hand, interpreting the attenuation as being due to flow between cracks gives an aspect ratio of $\sim 5 \times 10^{-3}$ through (18), which seems more reasonable. This interpretation is further supported by the nearness of the measured shear velocity to the value for a dry sample, which is characteristic of saturated isobaric behavior.

It should be pointed out that the actual presence of a peak in attenuation (which would indicate a narrow distribution of characteristic frequencies) in the data of Nur and Simmons [1969] is not clear, since it rests on two measurements with water and air in the cracks. The water-saturated sample may not have had the same crack structure as the glycerin-saturated samples, and for the sample with air-filled cracks the high compressibility of the air should be taken into account. (Nevertheless, a peak in attenuation near $\eta = 1$ P is clearly evident for measurements on a limestone sample by Nur [1971], where precise values of attenuation are reported.) If a more or less continuous distribution of aspect ratios were present in the sample, the attenuation for any given viscosity would reflect primarily only the effect of those cracks with aspect ratios in a limited range. Table 1 gives the aspect ratios that would be expected to dominate for different viscosities at the frequency used in the experiment by setting either ω_1 or ω_2 equal to 0.5 MHz in (18) (fluid flow) or (19) (viscous relaxation).

If the possibility of very thin cracks ($c/a < 10^{-5}$) is excluded, flow between cracks could account for most of the attenuation for viscosities of less than 1 P. For higher viscosities, both mechanisms would contribute, and for viscosities greater than ~ 1000 P, only viscous relaxation would contribute. If the peak in attenuation at 0.1 P is due to flow between cracks of aspect ratio 0.005, there should be another peak for viscous relaxation for viscosities of ~ 1000 P; such a peak is not apparent. For viscosities greater than 10^6 P the moduli should be nearly completely unrelaxed; the fact that the measured velocities continued to increase in this range may indicate incomplete saturation or the effect of the increase of the finite shear modulus of glycerol as the temperature was lowered.

A consistent interpretation of these results is that dissipation at low viscosities is due to fluid flow between cracks. The peak in attenuation may be a consequence of incomplete saturation

with glycerol, or it may be spurious owing to a change of the crack structure when the sample was saturated with water. At higher viscosities, attenuation was due to both fluid flow and viscous relaxation in cracks with a wide spectrum of aspect ratios.

Gordon [1974] reported measurements of a glycerin-saturated sample of Rhode Island granite that showed a very sharp relaxation peak for a viscosity of ~ 40 P at 50 kHz, with a corresponding change in the modulus. If the peak is due to the transition from saturated isobaric to saturated isolated behavior, the aspect ratios of the cracks must have a nearly discrete distribution at $c/a = 0.018$. This is consistent with the other measurements that we have discussed, although the sharpness of the relaxation is rather surprising. If the relaxation is interpreted as being due to viscous relaxation, the aspect ratio required is 1.3×10^{-5}, which is small but not completely implausible.

Gordon and Rader [1971] reported that saturating samples of amphibolite and Chester granite with water increased Young's modulus at high frequencies (10–100 kHz) but decreased the moduli at very low frequencies (0.01–0.1 Hz). Gordon [1974] has reported further similar results. This suggests that the upper frequency interval is in the range of the transition from saturated isobaric behavior (for which the shear modulus is the same as that for the dry case) to saturated isolated behavior, which is consistent with our interpretation of Nur and Simmons's [1969] data. The decrease in modulus at very low frequencies was interpreted by Gordon and Rader [1971] as being due to a lubricating effect of water, which would reduce dry friction across crack faces or otherwise enhance grain boundary sliding. An alternative explanation is that the process of saturation increases the crack density, possibly by the extension of cracks by the stresses arising from the fluid pressure gradients driving the infusing fluid. Such crack extension could presumably be enhanced by stress corrosion also. The increased crack density would not be noticed at higher frequencies owing to the increase in the effective modulus due to saturation in the regime between isobaric and isolated behavior. Only at very low frequencies, under completely isobaric conditions, would the increase in crack density be apparent. This interpretation is supported by O'Connell and Budiansky's

TABLE 1. Attenuation in Glycerol-Saturated Granite [Nur and Simmons, 1969]

$\log_{10} \eta$, P	c/a (equation (A6))	c/a (equation (25))
-2 (water)	$2.5E - 3$*	$9.5E - 8$
-1.6	$3.5E - 3$	$2.4E - 7$
-1.2	$4.7E - 3$	$6.0E - 7$
-0.8	$6.4E - 3$	$1.5E - 6$
-0.4	$8.7E - 3$	$3.8E - 6$
0	$1.2E - 2$	$9.5E - 6$
0.5	$1.7E - 2$	$3.0E - 5$
1.0	$2.5E - 2$	$9.5E - 5$
1.5	$3.7E - 2$	$3.0E - 4$
2.0	$5.5E - 2$	$9.5E - 4$
2.5	$8.1E - 2$	$3.0E - 3$
3.0	$1.2E - 1$	$9.5E - 3$
3.5	$1.7E - 1$	$3.0E - 2$
4.0	$2.5E - 1$	$9.5E - 2$
5.0	$5.5E - 1$	$9.5E - 1$
6.0	1.2	9.5

$K = 4.752 \times 10^{11}$ dyn/cm², $G = 3.305 \times 10^{11}$ dyn/cm², and $\omega = 0.5$ MHz.
*Read $2.5E - 3$ as 2.5×10^{-3}.

[1974] and *Hadley*'s [1976] interpretation of laboratory data that indicated that saturated rocks had higher crack densities than they had before saturation. If this interpretation is correct, it indicates that comparisons between the measured properties of dry and saturated rocks must take into account possible changes in the crack density and internal structure of the rocks caused by saturating them.

The conclusion is then that dissipation in water-saturated crystalline rocks containing cracks is primarily due to fluid flow between cracks, under most laboratory as well as seismic conditions. At seismic frequencies the wave velocities will be near those for saturated isobaric conditions, for which the shear wave velocity is independent of saturation. Rocks containing more equant pores, such as sedimentary rocks, will be close to the saturated isobaric condition at even higher frequencies, and models of their properties based on isolated fluid inclusions should not be expected to be accurate at seismic or laboratory frequencies.

Partially Molten Systems

Spetzler and Anderson [1968] reported measurements of changes of velocity and attenuation upon the appearance of a melt phase in frozen samples of water with 1 and 2% NaCl. Upon warming the samples through the eutectic temperature, discontinuous changes in velocity and attenuation were observed, apparently due to the formation of a liquid phase along the boundaries of the ice crystals. Their results are summarized in Table 2. From the observed changes in velocities one can infer crack densities of 0.15 and 0.4 for the two cases with 1 or 2% NaCl. From this and the reported melt concentration the 'average' aspect ratio of the cracks is ~0.05. This gives characteristic frequencies for flow between cracks and viscous relaxation of 600 and 200,000 MHz for 1% NaCl. These are far above the frequencies of 3–10 kHz used in the experiment [*Spetzler*, 1969]. However, if there were fluid pockets present with aspect ratios of 0.001, the frequencies for flow and viscous relaxation would be 4 kHz and 540 MHz. The presence of such low aspect ratio inclusions is certainly plausible and would account well for the changes in attenuation, which would then be due to flow between cracks. The frequency for viscous relaxation, though, would be far too high for that mechanism to contribute substantially. This interpretation is both plausible and consistent with the data; Figures 5 and 7 show that

the observed values of Q are of the same order as those predicted for the inferred crack densities of the samples, even if the crack aspect ratios are distributed over a considerable range. It may be noted that as melting proceeds in such experiments, the effective aspect ratio distribution may change; this causes changes in attenuation at a given frequency. Thus the changes in attenuation noted by Spetzler and Anderson during melting, which they partially attributed to stress-induced freezing and thawing, may have been due to this. Measurements over a wide frequency range would be necessary to resolve this question.

It should also be noted that although *Spetzler and Anderson* [1968] and *Anderson and Spetzler* [1970] interpreted the observed changes in velocity in terms of viscous relaxation, this mechanism was unable (by several orders of magnitude) to account for the observed values of Q [*Spetzler*, 1969]. This indication that the relaxation mechanism had been misidentified does not seem to have been widely noted.

Stocker and Gordon [1975] have reported measurements of extensional velocities and attenuation in partially molten copper-lead and copper-silver alloys. The melt phase in the CuAg samples formed thin films; hence our analysis should apply to these experiments. Table 3 shows the relevant data. The crack density was calculated from Young's modulus for the saturated isobaric case. The average aspect ratios are all small, around 0.005. The characteristic frequencies calculated from the average aspect ratios are around 10 MHz for flow between cracks and around 40,000 MHz for viscous relaxation in cracks; the latter is far above the frequencies of 100 kHz used in the experiments. An aspect ratio of 0.001 gives frequencies of 37 kHz and 6000 MHz. Thus it seems likely that fluid flow between cracks was a dominant dissipative mechanism in the experiments. The observed values of Q are consistent with this interpretation, but the samples would have required a narrow range of aspect ratios around 0.001 in order to have such low values of Q.

It should be noted that the measured Q includes the effects of dissipation in the solid phase as well, as is discussed by Stocker and Gordon. The background Q in Table 3 is their estimate of this component, presumably obtained from measurements taken immediately below the eutectic temperature. If grain boundary relaxation is an important dissipative mechanism in the solid, the appearance of a melt phase along grain

TABLE 2.   Properties of H_2O-NaCl Partial Melts [*Spetzler and Anderson*, 1968; *Spetzler*, 1969]

| | NaCl Concentration | | |
	1%	2%	Remarks
$\Delta V_P/V_P$	0.095	0.28	
$\Delta V_S/V_S$	0.135	0.40	
ΔQ_P	200* → 110.	···	fundamental
	150* → 80	···	overtone
	100* → 70	···	overtone
ΔQ_S	170* → 95	150* → 50	fundamental
	100* → 80	120* → 30	overtone
	···	110* → 22	overtone
Melt concentration	0.035	0.065	
$\epsilon(\Delta V_P)$	0.15	0.3	
$\epsilon(\Delta V_S)$	0.15	0.4	
c/a average	0.056	0.039	
ω_1, MHz	600	200	c/a average
ω_2, MHz	200,000	133,000	c/a average
ω_1, MHz	0.004	0.004	$c/a = 0.001$
ω_2, MHz	540	540	$c/a = 0.001$

*Background Q.

TABLE 3. Properties of CuAg Partial Melts [*Stocker and Gordon*, 1975]

	Melt Fraction			
	0.0014	0.0014	0.0110	0.0110
$\Delta V/V$	0.033	0.020	0.206	0.193
Q	61	91	17	45
Q (background)	179	175	83	62
ϵ	0.08	0.05	0.44	0.42
c/a (average)	0.004	0.007	0.006	0.006
ω_1, MHz	3	13	8	9
ω_2, MHz	28,000	47,000	40,000	42,000

$K = 1.4 \times 10^{12}$ dyn/cm², and $\mu = 0.01$ P.

boundaries may remove this mechanism. Thus care should be taken in assessing the contribution of partial melting to dissipation from the change in Q upon melting. For example, 1-mm grain boundaries 6 Å thick, with an effective viscosity of 1 P, result in $\omega_2 \sim 100$ kHz; thus grain boundary relaxation could be important at the frequencies of these experiments.

The Low-Velocity Zone

Partial melting has been proposed as the cause of the low-velocity zone (LVZ) of the upper mantle [*Anderson and Sammis*, 1970] in order to account for the low velocities and high attenuation there and to provide a source region for basaltic magmas. Other relaxation mechanisms could account for the first two of these, such as grain boundary relaxation [*Goetze*, 1969; *Jackson*, 1969] or dislocation-impurity interaction [*Gueguen and Mercier*, 1973], and it is perhaps only the sharp onset and termination of the LVZ that is diagnostic of partial melting, owing to the sudden appearance of a melt phase at the eutectic temperature.

The characteristics of a partially molten LVZ have been discussed by *Nur and Simmons* [1969], *Nur* [1971], *Solomon* [1972, 1973], and *Anderson and Spetzler* [1970] in terms of *Walsh's* [1969] model of viscous relaxation in crack and laboratory experiments [*Nur and Simmons*, 1969; *Spetzler and Anderson*, 1968]. As we have seen, the mechanism responsible for attenuation in the experiments was most probably fluid flow between cracks, and therefore a simple extrapolation from laboratory to mantle conditions based on Walsh's model is not valid. (Of course this does not rule out viscous relaxation in the LVZ.)

Figure 8 shows velocities and attenuation for a cracked solid taken as a model of a partially molten upper mantle. A wide distribution of crack aspect ratios has been chosen, uniform in log c/a from 10^{-4} to 10^{-1}. The melt concentrations for crack densities $\epsilon = 0.1, 0.3$, and 0.5 are 0.6, 1.8, and 3%, respectively, most of the contribution being from the cracks with $c/a \sim 0.1$. The viscosity of the melt is 10^6 P; the proper value for a thin film of melt at upper mantle conditions is uncertain and is discussed below. Nevertheless, this is probably an upper bound. It is apparent that this model can account for a Q that varies little with frequency over many decades. Accompanying this is a uniform velocity dispersion. The frequencies of seismic body waves occur in the region where relaxation from fluid flow is giving way to viscous relaxation at higher frequencies. The attenuation of surface waves and free oscillations will be dominated by fluid flow. In all cases the value of Q can be quite low. *Solomon* [1973] has reported a value of Q for shear waves of 10 beneath the mid-Atlantic ridge; such a value could be achieved with only a few percent melt if the crack density is

around 0.5. (Since Q changes little with frequency, the value of melt viscosity used is not critical.) A crack density of around 0.3 can account for Q values of ~ 30 in shear and ~ 100 for P waves over 10 decades in frequency.

The frequencies labeled m, d, and h in Figure 8 correspond to periods of 1 month, 1 day, and 1 hour, respectively. The dispersion of the velocities results in substantially lower shear velocities (and an even greater reduction in the shear modulus) at such long periods compared to the periods of seismic body waves. This difference may be important for the determination of static displacements associated with large earthquakes and Love numbers for long-period or static loads. In addition, the viscoelastic relaxation may give rise to slow deformation following large earthquakes, as is discussed by *Mavko and Nur* [1975].

Figure 9 shows the velocities and values of Q for a range of crack aspect ratios limited to 1 decade (10^{-3}–10^{-2}). The resulting Q values are lower than those in Figure 8 but are restricted to a narrower range of frequencies. The concentration of the liquid phase is very small, being less than 1% for all three crack densities. An extension of the crack aspect ratio distribution to smaller values (less than 10^{-3}) would not substantially increase the melt concentration but would smooth out the Q curves similar to those in Figure 8. Note that there is a low-dissipation 'window' in the frequency range of seismic body waves but that Q decreases markedly at free oscillation periods. This illustrates the possible hazards of comparing elastic properties and dissipation at different frequencies.

In Figures 8 and 9 we chose a value of 10^6 P for the melt viscosity, and with this value, dissipation at seismic frequencies was dominated by fluid flow between cracks. A higher viscosity, or smaller aspect ratios, would increase the contribution of viscous relaxation, and thus the choice of viscosity is important in determining which mechanism may dominate. Recent measurements of viscosities of tholeiitic and andesitic magmas [*Kushiro et al.*, 1976] indicate that the viscosities of melts in the upper mantle are significantly less than 10^6 P and that they decrease with depth. This decrease is due to the increased temperature along the liquidus and the smaller effect of increased pressure, which lowers the viscosity, possibly by changing the coordination of aluminum in the melt [*Waff*, 1975]. The resulting viscosities at ~ 100-km depth range from ~ 10 P for a tholeiitic or hydrous andesitic melt to ~ 1000 P for an anhydrous andesitic melt. Since the characteristic frequencies are linearly related to the fluid viscosities, Figures 8 and 9 can be scaled for these viscosities by shifting the frequency scale 3–5 decades to the left. If this is done, the seismic frequency range is clearly in the regime of dissipation due to fluid flow. Viscous relaxation would only be important for cracks with aspect ratios of less than $\sim 10^{-7}$; this is sufficiently small

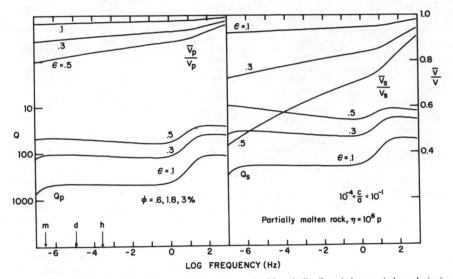

Fig. 8. Wave velocities and seismic Q for a partially molten rock with melt distributed along grain boundaries in cracks with aspect ratios from 10^{-4} to 10^{-1}. The melt viscosity is 10^6 P; a smaller (more realistic) viscosity would shift the frequency scale to the left. Q is relatively independent of frequency over a wide range. Dissipation at seismic frequencies and below is due to fluid flow between cracks. The frequencies marked m, d, and h correspond to periods of 1 month, 1 day, and 1 hour. The velocities at these frequencies can be considerably less than those measured at seismic frequencies.

that it most probably cannot be distinguished from grain boundary relaxation below the liquidus temperature.

The reduction of shear velocity in the low-velocity zone is ~10% or less [*Goetze*, 1977]. If this represents a relaxation from the intrinsic velocity of the solids, then a crack density of less than 0.2 can account for such a reduction in the regime of fluid flow. This would also account for shear Q values of 50–100, depending on the distribution of aspect ratios. The required melt concentration is very small and could be substantially less than 1%. In addition, a higher crack density would result in a much larger reduction of shear velocity than is observed. This places a restriction on the distribution of melt; coupled with knowledge of how the melt would wet grain

boundaries [*Stocker and Gordon*, 1975], it may also restrict the melt concentration.

Grain Boundary Relaxation

Relaxation of shear stresses across grain boundaries may be an important dissipative mechanism in the earth's mantle and has been proposed as a possible cause of the low-velocity zone [*Goetze*, 1969, 1977; *Jackson*, 1969; *Gueguen and Mercier*, 1973]. *Zener* [1941] analyzed this effect by considering the relaxation of shear stress on the boundary of a spherical grain, which should provide an upper bound for the magnitude of the relaxation, since it assumes that there is no geometric impediment to slip at the grain boundary. For polyhedral grains the

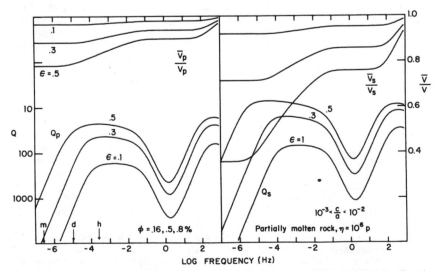

Fig. 9. Similar to Figure 8 but for a crack aspect ratio distribution between 10^{-3} and 10^{-2}. Note that the dependence of Q on frequency can be very variable, depending on the frequency.

edges and corners of the polyhedra would prevent slip [*Zener*, 1948], which would thus be confined to relatively planar portions of the grain boundaries. Each of these can be regarded as a crack filled with a viscous fluid, and we can estimate the magnitude of the relaxation from the analysis of saturated isolated cracks. Table 4 gives crack densities calculated from (1) by assuming that the grains are various polyhedra and that each face may be regarded as a crack. For each case we show the relaxed moduli found from (3) and (4) for a solid with an unrelaxed Poisson ratio of $\frac{1}{4}$ together with the associated velocities. Also shown is *Zener's* [1941] result, which gives substantially more relaxation than the others (except for the cube), which predict relaxation of the shear modulus by ~25%. This figure is of course approximate owing to the failure of real grains to be perfect polyhedra and to the limitations of the analysis, which assumes randomly oriented and located cracks. Nevertheless, it indicates the magnitude of the effect of irregularities of grain shape.

The magnitude of the dissipation can be seen in Figure 4. For a crack density of 0.3, Q can be as small as ~10 for shear and ~25 for P waves. Of course any distribution of characteristic frequencies over an interval will result in a smaller maximum dissipation.

The characteristic frequency for relaxation is given by (25), in which the thickness of the crack is taken to be the width of the disordered grain boundary in which viscous relaxation occurs. For simple solids this is usually one or two atomic dimensions [*Nowick and Berry*, 1972], but for silicates it may be larger, corresponding to the size of polymerized units such as SiO_4 tetrahedra. The effective viscosity will be governed by thermally activated motion of atoms and vacancies; one would expect it to be roughly similar to, but greater than, the viscosity of a melt of the crystals. In any event a distribution of relaxation frequencies would be expected in the mantle because of the range of sizes of the grain faces and the different types of grain boundaries in a multiphase aggregate. A typical characteristic frequency, ω_2, for 1-mm grain faces a few angstroms thick, with a viscosity of ~10^6 P, is in the seismic frequency band.

It has been mentioned before that grain boundary relaxation has not been ruled out as the cause of the low-velocity zone. The fact that the characteristic frequency for relaxation is near seismic frequencies raises another important possibility, namely, that the moduli appropriate to static displacements would be relaxed and less than those inferred from seismic velocities. Insofar as the effective grain boundary viscosity depends on temperature and pressure, the degree of relaxation at a given frequency would be a function of depth in the earth. Nevertheless, for sufficiently long periods the whole mantle would be relaxed. This possibility should be considered in calculations of static displacements from earthquakes and elastic Love numbers.

CONCLUSIONS

Two different dissipative mechanisms have been identified and analyzed for fluid-saturated solids in which the fluid fills cracklike voids in the solid. The first (viscous relaxation) is due to the relaxation of shear stresses in the viscous fluid in individual cracks, which was previously analyzed by *Walsh* [1969] for a sparse crack distribution. The analysis here extends this to larger crack densities and gives more simple expressions for the results. The second mechanism (fluid flow) results from the stress-induced flow of fluid between cracks at different orientations to the applied stress field. A similar mechanism (consolidation) arising from bulk flow of fluid out of regions of compressions has not been systematically analyzed here, but results for this mechanism follow simply from the analysis presented here.

A saturated cracked solid can be characterized by effective viscoelastic moduli, which depend on the crack density and the frequency of the applied stress field. Characteristic frequencies for each mechanism can be estimated from the geometric properties of the cracks and the fluid viscosity; that for fluid flow can also be estimated from the effective permeability for fluid flow between adjacent cracks, which may differ significantly from the bulk permeability of the whole body. In general, relaxation due to fluid flow occurs at lower frequencies than that due to viscous relaxation. For frequencies far removed from the characteristic frequencies, dissipation is small, and the body is nearly elastic. At low frequencies the pore pressure is uniform, and the shear modulus is unaffected by saturation. At high frequencies the viscous fluid prevents the cracks from responding to stress, and the moduli of the body are unaffected by the presence of the cracks. Intermediate to the characteristic frequencies for fluid flow and viscous relaxation, the pore pressure is inhomogeneous, and the moduli are those obtained by *Budiansky and O'Connell* [1976] for isolated cracks. Transitions between these limiting states give rise to dissipation and accompanying dispersion of phase velocities of elastic waves.

Interpretation of laboratory measurements of water-saturated rocks and partially molten solids indicated that the observed dissipation was most probably due to fluid flow between cracks. This accounts for the observation that the shear moduli of rocks are relatively unaffected by saturation with water. Dissipation in glycerol-saturated rocks was due to both effects, depending on the viscosity of the fluid (which was varied). At seismic frequencies, water-saturated crystalline rocks are most probably in the state with uniform pore pressure, and any dissipation is due to fluid flow between cracks.

Dissipation in a partially molten upper mantle is also most probably due to fluid flow between cracks unless the melt viscosity is significantly greater than that for magmas which have been observed in the field or measured in the laboratory.

TABLE 4. Relaxation From Grain Boundary Sliding

Grain Shape	ϵ	$\bar{\nu}$	$\bar{G}/G$	$\bar{E}/E$	$\bar{V}_P/V_P$	$\bar{V}_S/V_S$
Cube	0.48	0.34	0.60	0.64	0.91	0.77
Rhombic dodecahedron	0.28	0.30	0.76	0.79	0.94	0.87
Pentagonal dodecahedron	0.30	0.31	0.74	0.78	0.94	0.86
Cube octahedron	0.27	0.30	0.76	0.79	0.95	0.87
Sphere [*Zener*, 1941]	...	0.35	0.55	0.59	0.89	0.74

Viscous relaxation is important at seismic frequencies only for cracks so thin that the mechanism is equivalent to grain boundary relaxation. The seismic low-velocity zone can be accounted for by a fraction of a percent of melt distributed along a substantial fraction of the grain boundaries, or by extensive grain boundary relaxation. Improved knowledge of the detailed structure of the LVZ would allow discrimination between these alternatives, a sharp boundary of the LVZ favoring melting.

For both saturated and partially molten rocks, variability of crack shapes can result in a Q that varies little with frequency over many decades, with an accompanying uniform velocity dispersion. Superposition of the effects of grain boundary relaxation in a multiphase aggregate could give a similar effect. Relaxation at frequencies below the seismic frequency band could cause the static shear modulus of the mantle to be less than that measured by seismic techniques; this would influence estimates of static displacements from earthquakes and elastic Love numbers.

A consideration of the effects of polyhedral grain shape on grain boundary relaxation indicates that the relaxation of the shear modulus would be about half of that estimated by *Zener* [1941] for a spherical grain. In addition, variability of the size and shape of grain faces should spread the relaxation peak over wider range of frequency than that for a discrete relaxation frequency.

APPENDIX: DERIVATION OF VISCOELASTIC MODULI

Our aim here is to derive complex viscoelastic moduli for some simple states of stress and strain. We shall first derive equations governing the overall static elastic moduli of bodies containing thin ellipsoidal elastic inclusions. Subsequently, these equations will serve as a basis for the derivation of viscoelastic moduli for bodies containing fluid-filled cracks, viscous dissipation occurring either by viscous shear relaxation within the crack or by fluid flow between cracks.

Elastic Analysis

Consider an infinite isotropic homogeneous elastic solid subjected to stresses σ_{ij} at infinity. The decrease in potential energy due to the insertion of a single circular crack of radius a into this body is (BO)

$$\mathcal{E} = \frac{8(1 - \bar{\nu}^2)a^3}{3\bar{E}}\sigma^2 + \frac{16(1 - \bar{\nu}^2)a^3}{3(2 - \bar{\nu})\bar{E}}\tau^2 \qquad (A1)$$

where $\bar{E}$ and $\bar{\nu}$ are Young's modulus and Poisson's ratio of the body, $\sigma = \sigma_{ij}n_in_j$ is the resolved normal stress in the direction n normal to the plane of the crack, and τ is the maximum resolved shear stress parallel to the crack plane.

But now suppose that the crack is really a thin spheroidal inclusion, having semiaxes $c \ll a$, which contains material with bulk modulus $\bar{K}$ and shear modulus $\bar{G}$ (one or both of which will subsequently be given complex values). To avoid unnecessarily complicated expressions, we will restrict the analysis to the case in which

$$|\bar{G}| / |\bar{K}| \ll 1 \qquad (A2)$$

Then the state of stress in the inclusion (cf. (11)) consists essentially of a uniform hydrostatic stress

$$\tilde{\sigma}_{ij} = \tilde{\sigma}\delta_{ij} = \bar{K}\tilde{e}_{kk}\delta_{ij} \qquad (A3)$$

together with a uniform shear stress $\tilde{\tau}$ in the same orientation as τ [*Eshelby*, 1957]. The energy release (A1) must now be reduced by the inclusion strain energy

$$\left(\frac{\tilde{\sigma}^2}{2\bar{K}} + \frac{\tilde{\tau}^2}{2\bar{G}}\right)4\pi a^2c/3 \qquad (A4)$$

and also by the amount

$$\frac{8(1 - \bar{\nu}^2)a^3}{3\bar{E}}\tilde{\sigma}^2 + \frac{16(1 - \bar{\nu}^2)a^3}{3(2 - \bar{\nu})\bar{E}}\tilde{\tau}^2 \qquad (A5)$$

which (cf. (A1)) represents work that would be done on the body by the application of normal stress $\tilde{\sigma}$ and shear stress $\tilde{\tau}$ to the crack faces. The resulting energy decrease associated with each inclusion is then

$$\mathcal{E} = \frac{8(1 - \bar{\nu}^2)a^3}{3\bar{E}}\sigma^2\left[1 - \left(\frac{\tilde{\sigma}}{\sigma}\right)^2(1 + \gamma)\right]$$
$$+ \frac{16(1 - \bar{\nu}^2)a^3}{3(2 - \bar{\nu})\bar{E}}\tau^2\left[1 - \left(\frac{\tilde{\tau}}{\tau}\right)^2(1 + \beta)\right] \qquad (A6)$$

where

$$\gamma = \frac{3\pi}{4}\left(\frac{1 - 2\bar{\nu}}{1 - \bar{\nu}^2}\right)\frac{\bar{K}}{K}\frac{Kc}{\bar{K}a} \qquad (A7)$$

and

$$\beta = \frac{\pi}{4}\left(\frac{2 - \bar{\nu}}{1 - \bar{\nu}}\right)\frac{\bar{G}}{G}\frac{Gc}{\bar{G}a} \qquad (A8)$$

The stresses in the inclusion are obtained by equating the volume change and net shear of the inclusion to the corresponding quantities for the crack subject to effective stresses $\sigma - \tilde{\sigma}$ and $\tau - \tilde{\tau}$. Thus

$$\frac{\tilde{\tau}}{\bar{G}}v_c = \frac{32(1 - \bar{\nu}^2)a^3}{3(2 - \bar{\nu})\bar{E}}(\tau - \tilde{\tau})$$

which yields

$$\tilde{\tau}/\tau = (1 - \beta)^{-1} \qquad (A9)$$

Similarly,

$$\tilde{\sigma}/\sigma = (1 - \gamma)^{-1} \qquad (A10)$$

and (A6) becomes

$$\mathcal{E} = \frac{8(1 - \bar{\nu}^2)a^3}{3\bar{E}}D\sigma^2 + \frac{16(1 - \bar{\nu}^2)a^3}{3(2 - \bar{\nu})\bar{E}}C\tau^2 \qquad (A11)$$

where

$$D = (\sigma - \tilde{\sigma})/\sigma = \gamma/(\gamma + 1) \qquad (A12)$$

$$C = (\tau - \tilde{\tau})/\tau = \beta/(\beta + 1) \qquad (A13)$$

As in the work of BO we now invoke the self-consistent approximation by assuming that (A11) approximates the energy loss produced by each of very many randomly oriented and distributed inclusions as long as we identify $\bar{E}$ and $\bar{\nu}$ as the (as yet unknown) effective elastic moduli of the homogeneous isotropic composite body. Then by the procedure given by BO, expressions giving the effective moduli follow from (A11) as

$$\frac{\bar{K}}{K} = 1 - \frac{16}{9}\left(\frac{1 - \bar{\nu}^2}{1 - 2\bar{\nu}}\right)D\epsilon \qquad (A14)$$

$$\frac{\bar{G}}{G} = 1 - \frac{32}{45}\left(\frac{1 - \bar{\nu}}{2 - \bar{\nu}}\right)[(2 - \bar{\nu})D + 3C]\epsilon \qquad (A15)$$

$$\bar{\nu} = \nu + \frac{16}{45}\left(\frac{1 - \bar{\nu}^2}{2 - \bar{\nu}}\right)[2(1 - 2\nu)C$$
$$- (2 - \bar{\nu})(1 + 3\nu)D]\epsilon \qquad (A16)$$

Note that the dependence of D and C on the moduli requires the simultaneous solution of (A7), (A8), and (A12)–(A16) as well as the specification of the parameters $Kc/\bar{K}a$ and $Gc/\bar{G}a$ in (A7) and (A8).

The results for the static moduli given in the text follow from (A14)–(A16) with $C = 1$ and the appropriate value for D: dry, $D = 1$; saturated isolated, $D = 0$; saturated isobaric, $D = 0$ in (A14) and $D = 1$ in (A15) and (A16) with the bulk Poisson ratio calculated from (9). We also note that although (A14)–(A16) were derived here by an energy method, the same result can be obtained by calculation of inclusion strain after *Hill* [1965], but care must be taken during the approach to the singular limit of cracklike ellipsoidal inclusions of infinitesimal thickness.

Also the results should be entirely consistent, when the appropriate simplifying assumptions are incorporated and the right limits are taken carefully, with those of *Wu* [1966], who gives a general self-consistent derivation for arbitrary ellipsoids. *Walsh* [1969] did, in fact, extract such low aspect-ratio-limiting results from Wu's work but then abandoned the self-consistent approximation.

Viscoelastic Analysis

We consider the case in which the stress and strain vary sinusoidally with time. We will discuss how (and under which conditions) the elastic stiffnesses $\bar{K}$ and $\bar{G}$ of the inclusion must be modified in order to model viscous flow effects. For the case of cracks filled with a viscous fluid we replace $\bar{G}$ and $\bar{K}$ with complex moduli that take into account viscous relaxation of shear stresses in the inclusions and the flow of fluid between cracks.

Viscous shear relaxation. The response of a Newtonian viscous fluid in a crack to harmonic shear stress of amplitude $\tilde{t}$ in the plane of the crack can be immediately taken into account by replacing the shear modulus of an elastic inclusion, $\bar{G}$, in (A8) by the complex modulus $i\omega\eta$, where η is the viscosity of the fluid, ω is the frequency, and $i = (-1)^{1/2}$. Then the quantity C in (A13) becomes

$$C = \frac{-ie\omega_2/\omega}{1 - ie\omega_2/\omega} \qquad (A17)$$

where

$$e = \frac{\pi}{4}\left(\frac{2 - \bar{\nu}}{1 - \bar{\nu}}\right)\frac{\bar{G}}{G} \qquad (A18)$$

and the characteristic frequency for viscous shear relaxation is

$$\omega_2 = Gc/\eta a \qquad (A19)$$

Note that $\omega \ll \omega_2$ gives $C = 1$, corresponding to the limiting elastic case of saturated isolated cracks, and $\omega \gg \omega_2$, giving $C = 0$, is the glued case in which the crack faces undergo no relative displacement. The maximum dissipation of energy occurs for frequencies near ω_2. This mechanism is the same as that analyzed by *Walsh* [1968, 1969] for a dilute concentration of cracks.

Flow between cracks. The flow of fluid between cracks can be included in the viscoelastic analysis by incorporating an effective constitutive relation between the fluid pressure and the volume of the material in a crack. If no fluid is permitted to flow out of a crack (the elastic case), this relation remains $\theta = \tilde{\sigma}/\bar{K}$, where θ is the dilatation of the crack and $\tilde{\sigma}$ and $\bar{K}$ are the hydrostatic stress and bulk modulus of the inclusion. We assume that the rate at which fluid flows out of a crack, when

flow is permitted, is proportional to the fluid pressure in the crack and that the rate at which the crack volume changes owing to fluid flow can be written

$$dv_c/dt = v_c \, d\theta/dt = \lambda a^3 \tilde{\sigma}/\eta \qquad (A20)$$

where η is the fluid viscosity and λ is a dimensionless parameter. The total dilatation rate will be the sum of the elastic and viscous dilatation rates:

$$d\theta/dt = (d\tilde{\sigma}/dt) + \lambda a^3 \tilde{\sigma}/(\eta v_c)$$

For sinusoidal stress and strain variation the dilatation amplitude is then

$$\theta = \left(\frac{1}{\bar{K}} + \frac{\lambda a^3}{i\omega\eta v_c}\right)\tilde{\sigma} = \frac{\tilde{\sigma}}{\bar{K}_{VE}} \qquad (A21)$$

Replacing $\bar{K}$ by $\bar{K}_{VE}$ in (A7) gives

$$\gamma = \gamma_E + \frac{9}{16}\left(\frac{1 - 2\bar{\nu}}{1 - \bar{\nu}^2}\right)\frac{\bar{K}}{K}\frac{K\lambda}{i\omega\eta} \qquad (A22)$$

where γ_E is the elastic contribution given by (A7).

For thin cracks saturated with a 'stiff' fluid (e.g., water in cracks with $c/a \sim 10^{-3}$) the parameter $Kc/\bar{K}a$ will be small, and the term γ_E in (A22) may be neglected. Consequently, (A12) becomes

$$D = \frac{-id\omega_1/\omega}{1 - id\omega_1/\omega} \qquad (A23)$$

where

$$d = \frac{9}{16}\left(\frac{1 - 2\bar{\nu}}{1 - \bar{\nu}^2}\right)\frac{\bar{K}}{K} \qquad (A24)$$

and the characteristic frequency is

$$\omega_1 = K\lambda/\eta \qquad (A25)$$

Finally, we assess our initial assumption (A2). Since

$$\bar{G} = i\omega\eta = iG(\omega/\omega_2)(c/a)$$

and by (A21) and (A25)

$$\frac{1}{\bar{K}_{VE}} = \left(\frac{3}{4\pi iK}\right)\frac{\omega_1 a}{\omega c} + \frac{1}{\bar{K}}$$

then

$$\frac{\bar{G}}{\bar{K}_{VE}} = \frac{3}{4\pi}\frac{G}{K}\frac{\omega_1}{\omega_2} + i\frac{G}{K}\frac{\omega}{\omega_2}\frac{c}{a}\frac{K}{\bar{K}}$$

Since (cf. (18) and (19)) $\omega_1 \ll \omega_2$, the first term is small, and since $cK/a\bar{K} \ll 1$, the second term is small unless $\omega \gg \omega_2$. But when this occurs, the cracks are all glued ($D = C = 0$ in (13)), and the moduli are unaffected by the presence of the cracks.

Calculation of effective moduli. With the appropriate use of (A17) and (A23) for C and D, relations (A14)–(A16) can now serve as a basis for the construction of effective viscoelastic moduli. The calculation is not straightforward, however, because expression (A23) for D (together with (A24) for d) is not valid for all stress states. While fluid flow between cracks can occur under pure shear (when fluid flows from one half of the cracks into the other half), such flow does not take place at all for hydrostatic stress states. Correspondingly, intermediate stress states involve intermediate amounts of intercrack flow. Nevertheless, a valid calculation of the complex moduli can be made just by consideration of pure shear and hydrostatic stress.

Under hydrostatic compression, no fluid flow takes place, a situation corresponding to $D = 0$, and then (A14) gives

$$\bar{K}/K = 1 \qquad (A26)$$

which, of course, is the same as the static result for each of the saturated cases treated in the text.

Consider next the application of pure shear to the cracked body. Now fluid flow between cracks is permitted, and the shear response of the body is indistinguishable from that of a body in which fluid behavior, described by (A17) and (A23) for C and D, occurs under all stress conditions. Accordingly, to calculate $\bar{G}/G$, we use these values of C and D in the simultaneous solution of (A14)–(A16). Note that the values for $\bar{K}/K$ and $\bar{\nu}$ resulting from this calculation are not, of course, the correct ones; rather they are those of the fictitious comparison body in which fluid flow into or out of cracks occurs for all stress states. If we denote the fictitious $\bar{\nu}$ by ν', the solution for $\bar{G}/G$ is facilitated by writing (A24) as

$$d = \frac{9}{16}\left(\frac{1-2\nu}{1+\nu}\right)\frac{1}{1-\nu'}\frac{\bar{G}}{G} \qquad (A27)$$

This is obtained from the standard elastic relation $K/G = (2 + 2\nu)/(3 - 6\nu)$. The shear modulus will then be given by

$$\frac{\bar{G}}{G} = 1 - \frac{32}{45}\left(\frac{1-\nu'}{2-\nu'}\right)[(2-\nu')D + 3C]\epsilon \qquad (A28)$$

where ν' is the root of

$$\nu' = \nu + \frac{16}{45}\left(\frac{1-\nu'^2}{2-\nu'}\right)[2(1-2\nu)C$$
$$- (2-\nu')(1+3\nu)D]\epsilon \qquad (A29)$$

Finally, the correct value of $\bar{\nu}$ is found by using (A26) and the result for $\bar{G}/G$ in the standard relation given by (9).

Estimates of ω_1 for Flow Between Cracks

The parameter ω_1 in (A23) is related to the rate at which the transition from the saturated isolated to the saturated isobaric cases takes place. This transition involves flow of fluid between cracks. The rate at which fluid can flow out of a crack will depend on the details of the interconnections between cracks and is not susceptible to direct analysis. Nevertheless, we may estimate this rate and hence the characteristic frequency ω_1 from some simple models.

The first model is appropriate for a large crack density with many intersections between cracks, in which the rate of fluid flow will be limited by flow out of a crack rather than by flow between separated cracks. We consider two-dimensional flow of an incompressible fluid between two parallel plates of width $2a$ separated by a distance $2c$, approaching one another at constant velocity. The solution to this is given by *Jaeger* [1956, p. 140]. Relating the discharge from a length a of the plates to the average normal stress on the plates through (A20) gives

$$\lambda = 4(c/a)^3 \qquad \omega_1 = 4(K/\eta)(c/a)^3 \qquad (A30)$$

A second model, similar to the first, is that of a thin spherical shell of thickness $2c$ of fluid completely surrounding a solid sphere. Under the influence of a macroscopic shear stress field, fluid will flow from regions where the shell is compressed to regions where the shell is dilated. If the diameter of a crack, $2a$, is identified with one fourth of the circumference of the spherical shell, the analysis of this problem (R. J. O'Connell, unpublished data, 1976) gives $\omega_1 = \frac{1}{2}(K/\eta)(c/a)^3$. This agrees in form

with (A30) and may be regarded as being appropriate to a limiting case in which the solid is completely permeated with intersecting cracks or that in which a partial melt nearly completely surrounds a single grain. A reasonable approximation for ω_1, which we shall use, is obtained by setting the numerical factor in (A30) equal to unity, resulting in (18) in the text.

The third model is appropriate for widely separated cracks, where flow is limited by the interconnections between cracks. We consider the flow of fluid from a circular crack of radius a in an infinite medium of permeability k, in which the fluid velocity $\mathbf{v}$ is related to pressure p by $\mathbf{v} = -k/\eta\,\nabla p$. For an incompressible fluid, $\nabla\cdot\mathbf{v} = 0$, and the pressure satisfies Laplace's equation $\nabla^2 p = 0$. The solution for constant fluid pressure p_0 in a crack in the plane $z = 0$ at the origin of a cylindrical coordinate system may be found in the work by *Carslaw and Jaeger* [1959, p. 215]:

$$p(r, z) = \frac{2p_0}{\pi}\int_0^\infty \frac{1}{s}\exp\left(-s|z|\right)J_0(sr)\sin sa\,ds$$

where $J_0(sr)$ is the Bessel function of the first kind.

The rate of flow from the crack is

$$Q = \frac{-4\pi k}{\eta}\int_0^a \left(\frac{\partial p}{\partial z}\right)_{z=0} dr = \frac{8ak}{\eta}p_0$$

Hence

$$\lambda = 8k/a^2 \qquad \omega_1 = 8Kk/\eta a^2 \qquad (A31)$$

A fourth model along the lines of the third indicates the effects of fluid flowing to adjacent cracks and may be appropriate for moderate crack densities. For crack density $\epsilon = N\langle a^3\rangle$, regard each crack as being embedded in a sphere with volume a^3/ϵ. Approximate the crack as a smaller concentric sphere with the same surface area as that of the crack. The pressure on the outside of this shell is taken to be zero, and we calculate the flow through the permeable shell [cf. *Carslaw and Jaeger*, 1959, p. 146]. If a crack is oriented such that it has a positive fluid pressure, then half of its 'neighbors' will have positive pressures, and the remaining half, having negative pressures, will appear as sinks for the fluid flow. We consequently take the effective rate to be half of the flow out of the complete spherical shell:

$$Q = 2^{1/2}\,\pi ak/\eta(1 - s_0\epsilon^{1/3})$$

where

$$s_0 = (2^{1/2}\pi/3)^{1/3} = 1.14$$

Hence

$$\lambda = 4.4k/a^2(1 - 1.14\epsilon^{1/3})$$
$$\omega_1 = 4.4Kk/\eta a^2(1 - 1.14\epsilon^{1/3}) \qquad (A32)$$

This model gives the same result as does (A31) for $\epsilon = 0.06$ and gives a value of ω_1 that is 3 times greater than that given by (A31) for $\epsilon = 0.4$.

Finally, we note that *Mavko and Nur* [1975] give an estimated relaxation time for 'melt squirt' that is equivalent to

$$\omega_1 = \frac{4\pi}{3}\frac{K}{\eta}\left(\frac{c}{a}\right)^3\epsilon$$

In the discussion of fluid-saturated rocks and partial melts in the text the frequency for fluid flow was calculated by using (18) rather than any of the alternative estimates in the appendix. This was because it seemed easier to estimate crack aspect ratios than intercrack permeability (k in (A31)) and because the crack densities were of such magnitude to suggest that close connections between cracks were likely. Equations (18) and (A11) give the same frequencies for 100-μm cracks with $c/a = 10^{-4}$ if the permeability k is 10 ndarcies (1 darcy $\approx 10^{-8}$ cm^2); for 1 mm cracks of aspect ratio 10^{-3} the required permeability is 1 mdarcy. If it is recalled that this permeability is that for flow between adjacent cracks (at the proper orientation), these values seem consistent with the bulk permeabilities of crystalline rocks. A wide range of such intercrack permeabilities would be expected owing to their dependence on the details of the interconnections among cracks, which one would expect to be highly variable.

Acknowledgments. We appreciate helpful discussions with J. Rice, J. Rudnicki, and M. Cleary and comments from R. Brown and J. Korringa. The first author (R.J.O.) gratefully acknowledges the hospitality and support received while he was visiting the Department of Geodesy and Geophysics, Cambridge University, and the Institut de Physique du Globe, Université de Paris, where part of this research was done. This research was supported by the Earth Sciences Section, National Science Foundation, NSF grant EAR75-22433 and by the National Science Foundation Materials Research Laboratory Program, NSF grant DMR72-03020.

REFERENCES

Anderson, D. L., and C. Sammis, Partial melting in the upper mantle, *Phys. Earth Planet. Interiors, 3,* 41–50, 1970.

Anderson, D. L., and H. Spetzler, Partial melting in the low-velocity zone, *Phys. Earth Planet. Interiors, 4,* 62–64, 1970.

Budiansky, B., and R. J. O'Connell, Elastic moduli of dry and saturated cracked solids, *Int. J. Solids Struct., 12,* 81–97, 1976.

Carslaw, H. S., and J. C. Jaeger, *Conduction of Heat in Solids,* 510 pp., Oxford University Press, New York, 1959.

Eshelby, J. D., The determination of the elastic field of an ellipsoidal inclusion and related problems, *Proc. Roy. Soc. London, Ser. A, 241,* 376–396, 1957.

Futterman, W. I., Dispersive body waves, *J. Geophys. Res., 67,* 5279–5291, 1962.

Goetze, C., High temperature elasticity and anelasticity of polycrystalline salts, Ph.D. thesis, Harvard Univ., Cambridge, Mass., 1969.

Goetze, C., A brief summary of our present day understanding of the effect of volatiles and partial melt on the mechanical properties of the upper mantle, in *High Pressure Research: Applications to Geophysics,* edited by M. H. Manghnani and S. Akimoto, Academic, New York, 1977.

Gordon, R. B., Mechanical relaxation spectrum of crystalline rock containing water, *J. Geophys. Res., 79,* 2129–2131, 1974.

Gordon, R. B., and L. A. Davis, Velocity and attenuation of seismic waves in imperfectly elastic rock, *J. Geophys. Res., 73,* 3917–3935, 1968.

Gordon, R. B., and D. Rader, Imperfect elasticity of rock: Its influence on the velocity of stress waves, in *The Structure and Physical Properties of the Earth's Crust,* Geophys. Monogr. Ser., vol. 14, edited by J. G. Heacock, AGU, Washington, D. C., 1971.

Gueguen, Y., and J. M. Mercier, High attenuation and the low-velocity zone, *Phys. Earth Planet. Interiors, 7,* 39–46, 1973.

Hadley, K., Comparison of calculated and observed crack densities and seismic velocities in Westerly granite, *J. Geophys. Res., 81,* 3484–3494, 1976.

Hill, R., A self-consistent mechanics of composite materials, *J. Mech. Phys. Solids, 13,* 213–222, 1965.

Jackson, D. D., Elastic relaxation model for seismic wave attenuation in the earth, *Phys. Earth Planet. Interiors, 2,* 30–34, 1969.

Jaeger, J. C., *Elasticity, Fracture and Flow,* 208 pp., Methuen, London, 1956.

Knopoff, L., Q, *Rev. Geophys. Space Phys., 2,* 625–660, 1964.

Kushiro, I., H. S. Yoder, Jr., and B. O. Mysen, Viscosities of basalt and andesite melts at high pressures, *J. Geophys. Res., 81,* 6351–6356, 1976.

Liu, H. P., D. L. Anderson, and H. Kanamori, Velocity dispersion due to anelasticity: Implications for seismology and mantle composition, *Geophys. J. Roy. Astron. Soc., 47,* 41–58, 1976.

Mavko, G., and A. Nur, Melt squirt in the asthenosphere, *J. Geophys. Res., 80,* 1444–1448, 1975.

Nowick, A. S., and B. S. Berry, *Anelastic Relaxation in Crystalline Solids,* 677 pp., Academic, New York, 1972.

Nur, A., Viscous phase in rocks and the low-velocity zone, *J. Geophys. Res., 76,* 1270–1277, 1971.

Nur, A., Dilatancy, pore fluids and premonitory variations of t_s/t_p travel times, *Bull. Seismol. Soc. Amer., 62,* 1217–1222, 1972.

Nur, A., and G. Simmons, The effect of viscosity of a fluid phase on velocity in low porosity rocks, *Earth Planet. Sci. Lett., 7,* 99–108, 1969.

O'Connell, R. J., and B. Budiansky, Seismic velocities in dry and saturated cracked solids, *J. Geophys. Res., 79,* 5412–5426, 1974.

O'Connell, R. J., and B. Budiansky, Measures of attenuation in dissipative media, submitted to *Geophys. Res. Lett.,* 1977.

Pipkin, A. C., *Lectures on Viscoelastic Theory,* 180 pp., Springer, New York, 1972.

Rice, J. R., and M. P. Cleary, Some basic stress diffusion solutions for fluid-saturated elastic porous media with compressible constituents, *Rev. Geophys. Space Phys., 14,* 227–242, 1976.

Scholz, C. H., L. R. Sykes, and Y. P. Aggarwal, Earthquake prediction: A physical basis, *Science, 181,* 803–810, 1973.

Solomon, S. C., Seismic wave attenuation and partial melting in the upper mantle of North America, *J. Geophys. Res., 77,* 1483–1502, 1972.

Solomon, S. C., Shear wave attenuation and melting beneath the Mid-Atlantic Ridge, *J. Geophys. Res., 78,* 6044–6059, 1973.

Spetzler, H. A., The effect of temperature and partial melting on velocity and attenuation in a simple binary system, I, Ph.D. thesis, Calif. Inst. of Technol., Pasadena, 1969.

Spetzler, H. A., and D. L. Anderson, The effect of temperature and partial melting on velocity and attenuation in a simple binary system, *J. Geophys. Res., 73,* 6051–6060, 1968.

Stocker, R. L., and R. B. Gordon, Velocity and internal friction in partial melts, *J. Geophys. Res., 80,* 4828–4836, 1975.

Terzaghi, K., *Theoretical Soil Mechanics,* John Wiley, New York, 1943.

Waff, H. S., Pressure induced coordination changes in magmatic liquids, *Geophys. Res. Lett., 2,* 193–196, 1975.

Walsh, J. B., Attenuation in a partially melted material, *J. Geophys. Res., 73,* 2209–2216, 1968.

Walsh, J. B., New analysis of attenuation in partially melted rock, *J. Geophys. Res., 74,* 4333–4337, 1969.

Watt, J. P., G. F. Davies, and R. J. O'Connell, The elastic properties of composite materials, *Rev. Geophys. Space Phys., 14,* 541–563, 1976.

Whitcomb, J. H., J. D. Garmany, and D. L. Anderson, Earthquake prediction: Variation of seismic velocities before the San Fernando earthquake, *Science, 180,* 632–635, 1973.

Wu, T. T., The effect of inclusion shape on the elastic moduli of a two-phase material, *Int. J. Solids Struct., 2,* 1–8, 1966.

Zener, C., Theory of the elasticity of polycrystals with viscous grain boundaries, *Phys. Rev., 60,* 906–908, 1941.

Zener, C., *Elasticity and Anelasticity of Metals,* University of Chicago Press, Chicago, Ill., 1948.

(Received January 17, 1977;
revised June 27, 1977;
accepted June 28, 1977.)

Reprinted from Geophysics, v. 40, p. 224–232.

COMPUTED SEISMIC SPEEDS AND ATTENUATION IN ROCKS WITH PARTIAL GAS SATURATION

J. E. WHITE*

Calculations for an unconsolidated sand with partial gas saturation show a 20 percent increase in compressional wave velocity between 1 and 100 hz and attenuation of 27 db/1000 ft at 31 hz and 82 db/1000 ft at 123 hz.

Shear velocity and attenuation are not affected. Fluid-flow waves are shown to be responsible for the dispersion and attenuation at low frequencies; relations are derived by extending Gassmann's viewpoint to include coupling between fluid-flow waves and seismic body waves. This appears to be an important loss mechanism for heterogeneous porous rocks.

GENERAL FEATURES

Interest in partial saturation

At the gas-oil or gas-water contact in a homogeneous reservoir rock, capillary pressure is responsible for a transition zone in which the gas saturation varies through a wide range. When the reservoir rock is not homogeneous, it seems plausible that gas saturation may vary accordingly. Shale stringers may seal off local pockets of gas creating a multitude of gas-liquid contacts. During production of a field, gas may come out of solution and create distributed pockets of free gas. In view of these possibilities, a study of seismic wave propagation in porous rocks with mixed fluid saturation may well have practical application.

A mechanism and a model

Through the work of Frenkel (1944) and Biot (1956) and others, there is a good theoretical framework for treating seismic waves in porous rocks. The full expressions are complex, and for anything but the simplest geometry, their application is a formidable task. Gassmann

(1951) derived elastic constants for porous rocks by straightforward application of static elasticity, obtaining speeds which agree with Biot's low-frequency values. White (1965, p. 133) extended Gassmann's approach to calculate the effect of fluid flow on shear-wave attenuation, obtaining the same expression as Biot's low-frequency attenuation. The present paper can be looked at as a further application of the low-frequency analysis to show how pressure differences at a boundary may cause substantial fluid flow and how the shifting fluid couples back into the seismic wave to a significant degree. The approach will be described qualitatively, with brief derivations to be found in the Appendix.

As a compressional wave travels through a porous rock, pressure gradients in the fluid cause some flow relative to the rock skeleton and hence some loss of energy. Within a single homogeneous rock, these gradients are small and attenuation due to fluid flow is negligible at prospecting frequencies. If the rock has mixed saturation, with segregated pockets of gas, for instance, then pressure gradients are high near the inhomogeneities and fluid flow will be high

Paper presented at the 43rd Annual International SEG Meeting, October 24, 1973, Mexico City. Manuscript received by the Editor December 17, 1973; revised manuscript received April 9, 1974.
* University of Texas at El Paso, El Paso, Tex. 79968.

also. The loss of energy at these local spots may be averaged over the total volume and expressed as a large attenuation for the compressional wave.

For consideration of this loss mechanism mathematically, some idealized geometry must be adopted. Spherical gas pockets located at the corners of a cubic array are shown in Figure 1. Perhaps it should be repeated that the skeleton is uniform, and the pockets are spherical volumes saturated with gas in a rock which is elsewhere saturated with liquid. At low frequencies, an elementary cube with its enclosed sphere of gas is a typical volume, with average properties which are the same as the average properties of the composite medium. Even for computing average bulk modulus, however, the combination of a cube and a sphere is complicated. Hence the typical volume is considered to be equivalent to the concentric spheres shown

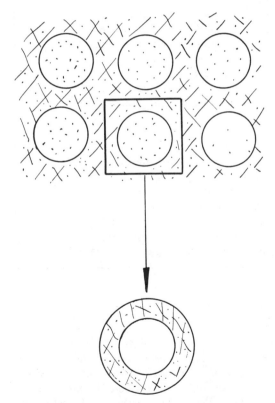

FIG. 1. Model of porous rock with mixed saturation—spherical pockets saturated with gas, intervening volume saturated with liquid. The typical volume considered in the calculations is the pair of concentric spheres shown in the lower part of the figure.

in the lower part of the figure, where the volume of the outer sphere is the same as the volume of the original cube. A specified fractional volume change is impressed, at a low frequency, on the outer surface, and the resulting pressure amplitude at the surface is computed; the effects of fluid flow in the spherical shell are included. The gas pocket is assumed to provide a complete pressure release. The ratio of pressure amplitude to the fractional volume change yields the complex bulk modulus. With known shear modulus and easily computed density, the complex bulk modulus yields the speed and attenuation of compressional waves.

Resulting average properties

It is easy to see that this model exhibits a relaxation phenomenon. There is a characteristic time, dependent on the size of the gas pockets and the distance between them. When the period of oscillation is long compared to this time, the fluid pressure is zero and the bulk modulus is independent of frequency. When the oscillation period is short, the bulk modulus is substantially the value computed as if there were no fluid flow, but with a phase angle which is decreasing with frequency. In the intermediate frequency range, as the modulus varies between these two limits, the medium is characterized by both dispersion and attenuation.

These features are illustrated by the numerical example, for an unconsolidated sandstone at a fairly shallow depth. The dispersion is obvious at a glance. The attenuation is drastic (27 db/1000 ft at 31 hz), perhaps 30 times greater than normal in the frequency range 20 to 50 hz. The round trip through a few hundred feet of this composite medium would just about wipe out a deep reflection.

The rock for this example was chosen to emphasize the effect of fluid content. More competent rocks at greater depths must have the same general features, to a lesser degree. In the example shown, size and spacing of the gas pockets were chosen to place the transition zone in the seismic prospecting band of frequencies. Presumably, random combinations of sizes and spacings would give a blurred-out transition zone which could span this frequency band. At any rate, the results shown here support the suggestion that fluid flow is indeed an important source of attenuation for compressional waves in rocks saturated with a mixture of liquid and gas. Attenuation of shear waves is not affected by the partial saturation.

MATHEMATICAL DERIVATION

Static stress with no fluid flow

We are dealing with such low frequencies that our small elementary volume is in equilibrium locally, and we can, therefore, use the equations of static elasticity. We apply them to two concentric spheres, the inner of radius a and the outer of radius b.

On the outer surface we impress a displacement $u_r(b) = U_0 e^{i\omega t}$, which creates a pressure on the outside of $P_0 e^{i\omega t}$, equal to $-p_{rr}(b)$. A list of symbols and their definitions is given. We express the displacement as a fractional change in volume

$$D_0 e^{i\omega t} = \frac{4\pi b^2}{(4/3)\pi b^3} U_0 e^{i\omega t},$$

$$\text{or} \quad D_0 = 3U_0/b. \quad (1)$$

The bulk modulus is the ratio of $-P_0$ to D_0,

$$k_0 = -P_0/D_0. \quad (2)$$

This is the average bulk modulus of the two concentric spherical bodies with no fluid flow. The expression for k_0 in terms of the elastic constants for the two media is derived in Appendix A.

For this composite medium with no fluid flow the elastic constants and velocities are

$$M_0 = k_0 + 4\mu_0/3,$$

$$\rho_0 = S_G \rho_1 + (1 - S_G)\rho_2, \quad \text{and} \quad (3)$$

$$C_{P0} = (M_0/\rho_0)^{1/2}.$$

The subscripts 0, 1, and 2 refer to the average properties of the concentric spheres, the inner sphere, and the concentric shell, respectively. Since we have the same skeleton in the entire volume and since fluid content does not affect rigidity,

$$\mu_0 = \mu_1 = \mu_2 = \bar{\mu}, \quad \text{and} \quad (4)$$

$$C_{S0} = (\mu_0/\rho_0)^{1/2}.$$

LIST OF SYMBOLS

$a_P{}^*, a_S{}^*$: attenuation of compressional and shear waves
a: radius of inner sphere
b: radius of outer sphere
$C_P, \bar{C}_P, C_P{}^*$: compressional-wave speed
$C_S, \bar{C}_S, C_S{}^*$: shear-wave speed
D_0: dilatation amplitude
E, E_1, E_2: Young's modulus
$k, k_S, \bar{k}, k^*, k_L$: bulk modulus
k_A, k_E: "bulk moduli", as defined
$M, \bar{M}, M_0, M^*$: plane-wave modulus
P_i, P_0: complex pressure amplitude
p_{rr}: normal stress
p_f: fluid pressure
q: frequency parameter, $(\omega/\omega_0)^{\frac{1}{2}}$
q': frequency parameter, $(1 - a/b)q$
r: radial spherical coordinate
S_G: fractional gas saturation, (a^3/b^3)
t: time
U_n: displacement amplitude
u_r: radial displacement
v: particle velocity in fluid relative to skeleton
x, z: complex variables
Z: acoustic impedance
α: complex constant, $(i\omega\eta/k_E \kappa)^{\frac{1}{2}}$
η: viscosity
ϕ: porosity
κ: permeability
$\mu, \bar{\mu}, \mu_1, \mu_2, \mu_0$: shear modulus
$\rho, \bar{\rho}, \rho_0, \rho_s, \rho_f, \rho^*$: density
$\sigma, \sigma_1, \sigma_2$: Poisson's ratio
$\theta_P{}^*$: phase angle
ω: angular frequency
ω_0: reference frequency, $(2\kappa k_E/\eta b^2)$
γ: complex propagation constant

Effect of fluid flow

The static stresses discussed above create a fractional volume change which is constant within the inner sphere and a different fractional volume change which is constant within the outer shell. Thus there is no tendency for fluid flow within either body when the boundary between the two is sealed. However, the values of fluid pressure (P_1 and P_2) created within the two bodies may be different; each is proportional to the impressed pressure P_0.

$$P_1 = R_1 P_0. \tag{5}$$

$$P_2 = R_2 P_0.$$

The constants of proportionality R_1 and R_2 are derived in Appendix B. The pressure difference causes a fluid flow across the boundary; the flow velocity is given by

$$v = \frac{(P_1 - P_2)}{(Z_1 + Z_2)} = \frac{P_0(R_1 - R_2)}{(Z_1 + Z_2)}. \tag{6}$$

The acoustic impedances Z_1 and Z_2 are derived in Appendix C. They are the impedances for fluid-flow waves (or diffusion waves) looking inward and outward, respectively, from the boundary. As fluid flows out of medium 1, for instance, "unloading" occurs and the volume occupied by the saturated rock decreases. Similarly, the (equal) flow of fluid into medium 2 causes it to expand. However, the amount of rock expansion for a given fluid volume is different in the two media, as derived in Appendix D. The constants of proportionality are Q_1 and Q_2.

Since we have chosen to impress a dilatation D_0 on the outer sphere and compute the resulting pressure, we are in effect using an infinite-impedance source. It is appropriate, therefore, to compute the change in pressure ΔP_0 on the surface of a rigidly held sphere when a small change in volume is introduced by expansion of a thin spherical shell. This pressure change is independent of the radius of the thin shell and is simply proportional to fractional volume change;

$$\Delta P_0 = k^* D. \tag{7}$$

We shall use the asterisk to indicate parameters of the case which includes fluid flow. Since the pressure change is independent of the radial position of the fluid which causes the expansion (or contraction) of each spherical shell, we only need to know the total fluid volume injected (or withdrawn) through the boundary. This volume is $(4\pi a^2 v/i\omega)$, leading to a fractional volume change in the medium of

$$D = (4\pi a^2 v/i\omega)(-Q_1 + Q_2)/(4\pi b^3/3). \tag{8}$$

Combining these expressions with equation (6) we obtain

$$\Delta P_0 = k^* \frac{(R_1 - R_2)(4\pi a^2)(-Q_1 + Q_2)P_0}{(Z_1 + Z_2)(i\omega)(4\pi b^3/3)},$$

or

$$\Delta P_0 = k^* W P_0, \tag{9}$$

where

$$W = \frac{3a^2(R_1 - R_2)(-Q_1 + Q_2)}{b^3 i\omega(Z_1 + Z_2)}.$$

Without flow,

$$P_0 = -k_0 D_0.$$

With flow,

$$P_0 + \Delta P_0 = -k^* D_0, \quad \text{or}$$

$$P_0 + k^* W P_0 = -k^* D_0, \quad \text{with} \tag{10}$$

$$k^* = [k_0/(1 - k_0 W)] = k_r^* + i k_i^*.$$

Since fluid content does not affect shear distortions, the shear modulus for the partial saturation is

$$\mu^* = \mu_0 = \mu = \bar{\mu}. \tag{11}$$

Density does not depend on fluid flow, so

$$\rho^* = \rho_0 = (1 - \phi)\rho_S + \phi(1 - S_G)\rho_f. \tag{12}$$

Plane-wave modulus is

$$M^* = (M_r^* + i M_i^*)$$

$$= (k_r^* + 4\mu^*/3 + i k_i). \tag{13}$$

Speeds and attenuation

For a plane compressional wave in this composite medium, we can write the wave equation

$$M^* \frac{\partial^2 u}{\partial x^2} = \rho^* \frac{\partial^2 u}{\partial t^2}$$

and its solution

$$u = U_0 e^{-\gamma x} e^{i\omega t}, \quad \text{where} \tag{14}$$

$$\gamma = a_P^* + i\omega/C_P^* = i\omega(\rho^*/M^*)^{1/2}.$$

The velocity and attenuation are

$$C_P^* = (|M^*|/\rho^*)^{1/2}/\cos(\theta_P^*/2), \quad \text{and}$$

$$a_P^* = \omega \tan(\theta_P^*/2)/C_P^*, \quad \text{where}$$

$$\theta_P^* = \tan^{-1}(M_i^*/M_r^*).$$

There is no attenuation of the shear wave due to partial saturation. We write for its velocity and attenuation

$$C_S{}^* = (\mu^*/\rho^*)^{1/2} \quad \text{and}$$

$$a_S{}^* = 0. \tag{15}$$

Partial gas saturation

It seems clear that the effect of fluid flow will be most pronounced when one medium is saturated with a very compressible fluid. We shall assume that the central sphere is saturated with a gas so light and compressible that R_1, Q_1, and Z_1 can all be neglected. In this case, equation (10) simplifies to

$$k^* = \frac{k_0}{1 + 3a^2 R_2 Q_2 k_0 / b^3 i\omega Z_2}. \tag{16}$$

Defining $(\text{Re} + i\,\text{Im}) = (2ak_{E2}/b^2 i\omega Z_2)$, we can write

$$k^* = \frac{k_0}{1 + (3k_0 R_2 Q_2 / 2k_{E2})(a/b)(\text{Re} + i\,\text{Im})}. \tag{17}$$

To simplify the expression for calculation, the following additional definitions are made:

$$\omega_0 = (2\kappa_2 k_{E2}/\eta_2 b^2);$$

$$\alpha_2 b = (1 + i)q \quad \text{and} \quad \alpha_2(b - a) = (1 + i)q';$$

$$q = (\omega/\omega_0)^{1/2}.$$

As a function of dimensionless frequency q,

$$(\text{Re} + i\,\text{Im})$$

$$= -\frac{i}{q^2}\left[\frac{(1 + q' + iq' - i2q^2 a/b) - (1 - q' - iq' - i2q^2 a/b)e^{2q'}(\cos 2q' + i \sin 2q')}{(1 + q + iq) - (1 - q - iq)e^{2q'}(\cos 2q' + i \sin 2q')}\right]. \tag{18}$$

Let us consider the limiting cases of zero gas saturation $(a/b \rightarrow 0)$ and full gas saturation $(a/b \rightarrow 1.0)$. In the first case, $(\text{Re} + i\text{Im})$ is not zero but it is multiplied by a/b in equation (17). Hence $k^* = k_0$, which is the bulk modulus for liquid-saturated rock. As a/b approaches unity, $(\text{Re} + i\text{Im})$ approaches zero, and again $k^* = k_0$. However, k_0 is now the bulk modulus for gas-saturated rock. Hence equation (17) gives the correct values for the bulk modulus at two limiting saturations.

At "high" frequencies (for which the wavelength of fluid-flow waves is small compared to sphere radius), $(\text{Re} + i\text{Im})$ approaches zero and $k^* = k_0$. Although the phase angle $\theta_P{}^*$ decreases with frequency, the attenuation (per unit distance) increases as the square root of frequency. At very low frequencies, $(\text{Re} + i\text{Im})$ approaches $+1$, and k^* approaches a value which depends on a/b. This sort of composite medium is a low-pass filter, with a transition frequency which is proportional to the inverse square of the typical dimension (b) of the inhomogeneity.

NUMERICAL EXAMPLE

Properties of skeleton

We start from measured speeds reported by Gardner et al (1964) for unconsolidated packings of sand under various confining pressures. At a pressure of 2500 psi on the skeleton, an air-filled sand of 30 percent porosity had a compressional speed of 1.5×10^5 cm/sec and a shear speed two-thirds as great. The solid material (quartz) has a density

$$\rho_S = 2.65 \text{ gm/cm}^3$$

and a bulk modulus

$$k_S = 35 \times 10^{10} \text{ dynes/cm}^2.$$

From this information we obtain the skeleton properties:

$$\phi = 0.30,$$

$$\bar{C}_P = 1.5 \times 10^5 \text{ cm/sec},$$

$$\bar{C}_S = 1.0 \times 10^5 \text{ cm/sec},$$

$$\bar{\rho} = (1 - \phi)\rho_S = 1.85 \text{ gm/cm}^3,$$

$$\bar{M} = \bar{\rho}\bar{C}_P{}^2 = 4.17 \times 10^{10} \text{ dynes/cm}^2,$$

$$\bar{\mu} = \bar{\rho}\bar{C}_S{}^2 = 1.85 \times 10^{10} \text{ dynes/cm}^2,$$

$$\bar{k} = \bar{M} - 4\bar{\mu}/3 = 1.71 \times 10^{10} \text{ dynes/cm}^2.$$

Properties of water-saturated rock

When the fluid in the pore space is pure water under normal conditions, then the constants of the fluid in region 2 are

$\rho_{f2} = 1.0$ gm/cm^3,

$k_{f2} = 2.25 \times 10^{10}$ dynes/cm^2,

$\eta_2 = 0.01$ gm/cm sec (one centipoise),

$\kappa_2 = 10^{-9}$ cm^2 (100 millidarcies).

The constants for the water saturated rock of region 2 are

$\rho_2 = (1 - \phi)\rho_S + \phi\rho_{f2} = 2.155$ gm/cm^3,

$k_2 = \bar{k} + \dfrac{(1 - \bar{k}/k_S)^2}{(\phi/k_{f2} + (1 - \phi)/k_S - \bar{k}/k_S{}^2)}$

$\qquad = 7.66 \times 10^{10}$ dynes/cm^2,

$\mu_2 = \bar{\mu} = 1.85 \times 10^{10}$ dynes/cm^2,

$M_2 = k_2 + 4\mu_2/3$

$\qquad = 10.12 \times 10^{10}$ dynes/cm^2,

$C_{P2} = (M_2/\rho_2)^{1/2} = 2.167 \times 10^5$ cm/sec,

and

$C_{S2} = (\mu_2/\rho_2)^{1/2} = 0.927 \times 10^5$ cm/sec.

Mixed saturation with no fluid flow

With $b/a = 2.0$, $a^3/b^3 = S_G = 0.125$ and for mixed saturation with no fluid flow we obtain the rock properties:

$\sigma_1 = \dfrac{\bar{M} - 2\bar{\mu}}{2(\bar{M} - \bar{\mu})} = 0.10$,

$E_1 = 2\bar{\mu}(1 + \sigma_1) = 4.07 \times 10^{10}$ dynes/cm^2,

$\sigma_2 = \dfrac{M_2 - 2\mu_2}{2(M_2 - \mu_2)} = 0.39$,

$E_2 = 2\mu_2(1 + \sigma_2) = 5.14 \times 10^{10}$ dynes/cm^2,

$K_1 = 1.158/(2.158 - S_G) = 0.570$,

$K_2 = (0.240/S_G) + 0.760 = 2.680$,

$K_3 = 1.872 \times 10^{10}(1/S_G - 1)$

$\qquad = 13.11 \times 10^{10}$ dynes/cm^2,

(see Appendix A)

$k_0 = 6.21 \times 10^{10}$ dynes/cm^2,

$M_0 = (k_0 + 4\mu_0/3)$

$\qquad = 8.68 \times 10^{10}$ dynes/cm^2,

$\rho_0 = S_G\bar{\rho} + (1 - S_G)\rho_2 = 2.11$ gm/cm^3,

$C_{P0} = (M_0/\rho_0)^{1/2} = 2.03 \times 10^5$ cm/sec,

and

$C_{S0} = (\mu_0/\rho_0)^{1/2} = 0.94 \times 10^5$ cm/sec.

Mixed saturation including fluid flow

For the case of mixed saturation including fluid flow, the pertinent constants are

$R_2 = 0.867$,

$k_{A2} = 6.58 \times 10^{10}$ dynes/cm^2,

$k_{E2} = 2.71 \times 10^{10}$ dynes/cm^2,

$Q_2 = 0.618$, and

$k^* = \dfrac{6.21 \times 10^{10}}{1 + 1.842(a/b)(\text{Re} + i\,\text{Im})}$.

For $a/b = 0.5$ and $q = (\pi/2)$, we obtain the complex quantities:

$(\text{Re} + i\,\text{Im}) = 0.684 - i0.535$,

$k^* = (3.45 + i1.056) \times 10^{10}$, and

$M^* = (5.91 + i1.056) \times 10^{10}$ dynes/cm^2.

Also we note that

$\theta_P{}^* = 10.13$ degrees and

$C_P{}^* = 1.69 \times 10^5$ cm/sec.

If we arbitrarily choose for the radius of the outer sphere $b = 8.3$ cm, $\omega_0 = 25\,\pi$. Then $q = \pi/2$ corresponds to 31 hz, and the attenuation

$a_P{}^* = 1.03 \times 10^{-4}$ cm^{-1}, or 27 db/1000 ft.

Numerical results for the same rock and dimensions and for various frequencies are shown in the table. These results for $C_P{}^*$ and $\theta_P{}^*$ are plotted against frequency in Figure 2.

Table 1. Results for numerical example

q	f hz	$C_P{}^*$ m/sec	$\theta_P{}^*$ radians	$a_P{}^*$ m^{-1}	$a_P{}^*$ db/1000 ft
0.02	.005	1620	0.004	—	—
1.57	31	1690	0.179	0.0103	27
3.14	123	1910	0.152	0.0307	82
6.28	495	1980	0.063	0.0487	130

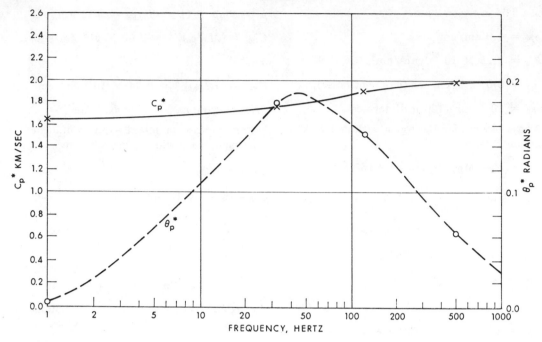

FIG. 2.   Phase angle $\theta_P{}^*$ and compressional wave speed $C_P{}^*$ for a porous rock with mixed saturation. Diameter of the gas pockets is 8.3 cm and the distance between centers is 13.3 cm.

ACKNOWLEDGMENTS

The author wishes to express his appreciation for the opportunity to complete this work at the Institute of Physics of the Earth in Moscow as an exchange scientist sponsored by the National Academy of Sciences and the Academy of Sciences of the USSR.

REFERENCES

Biot, M. A., 1956, Theory of propagation of elastic waves in a fluid-saturated porous solid: J. Acous. Soc. Am., v. 28, p. 168–191.
Frenkel, Ya. I., 1944, Theory of seismic and seismoelectric phenomena in damp soil: Izvestiya, Acad. Sci. USSR, Ser. Geogr. and Geoph., v. 8, no. 4, p. 133–150. (In Russian).
Gardner, G. H. F., Wyllie, M. R. J., and Droschak, D. M., 1964, Effects of pressure and fluid saturation on the attenuation of elastic waves in sands: J. Petr. Tech., February, p. 189–198.
Gassmann, F., 1951, Uber die Elastizität poröser Medien: Vierteljahrschr. Naturforsch. Ges., Zürich, v. 96, p. 1–23.
Lamb, Horace, 1960, Statics: New York, Cambridge University Press.
White, J. E., 1965, Seismic waves: New York, McGraw-Hill Book Co., Inc.

APPENDIX A

BULK MODULUS WITHOUT FLUID FLOW

We wish to calculate the average bulk modulus of a composite elastic body consisting of a sphere of radius a and a spherical shell of outer radius b. A text such as Lamb (1960, p. 342–43) gives for radially symmetric stress and displacement

$$p_{rr} = A + B/r^3$$

$$u_r = \frac{(1 - 2\sigma)}{E} Ar - \frac{(1 + \sigma)}{2E} \frac{B}{r^2}. \qquad \text{(A–1)}$$

Consider application of a pressure P_0 at the outer boundary, resulting in a pressure P_i at the inner boundary. Designating the inner sphere by subscript 1 and the shell by 2, we characterize the two media by Young's modulus E and Poisson's ratio σ. For the inner body, $B = 0$ and $A = p_{rr}(a) = -P_i$. Continuity of displacement at the inner boundary yields

$$-\frac{(1 - 2\sigma_1)}{E_1} aP_i =$$

$$\frac{(1 - 2\sigma_2)}{E_2} a \frac{(a^3 P_i - b^3 P_0)}{(b^3 - a^3)} \qquad \text{(A–2)}$$

$$-\frac{(1 + \sigma_2)}{2E_2} \frac{1}{a^2} \frac{a^3 b^3}{(b^3 - a^3)} (P_0 - P_i).$$

Displacement at the outer boundary is

$$U_0 = \frac{(1 - 2\sigma_2)}{E_2} \frac{(a^3 P_i - b^3 P_0)}{(b^3 - a^3)} b$$

$$- \frac{(1 + \sigma_2)}{2E_2} \frac{1}{b^2} \frac{a^3 b^3}{(b^3 - a^3)} (P_0 - P_i). \qquad \text{(A-3)}$$

From these equations we obtain the result

$$k_0 = K_3/(K_2 - K_1), \qquad \text{(A-4)}$$

where (with $a^3/b^3 = S_G$)

$$K_1 = \frac{3(1 - \sigma_2)/2}{(1 - 2\sigma_1)(1 - S_G)E_2/E_1 + S_G(1 - 2\sigma_2) + (1 + \sigma_2)/2},$$

$$K_2 = \frac{(1 - 2\sigma_2)/S_G + (1 + \sigma_2)/2}{3(1 - \sigma_2)/2}, \quad \text{and}$$

$$K_3 = \frac{E_2(1/S_G - 1)}{3(1 - 2\sigma_2) + 3(1 + \sigma_2)/2}.$$

APPENDIX B
FLUID PRESSURE DUE TO DILATATION

Assume that a pressure p is applied to the faces of a cube of saturated rock, creating dilatation D, fluid pressure p_f, and skeleton pressure $\bar{p}$. By definition, $p = -kD$. Also

$$D = -\phi p_f/k_f - (1 - \phi)p_f/k_s$$

$$- \bar{p}/k_S, \text{ and } p = p_f + \bar{p}.$$

From these relations we conclude that

$$p_f = -\frac{k_f}{\phi}\left[\frac{(1 - k/k_S)}{(1 - k_f/k_S)}\right] D. \qquad \text{(B-1)}$$

We need to apply this relation to the central sphere and to the spherical shell. For the same geometry as Appendix A, static analysis gives for the dilatation in the sphere $D_1 = -P_i/k_1 = -K_1P_0/k_1$ so that $p_{f1} = P_1 = R_1P_0$ with

$$R_1 = \frac{k_{f1}}{\phi}\left[\frac{(1 - k_1/k_{S1})}{(1 - k_{f1}/k_{S1})}\right]\frac{K_1}{k_1}. \qquad \text{(B-2)}$$

Dilatation in the shell is

$$D_2 = \frac{1}{k_2}\frac{(S_G P_i - P_0)}{(1 - S_G)} = \frac{1}{k_2}\frac{(S_G K_1 - 1)P_0}{(1 - S_G)},$$

and, hence, $p_{f2} = P_2 = R_2P_0$, with

$$R_2 = \frac{k_{f2}}{\phi k_2}\frac{(1 - k_2/k_{S2})(1 - S_G K_1)}{(1 - k_{f2}/k_{S2})(1 - S_G)}. \qquad \text{(B-3)}$$

In each fluid-saturated porous medium, we assume that the oscillatory particle velocity of the fluid relative to the skeleton is proportional to the pressure gradient (Darcy's law). For radial motion in spherical coordinates this relation is simply

$$v = -\frac{\kappa}{\eta}\frac{\partial p_f}{\partial r}. \qquad \text{(C-1)}$$

We also recognize that as fluid accumulates in any elementary volume of the medium (the rate of accumulation being proportional to the divergence of fluid velocity), the pressure increases. This linear relationship defines an "effective bulk modulus" k_E as follows:

$$p_f = k_E D_f, \qquad \text{(C-2)}$$

where D_f is the volume of fluid added divided by the volume of the element. This result may be derived through relations for an elementary cube in two steps. First, assume that the faces of the cube are held rigidly so that the volume of the injected fluid must be matched by compression of the fluid in the pore space ($p_{fB}\phi/k_f$), compression of the solid [$p_{fB}(1 - \phi)/k_s$], and expansion of the solid due to unloading of frame pressure ($-p_{fB}\bar{k}/k_s^2$). This defines an apparent bulk modulus k_A, yielding the pressure p_{fB} created by injection of fluid into a blocked elementary volume:

$$p_{fB} = k_A D_f, \qquad \text{(C-3)}$$

where

$$k_A = \left(\frac{\phi}{k_f} + \frac{1 - \phi}{k_s} - \frac{\bar{k}}{k_s^2}\right)^{-1}.$$

Because of shrinkage of the skeleton, skeleton pressure will take on the value $\bar{p}_B = -\bar{k}p_{fB}/k_s$, and the total rock pressure on the rigid container is $p_B = p_{fB} + \bar{p}_B = (1 - \bar{k}/k_s)p_{fB}$.

As the second step, we recognize that for a wave with displacement along only one coordinate, the blocked condition applies to the four faces parallel to the fluid flow, but the faces perpendicular to flow can move. At low frequencies, the constraining pressure on these faces is negligibly small. So consider a pressure $p_E = -p_B$ to be applied to these faces, the other four faces being blocked. The fluid pressure change is

$$p_{fE} = \frac{k_f}{\phi M} \frac{(1 - k/k_S)}{(1 - k_f/k_S)} p_E$$

$$= -\frac{k_f}{\phi M} \frac{(1 - k/k_S)(1 - \bar{k}/k_S)}{(1 - k_f/k_S)} p_{fB}. \qquad \text{(C-4)}$$

The combined fluid pressure from both steps is

$$p_f = p_{fB} + p_{fE} \qquad \text{(C-5)}$$

$$= \left[1 - \frac{k_f(1 - k/k_S)(1 - \bar{k}/k_S)}{\phi M(1 - k_f/k_S)}\right] k_A\ D_f.$$

$$p_2 e^{i\omega t} = (A/r)[e^{\alpha_2 r} + e^{2\alpha_2 b}(\alpha_2 b - 1)e^{-\alpha_2 r}/(\alpha_2 b + 1)]e^{i\omega t} \quad \text{and}$$

$$v_2 e^{i\omega t} = -(\kappa_2 A/\eta_2 r^2)[(\alpha_2 r - 1)e^{\alpha_2 r} - e^{2\alpha_2 b}(\alpha_2 b - 1)(\alpha_2 r + 1)e^{-\alpha_2 r}/(\alpha_2 b + 1)]e^{i\omega t}. \qquad \text{(C-11)}$$

The impedance looking outward is p_2/v_2, or

$$Z_2 = -\frac{\eta_2 a}{\kappa_2}\left[\frac{(\alpha_2 b + 1) + (\alpha_2 b - 1)e^{2\alpha_2(b-a)}}{(\alpha_2 b + 1)(\alpha_2 a - 1) - (\alpha_2 b - 1)(\alpha_2 a + 1)e^{2\alpha_2(b-a)}}\right]. \qquad \text{(C-12)}$$

Since $p_f = k_E D_f$, the "effective bulk modulus" is

$$k_E = \left[1 - \frac{k_f(1 - k/k_S)(1 - \bar{k}/k_S)}{\phi M(1 - k_f/k_S)}\right] k_A.$$

$$\text{(C-6)}$$

For radial fluid flow, the rate of fluid influx is $1/r^2 \partial/\partial r(r^2 v)$ so that

$$\frac{\partial p_f}{\partial t} = -k_E\left(\frac{\partial v}{\partial r} + \frac{2}{r}v\right). \qquad \text{(C-7)}$$

Using equation (C-1), we obtain

$$\frac{\partial^2 p_f}{\partial r^2} + \frac{2}{r}\frac{\partial p_f}{\partial r} = \frac{\eta}{\kappa k_E}\frac{\partial p_f}{\partial t}. \qquad \text{(C-8)}$$

In the central sphere,

$$p_1 e^{i\omega t} = (A/r)(e^{\alpha_1 r} - e^{-\alpha_1 r})\,e^{i\omega t} \quad \text{and}$$

$$v_1 e^{i\omega t} = \qquad \text{(C-9)}$$

$$-(\kappa_1 A/\eta_1 r^2)[(\alpha_1 r - 1)e^{\alpha_1 r} + (\alpha_1 r + 1) \cdot e^{-\alpha_1 r}]e^{i\omega t},$$

where $\alpha_1 = (i\omega \eta_1/\kappa_1 k_{E1})^{1/2}$.

The impedance looking inward is $-p_1/v_1$, or

$$Z_1 = \frac{\eta_1 a}{\kappa_1}\left[\frac{(1 - e^{-2\alpha_1 a})}{(\alpha_1 a - 1) + (\alpha_1 a + 1)e^{-2\alpha_1 a}}\right]. \qquad \text{(C-10)}$$

In the spherical shell,

APPENDIX D
SKELETON ADJUSTMENT TO FLUID INFLUX

As discussed in Appendix C, an elementary volume expands or contracts as a consequence of the fluid-flow wave. In the example, the fractional expansion required to achieve zero stress is

$$D_E = p_B/M = [(1 - \bar{k}/k_S)k_A/M]\ D_f,$$

$$= Q\ D_f,$$

where

$$Q = (1 - \bar{k}/k_S)k_A/M. \qquad \text{(D-1)}$$

Reprinted from Geophysics, v. 44, p. 161–178.

Wave attenuation in partially saturated rocks

Gerald M. Mavko* and Amos Nur*

A model is presented to describe the attenuation of seismic waves in rocks with partially liquid-saturated flat cracks or pores. The presence of at least a small fraction of a free gaseous phase permits the fluid to flow freely when the pore is compressed under wave excitation. The resulting attenuation is much higher than with complete saturation as treated by Biot. In general, the attenuation increases with increasing liquid concentration, but is much more sensitive to the aspect ratios of the pores and the liquid droplets occupying the pores, with flatter pores resulting in higher attenuation. Details of pore shape other than aspect ratio appear to have little effect on the general behavior provided the crack width is slowly varying over the length of the liquid drop.

INTRODUCTION

The velocity and attenuation of seismic waves in crustal materials are strongly dependent upon pore fluid content and the details of pore geometry. The degree of wave interaction with fluids is in general determined by the shape, and hence compliance, of the pores within the solid matrix of mineral grains. Rocks with flatter pores are more sensitive to the details of the fluid and its ability to support or transmit compressional and shear loads; rounder, more rigid pores are less sensitive to the presence of fluids. In the low frequency limit, pore fluids influence the system through their density, compressibility, and distinct lack of rigidity; at higher frequencies, viscous and inertial interactions are introduced.

This dependence of velocity and attenuation on pore geometry and fluid properties can, in principle, serve as a diagnostic of material structure both in situ and in the laboratory. Nur (1971) and Solomon (1972), using equations from Walsh (1969), made estimates of upper mantle partial melt configuration from the velocities of teleseismic compressional and shear waves. Related experimental work on the dependence of velocity on partial melt was reported by Anderson and Spetzler (1970). Nur (1973) interpreted observed temporal velocity anomalies as diagnostic of dilatant strain and varying pore water saturation in the crust prior to certain earthquakes.

In exploration geophysics, a substantial effort has been made to develop more detailed theoretical and experimental correlations between longitudinal wave velocities, rock type, and fluid content in the shallow crust. The goal in oil and gas exploration is to be able to distinguish reliably between gas, oil, and water in situ, as well as to infer their relative concentrations, rock type, porosity, and permeability. Similar problems in geothermal exploration concern assessing water and steam content, as well as permeability and state of fracture.

Work by Domenico (1974), Kuster and Toksöz (1974a, b), and Elliott and Wiley (1975) indicates that the velocity of a liquid-saturated rock can differ substantially from that in the same rock with a partial saturation of a free gaseous phase. The two rocks in contact can account for a large reflection coefficient and an observed "bright spot". However, the dependence of velocity on the amount of gas saturation is very weak over the range 10 to 90 percent. The contrast in velocity is an indicator of gas, but a poor quantitative measure of economic value.

In this paper we present a model of one particular mechanism for wave attenuation in partially saturated liquid-gas systems. The model predicts that for certain rocks with at least a small concentration of very flat pores, even a small amount of water can dramatically enhance the dissipation of energy of compressional waves. Furthermore, the level of attenuation is directly dependent on the actual concentrations of liquid and gas, as well as on the fluid viscosity and pore shape. This can, in principle, serve as an independent data point on the state of saturation of porous rocks.

Paper presented at the 46th Annual International SEG Meeting, October 27, 1976 in Houston, Texas. Manuscript received by the Editor September 12, 1977; revised manuscript received June 26, 1978.
*Rock Physics Project, Department of Geophysics, Stanford University, Stanford, CA 94305.

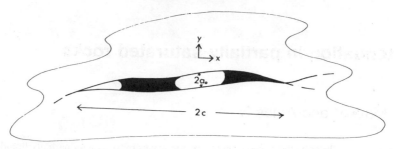

Fig. 1. Schematic view of partially saturated, two-dimensional pore. Thickness of pore perpendicular to the page is equal to d, an approximation of the third dimension.

In the first section that follows, we will discuss some general considerations of fluid attenuation and introduce our model of partial saturation. In the second section, mathematical formulas are derived for calculating attenuation of waves. The remainder of the paper gives a discussion of the model results including simplified expressions for attenuation for specific pore geometries, limiting expressions at high and low frequencies, the ratio of attenuation for P- and S-waves, and comparison with other models of fluid attenuation.

FLUID ATTENUATION

The primary source of fluid attenuation in porous media (ignoring nonmechanical effects) is relative motion between the solid and liquid. Such motion results in shearing stresses in the fluid and, consequently, viscous dissipation of mechanical energy. In our model of attenuation, we examine the details of flow and energy dissipation on the scale of the individual pores. We approximate the rock as a macroscopically isotropic elastic solid containing a distribution of partially saturated cracks or pores of the type shown schematically in Figure 1. We assume that the liquid is segregated into one or more discrete "drops" within each pore that flow as the pore is deformed. For mathematical convenience, the separate pores are treated as two-dimensional cracks in plane strain with width d (into the page) equal to some function of the half-length c. Furthermore, the separate pores are assumed not to interact, and only flat cracks with aspect ratio $\alpha_c < 0.1$ are considered (where $\alpha_c = a_0/c$ and a_0 is the maximum pore half-width in the plane of the page). Rigorously, this might limit our applications to low porosity igneous rocks. However, conceptually, the notions of dissipative flow that we will develop are quite applicable to more closely spaced, equidimensional pores. We will also find that flat pores give a much larger effect than the same volume of equant pores.

Our treatment of the details of flow in the individual pores will be reminiscent of Biot's (1956a, b) for the general porous solid except for the emphasis here on undersaturation. In this study, details of pore geometry will be retained, and both high and low frequencies are addressed.

MATHEMATICAL DERIVATION: NORMAL COMPONENT OF EXCITATION

Both P- and S-waves passing through our material exert oscillatory stresses which can be resolved into normal and shear components in the plane of each pore. In this section we find the attenuation resulting from the normal component of excitation on the individual pore; a later section treats the shear component. For a given wave and distribution of crack orientations, the total attenuation can then be found from a summation of both components over all pores. We will find, in fact, that the shear component of dissipation resulting from either compressional or shear waves gives a negligible contribution for most cases of interest involving water.

Seismic attenuation is estimated for a given fluid geometry by solving for the specific dissipation function Q^{-1} of the fluid-elastic composite, given by

$$Q^{-1} = \frac{\Phi}{2\pi W},$$

where Φ is the energy dissipated in the fluid phase during one cycle of sinusoidal oscillation and W is the peak energy stored during the same cycle. The spatial attenuation function can then be found from

$$\alpha_x = \frac{\omega}{2 C_P Q},$$

where α_x is the attenuation coefficient for plane wave amplitude decay with propagation and C_P is the compressional phase velocity at frequency ω. The method of solution will be to solve for the fluid flow field in each single partially filled pore resulting from a prescribed oscillation of the pore walls. The elastic energy of the rock surrounding the pore will then be obtained for the same oscillation.

For this study, pore geometries are limited to long narrow two-dimensional cracks as shown in Figure 1.

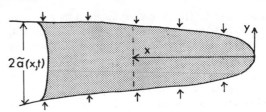

FIG. 2. Liquid drop subject to oscillations of pore walls.

LIST OF SYMBOLS

$a(x)$, $\tilde{a}$ = Pore half-width

a_0 = Maximum pore half-width

$A(x)$ = Fluid volume as a function of position in pore

c = Pore half-length

C_P = Compressional wave phase velocity

d = Pore dimension in third dimension

D = Drop length

E = Young's modulus

f = Frequency

K_f = Fluid bulk modulus

M = Modulus of rock

P = Fluid pressure

Q^{-1} = Specific dissipation function

$S(x)$ = Pore wall displacement as a function of position

u = Fluid velocity

U = Surface displacement of rock sample

V = Volume of rock sample

v = Volume of drop

V_P = Speed of sound in liquid

W, W_0, W_p = Elastic strain energy

α_c = Aspect ratio of pore

α_f = Aspect ratio of drop

α_x = Spatial attenuation function

β = Fractional volume of liquid in rock

δ = Maximum pore wall displacement

ε = Pore strain

η = Viscosity

λ = Lame's coefficient for rock material

μ = Shear modulus for rock material

ν = Kinematic viscosity

ξ = Poisson's ratio

ρ = Fluid density

σ = Applied stress

Σ = External surface of rock sample

τ = Viscous shear stress

Φ = Energy dissipated during one cycle

ψ = Porosity

ω = Frequency

Surface tension is neglected except for the assumption that, during seismic oscillation, each fluid drop remains intact rather than breaking up into smaller drops or losing contact with one or both of the pore walls. In our analysis we find that the induced gradients of pressure within each drop are important in determining dissipation, while the absolute pressure is not. Since the effect of surface tension is to superimpose an increment of essentially uniform pressure within the drop, we ignore only a slight perturbation in the estimation of W. A much more important effect which we consider later is the effect of surface tension on the distribution of liquid throughout the rock. Because the pores are undersaturated, the flow is incompressible for most frequencies and geometries of interest (see Appendix A). (An extension to compressible flow at very high frequencies is discussed in the next section and in Appendix B.) Furthermore, by assuming that the crack half-width $a(x)$ is a slowly varying function of x, the flow is essentially one-dimensional for small oscillations of the pore walls about the static position. Hence, the approximate equation of motion governing the fluid flow reduces to

$$\rho \frac{\partial u}{\partial t} = -\frac{\partial P}{\partial x} + \eta \frac{\partial^2 u}{\partial y^2}, \qquad (1)$$

where u is the x-component of velocity, ρ is the density, η is the viscosity, and P is the pressure of the fluid. [A discussion of the assumptions involved in obtaining the equation of motion (1) is given in Appendix A.]

The boundary conditions for the fluid are stated in terms of pore wall displacements (Figure 2). (The analysis is not limited to drops at crack tips. For drops away from the tips, the figure can be thought of as showing half the drop.) For very small strains, the pore half-width $\tilde{a}(x, t)$ is assumed to oscillate about the static shape $a(x)$ as

$$\tilde{a}(x, t) = a(x)[1 + \varepsilon e^{i\omega t}], \qquad (2)$$

where $\epsilon \ll 1$. The pore strain is given by $\epsilon = \delta/a_0$ where δ is the maximum pore wall displacement. Although the motion in (2) is in the y-direction, the y-component of fluid velocity is neglected in the one-dimensional approximation. However, the oscillation in (2) results in a volumetric oscillation which causes a pressure gradient and a lateral displacement or flow of the fluid.

At any station x, with local pressure gradient $\partial P/\partial x$, the solution of equation (1) for (approximately) parallel-plate flow is given by

$$u(x, y) = \frac{-\partial P(x)}{\partial x} \frac{1}{i\omega\rho} \cdot$$

$$\cdot \left[1 - \frac{\cosh\left(\sqrt{\frac{i\omega}{\nu}}\, y\right)}{\cosh\left(\sqrt{\frac{i\omega}{\nu}}\, a(x)\right)} \right], \quad (3)$$

where $\nu = \eta/\rho$. The weak x-dependence is contained in $P(x)$ and $a(x)$. The explicit time dependence is dropped here and in the remaining derivation since all quantities vary as $e^{i\omega t}$. Since the flow is incompressible, the net flow at x must be equal to the rate of change of volume $\mathring{A}(x)$ of the portion of the pore to the right of x. Hence,

$$\int_{-a(x)}^{a(x)} u(x, y)\, dy = -\mathring{A}(x), \quad (4)$$

and from (3) and (4) we can solve for the pressure gradient

$$\frac{\partial P(x)}{\partial x} = \mathring{A}(x) \cdot$$

$$\cdot \frac{i\omega\rho}{2a\left[1 - \frac{1}{a}\sqrt{\frac{\nu}{i\omega}}\tanh\sqrt{\frac{i\omega}{\nu}}\, a\right]},$$

where $\quad (5)$

$$\mathring{A}(x) = 2\int_0^x \frac{\partial}{\partial t}\tilde{a}(x, t)\, dx.$$

By combining (3) and (5), the flow field is obtained,

$$u(x, y) = \frac{-\mathring{A}(x)}{2a(x)} \cdot$$

$$\cdot \frac{1 - \frac{\cosh\sqrt{\frac{i\omega}{\nu}}\, y}{\cosh\sqrt{\frac{i\omega}{\nu}}\, a}}{1 - \frac{1}{a}\sqrt{\frac{\nu}{i\omega}}\tanh\sqrt{\frac{i\omega}{\nu}}\, a} \cdot$$

$$(6)$$

The energy dissipation Φ for the entire pore during one cycle is found by solving for the shear stress $\tau = \eta(\partial u/\partial y)$, and integrating the energy dissipation density over the period and the volume of fluid,

$$\Phi = \int_V \int_T \frac{1}{2\eta} \tau \cdot \tau^* \, dt\, dv$$

$$= \frac{3}{2}\frac{\pi\eta d}{\omega}\int_0^D \frac{\mathring{A}\mathring{A}^*}{a^3(x)} R(z)\, dx, \quad (7)$$

where $*$ means complex conjugate, $z = a(x)(\omega/2\nu)^{1/2}$ and $R(z)$ is given in Appendix B.

To find the peak mechanical energy stored in the rock-fluid system we consider a uniform oscillatory stress $\sigma = \sigma_0 e^{i(\omega t + \phi)}$ applied to a sample of the material and the resulting surface displacement $U = U_0 e^{i(\omega t + \theta)}$. If the difference in phase between displacement and stress is small (i.e., $\theta - \phi \approx Q^{-1} \ll 1$), then the peak energy is approximately

$$W \cong \frac{1}{2}\iint_\Sigma \sigma_0 U_0\, ds$$

$$\cong \mathrm{Re}\,\frac{1}{2}\iint_\Sigma \sigma^* U\, ds,$$

where the integral is over the external surface Σ of the sample and Re refers to the real part. To find the product

$$\frac{1}{2}\iint_\Sigma \sigma^* U\, ds,$$

and hence the peak energy, we use the Betti-Rayleigh reciprocity theorem. The theorem states that for a body acted upon separately by two sets of tractions, the work done by the first set of tractions acting through the displacement produced by the second set of tractions is equal to the work done by the second set of tractions acting through the displacements produced by the first set of tractions.

In the derivation that follows, we divide the rock into elements of volume, each containing a single pore. A uniform traction is applied to the boundary of each volume corresponding to the stress from the propagating wave. For cracks oriented at some arbitrary angle with the direction of propagation, the applied traction can be resolved into normal and shear components of stress in the plane of the crack. The crack compression from the normal component results in the fluid flow and attenuation derived above. The crack shearing from the shear component of stress results in the flow problem to be treated in a later section. (In most cases of interest, the shear dissipation will be negligible.) To find the stored energy for the arbitrarily oriented crack, both components of stress must be considered. Jaeger and Cook (1969, p. 313) show in detail how this energy varies with orientation for the simple case of a dry crack and how to sum over a random distribution of orientations. In our problem, the crack has a non-uniform internal fluid pressure distribution given by equation (5) which will modify the energy from the simple dry case. We will only solve in detail for the case of cracks oriented perpendicular to the principal stresses of excitation, in order to determine the first-order fluid effects on propagation. However, in principle, variation with orientation can be found by repeating the derivation which follows, keeping careful note of the resolved components of stress, in a fashion analogous to Jaeger and Cook. We expect that this would result in only a small perturbation of our results. In a later section we do, however, estimate the P- and S-wave attenuation for random crack orientations at low frequencies, for which the expressions for strain energy simplify.

Walsh (1965a, b) discusses the difficulties and uncertainties associated with choosing the uniform stress versus a uniform strain boundary condition, as well as the difference between the penny-shaped crack and a two-dimensional crack in plane strain or plain stress. He concludes, as we do, that the differences resulting from these various assumptions are negligible compared to the overall level of approximation in the analysis.

To apply the reciprocity theorem, consider the two sets of tractions shown in Figure 3. The system on the left is loaded by an externally applied stress corresponding to the peak stress of the seismic wave. The pore wall displacement $S(x)$ is due to both the external stress σ, tending to close the crack, and the instantaneous internal fluid pressure $P(x)$ tending to keep the crack open. The system on the right has the same external stress σ^* applied to both the external and internal surfaces. In this case, the system behaves like a solid elastic block without the cavity or fluid. Applying the reciprocity theorem, we can write

$$-2d \int_{-c}^{c} P(x) \frac{\sigma^*}{M} a(x)\,dx + \frac{\sigma\sigma^*}{M} V$$
$$= \iint_{\Sigma} \sigma^* U\,ds - 2\sigma^* d \int_{-c}^{c} S(x)\,dx, \quad (8)$$

where V is the volume of the block, M is the elastic modulus of the solid rock, Σ is the external surface of the volume (excluding pore surfaces), and dimensions c and d are as given in Figure 1. Since

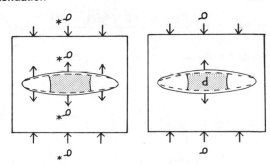

FIG. 3. Applying the reciprocity theorem to a rock under two sets of applied stress. On the left, only the induced pore pressure is applied to the pore. On the right, the applied stress σ^* is also applied to the pore, making the rock deform as though uncracked.

the theorem applies to elastic bodies it is necessary to treat the pores as external surfaces and the pore pressure as an externally applied load. The integral on the left is negligibly small and can be dropped (see Appendix D). For the case of plane wave propagation of a compressional wave, $M = \lambda + 2\mu$ where μ is the shear modulus and λ is Lame's coefficient for the intrinsic rock material. Rearranging equation (8) and taking the real part, we get

$$\mathrm{Re}\, \frac{1}{2} \iint_{\Sigma} \sigma^* U\,ds = \frac{1}{2} \frac{|\sigma|^2}{M} V + \mathrm{Re}\, \sigma^* d \cdot$$
$$\cdot \int_{-c}^{c} S(x)\,dx,$$

or

$$W = W_0 + W_p. \quad (9)$$

The term on the left is recognized as the approximate peak mechanical energy of the composite system W, which we are seeking. The terms on the right are the peak strain energy W_0 of a comparable rock without porosity, plus the energy contribution W_p due to the presence of the pore.

In order to relate the displacement $S(x)$ to the applied stress σ, it is convenient to consider a crack with uniform pore pressure $\bar{P}$ (see Appendix C). For very small stresses σ and $\bar{P}$ (as for a wave), the resulting small pore displacement and its integral W_p can be written as

$$S(x) \simeq \frac{2(\sigma - \bar{P})c(1 - \xi^2)}{E} \cdot$$
$$\cdot \left[1 - \left(\frac{x}{c}\right)^2\right]^{1/2}, \quad (10)$$

$$W_p \simeq \mathrm{Re}\ \frac{\pi \sigma^*(\sigma - \bar{P})c^2 d(1 - \xi^2)}{E}, \quad (11)$$

where ξ is Poisson's ratio and E is Young's modulus. With nonuniform pressure $P(x)$, the energy W_p can still be written in the form (11) if the pressure $\bar{P}$ is chosen as

$$\bar{P} = \frac{2}{\pi c}\int_{-c}^{c} P(x)\left[1 - \left(\frac{x}{c}\right)^2\right]^{1/2} dx, \quad (12)$$

where $P(x)$ is obtained from equation (5) (see Appendix C). Setting $S(0) \simeq a_0\varepsilon$, we can solve for σ in terms of $\bar{P}$ and ε or $\sigma = \bar{P} + \varepsilon\alpha_c E/2(1 - \xi^2)$. Finally, combining this with equations (9) and (11) and setting $\lambda = \mu$ and $\xi = 1/4$, we arrive at the total strain energy in terms of ε and $\bar{P}$:

$$W = \frac{8}{27}\ \mu V\alpha_c^2\varepsilon^2 \left(1 + \frac{9}{4}\ \frac{\psi}{\alpha_c}\right)$$
$$+ \frac{\mu V}{6}\left[\left(\frac{|\bar{P}|}{\mu}\right)^2\right.$$
$$\left. + \frac{8}{3}\ \alpha_c\varepsilon\left(1 + \frac{9}{8}\ \frac{\psi}{\alpha_c}\right)\mathrm{Re}\left(\frac{\bar{P}}{\mu}\right)\right]. \ (13)$$

Here ψ is the porosity and α_c is the crack aspect ratio.

The specific dissipation function Q^{-1} now can be given by combining equations (7) and (13),

$$Q^{-1} = \frac{\Phi}{2\pi W} = \cfrac{\dfrac{3}{2}\ \pi\eta\dfrac{d}{\omega}\displaystyle\int_0^D \dfrac{\mathring{A}\mathring{A}^* R(z)}{a^3(x)}\ dx}{\dfrac{8}{27}\ \mu V\alpha_c^2\varepsilon^2\left(1 + \dfrac{9}{4}\ \dfrac{\psi}{\alpha_c}\right) + \dfrac{\mu V}{6}\left[\left(\dfrac{|\bar{P}|}{\mu}\right)^2 + \dfrac{8}{3}\ \alpha_c\varepsilon\left(1 + \dfrac{9}{8}\ \dfrac{\psi}{\alpha_c}\ \mathrm{Re}\left(\dfrac{\bar{P}}{\mu}\right)\right)\right]}. \quad (14)$$

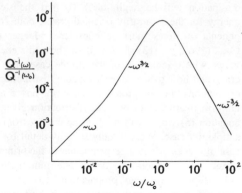

FIG. 4. The frequency dependence of Q^{-1} for the parallel walled pore in compression.

where $Y(z)$ is given in Appendix B. Using equation (12), we get finally the equivalent uniform pressure that appears in the formula for W,

$$|\bar{P}| = \frac{\omega^2\varepsilon\rho}{\pi}\ c^2\gamma Y(z),$$

$$\mathrm{Re}\ \bar{P} = \frac{\omega^2\varepsilon\rho}{\pi}\ c^2\gamma Z(z), \quad (17)$$

where γ and $Z(z)$ are also given in Appendix B. The attenuation Q^{-1} can be found by placing these simplified forms in equation (14).

A plot of Q^{-1} versus ω for the parallel walled pore is given in Figure 4. The most striking feature is the sharp peak at $\omega = \omega_0$ (ω_0 is the frequency at maximum attenuation), and the rapid fall off at both high frequencies, where $Q^{-1} \simeq \omega^{-3/2}$, and low frequencies, where $Q^{-1} \simeq \omega$. The peak and high-frequency fall-off result from the fluid compressibility which becomes important at very high pressures. We include this effect by modifying the continuity condition, equation (4), in the above derivation (see Appendix B). For most applications, however, the frequencies of interest are well within the limits of incompressible flow.

Let us examine more closely the asymptotic behavior. Table 1 gives the high and low frequency

RESULTS: THE PARALLEL-WALLED PORE

Consider the specific example of a pore where $a(x)$ is approximately a constant (i.e., $a(x) = a_0$) over the region occupied by the drop of length $2D$. Here, $R(z)$ is independent of x and

$$\mathring{A}(x) = +2x\varepsilon a_0 i\omega e^{i\omega t}.$$

Equation (7) can easily be evaluated to give the energy dissipated per cycle,

$$\Phi = \frac{4\pi\eta\omega d\varepsilon^2 D^3}{a_0}\ R(z). \quad (15)$$

The pressure distribution $P(x)$ is obtained by integrating equation (5) and taking the modulus

$$|P(x)| = \left|\int \frac{\partial P}{\partial x}\ dx\right|$$
$$= \frac{\omega^2\varepsilon\rho}{2}\ (D^2 - x^2)\ Y(z), \quad (16)$$

Table 1. Limiting forms of frequency dependent terms.

	$\omega \ll \dfrac{2\nu}{a^2}$ (Poiseuille)	$\omega \gg \dfrac{2\nu}{a^2}$ (inertial)	
		$\omega \ll \dfrac{1}{D}\sqrt{\dfrac{K_f}{\rho}}$ (incompressible)	$\omega \gg \dfrac{1}{D}\sqrt{\dfrac{K_f}{\rho}}$ (compressible)
R	1	$\dfrac{a_0}{3}\sqrt{\dfrac{\omega}{2\nu}}$	
Y	$\dfrac{3\eta}{\rho\omega a_0^2}$	1	—
Z	$\dfrac{6}{5}$	1	—
Φ	$\dfrac{4\pi\eta\varepsilon^2 dDa\omega}{\alpha_f^2}$	$\dfrac{4\pi\eta\omega\varepsilon^2 D^3 d}{3}\sqrt{\dfrac{\omega}{2\nu}}$	$\sqrt{2\eta\rho}\,\dfrac{\pi d\varepsilon^2}{\omega^{3/2}}\left(\dfrac{K}{\rho}\right)^{3/2}$
$\lvert P\rvert$	$\dfrac{3\eta\varepsilon\omega}{2a_0^2}(D^2-x^2)$	$\dfrac{\omega^2\varepsilon\rho}{2}(D^2-x^2)$	$\varepsilon K_f\left[1-e^{-\omega\sqrt{\rho/K_f}\,(D-x)}\right]$
W	$\dfrac{8}{27}\mu V\alpha_c^2\varepsilon^2\left(1+\dfrac{9}{4}\dfrac{\psi}{\alpha_c}\right)$	$\dfrac{8}{27}\mu V\alpha_c^2\varepsilon^2\left(1+\dfrac{9}{4}\dfrac{\psi}{\alpha_c}\right)$	$\dfrac{8}{27}\mu V\alpha_c^2\varepsilon^2\left(1+\dfrac{9}{4}\dfrac{\psi}{\alpha_c}\right)+\dfrac{\mu V}{6}\left[\dfrac{8}{3}\alpha_c\varepsilon\left(1+\dfrac{9}{8}\dfrac{\psi}{\alpha_c}\right)\left|\dfrac{P}{\mu}\right|+\left(\left|\dfrac{P}{\mu}\right|^2\right)\right]$
Q^{-1}	$\dfrac{27}{16}\dfrac{\eta\omega\beta}{\mu\alpha_c^2\alpha_f^2\left(1+\dfrac{9}{4}\dfrac{\psi}{\alpha_c}\right)}$	$\dfrac{9}{16}\dfrac{\eta\omega\beta a_0}{\mu\alpha_f^2\alpha_c^2\left(1+\dfrac{9}{4}\dfrac{\psi}{\alpha_c}\right)}\sqrt{\dfrac{\omega}{2\nu}}$	$\propto \omega^{-3/2}$

forms of Φ, P, W, and Q^{-1} based on the limiting forms of $R(z)$, $Y(z)$, and $Z(z)$. The new parameters are β, the volume concentration of liquid in the rock, K_f, the fluid bulk modulus, and $\alpha_f = a_0/D$, the aspect ratio of the drop.

Physically, the low-frequency expressions ($\omega \ll 2\nu/a^2$) correspond to Poiseuille flow. That is, the inertial terms are negligible, and the flow is governed by the balance between viscous shear forces and the pressure gradient driving the flow. These viscous shear stresses are proportional to ω; the average dissipation goes as ω^2; and the dissipation during one period, $T = 2\pi/\omega$, varies as ω.

The fluid pressure P, varies as $\eta\omega\varepsilon D^2/a^2$ at low frequencies. The pressure-shear balance is obvious here since the peak velocity is approximately $\omega\varepsilon D$, the shear stress is approximately $\eta\omega\varepsilon D/a$, and the integrated shear force over the drop length is approximately $\eta\omega\varepsilon D^2/a$. This equals a change in pressure force of approximately Pa over the length of the drop. For the case of water, the low-frequency pressure is usually small compared with the stress in the rock. Hence, at low frequencies (incompressible flow), the energy W is approximately the energy of a dry porous rock, independent of ω.

The condition for Poiseuille flow, and therefore a condition on the low-frequency form of Φ in Table 1, corresponds to $z < 1$, $\omega < 2\nu/a^2$, or $f < \nu/\pi a^2$. For a water system $\nu = 0.01$ cgs, so that $f < 0.003/a$, where a is in centimeters and f is in Hz. For the case of joints or fractures where $a \simeq 1$ mm, the low-frequency expressions are restricted to frequencies below 0.3 Hz; for pores or microcracks where $a \simeq 0.1$ mm, the restriction is $f < 30$ Hz. It would seem that for exploration frequencies the low-frequency approximation is generally adequate. However, where unusually wide cracks or high frequencies are encountered (in the laboratory, for example) the other forms should be used.

The viscosity always appears with frequency as the product $\eta\omega$ in the Poiseuille flow expressions (in Table 1). This is a characteristic of viscous flow that can be particularly useful. For example, in the laboratory, we can define $\eta\omega$ as an effective frequency $\omega' = \eta\omega$ and measure the frequency dependence of liquid-solid systems by varying either the real frequency ω or the fluid viscosity.

Computed values of Q^{-1} for some specific rocks, Boise sandstone, Bedford limestone, and Troy granite at 50 percent water saturation ($\beta = \psi/2$) and $f = 1.6$ Hz, are shown in Table 2. These are computed using the low-frequency formula and aspect ratio distributions given by Toksöz et al (1976). We assume for simplicity and for comparison of aspect ratios that all pores are oriented alike, and that each pore is exactly half-filled with water. The contributions from the first two pore shapes from each rock type are not computed because the quantitative reliability of the model is low for such large aspect ratios. However, the trend indicates that smaller aspect ratios dominate the overall behavior, almost independent of the porosity of each aspect ratio group. Hence, the total rock attenuation is insensitive to the assumed distribution of fluid except in the flattest pores. All three rock types show a similar sensitivity to saturation and a dominance of flatter pores or cracks in determining the overall attenuation. Although the three rocks are very different in nature, the population of small aspect ratio cracks is somewhat similar in each. It should be noted that if smaller aspect ratios or higher frequencies are to be considered, the higher frequency expressions in Table 1 must be used.

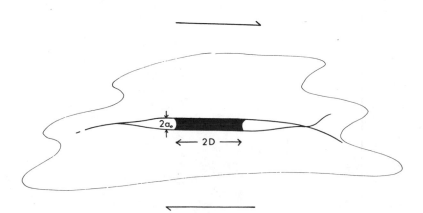

FIG. 5. Nearly parallel-walled pore in shear.

Table 2. Numerical examples of attenuation for three rock types at $f = 1.6$ Hz and uniform 50 percent water saturation.

ψ	α_c	α_f	β	Q^{-1}
Boise sandstone				
1.8×10^{-1}	1			
6.9×10^{-2}	$1. \times 10^{-1}$			
$1. \times 10^{-4}$	2.5×10^{-3}	$5. \times 10^{-3}$	$5. \times 10^{-5}$	5.4×10^{-7}
$1. \times 10^{-4}$	$2. \times 10^{-3}$	$4. \times 10^{-3}$	$5. \times 10^{-5}$	1.3×10^{-6}
1.5×10^{-4}	1.5×10^{-3}	$3. \times 10^{-3}$	7.5×10^{-5}	6.2×10^{-6}
$2. \times 10^{-4}$	$1. \times 10^{-3}$	$2. \times 10^{-3}$	$1. \times 10^{-4}$	4.2×10^{-5}
$1. \times 10^{-4}$	$5. \times 10^{-4}$	$1. \times 10^{-3}$	$5. \times 10^{-5}$	3.3×10^{-4}
$2. \times 10^{-5}$	$1. \times 10^{-4}$	$2. \times 10^{-4}$	$1. \times 10^{-5}$	$4. \times 10^{-2}$
Bedford limestone				
$1. \times 10^{-1}$	$1.$			
$2. \times 10^{-2}$	$1. \times 10^{-1}$			
2.5×10^{-3}	$1. \times 10^{-2}$	$2. \times 10^{-2}$	1.2×10^{-3}	$5. \times 10^{-8}$
$5. \times 10^{-4}$	$5. \times 10^{-3}$	$1. \times 10^{-2}$	2.5×10^{-4}	1.7×10^{-7}
$5. \times 10^{-4}$	$4. \times 10^{-3}$	$8. \times 10^{-3}$	2.5×10^{-4}	4.1×10^{-7}
$7. \times 10^{-4}$	$3. \times 10^{-3}$	$6. \times 10^{-3}$	3.5×10^{-4}	1.8×10^{-6}
$6. \times 10^{-4}$	$2. \times 10^{-3}$	$4. \times 10^{-3}$	$3. \times 10^{-4}$	7.9×10^{-6}
$4. \times 10^{-4}$	1.5×10^{-3}	$3. \times 10^{-3}$	$2. \times 10^{-4}$	1.7×10^{-5}
$5. \times 10^{-4}$	$1. \times 10^{-3}$	$2. \times 10^{-3}$	2.5×10^{-4}	$1. \times 10^{-4}$
Troy granite				
$1. \times 10^{-3}$	$1.$			
$5. \times 10^{-4}$	$1. \times 10^{-1}$			
1.5×10^{-3}	$1. \times 10^{-2}$	$2. \times 10^{-2}$	7.5×10^{-4}	$3. \times 10^{-8}$
$5. \times 10^{-5}$	$5. \times 10^{-4}$	$1. \times 10^{-3}$	2.5×10^{-5}	1.7×10^{-4}
7.5×10^{-5}	$4. \times 10^{-4}$	$8. \times 10^{-4}$	3.5×10^{-5}	5.8×10^{-4}
$1. \times 10^{-4}$	2.5×10^{-4}	$5. \times 10^{-4}$	$5. \times 10^{-5}$	5.4×10^{-3}
$5. \times 10^{-5}$	$1. \times 10^{-4}$	$2. \times 10^{-4}$	2.5×10^{-5}	$1. \times 10^{-1}$

The higher frequency behavior of dissipation and attenuation, Table 1, is governed by the increasing importance of fluid inertial stresses with respect to viscous stresses. In particular, the pressure at intermediate frequencies $(2\nu/a^2 \ll \omega \ll (1/D)\sqrt{K_f/\rho})$ is a balance between pressure gradient $\approx P/D$ and inertial stress $\approx \rho(\omega^2 \varepsilon D)$, where $\omega^2 \varepsilon D$ gives the acceleration. At very high frequencies $(\omega \gg (1/D)\sqrt{K_f/\rho})$, inertial forces greatly stifle the flow so that it becomes easier to compress the fluid (with pressure $\sim \varepsilon K_f$) than to accelerate it along the pore. The transition between incompressible and compressible flow, and hence the peak in the attenuation curve, occurs when the fluid inertial stress $\sim \omega^2 \varepsilon \rho D^2$ is comparable with the fluid compressive stress $\sim \varepsilon K_f$ or $\omega \sim (1/D)\sqrt{K_f/\rho}$.

SHEAR COMPONENT OF EXCITATION

Consider once again the nearly parallel-walled pore of width $2a_0$ over the region occupied by the drop of length $2D$. This time the pore is excited by a pure shear in the plane of the pore as shown in Figure 5. This is essentially the same mechanism considered by Walsh (1969) for saturated pores. The equation of motion is

$$\frac{\partial u}{\partial t} = \nu \frac{\partial^2 u}{\partial y^2},$$

with the boundary conditions on velocity of:

$$u = \begin{cases} i\omega\delta e^{i\omega t} & y = +a_0 \\ -i\omega\delta e^{i\omega t} & y = -a_0 \end{cases},$$

at the upper and lower pore walls, respectively. The flow field can easily be found:

$$u(y,t) = \frac{i\omega\delta \sinh\left(y\sqrt{\dfrac{i\omega}{\nu}}\right)}{\sinh\left(a_0\sqrt{\dfrac{i\omega}{\nu}}\right)} e^{i\omega t}. \quad (18)$$

The energy dissipated in the entire pore during one cycle is found exactly as in the compressional problem and is given by

$$\Phi_{\text{shear}} = \frac{2\omega^2 \pi \eta \delta^2 Dd}{\nu}\sqrt{\frac{\nu}{2\omega}}.$$

$$\cdot \frac{\sinh \sqrt{\frac{2\omega}{\nu}} \, a + \sin \sqrt{\frac{2\omega}{\nu}} \, a}{\sinh^2 \sqrt{\frac{\omega}{2\nu}} \, a \, \cos^2 \sqrt{\frac{\omega}{2\nu}} \, a + \cosh^2 \sqrt{\frac{\omega}{2\nu}} \, a \, \sin^2 \sqrt{\frac{\omega}{2\nu}} \, a}, \qquad (19)$$

or, in the limits of high and low frequency,

$$\Phi_{\text{shear}} = \begin{cases} 4\pi\omega\eta\varepsilon^2 Dad & \omega \ll 2\nu/a^2 \\[2mm] \dfrac{4\pi\omega^2\eta\delta^2 Dd}{\nu} \sqrt{\dfrac{\nu}{2\omega}} & \omega \gg 2\nu/a^2. \end{cases} \qquad (20)$$

The relative importance of dissipation in shear and compression for this geometry is found by taking a simple ratio Φ (compression)$/\Phi$ (shear) with identical pores and fluid and equal amplitudes of pore displacements. Using equation (20) and Table 1, we have

$$\frac{\Phi_{\text{compression}}}{\Phi_{\text{shear}}}$$

$$= \begin{cases} 1/\alpha_f^2 & \omega \ll 2\nu/a^2 \\[2mm] 1/3\,\alpha_f^2 & 2\nu/a^2 \ll \omega \ll \dfrac{1}{D}\sqrt{K_f/\rho}. \end{cases} \qquad (21)$$

It is clear that for small aspect ratios, the dissipation in shear will be many orders of magnitude smaller than for compression in either limit. The interpretation of this is simple. The fluid velocity in the shear case is approximately the pore wall velocity $\simeq \omega\delta$. This gives an average fluid shear stress of $\eta\omega\delta/a$. On the other hand, the fluid velocity in the compressional case is on the order of $\simeq \omega\delta/\alpha_f^2$, with a fluid shear stress $\simeq \eta\omega\delta/\alpha_f^2 a$. The flow in compression is amplified by the aspect ratio.

We should emphasize that these analyses of dissipation in shear and compression refer to the shear and compressional components of stress resolved on each individual crack. The results are not identically the dissipation that would be associated with shear (S) and compressional (P) wave propagation. For either type of wave, each crack depending on its orientation, would have both a shear and compressional contribution to energy dissipation in the fluid and strain energy stored in the elastic matrix. These are discussed in the next section.

P- AND S-WAVE ATTENUATION

To estimate low-frequency P- and S-wave attenuation we assume a random distribution of crack orientations and use the results of the parallel-walled pore. A crack at arbitrary orientation to the principal stresses encounters both shear and normal excitation. However, based on the results of the previous sections only the normal component of fluid excitation contributes significantly to dissipation. In addition, at low frequencies, the induced fluid pressure is small enough that the pores deform under stress as though dry. Hence, the maximum pore strain ε, using equation (10), is approximately related to the resolved normal stress σ_n by

$$\varepsilon = \frac{2(1-\xi^2)}{E\,\alpha_c}\,\sigma_n.$$

Substituting this into the expression for low-frequency dissipation shown in Table 1 gives

$$\Phi = \frac{16\pi\eta\omega dD^3(1-\xi^2)^2}{a\alpha_c^2 E^2}\,\sigma_n^2. \qquad (22)$$

The crack orientation is defined by spherical coordinates θ and ϕ. θ is the angle between the crack normal and greatest principal stress axis. ϕ is the azimuthal angle of rotation about the stress axis.

For a P-wave, the normal stress is rotationally symmetric about the direction of propagation and is given by

$$\sigma_n = \sigma[\cos^2\theta + \xi'\sin^2\theta/(1-\xi')].$$

Here ξ' is the effective Poisson's ratio of the porous rock and σ is the normal stress in the direction of propagation (Walsh, 1966). The total dissipation Φ_P is given by the sum of the expression (22) over all N cracks:

$$\Phi_P = \frac{16\pi\eta\omega(1-\xi^2)^2}{E^2}\,\sigma^2 \sum_N \frac{dD^3}{a\alpha_c^2} \cdot$$

$$\cdot \left[\cos^2\theta + \frac{\xi'}{1-\xi'}\,\sin^2\theta\right]^2. \qquad (23)$$

For very large N, the summation in (23) can be

approximated with an integral over a continuous distribution of crack orientation. In addition, we assume that the dimensions d, D, a, and α now represent appropriate averages for the distribution:

$$\Phi_P = \frac{16\pi\eta\omega(1 - \xi^2)^2 \sigma^2 dD^3}{a\alpha_c^2 E^2} \frac{N}{4\pi} \cdot$$

$$\cdot \int_0^{2\pi} \int_0^{\pi} \left[\cos^2\theta + \frac{\xi'}{1 - \xi'} \cdot \right.$$

$$\left. \cdot \sin^2\theta \right]^2 \sin\theta \, d\theta \, d\phi. \qquad (24)$$

The peak strain energy in a volume of rock V is

$$W = \frac{\sigma^2 V}{2E'} \frac{(1 - 2\xi')(1 + \xi')}{(1 - \xi')},$$

where E' is the effective Young's modulus of the porous rock. Finally, the P-wave attenuation becomes

$$Q_P^{-1} = \frac{16\eta\omega dD^3}{5\, a\alpha_c^2 E} \left(\frac{N}{V} \right) \left(\frac{E'}{E} \right) \cdot$$

$$\cdot \frac{(1 - \xi^2)^2 \left[1 + \frac{4}{3} \frac{\xi'}{1 - \xi'} + \frac{8}{3} \left(\frac{\xi'}{1 - \xi'} \right)^2 \right](1 - \xi')}{(1 + \xi')(1 - 2\xi')}. \qquad (25)$$

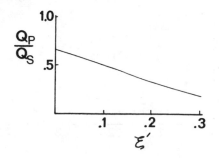

FIG. 6. The ratio Q_p/Q_s versus effective Poisson's ratio ξ' at low frequencies.

Similarly, for an S-wave the normal stress on a crack is given by (Walsh, 1966)

$$\sigma_n = \sigma(\cos^2\theta - \cos^2\phi \sin^2\theta)$$

where σ is the principal stress. Combining this with equation (22) and summing over all cracks gives the S-wave dissipation,

$$\Phi_S = \frac{16\pi\eta\omega(1 - \xi^2)^2 \sigma^2 dD^3}{a\alpha_c^2 E^2} \frac{N}{4\pi} \cdot$$

$$\cdot \int_0^{2\pi} \int_0^{\pi} [\cos^2\theta - \cos^2\phi \sin^2\theta]^2 \cdot$$

$$\cdot \sin\theta \, d\theta \, d\phi$$

The peak strain energy is given by

$$W = \frac{\sigma^2 V}{E'}(1 + \xi').$$

The S-wave attenuation becomes

$$Q_S^{-1} = \frac{32\eta\omega dD^3}{15\, a\alpha_c^2 E} \left(\frac{N}{V} \right) \left(\frac{E'}{E} \right) \frac{(1 - \xi^2)^2}{(1 + \xi')}. \qquad (26)$$

Although it is difficult to test these absolute values of Q_P and Q_S, it is interesting to take their ratio:

The ratio of Q_P/Q_S is plotted versus ξ' in Figure 6. The fact that $Q_P/Q_S < 1$ emphasizes that this is physically a local compressional mechanism of attenuation. Although the expressions (25) and (26) apply only at low frequencies, the ratio in equation (27) should be at least approximately correct to much higher frequencies.

Laboratory measurements on nominally dry rocks usually show $Q_P/Q_S \cong .5$, in fair agreement with our prediction in Figure 6. For example, for dry Quincy granite at low stress $\xi' \cong .1$ (Birch et al, 1942, p. 73), while $Q_P/Q_S \cong .5$ (Birch and Bancroft, 1938). Similarly, for Solenhofen limestone $\xi' \cong .25$ (Birch et al, 1942, p. 76) while $Q_P/Q_S \cong 0.6$ (Peselnick and Zietz, 1959). One could argue that dry rock measurements should not be compared with these results. However, because of capillarity some nominally dry rocks might contain small amounts of water in flat cracks. For example, Born (1941) found that when previously dried samples of Amherst sandstone were exposed to air their attenuation

$$\frac{Q_P}{Q_s} = \frac{2(1 - 2\xi')}{(1 - \xi')\left[3 + 4\left(\frac{\xi'}{1 - \xi'} \right) + 8\left(\frac{\xi'}{1 - \xi'} \right)^2 \right]}. \qquad (27)$$

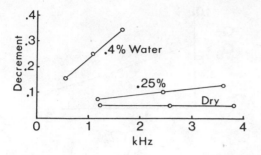

FIG. 7. Logarithmic decrement ($\sim Q^{-1}$) versus frequency for Amherst sandstone (Born, 1941).

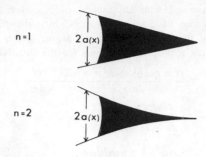

FIG. 8. Two simple drop geometries. The upper drop is a linear or triangular wedge. The lower drop has parabolic shape.

increased due to absorbed moisture. The added increment of attenuation was frequency-dependent ($\Delta Q^{-1} \propto \omega$), as expected from shear losses in a fluid phase, whereas Q^{-1} for the dried samples was independent of frequency. This is shown in Figure 7. A core of gypsum that was only air dry showed a similar frequency-dependent Q (Born, 1941).

Winkler (1977, personal communication) found that $Q_P/Q_S > 1$ for dry Massilon sandstone, but when almost fully saturated $Q_P/Q_S < 1$.

OTHER GEOMETRIES

The parallel pore is particularly useful in understanding the characteristics of the model because the integral in equations (6) and (13) for compressional dissipation can very easily be evaluated. For other simple pore geometries, the integral can also be evaluated if we work in the low frequency regime such that $R(z) \simeq 1$.

Consider the two geometries shown in Figure 8. In each case the pore half-width is given by $a(x) = a_0(x/c)^n$, $n = 1, 2$, and the maximum pore displacement is δ. Using formula (6), we first note that

$$A(x) = \frac{2a_0}{c^n} \frac{x^{n+1}}{n+1} e^{i\omega t},$$

or   (28)

$$\mathring{A}\mathring{A}^* = \frac{4\omega^2 \delta^2}{c^{2n}} \frac{x^{2n+2}}{(n+1)^2}.$$

The dissipation Φ is then

$$\Phi = \begin{cases} \dfrac{3\pi\eta\omega\varepsilon^2 v}{\alpha_f^2} & n = 1 \\[2em] \dfrac{4\pi\eta\omega\varepsilon^2 v}{\alpha_f^2} & n = 2 \end{cases}, \qquad (29)$$

where v is the volume of the drop. This result compared with Table 1 for the parallel pore suggests that for low frequencies the fluid dissipation can be approximated for a variety of geometries by

$$\Phi \simeq g \frac{n\omega\varepsilon^2 v}{\alpha_f^2}, \qquad (30)$$

where g is a geometric factor which we have found to vary from π to 4π for the specific geometries studied. This rather weak dependence on geometry suggests that the results derived for the parallel pore have general applicability, particularly for order-of-magnitude behavior.

COMPARISON WITH OTHER MODELS

Perhaps the most comprehensive single treatment of wave propagation incorporating dissipative fluid motion was presented by Biot (1956a, b). His formulation assumes a fully saturated porous material and includes the effects of fluid compressibility and coupled fluid and solid stress. At low frequencies, the relative fluid flow is assumed to resemble Poiseuille flow in a flat or circular duct, while at higher frequencies inertial terms are also included (laminar flow is always assumed). This microscopic flow field is used only to establish a frequency-dependent proportionality between the average flow and the stresses transmitted to the solid rock. Details of pore shape and local flow are neglected thereafter, and are lumped into parameters which relate only the averaged solid and fluid motions on a scale much greater than the pore size.

The elimination of local flow, and the condition of complete liquid saturation treated by Biot, severely limit the amount of flow and dissipation that can occur under wave excitation. White (1965, p. 133) shows that for the saturated medium, the only sources of relative fluid motion from the passing wave are viscous drag due to acceleration of the solid with respect to the fluid and pressure gradients between the peak and trough of the passing wave. His numerical example of a water-saturated sandstone shows negligible dissipation.

In our model we specifically neglect the large-scale acceleration and diffusion and look only at the local flow. In this sense our mechanism of dissipation is more a point property. In addition, the undersaturation provides a strong local heterogeneity which causes locally high pressure gradients and flow.

At low frequencies, Q^{-1} is proportional to ω in both Biot's model and ours, although the magnitude of attenuation is much greater in ours. At higher frequencies the dependence is quite different. Biot's expression for Q^{-1} varies as $\omega^{-1/2}$, while ours varies in a complicated fashion toward $\omega^{-3/2}$ at very high frequencies.

Another model of attenuation in porous media is by White (1975). His treatment resembles Biot's in that lumped parameters describe the material and flow properties on a scale much greater than a pore dimension. Only low frequency viscous flow is considered. White includes undersaturation by considering regions of dry rock containing many pores imbedded in regions of saturated rock also containing many pores. High pressure gradients and flow occur at the contact between wet and dry rock and result in large attenuation. At low frequencies White's expression for Q^{-1} varies as ω, while at higher frequencies (still within the Poiseuille range) Q^{-1} varies as ω^{-1}. The only physical comparison we can make between the two models is to suggest that the contact region between wet and dry in White's model is somewhat similar to our description of the individual partially saturated pores.

DISCUSSION

We have presented a simple model to describe the attenuation of seismic waves in rocks with partially liquid saturated flat cracks or pores. In this study, the presence of at least a small fraction of a highly compressible gaseous phase permits the fluid to flow freely when the pore is compressed under seismic excitation. This leads to viscous shearing in the fluid and high energy dissipation. In general, the attenuation increases with increasing liquid concentration. However, only partially saturated pores contribute, as well as saturated pores connected to open pores. As successive regions saturate, the attenuation will fall off. In the limit of complete saturation, the flow is stifled and the attenuation is expected to essentially

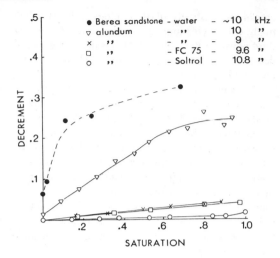

FIG. 9. Attenuation versus saturation for several materials and fluids. Berea sandstone from Gardner et al (1964); Alundum rods from Wyllie et al (1962). All samples excited in extensional oscillation, except for sample with FC-75, which is in torsion.

disappear as is predicted by Biot's formulation. Likewise, the dissipation from pure shear distortion of the pores, whether completely saturated or not, is negligible.

The attenuation or Q^{-1} is found to be extremely sensitive to the aspect ratios of the pores and the liquid droplets occupying the pores, with flatter pores and drops resulting in higher attenuation. Details of pore shape other than aspect ratio appear to have little effect on the general behavior, provided the crack width is slowly varying over the length of the liquid drop.

An important factor in interpreting attenuation measurements is the actual distribution of liquid within the rock. Since only the flattest cracks are important, it is the state of flat crack saturation, rather than the total rock saturation, that determines attenuation. The distribution will depend, for the most part, on capillarity. Wetting fluids will tend to occupy the flattest pores, while nonwetting fluids might be more randomly distributed. The distribution and, hence, attenuation might also depend on saturation history. For example, if one flat crack after another is sequentially wetted as fluid is added to a rock, we might expect the low frequency attenuation Q^{-1} to increase linearly with fluid content (see Table 1). This is due to the linear increase in the fluid content β, while the average aspect ratios of wetted cracks and droplets do not change. On the other hand, if all flat cracks are gradually filled simultaneously, then we might expect Q^{-1} initially to increase roughly as the third power of fluid content. This is due to the steady decrease of droplet aspect ratio in addition to the increase of β as fluid is added.

Measurements of attenuation under partially saturated conditions have been reported by a number of authors (Born, 1941; Wyllie et al, 1962; Gardner et al, 1964; Usher, 1962). Some of the results are shown in Figure 9. All these curves show a monotonic increase of attenuation with increasing saturation, as we might expect. The data for water in sandstones show a faster, nonlinear increase with saturation than for the nonaqueous liquids in alundum rods. This is perhaps due to a greater number of very thin cracks in the sandstones, and greater wetting with water.

Since small amounts of water can decrease Q, it is interesting to consider also the effect on elastic moduli. At low frequencies, a partially saturated rock behaves as though dry (neglecting the effects of surface tension and chemical softening), while at extremely high frequencies droplets of water behave as rigid bridges, effectively shortening flat cracks and stiffening the rock. Comparison of measured static and dynamic moduli of nominally dry rocks shows that high-frequency moduli can be as much as 100 percent greater (Simmons and Brace, 1965), particularly on igneous rocks which have a large relative population of flat cracks. We might expect at least part of the difference to result from the fluid stiffening. To estimate the magnitude of the effect, we can use the dispersion relation given by Kjartansson (1977),

$$\frac{M_1}{M_2} \cong \left(\frac{f_1}{f_2}\right)^{2/\pi Q} \qquad (31)$$

where M_1 and M_2 are two moduli measured at two frequencies f_1 and f_2. Using this relation, we assume that Q stays roughly the same over the band of interest. For static measurements at $f_2 < .1$ Hz and dynamic measurements $f_1 \cong 10^6$ Hz, this gives $f_1/f_2 > 10^7$. For $Q = 100$, this gives $M_1/M_2 \cong 1.1$. For $Q \cong 50$, $M_1/M_2 \cong 1.2$. These are less than some of the observed differences, but in the right direction. This estimate, as obtained from equation (31), is a conservative one, based on the assumption of frequency-independent Q. To further investigate the static-dynamic discrepancy, it will be useful and necessary to measure the moduli as a function of both frequency and degree of partial saturation. Such measurements could well unravel the physical mechanism of attenuation in partially saturated porous rocks.

ACKNOWLEDGMENTS

We thank the reviewers for constructive comments. This work was supported by grants from the National Science Foundation, Earth Sciences Division grant no. EAR76-22501; and ERDA, Division of Basic Energy Sciences contract no. EY76-S-0326 PA no. 45. The author was supported in part by NSF graduate and postdoctoral Fellowships.

REFERENCES

Anderson, Don L. and Spetzler, Harmut, 1970, Partial melting and the low velocity zone: Phys. Earth Planet Interiors, v. 4, p. 62–64.
Batchelor, G. K., 1967, An introduction to fluid dynamics: Cambridge University Press.
Biot, M. A., 1956a, Theory of propagation of elastic waves in a fluid-saturated porous solid. I. Low-frequency range: J. Acoustical Soc. of Am., v. 28, p. 168–178.
——— 1956b, Theory of propagation of elastic waves in a fluid-saturated porous solid. II. Higher frequency range: J. Acoustical Soc. Am., v. 28, p. 179–191.
Birch, F., Schairer, J. F. and Spicer, H. C., 1942, Handbook of Physical Constants, Geol. Soc. Am. Spec. Paper 36, 325 p.
Birch, F. and Bancroft, D., 1938, Elasticity and internal friction in a long column of granite: Bull. Seis. Soc. Amer., v. 28, p. 243–254.
Born, W. T., 1941, The attenuation constant of earth materials: Geophysics, v. 6, p. 132–148.
Domenico, S. N., 1974, Effect of water saturation on seismic reflectivity of sand reservoirs encased in shale: Geophysics, v. 39, p. 759–769.
Elliot, S. E. and Wiley, B. F., 1975, Compressional velocities of partially saturated, unconsolidated sands: Geophysics, v. 40, p. 949–954.
Gardner, G. H. F., Wyllie, M. R. J., and Droschak, D. M., 1964, Effects of pressure and fluid saturation on the attenuation of elastic waves in sands : Jour. of Pet. Tech., v. 16, p. 189–198.
Jaeger, J. C. and Cook, N.G.W., 1969, Fundamentals of rock mechanics: London, Metheun and Co. Ltd.
Kjartansson, E., 1977, Constant Q: Wave propagation and attenuation: EOS, Trans. AGU, v. 58, p. 1183.
Kuster, G. T. and Toksöz, M. N., 1974a, Velocity and attenuation of seismic waves in two-phase media: Part I. Theoretical formulations: Geophysics, v. 39, p. 587–606.
——— 1974b, Velocity and attenuation of seismic waves in two-phase media: Part II. Experimental results: Geophysics, v. 39, p. 607–618.
Mavko, G. and Nur, A., 1977a, The effect of non-elliptical cracks on the compressibility of rocks: (in press), J. Geograph. Res.
——— 1977b, Unpublished manuscript.
Nur, A., 1971, Viscous phase in rocks and the low-velocity zone: J. Geophys. Res., v. 76, 5, p. 1270–1277.
——— 1973, Role of por fluids in faulting: Trans. Phil. R. Soc. London A. v. 274, p. 297–304.
Peselnick, L., and Zietz, I., 1959, Internal friction of fine-grained limestones at ultrasonic frequencies: Geophysics, v. 24, p. 285–296.
Simmons, G., and Brace, W. F., 1965, Comparison of static and dynamic measurements of compressibility of rocks: J. Geophys. Res., v. 70, p. 5649–5656.

Solomon, S. C., 1972, Seismic wave attenuation and partial melting in the upper mantle of North America: J. Geophys. Res., v. 77, p. 1483–1502.

Toksöz, M. N., Cheng, C. H., and Timur, A., 1976, Velocities of seismic wave in porous rocks: Geophysics, v. 41, p. 621–645.

Usher, M. J., 1962, Elastic behaviour of rocks at low frequencies: Geophys. Prosp., v. 10, p. 119–127.

Walsh, J. B., 1965a, The effect of cracks on the compressibility of rock: J. Geophys. Res., v. 70, p. 381–389.

—— 1965b, The effect of cracks on the uniaxial elastic compression of rocks: J. Geophys. Res., v. 70, p. 399–411.

—— 1966, Seismic wave attenuation in rock due to friction: J. Geophys. Res., v. 71, p. 2591–2599.

—— 1969, New analysis of attenuation in partially melted rocks: J. Geophys. Res., v. 74, p. 4333–4337.

White, J. E., 1965, Seismic waves: New York, McGraw-Hill Book Co., Inc.

—— 1975, Computed seismic speeds and attenuation in rocks with partial gas saturation: Geophysics, v. 40, p. 224–232.

Wyllie, M. R. J., Gardner, G. H. F., and Gregory, A. R., 1962, Studies of elastic wave attenuation in porous media: Geophysics, v. 27, p. 569–589.

APPENDIX A
THE FLUID EQUATION OF MOTION

In most of our treatment of fluid flow within the individual pores of a rock, we assume that the fluid behaves as though incompressible. Usually the principal requirement for this to be approximately true is that the fluid speed be small compared to the speed of sound V_p (Batchelor, 1967, p. 174). Our results indicate that the peak fluid velocity is on the order of $\omega\varepsilon D$. Then for incompressible flow, we demand

$$\omega\varepsilon D \ll V_p = \left[\frac{K_f}{\rho}\right]^{1/2}.$$

For water $V_p \simeq 10^3$ m/sec and for strains $\varepsilon < 10^{-4}$, this becomes

$$\omega D \ll 10^7 \text{ m/sec.} \qquad (A-1)$$

A second requirement we can impose is that the volumetric strain due to fluid compression be small compared with the volumetric strain imposed as a boundary condition, i.e.,

$$\frac{P}{K_f} \ll \frac{\delta}{a_0} = \varepsilon.$$

Our results give a peak fluid pressure of

$$P \simeq \begin{cases} \dfrac{3}{2}\dfrac{\eta\omega\varepsilon D^2}{a_0^2} & \omega \text{ small} \\[3mm] \dfrac{1}{2}\omega^2\rho D^2\varepsilon & \omega \text{ large} \end{cases}.$$

Then the requirement becomes

$$\frac{\eta\omega\varepsilon D^2}{a_0^2 K_f} \ll \varepsilon \Rightarrow \frac{\eta\omega D^2}{a_0^2} \ll K_f \qquad \omega \text{ small,}$$

$$\frac{\omega^2\rho D^2\varepsilon}{K} \ll \varepsilon \Rightarrow \omega^2 D^2 \ll \frac{K_f}{\rho} = V_p^2 \qquad \omega \text{ large.}$$

These are more restrictive than (A–1). For water $K_f = 2 \cdot 10^{10}$ dyne/cm^2 and $\eta \simeq 10^{-2}$ dyne sec/cm^2, we require:

$$\frac{\omega}{\alpha_f^2} \ll 2 \cdot 10^{12} \text{ sec}^{-1} \qquad \omega \text{ small,}$$

$$\omega D \ll 10^5 \text{ cm/sec} \qquad \omega \text{ large.}$$

Within these bounds of incompressibility, the equation of motion for the fluid is (Batchelor, 1967)

$$\rho\frac{Du_i}{Dt} = -\frac{\partial P}{\partial x_i} + \eta\nabla^2 u_i,$$

where $D/Dt = \partial/t + u\cdot\nabla$. We will assume that η is constant with time and position.

For a two-dimensional pore, $u_z = \partial/\partial z = 0$. Furthermore, if the pore width is slowly varying, i.e., $1/a\cdot\partial a/\partial x \ll 1$ and the pore strains are very small, then the flow is approximately unidirectional, and we can neglect the terms $u(\partial u/\partial x)$ and $\partial^2 u/\partial x^2$. Hence, the equation of motion becomes

$$\rho\frac{\partial u}{\partial t} = -\frac{\partial P}{\partial x} + \eta\frac{\partial^2 u}{\partial y^2}.$$

APPENDIX B
FREQUENCY-DEPENDENT EXPRESSIONS AND THEIR LIMITING FORMS

The frequency dependence of Φ and $P(x)$ is found as follows. The flow field is given by equation (6):

$$u(x, y) = \frac{-\mathring{A}(x)}{2a(x)}.$$

$$\cdot \; \frac{1 - \dfrac{\cosh \sqrt{\dfrac{i\omega}{\nu}}\, y}{\cosh \sqrt{\dfrac{i\omega}{\nu}}\, a}}{1 - \dfrac{1}{a}\sqrt{\dfrac{\nu}{i\omega}}\, \tanh \sqrt{\dfrac{i\omega}{\nu}}\, a} \cdot$$

The shear stress $\tau = \eta(\partial u/\partial y)$ is then

$$\tau(x, y) = \frac{\eta \mathring{A}(x)}{2a} \cdot$$

$$\cdot \; \frac{\sqrt{\dfrac{i\omega}{\nu}}}{\left[1 - \dfrac{1}{a}\sqrt{\dfrac{\nu}{i\omega}}\, \tanh \sqrt{\dfrac{i\omega}{\nu}}\, a\right]} \cdot$$

$$\cdot \; \frac{\sinh \sqrt{\dfrac{i\omega}{\nu}}\, a}{\cosh \sqrt{\dfrac{i\omega}{\nu}}\, a} \cdot$$

The average dissipation rate per unit volume of fluid is given by

$$\frac{1}{2\eta}\tau\tau^* = \frac{\omega\rho}{8a^2}\frac{\mathring{A}\mathring{A}^*}{T(z)} \cdot \left[\frac{\sinh^2\sqrt{\dfrac{\omega}{2\nu}}\, y \cos^2\sqrt{\dfrac{\omega}{2\nu}}\, y + \cosh^2\sqrt{\dfrac{\omega}{2\nu}}\, y \sin^2\sqrt{\dfrac{\omega}{2\nu}}\, y}{\cosh^2 z \cos^2 z + \sinh^2 z \sin^2 z}\right],$$

where

$$T(z) = 1 - \frac{1}{z}\cdot\left[\frac{\tanh z + \tanh z \tan^2 z - \tanh^2 z \tan z + \tan z}{1 + \tanh^2 z \tan^2 z}\right] + \frac{1}{2z^2}\left[\frac{\tanh^2 z + \tan^2 z}{1 + \tanh^2 z \tan^2 z}\right],$$

$$z = a\sqrt{\frac{\omega}{2\nu}}\,.$$

Finally, integrating over the volume and period gives

$$\Phi = \int_0^D \frac{\pi d\eta \mathring{A}\mathring{A}^* z}{4a^3 \omega T(z)} \cdot$$

$$\cdot \left[\frac{\sinh 2z - \sin 2z}{\cosh^2 z \cos^2 z + \sinh^2 z \sin^2 z}\right]dx$$

$$= \frac{3}{2}\frac{\pi\eta d}{\omega}\int_0^D \frac{\mathring{A}\mathring{A}^*}{a^3}R(z)\,dx,$$

$$R(z) = \frac{z}{6T(z)} \cdot$$

$$\cdot \left[\frac{\sinh 2z - \sin 2z}{\cosh^2 z \cos^2 z + \sinh^2 z \sin^2 z}\right].$$

The asymptotic expressions given in Table 1 are

found by noting that at large z, $\sinh z \simeq \cosh z$ and at small z by expanding the trigonometric and hyperbolic functions in power series and dropping higher order terms. Hence the definition of large and small z is $z \gg 1$ and $z \ll 1$, respectively, or in terms of ω: $\omega \gg 2\nu/a$, $\omega \ll 2\nu/a$.

The complex pressure $P(x)$ is obtained by taking the integral of equation (5):

$$\frac{\partial P}{\partial x} = \frac{\mathring{A}(x)i\omega\rho}{2a\left[1 - \dfrac{1}{a}\sqrt{\dfrac{\nu}{i\omega}}\, \tanh \sqrt{\dfrac{i\omega}{\nu}}\, a\right]},$$

$$P(x) = P_0 + \int_0^x \frac{\partial P}{\partial x}\,dx.$$

For the parallel drop, we set $\mathring{A}(x) = 2x\varepsilon a_0 i\omega$, and assume for simplicity that the pressure at the edge of the drop, $x = D$, is zero. This gives

$$P(x) = \frac{\omega^2 \varepsilon \rho}{2}(D^2 - x^2) \cdot$$

$$\cdot \left[1 - \frac{1}{a}\sqrt{\frac{\nu}{i\omega}}\, \tanh \sqrt{\frac{i\omega}{\nu}}\, a\right]^{-1}$$

$$= \frac{\omega^2 \varepsilon \rho}{2}(D^2 - x^2)Y(z).$$

The modulus and real part of P are

$$|P| = \frac{\omega^2 \varepsilon \rho}{2}(D^2 - x^2)Y(z),$$

$$\text{Real }(P) = \frac{\omega^2 \varepsilon \rho}{2}(D^2 - x^2)Z(z),$$

where

$$Y(z) = \left[(\cos^2 z + \sinh^2 z)^2 - \frac{1}{2z} \cdot \right.$$
$$\cdot (\cos^2 z + \sinh^2 z) \cdot$$
$$\cdot (\sinh 2z + \sin 2z)$$
$$\left. + \frac{1}{8z^2}(\sinh^2 2z + \sin^2 2z)\right]^{1/2}\Big/ F(z).$$

$$Z(z) = \left[\cos^2 z + \sinh^2 z\right.$$
$$\left. - \frac{1}{4z}(\sinh 2z + \sin 2z)\right] \Big/ F(z),$$

$$F(z) = \left[\cosh z \cos z - \frac{1}{2z} \cdot\right.$$
$$\left. \cdot (\sinh z \cos z + \cosh z \sin z)\right]^2$$
$$+ \left[\sinh z \sin z - \frac{1}{2z} \cdot\right.$$
$$\left. \cdot (\cosh z \sin z - \sinh z \cos z)\right]^2,$$

$$z = a\sqrt{\frac{\omega}{2\nu}}.$$

To find the equivalent pressure $\bar{P}$ for the case of a drop centered in the pore, we substitute these values into equation (12):

$$\begin{Bmatrix} |P| \\ \text{Real } P \end{Bmatrix} = \frac{\omega^2 \varepsilon \rho}{2} \begin{Bmatrix} Y(z) \\ Z(z) \end{Bmatrix} \frac{2}{\pi c} \cdot$$
$$\cdot \int_{-D}^{D} (D^2 - x^2)\left[1 - \left(\frac{x}{c}\right)^2\right]^{1/2} dx$$
$$= \frac{\omega^2 \varepsilon \rho c^2}{\pi} \gamma \begin{Bmatrix} Y(z) \\ Z(z) \end{Bmatrix},$$

where

$$\gamma = \left[\left(\frac{1}{4} + \frac{1}{2}\frac{D^2}{c^2}\right)\frac{D}{c}\sqrt{1 - \left(\frac{D}{c}\right)^2}\right.$$
$$\left. + \left(\frac{D^2}{c^2} - \frac{1}{4}\right)\sin^{-1}\frac{D}{c}\right].$$

The high and low frequency limits of $Y(z)$ and $Z(z)$ are found in the same manner as $R(z)$.

We include the effect of fluid compressibility in the parallel pore by using a modified form of the continuity condition given by equation (4):

$$\int_{-a}^{a} U(x,y)\,dy = -\left[\mathring{A}(x) - \frac{2a_0}{K_f}\int_0^x \mathring{P}(x)\,dx\right],$$

where K_f is the fluid bulk modulus. Now the total rate of flow past a given plane (x = const) is equal to the difference between the volume change of the pore

and the fluid volume change due to compression. Substituting for U from equation (3), this gives.

$$\frac{2a_0}{i\omega\rho}\left[1 - \frac{1}{a_0}\sqrt{\frac{\nu}{i\omega}}\tanh\sqrt{\frac{i\omega}{\nu}}\,a_0\right]\frac{\partial P}{\partial x}$$
$$= 2x\varepsilon a_0 i\omega - \frac{2a_0 i\omega}{K_f}\cdot$$
$$\cdot \int_0^x P\,dx. \qquad (B-1)$$

For frequencies well above the Poisseuille range ($\omega \gg 2\nu/a^2$), this becomes

$$\frac{\partial P}{\partial x} = -x\varepsilon\omega^2\rho + \frac{\omega^2\rho}{K_f}\int_0^x P(x)\,dx,$$

with solution (corresponding to zero pressure at the edge of the drop):

$$P = \varepsilon K_f\left[1 - \frac{\cosh \alpha x}{\cosh \alpha D}\right]$$
$$\simeq \begin{cases} \dfrac{\omega^2 \varepsilon \rho (D^2 - x^2)}{2}, & \alpha D \ll 1 \text{ (incompressible)} \\ \varepsilon K_f(1 - e^{-\alpha(D-x)}), & \alpha D \gg 1 \end{cases},$$
$$(B-2)$$

where $\alpha = \omega\sqrt{\rho/K_f}$. Viscous effects can be retained if the term in brackets [ ] in (B-1) is a weak function of x, giving a modified form of (B-2) with

$$\alpha \sim \omega \cdot$$
$$\cdot \left[\rho/K_f\left(1 - \frac{1}{a_0}\sqrt{\nu/i\omega}\tanh\sqrt{\frac{i\omega}{\nu}}\,a_0\right)\right]^{1/2}.$$

Substituting the pressure gradient into (3), we obtain the flow field and, finally, the dissipation:

$$\Phi = \frac{2\pi\eta d\varepsilon^2 K_f}{\omega\rho(\cosh 2\alpha D + 1)}\cdot$$
$$\cdot \sqrt{\frac{\omega}{2\nu}}\left[\frac{\sinh 2\alpha D}{\alpha} - 2D\right]$$
$$\simeq \begin{cases} \dfrac{4\pi\eta\omega\varepsilon^2 dD^3}{3}\sqrt{\dfrac{\omega}{2\nu}} & \omega \ll \alpha D \\ \sqrt{2\eta\rho}\pi d\varepsilon^2\left(\dfrac{K}{\rho\omega}\right)^{3/2} & \omega \gg \alpha D. \end{cases}$$

APPENDIX C
SHAPE APPROXIMATIONS IN CALCULATING ENERGIES

The separate calculation of Φ and W assumes a certain degree of uncoupling between the details of the fluid flow field and strain field in the rock around

the pore. In a rigorous approach the pore strain ε in equation (2) is a function of x and is coupled to the fluid pressure distribution $P(x)$.

The general form for the dissipation Φ is given by equation (7) as a function of $\mathring{A}$ where

$$\mathring{A}(x) = 2 \int_0^x \frac{\partial}{\partial t} \tilde{a}(x, t)\, dt.$$

This can be rewritten using (2) as

$$\mathring{A} = 2i\omega e^{i\omega t} \int_0^x a(x)\, \varepsilon(x)\, dx.$$

In our calculations for specific geometries, we simplify the integral by replacing $\varepsilon(x)$ by an appropriate average constant value ε, such that

$$\mathring{A} \simeq 2i\omega \varepsilon e^{i\omega t} \int_0^x a(x)\, dx.$$

We further assume that this constant value of ε is approximately the strain that would occur for a uniformly pressurized pore deforming as equation (10).

In general, the pore wall displacement $S(x)$ resulting from uniform confining pressure and uniform pore pressure will depend on the details of crack shape. However, Mavko and Nur (1977a) have shown for a broad class of two-dimensional flat crack shapes that, for very small increments of loading as we might expect from a passing wave, the incremental change of pore shape is elliptical in form, i.e.,

$$\frac{\partial a(x)}{\partial \sigma} = \frac{2c(1 - \xi^2)}{E} \sqrt{1 - \left(\frac{x}{c}\right)^2}.$$

Then we can approximate $S(x)$ by

$$S(x) = \frac{\partial a}{\partial \sigma}(\sigma - \bar{P}) = \frac{2(\sigma - \bar{P})c(1 - \xi^2)}{E} \cdot$$
$$\cdot \sqrt{1 - \left(\frac{x}{c}\right)^2},$$

which is equation (10). The maximum displacement is $S(0) = 2(\sigma - \bar{P})c(1 - \xi^2)/E$. Comparing with equation (2), the maximum displacement is $a_0\varepsilon$ so that we set

$$\frac{2(\sigma - P)c(1 - \xi^2)}{E} \simeq a_0\varepsilon.$$

Equation (8) shows that the energy W depends only on the integral of crack displacement

$$\bar{S} = \int_{-c}^{c} S(x)\, dx.$$

Hence, for calculating W any approximation of pore wall deformation $S(x)$ is valid as long as it has the correct average value $\bar{S}$. Mavko and Nur (1977b) have shown that for a broad class of two-dimensional flat cracks, the integrated displacement due to pressure distribution $P(x)$ is exactly the same as for a uniform pressure $\bar{P}$, where $\bar{P}$ is an appropriately weighted average of $P(x)$, equation (12).

APPENDIX D
PORE PRESSURE APPROXIMATION IN
CALCULATING ENERGY

The pore pressure term in equation (8) can be shown negligible as follows. Since the bulk modulus of water K_f is much less than M, the modulus of rock, then

$$2d \int_{-c}^{c} P(x)\, \frac{\sigma^*}{M}\, a(x)\, dx \ll 2d \cdot$$
$$\cdot \int_{-c}^{c} P(x)\, \frac{\sigma^*}{K_f}\, a(x)\, dx.$$

In Appendix A, we require for the incompressible analysis to be valid that $P \ll \varepsilon K_f$, then

$$2d \int_{-c}^{c} \frac{P(x)}{K_f}\, \sigma^* a(x)\, dx \ll 2d \int_{-c}^{c} \sigma^* \varepsilon\, a(x)\, dx.$$

(If we allow for compressible flow, $P < \varepsilon K_f$, and this inequality still holds.) But by equation (2), $\varepsilon a(x) \simeq S(x)$:

$$2d \int_{-c}^{c} \sigma^* \varepsilon\, a(x)\, dx \simeq 2d \sigma^* \int_{-c}^{c} S(x)\, dx.$$

Therefore,

$$2d \int_{-c}^{c} P(x)\, \frac{\sigma^*}{M}\, a(x)\, dx \ll 2d\sigma^* \int_{-c}^{c} S(x)\, dx.$$

Reprinted from Geophysics, v. 44, p. 691–711.

Attenuation of seismic waves in dry and saturated rocks: II. Mechanisms

D. H. Johnston*, M. N. Toksöz*, and A. Timur‡

Theoretical models based on several hypothesized attenuation mechanisms are discussed in relation to published data on the effects of pressure and fluid saturation on attenuation. These mechanisms include friction, fluid flow, viscous relaxation, and scattering. The application of these models to the ultrasonic data of Toksöz et al (1979, this issue) indicates that friction on thin cracks and grain boundaries is the dominant attenuation mechanism for consolidated rocks under most conditions in the earth's upper crust. Increasing pressure decreases the number of cracks contributing to attenuation by friction, thus decreasing the attenuation. Water wetting of cracks and pores reduces the friction coefficient, facilitating sliding and thus increasing the attenuation. In saturated rocks, fluid flow plays a secondary role relative to friction. At ultrasonic frequencies in porous and permeable rocks, however, Biot-type flow may be important at moderately high pressures. "Squirting" type flow of pore fluids from cracks and thin pores to larger pores may be a viable mechanism for some rocks at lower frequencies. The extrapolation of ultrasonic data to seismic or sonic frequencies by theoretical models involves some assumptions, verification of which requires data at lower frequencies.

INTRODUCTION

To reasonably evaluate and interpret laboratory measurements generally made at ultrasonic frequencies (~1 MHz) and, more importantly, to extrapolate these results to seismic frequencies, a precise definition of the mechanisms involved in attenuation along with their pressure and frequency dependence is needed. Numerous mechanisms have been proposed and each may be considered to have a greater degree of importance to the overall attenuation under certain physical conditions. These mechanisms include: matrix anelasticity, including frictional dissipation due to relative motions at the grain boundaries and across crack surfaces (Walsh, 1966); attenuation due to fluid flow, including relaxation due to shear motions at pore-fluid boundaries (Walsh, 1968, 1969; Solomon, 1973); dissipation in a fully saturated rock due to the relative motion of the frame with respect to fluid inclusions (Biot, 1956a,b; Stoll and Bryan, 1970); squirting phenomena (Mavko and Nur, 1975; O'Connell and Budianski, 1977); partial saturation effects such as gas pocket squeezing (White, 1975); energy absorbed in systems undergoing phase changes (Spetzler and Anderson, 1968); and a large category of geometrical effects, including scattering by small pores and large irregularities and selective reflection from thin beds (O'Doherty and Anstey, 1971; Spencer et al, 1977). Of these mechanisms listed, all except the geometrical effects are dependent upon intrinsic rock properties and will be considered here. It is our purpose to evaluate these mechanisms in terms of experimental data in order to determine under what conditions one or more may be dominant in causing the overall attenuations of both P- and S-waves.

We begin by examining the published data on seismic wave attenuation in rocks under varying physical conditions, highlighting important features that contribute to our understanding of the mechanisms involved. In the second section, attenuation mechanisms along with available theoretical formulations are presented. In the final section, the theoretical models are compared to the data obtained by Toksöz et al (1979, this issue). Our main emphasis in this part will be to determine the relative importance of the mechanisms in contributing to the overall attenuation and to examine to what extent laboratory data may be used to infer rock properties from seismic

Manuscript received by the Editor August 30, 1977; revised manuscript received August 25, 1978.
*Department of Earth and Planetary Sciences, M.I.T., Cambridge, MA 02139.
‡Chevron Oil Field Research Co., Box 446, La Habra, CA 90631.

LIST OF SYMBOLS

a, a' = Biot flow structure constants
$c(\alpha_m)$ = Pore-crack porosity
f = Frequency
k = Wavenumber
l = Crack half-length
q = Volume flow
r = Radius of scattering inclusions
K, E, μ, ρ = Matrix moduli and density
K', E', μ', ρ' = Inclusion moduli and density
K^*, E^*, μ^*, ρ^* = Effective moduli and density
$K_{\frac{*}{A}}$ = Frame bulk modulus
N/V_0 = Number of cracks per volume
P = Pressure
P_c, P_f, P_d = Confining fluid and differential pressures
Q, Q^{-1} = Quality and dissipation factors
V = Seismic velocity
α = Attenuation coefficient
α_m = Aspect ratio
δ = Log decrement
ε = Crack/pore porosity ratio
η = Viscosity
θ = Dilatation
κ = Coefficient of friction
σ, σ^* = Matrix and effective Poisson's ratios
τ = Relaxation time
ϕ = Total porosity
χ = Permeability
ω = Angular frequency
ω_c, ω_d = Critical frequencies for squirting flow and shear relaxation

data obtained in the field. Our discussions will be limited primarily to fully saturated and completely dry competent rocks under pressures up to a few kilobars and at the relatively low temperatures that might be encountered in geophysical exploration.

ATTENUATION DATA

Seismic body wave attenuation has been measured for many rock types over wide ranges of physical conditions and frequencies, and by many techniques. Unfortunately, the systematics of attenuation behavior with pressure, temperature, and saturation conditions has not been adequately measured, nor is it well understood. With some exceptions, laboratory determinations of attenuation are limited to specific rocks under one physical state. However, an overview of these data can provide useful information on the nature and mechanics of attenuation in upper crustal rocks. In this section, individual determinations of attenuation will be briefly summarized, followed by a more extensive review of pertinent data.

The data examined here have been obtained by numerous experimental techniques, including pulse transmission of several types, resonant bars, and slow stress cycles. Each method determines a different measure of attenuation. The most commonly found quantities in the literature are the attenuation coefficient α, for a plane propagating wave in an infinite medium; the logarithmic decrement δ; and the dissipation factor Q^{-1}, or its inverse, the quality factor Q. The relationships among these are given by:

Table 1. Measured body wave Q for several rock types.

Rock	Q	Frequency, Hz	Method	References
Quincy Granite	125	$(.14-4.5) = 10^3$	long resonance	Birch and Bancroft
(air dry)	166		tors. resonance	(1938)
Solenhofen Limestone	112	$(3-15) \times 10^6$	P wave pulses	Peselnick and Zietz
(air dry)	188		S wave pulses	(1959)
I-1 Limestone	165	$(5-10) \times 10^6$	P wave pulses	Peselnick and Zietz
(air dry)				(1959)
Hunton Limestone	65	$(2.8-10.6) \times 10^3$	long. resonance	Born (1941)
(oven dry)				
Amherst Sandstone	52	$(.930-12.8) \times 10^3$	long. resonance	Born (1941)
(oven dry)				
Berea Sandstone	10	$(.2-.8) \times 10^6$	P and S wave pulses	Toksöz et al. (1979)
(brine saturated)				
Navajo Sandstone				
(air dry)	21	50-120	flexural vibrations	Bruckshaw & Mahanta
				(1954)
(water saturated)	7	$(.2-.8) \times 10^6$	P and S wave pulses	Toksöz et al. (1979)
Pierra Shale	32	50-450	P wave in situ	McDonel et al. (1958)
(in situ)	10		S wave in situ	

$$Q = \frac{\pi f}{\alpha V} = \frac{\pi}{\delta}, \qquad (1)$$

where V = velocity and f = frequency. We will deal with the parameters Q^{-1}, Q, and α exclusively.

A representative sample of individual attenuation measurements is listed in Table 1 along with other pertinent parameters. A summary of individual measurements taken from the compilation by Bradley and Fort (1966) is shown graphically in Figure 1, where Q as a function of rock type and rock porosity is plotted. The values taken are generally at surface pressure, although they cover wide frequency and saturation ranges. Figure 1 shows the wide variability of attenuation in rocks and a general trend of Q inversely proportional to porosity. As noted by many investigators (Wyllie et al, 1962; Knopoff, 1964; Bradley and Fort, 1966; and others), the accumulation of individual attenuation measurements has led to a series of generalities that may be applied to the nature of Q in crustal rocks. These are summarized below, with references to later sections of this paper where certain effects are discussed in more detail.

(1) **Frequency dependence.**—Laboratory experiments show that Q may be independent of frequency (α proportional to frequency) over a broad frequency range (10^{-2}–10^7 Hz), especially for some dry rocks (Birch and Bancroft, 1938; Born, 1941; McDonal et al, 1958; Peselnick and Outerbridge, 1961; Attewell and Ramana, 1966; Pandit and Savage, 1973; and others). Q^{-1} in liquids, however, is proportional to frequency (Pinkerton, 1947), so that in some highly porous and permeable rocks the total Q^{-1} may contain a frequency dependent component (Born, 1941; Wyllie et al, 1962). This component may be negligible at seismic frequencies, even in unconsolidated marine sediments (Hamilton, 1972).

(2) **Strain amplitude.**—Attenuation appears to be independent of strain amplitude for low strains such as those associated with seismic waves (Mason, 1958; Gordon and Davis, 1968). Data exist that show attenuation rapidly increasing above strains of about 10^{-6} (Winkler and Nur, 1978; Johnston and Toksöz, 1978; and others).

(3) **Fluid saturation.**—Attenuation for fluid-saturated rocks is higher than for dry rocks and depends on the degree of saturation, fluid type, and frequency in a complicated way. For rocks fully saturated with a low viscosity fluid (e.g., water, oil), it is generally found that at ultrasonic frequencies $Q_p \gtrsim Q_s$. This topic will be further discussed in the next section.

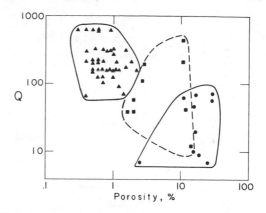

FIG. 1. Q as a function of porosity. Data for igneous and metamorphic rocks (triangles), limestones (squares), and sandstones (circles) are taken from Bradley and Fort (1966) and cover wide-frequency and saturation ranges.

(4) **Pressure and stress dependence.**—Observations show that attenuation decreases with increasing confining pressure. This is usually considered to be due to the closing of cracks in the rock matrix. Data supporting this and theoretical models of the pressure effects will be discussed later. For applied nonhydrostatic stress, the attenuation appears to be anisotropic (Merkulova et al, 1972; Walsh et al, 1970). For shear (S) waves polarized normal to the axis of maximum compression, attenuation is lowest due to the closure of cracks with faces normal to this axis (Lockner et al, 1977). At high differential stresses, the onset of dilatancy increases the attenuation (Lockner et al, 1977).

(5) **Temperature dependence.**—The small amount of data on this topic (Volarovich and Gurevich, 1957; Gordon and Davis, 1968) indicates that Q is generally independent of temperature at temperatures low relative to the melting point. An increase of attenuation in quartzite with temperatures above 150°C noted by Gordon and Davis (1968) may be due to thermal cracking of the rock. Near the boiling temperatures of pore fluids, attenuation may be affected strongly by temperature.

We will now consider in more detail data pertaining to the roles of fluid saturation and hydrostatic pressure in determining the attenuation of seismic waves in crustal rocks. In some cases, the absolute determinations of attenuation reported by investigators are unreliable, yielding unreasonable values. Therefore, we generally present the data in terms of the relative change in α or Q^{-1}.

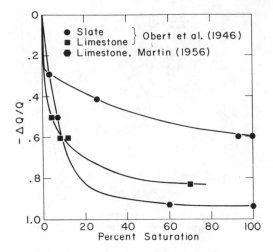

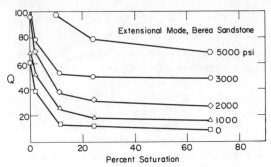

FIG. 4. Q as a function of saturation and differential pressure in Berea sandstone, extensional mode. Data from Gardner et al (1964).

FIG. 2. Change in Q as a function of saturation. Data from Obert et al (1946) and Martin (1956).

Attenuation as a function of saturation conditions

Although of great interest to the exploration community, relatively little experimental work has been done on the nature of attenuation as a function of saturation conditions. Even published data must be examined critically due to the inherent difficulties involved in partial saturation work. Unfortunately, little or no detailed description is given in the experimental literature about the techniques of fluid satura-

tion. An important, yet experimentally difficult, task is maintaining a homogeneous distribution of the saturant in the bulk of the rock. We must also address the question of what constitutes a "dry" rock. In most cases, samples are oven-dried prior to fluid injection. Heating the sample will cause some alterations of the matrix structure. In any event, it is nearly impossible to remove the fluid completely; at least a mono-molecular layer of fluid will probably remain in the thinnest cracks.

The degree of saturation and the type of saturant, characterized primarily by viscosity, appear to play an important role in attenuation. Studies of the effect of partial saturation by various fluids have been reported in Born (1941), Obert et al (1946), Collins and Lee (1956), Wyllie et al (1962), and Gardner et al (1964). A summary of these results is shown in Figures 2–5, where Q or the fractional change in Q is plotted as a function of percent saturation. As pointed out in the preceding section, the overall Q of the rock may be considered to contain a frequency-independent component plus a frequency-dependent component due to the fluid inclusions. Thus the effect of partial saturation may be frequency-dependent (Born, 1941). However, since the curves shown here were taken over a wide range of frequencies but exhibit similar behavior, fluid losses may not dominate frequency-independent losses in most rocks at surface pressures.

All of the samples shown in Figures 2–5 are saturated with water, chemically active with intergranular material, except alundum (Al_2O_3), which is saturated with soltrol, a relatively inert petroleum naptha, shown in Figure 3. The behavior of attenuation as a function of water saturation is similar for all rocks. Q is sharply reduced at low saturations,

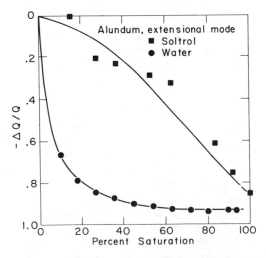

FIG. 3. Change in Q as a function of soltrol and water saturation in alundum at about 10 kHz. Data from Wyllie et al (1962). Samples 7915-B and 7928-B for soltrol and water, respectively.

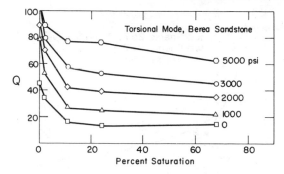

FIG. 5. Q as a function of saturation and differential pressure in Berea sandstone, torsional mode. Data from Gardner et al (1964).

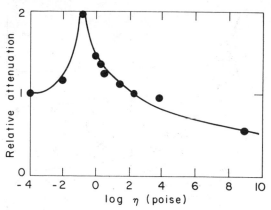

FIG. 6. Relative attenuation of S-waves as a function of pore fluid viscosity in Barre granite. Data from Nur and Simmons (1969a).

presumably due to the wetting effect of water entering the fine cracks, possibly reacting with intergranular material and softening the rock. Also evident is that the effect of pressure is to reduce the effect of saturation for both P- and S-waves, as shown in Figures 4 and 5, since the finer cracks are closed. In the case of soltrol saturation (Figure 3), the change in Q observed for water sautration is not seen. This implies that the effect observed for water saturation is primarily due to either chemical alteration of the intergranular material or a wetting phenomenon. Since it is unlikely that water reacts strongly with alundum, we favor the explanation that different wetting properties cause the different saturation effects observed. In real rocks, of course, a combination of the two mechanisms is likely.

The effect of fluid type, i.e., viscosity, has been discussed in detail by Wyllie et al (1962) and Nur and Simmons (1969a). The dependence of attenuation on fluid viscosity is complicated and not at all obvious from results presented by Wyllie et al (1962). Taking these data at face value, it would appear that very large viscosity fluids (e.g., glycerol) result in small fluid contributions to attenuation. This is reasonable for some attenuation mechanisms such as fluid flow, in that higher viscosity fluids decrease the effective permeability. However, Nur and Simmons (1969a) have shown that the viscosity effect is frequency dependent, consistent with a relaxation-type mechanism. In their experiment, a Barre granite (porosity = 0.6 percent) was saturated with glycerol which has a viscosity extremely dependent on temperature. Thus, by varying the temperature of the saturated sample, the effect of viscosity on velocities and relative attenuation of P- and S-waves is measured. The attenuation of S-waves as a function of pore fluid viscosity is shown in Figure 6. The relaxation peak occurs at a viscosity where the charac-

teristic relaxation time is equal to the wave period. An experiment reported by Gordon (1974) shows similar results.

Attenuation as a function of pressure

The pressure dependence of attenuation generally has been neglected by most investigators, yet the behavior of Q with pressure can yield as much information about mechanisms as does the frequency dependence. When a rock is subjected to hydrostatic pressure such as overburden pressure, its elastic and anelastic properties will change. The behavior of elastic properties under pressure is well known, and a theoretical treatment of it may be found in Toksöz et al (1976). The most important factor causing changes in velocity is the change of porosity with pressure; in particular, the closing of thin cracks. This also holds true for changes in attenuation, as will be discussed in the next section. In all cases, attenuation decreases (Q increases) with increasing pressure. Experimental data verifying this are found in Gardner et al (1964), Klima et al, (1964), Levykin (1965), Gordon and Davis (1968), Al-Sinawi (1968), Walsh et al (1970), and Toksöz et al (1979). For these data and the theoretical models to be presented, the pressure given is the differential pressure, $P_d = P_c - P_f$, where P_c is the confining pressure and P_f is the pore fluid pressure. This relationship holds for all rocks, as demonstrated by laboratory tests (Wyllie et al, 1958; Nur and Simons, 1969b).

The attenuation of P-waves in diabase and graywacke was measured by Klima et al (1964) up to a pressure of 4 kilobars (kb) by a pulse transmission method with a prevailing frequency of 0.9 MHz. Although not stated explicitly, the samples are assumed to be air-dry. The results of this experiment are shown

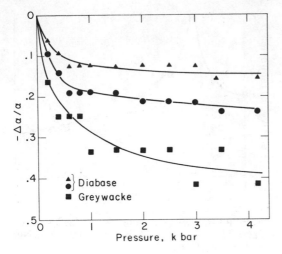

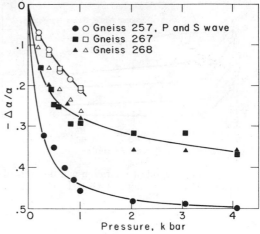

FIG. 7. Change in the attenuation coefficient as a function of pressure for diabase and graywacke samples. Data from Klima et al (1964).

FIG. 8. Change in the attenuation coefficients of P- and S-waves as functions of pressure for several gneisses. Data from Levykin (1965).

in Figure 7, which plots the change in the attenuation coefficient α as a function of pressure. In all cases, a clear decrease in α is observed up to about 1 kb. The fractional changes in attenuation are greater than those of the velocity measured in the same type rocks under the same conditions (Pros et al, 1962).

Levykin (1965) investigated the attenuation of both P- and S-waves in several igneous and metamorphic rock types up to pressures of about 4 kb. A pulse-echo technique at a frequency of 1 MHz was used. Samples were air-dry. The results of these experiments for several gneiss samples are shown in Figure 8. Again, the attenuation decreases rapidly with increasing pressure, leveling off after about 1 kb. Levykin attributes the differing extent to which attenuation is changed under pressure to be due to differences in the weathering of the rocks.

Gordon and Davis (1968) studied the effect of pressure (up to 4 kb) on a fluid-saturated granite using slow stress cycles ($f = 10$ mHz). Their data are reproduced in Figure 9. The same features as seen in the previous works are evident here.

So far, we have considered data only for low-porosity rocks, either dry or completely saturated. However, the pressure effect for a partially saturated Berea sandstone has been studied by Gardner et al (1964). Both extensional and torsional Q values were determined using resonance techniques at frequencies up to 30 kHz. External influences on the sample, such as losses into the pressure medium, were considered. These data are shown in Figures 4 and 5.

The same general behavior is seen for the data in Toksöz et al (1979) for dry, methane-, and water-saturated Berea sandstone at ultrasonic frequencies using the pulse transmission technique. The Q, however, levels off at a lower pressure than for the igneous and metamorphic rocks.

The variation of attenuation with pressure for P- and S-waves was also studied for a variety of rock types by Al-Sinawi (1968). A pulse transmission technique using 122 kHz transducers was used and the confining pressures for which measurements were taken were 0.5, 1, and 2 kb. All of the rocks studied were sedimentary except a granite gneiss and a volcanic tuff. Al-Sinawi found, as observed before, that both α_p and α_s decreased with pressure. In some rocks, particularly limestones, the pressure effect is different; however, this is not completely described.

ATTENUATION MECHANISMS

As a first approximation, we will assume that attenuation mechanisms are independent of each other. Thus, we may consider each mechanism separately and then combine the results to determine the overall attenuation. More specifically, we will consider separately the relative effects of the matrix anelasticity, the viscosity and flow of saturating fluids, and scattering from inclusions. The pressure dependence of these effects will be included. In all these cases, the available theoretical formulations are not very rigorous. They are guided primarily by experimental observations and, as a whole, should be treated as empirical relationships. The calculation of effective elastic properties, necessary for the determination of attenuation, is discussed in Appendix A. The method used is that of Kuster and Toksöz (1974) and Toksöz et al (1976).

Attenuation due to matrix anelasticity

Attenuation of seismic waves in a rock matrix can be attributed to two factors: (1) intrinsic anelasticity of matrix minerals, and (2) frictional dissipation due to relative motions at the grain boundaries and across crack surfaces. The intrinsic anelasticity of minerals is generally small. In individual crystals, Q values are generally higher than a few thousand, while in the whole rock, Q values are normally lower than a few hundred. Thus, in considering matrix attenuation, it is reasonable to neglect the intrinsic attenuation in minerals and to consider only the attenuation across grain surfaces and thin cracks.

The importance of frictional dissipation is supported by the observation that Q is generally independent of frequency, as predicted by this mechanism. However, friction across crack surfaces cannot account for all the anelasticity of the matrix. As pointed out by Walsh (1966), rocks subjected to confining pressures high enough to close all cracks still exhibit nonzero attenuation. Thus, it is necessary to consider, in addition to dissipation across crack surfaces, an "intrinsic" anelasticity of the aggregate minerals.

The exact mechanism of grain boundary and crack dissipation is not known, but frictional dissipation due to relative motions of the two sides may be the major factor (Walsh, 1966). If this is the case, then the attenuation should depend very strongly on the surface conditions that affect friction between grains. Among these are whether rocks are saturated or dry, the properties of saturating fluids, and the amount of clay or other soft components in the matrix.

From experience in laboratory experiments and with lunar rock samples, it is found that granular materials exhibit very high Q values when totally dry and in a vacuum. In the absence of atmosphere and water, the Coulomb forces across grains are very strong and friction coefficients are high; hence, no sliding motion can take place across the surfaces. This accounts for very high Q values measured for seismic waves in the moon ($Q = 2000-5000$: Dainty et al, 1976; Nakamura et al, 1974; Latham et al, 1974; Toksöz et al, 1974) and in the laboratory under hard vacuum conditions (Pandit and Tozer, 1970; Warren et al, 1974; Tittmann et al, 1972, 1975). In the laboratory, when a little water vapor was

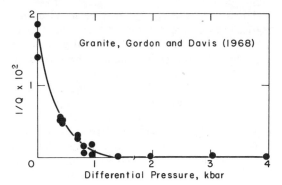

FIG. 9. Q^{-1} as a function of differential pressure in a granite. Data from Gordon and Davis (1968).

introduced into the vacuum chamber, Q values decreased significantly.

It is difficult to formulate attenuation due to grain boundary and "frame anelasticity" effects since this requires detailed knowledge of crack and grain boundary properties. Walsh (1966) formulated the problem by approximating the cracks as ellipsoids in plane strain. For random orientation of cracks, the Q values for compressional and shear waves were computed using the friction coefficient κ, effective Poisson's ratio σ^*, and matrix and effective rock moduli as parameters. The resulting expression for P-waves in an infinite medium is too complicated to be presented conveniently but has the following form:

$$Q_p^{-1} = \frac{E^*}{E} \frac{(1 - \sigma^*)}{(1 - 2\sigma^{*2})} \frac{l^3 N}{V_0} F(\kappa, \sigma^*), \quad (2)$$

when E^* and E are the effective and matrix Young's moduli, respectively, and N is the number of cracks with half-length l in a volume V_0. The function $F(\kappa, \sigma^*)$ is implicitly dependent on the angle between the normal to the crack plane and the direction of wave propagation. Only cracks of certain orientations, determined by κ and σ^*, will contribute to the attenuation.

A closed-form solution for the attenuation of S-waves is impossible to obtain but, again from the Walsh (1966) formulation we may write the general form as

$$Q_s^{-1} = \frac{E^*}{(1 + \sigma^*)E} \frac{l^3 N}{V_0} F(\kappa), \quad (3)$$

where $F(\kappa)$ is a function of the friction coefficient.

For reasonable values of the friction coefficient

and Poisson's ratio, Q_p/Q_s may be found by numerically evaluating equations (2) and (3) (Walsh, 1966). For κ between 0.0 and 0.5, and σ^* between 0.15 and 0.25, Q_p/Q_s is found to be between about 0.4 and 1.5. For most dry rocks, $Q_p/Q_s < 1$, while for saturated rocks $Q_p/Q_s \sim 1.0$ at surface pressure.

Many data (Peselnick and Outerbridge, 1961; Peselnick and Zietz, 1959; Knopoff, 1964) can be explained by the frictional dissipation mechanism. This mechanism, which yields a constant Q with frequency, also explains the frame anelasticity incorporated in Biot's (1956 a, b) formulations.

Although friction explains much of the observed behavior of attenuation in rocks, the calculation of absolute values requires the specification of too many unknown parameters (e.g., friction coefficients and the number and radii of cracks whose surfaces are in contact). Furthermore, these parameters most likely will change with saturation conditions. However, the Walsh formulation [equations (2) and (3)] is useful in determining the effect of pressure on the frictional mechanism.

To formulate this pressure dependence, we assume:

(1) The cracks and grain boundaries that contribute to friction can be characterized by very thin (oblate) spheroids with a small aspect ratio, α_m (where $\alpha_m = $ thickness/diameter). From equation (A–3), the fractional change of volume c for this family of cracks as a function of differential pressure is

$$\frac{dc}{c} = \frac{-P}{K_A^*}\left[\frac{4}{3\pi\alpha_m}\frac{(1-\sigma^2)}{(1-2\sigma)}\right], \qquad (4)$$

where σ is the matrix Poisson's ratio and K_A^* is the effective static or frame bulk modulus.

(2) The effective coefficient of friction κ is constant with pressure. Thus, $F(\kappa)$ in equation (3) is a constant. If we assume that the effective Poisson's ratio σ^* varies more slowly with pressure than c, then $F(\kappa, \sigma^*)$ in equation (2) is essentially a constant, also.

Since the fractional volume of cracks with aspect ratio α_m is

$$c(\alpha_m) = \frac{4\pi\alpha_m}{3}\frac{N(\alpha_m)l^3}{V_0}, \qquad (5)$$

equation (2) may be written

$$Q_p^{-1} = \frac{3}{4}\frac{E^*}{E}\frac{(1-\sigma^*)}{(1-2\sigma^{*2})}\frac{c(\alpha_m)}{\pi\alpha_m}F(\kappa, \sigma^*), \qquad (6)$$

with a similar change for equation (3). Then

$$\frac{dQ_p^{-1}}{Q_p^{-1}} = \frac{dE^*}{E^*} + \frac{dc}{c} + \Sigma, \qquad (7)$$

where Σ is a small quantity and includes variations in σ^* and $F(\kappa, \sigma^*)$. Using assumption 2, $\varepsilon \to 0$. Substituting equation (4) into (7) and then integrating, we finally obtain

$$Q_p^{-1} = Q_{p0}^{-1}\frac{E^*}{E_0^*}e^{-AP/K_A^*}, \qquad (8)$$

where $A = [4/(3\pi\alpha_m)] \cdot [(1-\sigma^2)/(1-2\sigma)] = $ constant. A similar expression is obtained for the attenuation of S-waves. Q_p^{-1} and Q_s^{-1} at $P = 0$ are found empirically and, thus, the imaginary parts of the matrix moduli can be set as described in Appendix A. In fact, at each pressure, the imaginary parts (subscript I) are given in terms of the real parts (subscript R) by

$$K_I = (K_R + 4/3\mu_R)Q_p^{-1} - 4/3\mu_R Q_s^{-1},$$

and $\qquad\qquad\qquad\qquad\qquad\qquad\qquad (9)$

$$\mu_I = \mu_R Q_s^{-1}.$$

These results can then be used in equations (A–1) and (A–2) to determine the effective moduli, velocities, and attenuation. Since α_m is arbitrary, the constant A is a free parameter and must be found empirically.

At first glance, the exponential decay of Q^{-1} with pressure predicted by equation (8) may not seem reasonable. As stated before, the attenuation of many rocks at high pressure is nonzero. However, equation (8) describes only the effects of cracks which control the behavior of the elastic and anelastic properties at relatively low pressures. If one considers a rock with an extremely low total porosity but moderate crack porosity such as a granite, then equation (8) may truly represent the pressure dependence of Q^{-1}. This is indeed observed in the data from Gordon and Davis (1968) shown in Figure 9. For rocks such as sandstones, however, we must consider the intrinsic aggregate anelasticity to contribute to the observed attenuation at pressures where the cracks are closed. In our models, this is determined empirically and assumed to be constant with pressure.

One further consideration is the difference between surface (atmospheric) pressure Q values for the dry Berea sandstone determined by the ultrasonic pulse method (Toksöz et al, 1979) and values obtained by dynamic resonance (Gardner et al, 1964, and unpublished data by the authors). Compared on a common basis, the Q_p value for the pulse technique is about 20, while for the resonance method it is higher than 50. The discrepancy is smaller for the saturated case. Two explanations are possible: either the friction mechanism as we understand it does not provide a frequency-independent Q, or the attenuation is dependent on strain amplitude. Some evidence favors the latter. Winkler and Nur (1978) reported that Q rapidly decreases with increasing strain at a strain of about 10^{-6}. This may be due to the presence of asperities in the cracks which inhibit sliding until a threshold amplitude is exceeded. The higher amplitude ultrasonic pulses (strain $>10^{-6}$) may thus be able to cause sliding on these rough surfaces and result in a higher attenuation. In the saturated case, crack surfaces are lubricated and the threshold amplitude is lower. Our own resonance data on the Berea sandstone and on Plexiglas (Johnston and Toksöz, 1978) corroborate the amplitude threshold theory. A discontinuous increase in attenuation was observed in the sandstone. However, no such increase was observed in the crack- and grain-boundary-free Plexiglas. This result strengthens the idea that a crack or grain boundary friction mechanism may result in amplitude attenuation. It may be valid, therefore, to compare experimental and in-situ results only when assured that factors such as strain amplitude are equivalent.

Attenuation due to viscosity and flow of saturating fluids

All rocks in the upper crust are partially or completely saturated with some fluid. It is of special interest, then, to consider the effect of viscous fluids in a solid rock matrix. Some mechanisms by which fluids contribute to attenuation are illustrated in Figure 10. Fluids in elongated pores and fine cracks contribute to attenuation in a complex manner. First, attenuation peaks due to viscous relaxation will develop at frequencies dependent both on pore geometry and fluid viscosity. For a rock with a wide spectrum of pore aspect ratios, the attenuation spectrum is of a complicated form. This problem has been discussed by Walsh (1968, 1969), Solomon (1973), and Kuster and Toksöz (1974) for spheroidal pores.

Second, fluid flow between pores, induced by the stress (seismic) wave, may cause attenuation. These flow mechanisms fall into two categories: inertial flow (Biot, 1956a, b), important at ultrasonic frequencies; and squirting flow (Mavko and Nur, 1975; O'Connell and Budiansky, 1977), more prominent at lower frequencies. We will consider each separately,

Effects of Fluid Saturation

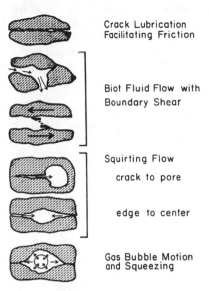

Crack Lubrication Facilitating Friction

Biot Fluid Flow with Boundary Shear

Squirting Flow

crack to pore

edge to center

Gas Bubble Motion and Squeezing

FIG. 10. Schematic illustration of several proposed attenuation mechanisms for saturated and partially saturated rocks.

and our analysis of squirting flow will also include the formulation for viscous relaxation.

In highly porous and permeable rocks, relative motion may take place between the rock frame and the saturating fluid as seismic waves propagate. Biot (1956a, b and 1962a, b) derived a theory for acoustical wave propagation in an isotropic solid with interacting pores. This theory can be used to calculate both velocity and attenuation. Biot theory predicts the existence of three types of body waves, two dilatational and one shear. One dilatational wave is highly attenuated and resembles a diffusion wave. The other is the P-body wave that travels with little attenuation or dispersion. A formulation of Biot's theory has been developed by Stoll and Bryan (1970) and Stoll (1974, 1977) and has been adopted for this study.

As with attenuation due to viscous shear relaxation, the viscous resistance to fluid flow is frequency-dependent for oscillating motion. Below a certain frequency, dependent on the fluid properties and pore characteristics, this resistance is given by the ratio of the fluid viscosity η to the physical permeability χ and may be considered approximately constant, describing Poiseuille flow. At higher frequencies, turbulent flow develops in which the effects of viscosity are felt only in a thin boundary layer.

For frequencies at which Poiseuille flow is valid, the attenuation coefficient α for the P-type body wave varies as the square of the frequency ($Q^{-1} \propto f$). At higher frequencies, Biot derived a correction factor to the fluid viscosity and found that α is proportional to $f^{1/2}$ ($Q^{-1} \propto f^{-1/2}$). Shear wave attenuation involves only the moving solid frame dragging the viscous fluid with it. Since the fluid motion is due only to inertial stresses, this mechanism must be treated in addition to the viscous relaxation model. The Biot-type loss mechanisms, pressure gradient flow and viscous drag, are schematically illustrated in Figure 10.

Biot theory and the numerical model of Stoll and Bryan (1970) require the following parameters: (following previous notations) the elastic moduli for the frame K_f^* and μ_f^*, and matrix K and μ; the bulk modulus and absolute viscosity of the fluid inclusions K' and η; the porosity ϕ; the physical permeability χ; and the densities of the matrix and fluid ρ and ρ'. All of these parameters are either known or can be calculated using the technique described by Kuster and Toksöz (1974), and Toksöz et al (1976) (Appendix A). Two other free constants derived from Biot theory, a pore size parameter a and a structure constant a' must be appropriately chosen or experimentally found for the material being considered. The choice of these values is discussed by Stoll and Bryan (1970).

The attenuation formulations for the two dilatational waves, after lengthy algebra, reduce to the solution of the period equation (Stoll, 1974)

$$\begin{vmatrix} Hk^2 - \rho\omega^2 & \rho'\omega^2 - Ck^2 \\ Ck^2 - \rho'\omega^2 & m\omega^2 - Mk^2 - \dfrac{i\omega F\eta}{\chi} \end{vmatrix} = 0,$$
(10)

where ω is the angular frequency and k is the wavenumber. $m = a'\rho'/\phi$ (with $a' \geq 1$). H, C, and M are operators which are functions of the frame, matrix, and fluid moduli, and F is a complex, high-frequency correction factor derived by Biot (1956b). The attenuation coefficient is obtained by solving for the complex roots ($k = k_R + ik_I$) of the period equation and using the imaginary part of the wavenumber k_I. One root represents the diffusion wave and the other the propagating P-wave. Another, more simple period equation for k may be found for the S-wave. Viscous drag at the pore-fluid interface results in greater loss than flow induced by pressure gradients. Thus, the model predicts that the attenuation of S-waves is greater than for the P-waves in

the case of the fluid flow mechanism.

In general, the elastic moduli of the frame in this formulation may be complex, allowing for the anelasticity of the frame. Since this effect is considered separately in this study, the imaginary parts of the frame moduli are set to zero. Numerical calculations carried out by Stoll and Bryan (1970) indicate that frame anelasticity dominates the fluid flow effects at lower frequencies ($f \leq 10^4$ Hz). At high frequencies, the fluid flow contribution could be detected for high-porosity rocks if the permeability is also high. In this case, the frequency dependence of the attenuation coefficient is f^2 at lower frequencies and $f^{0.5}$ at higher ($f \geq 10$ kHz) frequencies. For most sedimentary rocks saturated with water, the effects of fluid flow are small at seismic frequencies ($f = 10$–200 Hz), but could become important at ultrasonic frequencies.

The pressure dependence of attenuation due to fluid flow depends primarily on the change in permeability in the rock due to compaction and pore collapse. The elastic moduli and total porosity are easily obtained as functions of pressure using the method of Toksöz et al (1976). Furthermore, we may assume that the viscosity of the fluid inclusion remains relatively constant in the pressure range of interest.

Experimental determinations of permeability as a function of confining hydrostatic pressure have been made for several sandstones (Fatt and Davis, 1952), Westerly granite (Frangos, 1967), and Ottawa sand (Zoback and Byerlee, 1976). In general, permeability decreases with increasing pressure, but the rate of decrease depends on the total porosity and fraction of crack porosity. In highly porous and permeable consolidated rocks, the bulk of the porosity and permeability is contained in the large aspect ratio pores which do not close under pressure. Fatt and Davis (1952) found a maximum reduction in permeability of 25 percent at 350 bars for the sandstones, while for a granite the reduction may be as much as an order of magnitude (Frangos, 1967). However, since the effect of fluid flow is negligible in all but the highly permeable rocks, we need only consider data on that type. Measurements of permeability in unconsolidated Ottawa sand (Zoback and Byerlee, 1976) show a slow reduction up to 800 bars, where it drops off rapidly to level off again between 2000 and 3000 bars. The acceleration in permeability loss at 800 bars is presumably due to grain crushing and pore collapse. However, the applicability of this study to consolidated rocks is uncertain, and it could not easily be modeled. We shall assume that the permeability of highly porous rocks is constant with pressure. The effect of this is to give an upper bound on the contribution due to fluid flow on attenuation.

Several investigators have proposed attenuation mechanisms by which flow is induced between two adjacent connected cracks due to the relative volume change caused by the stress wave (Mavko and Nur, 1975; O'Connell and Budiansky, 1977). These are commonly known as squirting mechanisms, and while they are not important at ultrasonic frequencies, they may be so at sonic or seismic frequencies. The elastic model of Toksöz et al (1976) is particularly useful in treating these mechanisms in that a distribution of crack aspect ratios is uniquely determined and pressure gradients between cracks may be readily calculated.

Flow in any squirting mechanism is generally from small aspect ratio (thin cracks) to large aspect ratio (pores) voids. Thus, the flow field within the crack may be approximated by the flow between two infinite plates, as is done by Mavko and Nur (1975) and O'Connell and Budiansky (1977). Here we consider an approach to the problem consistent with the concepts and formulations introduced by Toksöz et al (1976). The details of the calculations are given in Appendix B. We assume that flow will take place between very thin cracks with $\alpha_m \approx 0$ and pores with $\alpha_m = 1$ due to a differential volume change induced by the stress wave. The pressure difference, the equalized pressure after flow, the instantaneous flow q, and the total flow q_T can be easily calculated. Assuming a relaxation of the form

$$q_T = q \int_0^\infty e^{-t/\tau} dt = q\tau, \qquad (11)$$

where τ is the relaxation time, we find that

$$\tau = 8\eta / \alpha_m^2 K' (1 + \varepsilon), \qquad (12)$$

where η is the viscosity, α_m is the aspect ratio, K' is the fluid bulk modulus, and ε is the ratio of connected crack volume to pore volume. We can make the approximation $\varepsilon \approx 0$ for porous rocks. Taking $K' = 2 \times 10^{10}$ dynes/cm^2, $\eta = 10^{-2}$ poise, with α_m ranging from 10^{-3} to 10^{-4}, we obtain relaxation times ranging from 4×10^{-6} to 4×10^{-4} sec.

The formulation of this mechanism in terms of complex moduli yields an expression that also includes the viscous relaxation mechanism in pores as discussed earlier. This is a result of applying the correspondence principle for the shear modulus $\mu' = i\omega\eta$ and expressing the bulk modulus as $K' = K_R' + i\omega g$, where g is considered an unknown to be

determined from the relaxation time for the squirting flow. While this is a good approximation for high frequencies, at very low frequencies (<0.1 Hz), the fluid offers little resistance to flow and thus $K_R \approx 0$ (O'Connell and Budiansky, 1977). It is shown in Appendix B that equations (A–1) and (A–2) for the effective moduli can be written in terms of two characteristic frequencies: $\omega_c = K/g$ and $\omega_d = 3K/4\eta$ [equation (B–13)]. ω_d is recognized as the characteristic frequency for viscous relaxation (Walsh, 1969), and ω_c is the characteristic frequency for fluid flow from cracks. From the estimate of the relaxation time for this mechanism, we obtain

$$g = \frac{8\eta}{\alpha_m^2(1 + \varepsilon)} \frac{K}{K_R'}. \qquad (13)$$

For example, with $\varepsilon = 0$, $\alpha_m = 10^{-3}$, $\eta = 10^{-2}$ poise, $K = 4 \times 10^{11}$ dynes/cm^2 and $K_R' = 2 \times 10^{10}$ dynes/cm^2, we find that $g = 1.6 \times 10^6$ poise or, more generally, $g = 1.6/\alpha_m^2$ poise. This mechanism is readily included in the elastic moduli formulations by finding g from equation (13) and then substituting to find K' to be used in the elastic moduli calculations.

Other sources of attenuation

In many cases, rocks in the crust are partially saturated by two or more fluids—air and water, oil and brine, gas and oil, to name a few. The effect of partial saturation on velocity is fairly well known; however, its effect on attenuation is not as well understood. Apparent attenuation of seismic waves propagated through gas-sands imply that the effect can be large. One problem encountered is the distribution of the saturants in the rock frame. Not only are large-scale irregularities in partial saturation found in rock formations, but the distribution on a smaller scale, pore to pore, may change. Gas bubbles in water or oil are more likely to occupy space in pores with larger aspect ratios than in the finer cracks, where the friction and relaxation mechanisms are more important. The latter effect is evident from the data at low saturations discussed earlier and shown in Figures 2–5.

Several mechanisms involving the presence of free gas in the pores may contribute to the attenuation in partially saturated rocks. This is illustrated in Figure 10. Gas bubbles have several effects. First, the pore fluid bulk modulus is reduced, facilitating flow even under very small pressure gradients. (Stoll (1977) suggested that in this case, conversion to Biot diffusion-type waves at an interface can result in substantial energy loss. Squirting flow would also be enhanced. Secondly, bubble squeezing and moving in particular may contribute to a decreased Q. Thus, in partially saturated rocks, the attenuation may be greater than in the fully saturated case. A small amount of fluid is required to lubricate cracks and grain boundaries to facilitate sliding and energy loss due to friction. The presence of gas bubbles, on the other hand, enhances energy dissipation mechanisms operative at full saturation, and further loss may result from motions of the bubbles themselves. This enhanced attenuation, particularly, concerns mechanisms dependent on pressure gradients induced by P-waves.

An attenuation model describing the effects of large-scale irregularities (on the order of 10 cm) in saturation conditions has been proposed by White (1975). The porous rock model contains spherical pockets saturated with gas with the rest of the volume saturated with liquid. Loss due to fluid flow is enhanced at the gas-liquid interfaces. White showed that for the particular model chosen, attenuation due to this mechanism can be important at seismic frequencies. There is some debate, however, as to the occurrence of the saturation irregularities.

Several other mechanisms for attenuation have been proposed, although their applicability to upper crustal rocks is debatable. Several of these mechanisms may be operable in the upper mantle, however, such as grain boundary relaxation, relaxation caused by a phase change, and a ''high temperature background'' attenuation probably related to Nabarro diffusion (Jackson and Anderson, 1970). Experimental evidence suggests little change in attenuation as a function of temperature at relatively low temperatures (Volarovich and Gurvich, 1957) when rock is not cracked and saturating fluids are not altered. However, near phase changes, attenuation could change rapidly with temperature. High attenuation has been observed at critical points in multicomponent systems (Spetzler and Anderson, 1968; Wang and Meltzen, 1972). Energy is absorbed by a medium whose equilibrium is disturbed by a stress wave. The frequency at which this occurs is dependent on the rate at which phase equilibrium can follow the changes imposed on it by the wave (Spetzler and Anderson, 1968). This mechanism may result in high attenuation in certain geothermal areas.

Finally, we consider the effective attenuation due to scattering by inclusions in the rock. Although this is a geometrical effect, it can, in some cases, affect the observed attenuation. Yamakawa (1962) has analyzed the scattering of compressional waves by spherical pores. The equivalent attenuation coefficient α is given by

$$\alpha = \phi \frac{12\pi^4 f^4 r^3}{V_p^4} \left[2B_0^2 + \frac{2}{3}(1+\nu^3) B_1^2 + \frac{(2+3\nu^5)}{5} B_2^2 \right], \qquad (14)$$

where

$$B_0 = \frac{K - K'}{3K' + 4\mu},$$

$$B_1 = (\rho - \rho')/3\rho,$$

$$B_2 = \frac{-20\mu^3}{3\mu(9K + 8\mu)},$$

and $\nu = V_p/V_s$, r = radius of inclusions, f = frequency. In the above, primed coefficients represent the inclusion properties. Although the effective attenuation of incident plane S-waves has not been calculated, we may estimate this effect by noting that the energy loss due to SP reflections is equivalent to PS reflections because of the reciprocal theorem. While losses due to SS reflections are not the same as PP, they are close, and we can reevaluate equation (19) for incident S-waves assuming $SS \equiv PP$. Doing so, the only changes in the equation are that V_p is replaced by V_s and $\nu = V_s/V_p$. Attenuation due to scattering is strongly dependent on frequency ($\propto f^4$). As will be shown in the next section, scattering effects can be important, if not dominant, at high ultrasonic frequencies ($f > 1$ MHz). At seismic frequencies, scattering due to pores is negligible.

Another geometric effect is the apparent attenuation due to selective reflection of the short wavelength component of seismic waves in thin beds. Although of little importance with respect to laboratory measurements, this mechanism may, under certain conditions, contribute to observed amplitude loss in seismic sections. O'Doherty and Anstey (1971), Schoenberger and Levin (1974), and Spencer et al (1977) have examined these cases in detail. In general, selective reflection due to cyclic stratification contributes a small but important part to the overall attenuation. If high reflection coefficients occur, the apparent attenuation can be high.

INTERPRETATION OF LABORATORY DATA

We shall now consider in more detail the relative effects of the various attenuation mechanisms in dry and saturated porous rocks. The methods and tech-

Table 2. Physical properties used for modeling the Berea sandstone.

Matrix
  $K = 35 \times 10^{10}$ dynes/cm²
  $\mu = 25 \times 10^{10}$ dynes/cm²
  $\rho = 2.61$ g/cm³
Inclusion
  $K' = 2.6 \times 10^{10}$ dynes/cm²
  $\eta = 4 \times 10^{-2}$ poise
  $\rho' = 1.0$ g/cm³
Frame
  $\phi \approx 0.16$
  $\chi = 75$ md
Fluid flow
  Structure constants:
  $a = 1.0 \times 10^{-4}$
  $a' = 3.0$

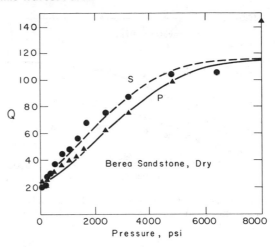

FIG. 11. Model fit (solid and dashed lines) to dry Berea sandstone data from Toksöz et al (1979, this issue, Figure 7). Pore pressure is assumed to be 1 bar (14.7 psi).

niques discussed in the previous section will be applied to model the behavior of attenuation as a function of differential pressure for the ultrasonic data on the Berea sandstone presented in Toksöz et al (1979). These models will then be extrapolated to other frequencies. The application of these models to previously reported data is difficult, because absolute values of attenuation appear to be unreliable in some cases, and parameters needed in the calculations are unavailable in other cases.

The procedure taken involves first modeling the attenuation in the dry rock in order to establish the needed parameters for the friction mechanism and intrinsic attenuation in the absence of fluid-associated mechanisms. These parameters will then be used in the modeling of the saturated sample data. An important (but probably valid) assumption made here is that all attenuation mechanisms that occur in dry rocks also occur in wet ones. Given the parameters obtained from the dry case, we may examine in more detail the relative importance of the mechanisms contributing to the attenuation in the brine-saturated case as a function of pressure. In particular, since the attenuation due to Biot-type fluid flow, squirting, and scattering are readily calculable, it remains to be seen what the contribution due to the presence of pore fluid is in terms of the friction mechanism and intrinsic aggregate anelasticity. The approach taken here is empirical, and thus the models presented have no absolute predictive ability.

The elastic moduli, fluid, and frame properties used in modeling the Berea sandstone are listed in Table 2. The bulk modulus of brine as a function of pore pressure is given by Adams (1931) and Long and Chierici (1961). For the dry case, the bulk modulus of air is taken to be 1 bar, and the pore pressure is assumed to be constant at 1 bar.

The pore aspect ratio distribution at surface (atmospheric) pressure in Table 3 is determined by fitting theoretically calculated elastic properties [equations (A–1) and (A–2)] to the P- and S-wave velocity versus differential pressure data for both saturated and dry cases, as described by Toksöz et al (1976). The frequency is taken to be 0.5 MHz.

The contributions to attenuation in the dry case are assumed to be due to friction and the intrinsic aggregate attenuation only. Zero-pressure Q's were taken as 23 for P-waves and 26 for S-waves, based on the data from Toksöz et al (1979). The pressure dependence of Q for the dry Berea sandstone may be reasonably modeled with $A = 0.2 \times 10^4$ [equation (8)] and an intrinsic aggregate Q for both P- and S-waves of 120. The possible variations in the parameter A are not as wide as one might expect, ranging from 0.15×10^4 to 0.25×10^4. The results of this empirical model fitted to the data are shown in Figure 11.

The introduction of brine as the pore saturant results in no change in the parameter A, since the crack-closing rate is the same as for the dry case, determined by the static rather than the dynamic effective bulk modulus.

In the preceding section, the role of fluids in determining the attenuation was discussed. In particular, water may soften and lubricate the matrix, resulting in a higher attenuation due to a friction type mechanism, especially for S-waves. Since the contributions

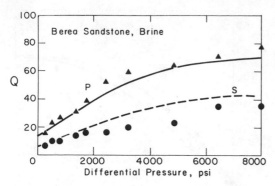

FIG. 12. Model fit to brine-saturated Berea sandstone data from Toksöz et al (1979, this issue, Figure 11).

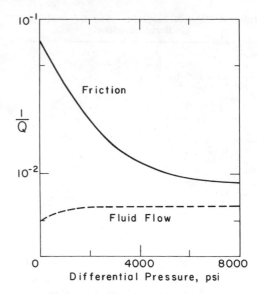

FIG. 13. Relative contributions of the friction and fluid flow mechanism for P-waves from the model of Figure 12 as a function of differential pressure.

due to Biot fluid flow, squirting flow, viscous shear relaxation, and scattering are fairly well determined from the properties listed in Table 2, the contribution due to friction remains to be seen in modeling the saturated data. This must be determined empirically. One important constraint, however, is the low Q, especially Q_s, at high pressures. This implies that a mechanism which is relatively independent of pressure, such as Biot fluid flow, is required under those conditions.

The fluid flow contributions to the attenuation are calculated as described in the previous section. Given the attenuation due to all the mechanisms other than friction, it is found that to fit the data, one must choose a zero pressure Q_p of 15 and a Q_s of 10 for friction. These low values of Q relative to the dry case indicate that brine saturation increases the attenuation due to friction by almost a factor of two. Although the data may be fit with a fluid viscosity of 1 cp, a better fit is obtained by allowing the effective viscosity to be 4 cp. This might be expected from experimental measurements of the viscosity of water in clay-water systems (Low, 1959). Such an effect would predict a higher attenuation in rocks with higher clay content. Furthermore, while not necessary, the best fit to the data shown in Figure 12 is obtained by reducing the intrinsic aggregate Q for shear waves by 5 percent. It is perhaps no coincidence that the seismic velocities are best fit in the saturated case by reducing the matrix shear modulus 5 percent relative to the dry case. This may reflect the possibility of increased shear and thus higher attenuation at grain boundaries due to the presence of water, as discussed earlier.

The relative contributions of the two important mechanisms, friction and Biot-type fluid flow in the brine-saturated case, are easily seen in Figure 13 showing Q_p^{-1} for each mechanism as a function of pressure. The small increase in the fluid flow contribution at low pressures is an artifact of the calculations. As would be expected, friction across cracks and grain boundaries is dominant at low pressures but becomes less important as cracks close. Since the bulk of the porosity and permeability is unaffected under the pressure conditions of interest, the fluid flow contribution to attenuation remains relatively constant with pressure and becomes an increasingly important mechanism. Obviously, at some pressure the porosity and permeability of the rock will break down, and one should expect a rapid increase in Q.

Using the Berea sandstone properties from model calculations, we shall now examine in more detail the individual contributions of each mechanism for the fully saturated case and extrapolate these results to other frequencies. The interpretation of these models must remain strictly within the confines imposed upon them. That is, it is assumed that strain amplitudes are equivalent to those in the laboratory experiment and that no other mechanisms contribute to attenuation at frequencies other than those at 0.5 MHz.

A theoretical overview of the relative contribution of each mechanism considered is shown in Figure 14. Here, the P-wave attenuation coefficients are plotted as functions of frequency for a surface pressure con-

Table 3. Aspect ratio distributions

Concentration, c	Aspect ratio, α
Surface	
0.12	1.00
0.04	0.10
0.10×10^{-3}	0.17×10^{-2}
0.10×10^{-3}	0.14×10^{-2}
0.20×10^{-3}	0.10×10^{-2}
0.15×10^{-3}	0.60×10^{-3}
0.75×10^{-4}	0.30×10^{-3}
0.30×10^{-4}	0.10×10^{-3}
0.90×10^{-5}	0.30×10^{-4}
0.30×10^{5}	0.10×10^{-4}
10,000 ft	
0.119	1.00
0.395×10^{-1}	0.98×10^{-1}
0.152×10^{-1}	0.258×10^{-3}

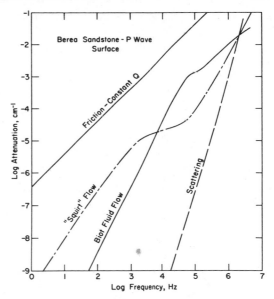

FIG. 14. *P*-wave attenuation coefficients at surface pressure as functions of frequency for several mechanisms considered in the saturated Berea sandstone model. Model parameters are listed in Tables 2 and 3 and in the text. The viscous shear relaxation mechanism is included on the line labeled "squirt" flow.

dition. Figure 14 was obtained by fixing the attenuation at 0.5 MHz based on the theoretical model of the pressure data (Figure 12). The resulting curves are theoretical extrapolations. A constant Q mechanism for friction is assumed. The same model is shown in Figure 15, except that the attenuation coefficients are calculated for a differential pressure equivalent to a depth of about 10,000 ft. The corresponding aspect ratio distribution is listed in Table 3.

Figures 14 and 15 clearly show the relative effects of friction, fluid flow, shear relaxation, and scattering on the attenuation of *P*-waves. Similar results are obtained for *S*-waves. If friction is indeed a frequency-independent attenuation mechanism, then it dominates the other mechanisms for this case. However, as seen before, friction is of somewhat less importance at higher pressures. As assumed in our models, the contribution of Biot fluid flow remains essentially unchanged between Figures 14 and 15. While never dominating in this case, it is of importance at about 10^5 Hz where Poisseuille flow breaks down. A striking change in the squirt flow and shear relaxation mechanism is apparent, however. For surface conditions, the contribution due to these mechanisms is readily seen from Figure 16. Here, Q^{-1} for both *P*- and *S*-waves is shown for the squirting and shear relaxation mechanisms only. Two peaks are evident, the lower frequency one corresponding to the flow mechanism and the other to viscous relaxation. The shape of the relaxation peaks are complicated, reflecting the spectrum of pore and crack shapes. The transition from flow to viscous relaxation takes place at about 50 kHz, below which $Q_p^{-1} > Q_s^{-1}$ and above which $Q_s^{-1} > Q_p^{-1}$. Even though viscous relaxation peaks at $f = 10^9$ Hz, it is clear from Figures 14 and 16 that the contribution of these mecha-

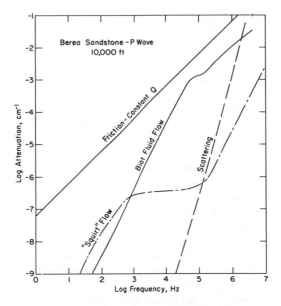

FIG. 15. *P*-wave attenuation coefficients as functions of frequency for the saturated Berea model as in Figure 14. Here, the contributions for each mechanism are calculated at a differential pressure equivalent to about a 10,000-ft depth.

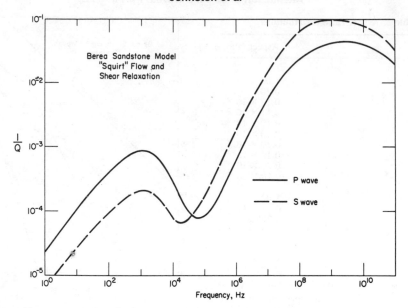

FIG. 16. Q_p^{-1} and Q_s^{-1} for the squirt flow and viscous shear relaxation mechanisms in the saturated Berea sandstone model at surface pressure as functions of frequency.

nisms to the attenuation in the Berea sandstone is small in the frequency band of interest, even at surface pressure. Furthermore, the effect of pressure, as seen in Figure 15, is to close cracks contributing to both the squirt flow and viscous relaxation, thus lowering even further their associated attenuations.

Scattering by inclusions produces a negligible effect, except at very high frequencies, where this mechanism clearly dominates. A larger scatterer radius will shift this curve to lower frequencies.

We finally combine both the frequency and pressure behavior of attenuation in our saturated Berea sandstone model in Figure 17, where the total Q_p of the rock is shown. For low differential pressures, Q_p remains essentially unchanged as a function of frequency, reflecting the importance of the friction mechanism. Q_p increases with differential pressure, and at high pressures and low frequencies ($<10^4$ Hz), Q_p is greater than 100. Q_p decreases with increasing frequency at higher pressures due to the increasing contribution of Biot flow. Finally, at very high frequencies (10^7 Hz), Q_p decreases sharply because of scattering.

While the ultrasonic attenuation data may be understood and modeled by several mechanisms, some problems exist in the extrapolation of these data to lower frequencies. As discussed earlier, the frequency dependence of the important friction mechanism is

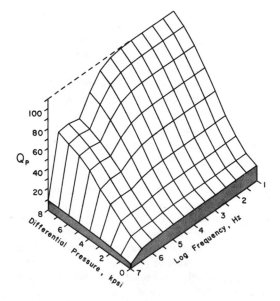

FIG. 17. Total Q_p for the saturated Berea model as a function of frequency and differential pressure based on the results presented in Figures 12, 14, and 15.

not clearly established. Furthermore, strain amplitudes at seismic exploration frequencies may not be the same as those used in ultrasonic measurements, thus invalidating the absolute estimates of Q. However, theoretical models such as the ones presented in this paper, provide a method of comparing laboratory data taken under controlled conditions with in-situ data. One has to be aware, however, that the contributions of mechanisms that may be important at low frequencies are difficult to establish from ultrasonic data unless supplementary information is available. Furthermore, it is obvious that from these models one would obtain only a point property of the rock. For in-situ data, the intrinsic attenuation must be isolated from other amplitude-reducing mechanisms such as scattering, spreading, or multiple reflections, before comparison with laboratory data.

CONCLUSIONS

The investigation of both published and new laboratory data on the attenuation of seismic waves in rocks, particularly sandstones, has shown that many of the same properties and processes that affect velocity also affect attenuation, many times to a greater extent. These properties include the number and distribution of cracks, the type and amount of fluid saturation, and the mechanical properties of the rock matrix.

We have approached the problem of attenuation in dry and completely saturated rocks by examining a number of hypothesized mechanisms for which numerical models may be applied. The formulation of the pressure dependence of these models enables us to reasonably fit ultrasonic data for Q_p and Q_s in a Berea sandstone. The models for attenuation require the specification of several free parameters and thus limit their predictive abilities. Furthermore, assumptions involving the frequency and amplitude behavior of the friction mechanism must be considered if laboratory data are compared to in-situ data. However, given the limitations of the models, several conclusions regarding the attenuation of seismic waves in rocks are possible:

1) Under the conditions studied in this paper, the primary mechanism for attenuation is friction on grain boundaries and thin cracks.
2) Increasing differential pressure decreases the number of cracks contributing to attenuation by friction. Since frictional loss depends on the number of cracks, the attenuation decreases with increasing pressure and eventually approaches a

limiting value we call the intrinsic aggregate anelasticity. This is probably due to grain boundaries and fine structure relatively unaffected by pressure.
3) In totally dry rocks, the attenuation is less than in wet or saturated rocks. The introduction of fluid into a dry rock will wet crack surfaces and grain boundaries. By this crack lubrication, frictional sliding is facilitated and the attenuation increases.
4) In a saturated porous rock, attenuation due to fluid flow plays a secondary role relative to friction. At low frequencies, squirting flow may be a viable mechanism, especially in the case of partial saturation. At ultrasonic frequencies, the Biot-type fluid flow mechanism, while not necessarily dominating, plays an important role in the overall attenuation at high pressures.

ACKNOWLEDGMENTS

The authors thank Dr. Joseph Walsh and Dr. C. H. Cheng of M. I. T. for their valuable assistance in preparing this manuscript. Prof. Jan Korringa and Dr. Jim Spencer of Chevron Oil Field Research Co. provided beneficial discussions and comments on the paper. This research was supported in part by the Advanced Research Projects Agency, monitored by the Air Force Office of Scientific Research under contract F44620-75-C-0064. David Johnston was supported in part by a Chevron fellowship.

REFERENCES

Adams, L. H., 1931, Equilibrium in binary systems under pressure; I. An experimental and thermodynamic investigation of the system, NaCl—H_2O at 20°: Am. Chem. Soc. J., v. 53, p. 2769–3785.
Al-Sinawi, S., 1968, An investigation of body wave velocities, attenuation on elastic parameters of rocks subjected to pressure at room temperature: Ph.D. thesis, St. Louis University.
Anderson, D. L., Ben-Menahem, A., and Archambeau, C. B., 1965, Attenuation of seismic energy in the upper mantle: J. Geophys, Res., v. 70, p. 1441–1448.
Attewell, P. B., and Ramana, Y. V., 1966, Wave attenuation and internal friction as functions of frequency in rocks: Geophysics, v. 31, p. 1049–1056.
Biot, M. A., 1956a, Theory of propagation of elastic waves in a fluid-saturated porous solid. I. Low frequency range: J. Acoust. Soc. Am., v. 28, p. 168–178.
——— 1956b, Theory of propagation of elastic waves in a fluid-saturated porous solid. II. High frequency range: J. Acoust. Soc. Am., v. 28, p. 179–191.
——— 1962a, Mechanics of deformation and acoustic propagation in porous media: J. Appl. Phys., v. 33, p. 1482–1498.
——— 1962b, Generalized theory of acoustic propagation in porous dissipative media: J. Acoust. Soc. Am., v. 34, p. 1254–1264.
Birch, F., and Bancroft, D., 1938, Elasticity and internal friction in a long column of granite, Bull. SSA, v. 28, p. 243–254.
Bland, D. R., 1960, The theory of linear viscoelasticity:

New York, Pergamon Press.

Born, W. T., 1941, Attenuation constant of earth materials: Geophysics, v. 6, p. 132–148.

Bradley, J. J., and Fort, A. N., Jr., 1966, Internal friction in rocks, *in* Handbook of physical constants: S. P. Clark, Jr., Ed., GSA. Pub., p. 175–193.

Collins, F., and Lee, C. C., 1956, Seismic wave attenuation characteristics from pulse experiments: Geophysics, v. 21, p. 16–40.

Dainty, A. M., Goins, N. R., and Toksöz, M. N., 1976, Seismic investigation of the lunar interior: Lunar Science VII, Lunar Science Institute, Houston, p. 181–183.

Fatt, I., and Davis, D. H., 1952, Reduction in permeability with overburden pressure: Trans. AIME, Pet. Branch, v. 195, p. 329.

Frangos, W. T., 1967, The effect of continuing pressure on the permeability of westerly granite: B.S. thesis, M.I.T.

Gardner, G. H. F., Wyllie, M. R. J., and Droschak, D. M., 1964, Effects of pressure and fluid saturation on the attenuation of elastic waves in sands: J. Petroleum Tech., v. 16, p. 189–198.

Gordon, R. B., 1974, Mechanical relaxation spectrum of crystalline rock containing water: J. Geophys. Res., v. 79, p. 2129–2131.

Gordon, R. B., and Davis, L. A., 1968, Velocity and attenuation of seismic waves in imperfectly elastic rock: J. Geophys. Res., v. 73, p. 3917–3935.

Hamilton, E. L., 1972, Compressional wave attenuation in marine sediments: Geophysics, v. 37, p. 620–646.

Jackson, D. D., and Anderson, D. L., 1970, Physical mechanisms of seismic wave attenuation: Rev. Geophys. Space Phys., v. 8, p. 1–63.

Johnston, D. H., and Toksöz, M. N., 1978, Cracks and Q: presented at the 48th Annual International SEG Meeting, November 1, in San Francisco.

Klima, K., Vanek, J., and Pros, Z., 1964, The attenuation of longitudinal waves in diabase and graywacke under pressures up to 4 kilobars: Studia Geoph. et Geod., v. 8, p. 247–254.

Knopoff, L., 1964, *Q*: Rev. Geophys., v. 2, p. 625–660.

Kuster, G. T., and Toksöz, M. N., 1974, Velocity and attenuation of seismic waves in two-phase media. Part I. Theoretical formulations: Geophysics, v. 39, p. 587–606.

Latham, G. V., Nakamura, Y., Lammlein, D., Dorman, J., and Duennebier, F., 1974, Structure and state of the lunar interior based upon seismic data: (abstr.) Lunar Science V, Lunar Science Institute, Houston, p. 434.

Levykin, A. I., 1965, Longitudinal and transverse wave absorption and velocity in rock specimens at multilateral pressures up to 4000 km/cm²: USSR Geophys. Ser. (Engl. transl.), v. 1, Physics of the Solid Earth, p. 94–98.

Lockner, D., Walsh, J. B., and Byerlee, J., 1977, Changes in seismic velocity and attenuation during deformation of granite: submitted to J. Geophys. Res., v. 82, p. 5374–5378.

Long, G., and Chierici, G., 1961, Salt content changes compressibility of reservoir brines: Pet. Engineer, July, B-25, B-26, B-31.

Low, P. H., 1959, Viscosity of water in clay systems: Eighth Nat. Conf. Clays and Clay Minerals, New York, Pergamon Press, p. 170–182.

Mason, W. P., 1958, Physical acoustics and the properties of solids: Princeton, Van Nostrand Co.

Mavko, G., and Nur, A., 1975, Melt squirt in the aesthenosphere: J. Geophys. Res., v. 80, p. 1444–1448.

McDonal, F. J., Angona, F. A., Mills, R. L., Sengbush, R. L., Van Nostrand, R. G., and White, J. E., 1958, Attenuation of shear and compressional waves in Pierre shale: Geophysics, v. 23, p. 421–439.

Merkulova, V. M., Pigulevskiy, E. D., and Tsaplev, U. M., 1972, Sound absorption measurements in uniaxially compressed rocks: USSR, Physics of the solid earth, v. 3,

p. 166–167.

Nakamura, Y., Dorman, J., Duennebier, F., Ewing, M., Lammlein, D., and Latham, G., 1974, High frequency lunar teleseismic events: Geochim. Cosmochim. Acta, Suppl. 5, p. 2883–2890.

Nur, A., and Simmons, G., 1969a, The effect of viscosity of a fluid phase on velocity in low porosity rocks: Earth Planet. Sci. Lett., v. 7, p. 99–108.

——— 1969b, The effect of saturation on velocity in low porosity rocks: Earth Planet. Sci. Lett., v. 7, p. 183–193.

Obert, L., Windes, S. L., and Duvall, W. I., 1946, Standardized tests for determining the physical properties of mine rock: U.S. Bur. Mines, R.I. 3891.

O'Connell, R. J., and Budiansky, B., 1977, Viscoelastic properties of fluid saturated cracked solids: submitted to J. Geophys. Res., v. 82, p. 5719–5736.

O'Doherty, R. F., and Anstey, N. A., 1971, Reflections on amplitudes: Geophys. Prosp., v. 19, p. 430–458.

Pandit, B. I., and Savage, J. C., 1973, An experimental test of Lomnitz's theory of internal friction in rocks: J. Geophys. Res., v. 78, p. 6097–6099.

Pandit, B. I., and Tozer, D. C., 1970, Anomalous propagation of elastic energy within the Moon: Nature, v. 226, p. 335.

Peselnick, L., and Zietz, I., 1959, Internal friction of fine-grained limestones at ultrasonic frequencies: Geophysics, v. 24, p. 285–296.

Peselnick, L., and Outerbridge, W. F., 1961, Internal friction and rigidity modulus of Solenhofen limestone over a wide frequency range: U.S.G.S. Prof. paper no. 400B.

Pinkerton, J. M. M., 1947, A pulse method for the measurement of ultrasonic absorption in liquids, results for water: Nature, v. 160, p. 128–129.

Pros, Z., Vanek, J., and Klima, K., 1962, The velocity of elastic waves in diabase and graywacke under pressures up to 4 kilobars: Studia Geoph. et Geod., v. 6, p. 347–367.

Schoenberger, M., and Levin, F. K., 1974, Apparent attenuation due to intrabed multiples: Geophysics, v. 39, p. 278–291.

Solomon, S. C., 1973, Shear wave attenuation and melting beneath the mid-Atlantic Ridge: J. Geophys. Res., v. 78, p. 6044–6059.

Spencer, T. W., Edwards, C. M., and Sonnad, J. R., 1977, Seismic wave attenuation in non-resolvable cyclic stratification: Geophysics, v. 42, p. 939–949.

Spetzler, H., and Anderson, D. L., 1968, The effect of temperature and partial melting on velocity and attenuation in a simple binary system: J. Geophys. Res., v. 73, p. 6051–6060.

Stoll, R. D., 1974, Acoustic waves in saturated sediments: *in* Physics of sound in marine sediments: L. Hampton, Ed., New York, Plenum Press.

——— 1977, Acoustic waves in ocean sediments: Geophysics, v. 42, p. 715–725.

Stoll, R. D., and Bryan, G. M., 1970, Wave attenuation in saturated sediments: J. Acous. Soc. Am., v. 47, p. 1440–1447.

Tittmann, B. R., Abdel-Gawad, M., and Housley, R. R., 1972, Elastic velocity and Q factor measurements on Apollo 12, 14, and 15 rocks: Proc. 3rd Lunar Sci. Conf., v. 3, p. 2565–2575.

Tittmann, R. B., Housley, R. M., and Abdel-Gawad, M., 1975, Internal friction quality factor–3100 achieved in lunar rock 70215,85: Lunar Science VI, Lunar Science Institute, Houston, p. 812–814.

Toksöz, M. N., Dainty, A. M., Solomon, S. C., and Anderson, K. R., 1974, Structure of the moon: Rev. Geophys. Space Phys., v. 12, p. 539–567.

Toksöz, M. N., Cheng, C. H., and Timur, A., 1976, Velocities of seismic waves in porous rocks: Geophysics,

v. 41, p. 621–645.

Toksöz, M. N., Johnston, D. H., and Timur, A., 1979, Attenuation of seismic waves in dry and saturated rocks. I. Laboratory measurements: Geophysics, this issue, p. 681–690.

Volarovich, M. P., and Gurevich, G. I., 1957, Investigation of dynamic moduli of elasticity for rocks in relation to temperature: Bull. Acad. Sci. USSR, Geophys., v. 4, p. 1–9.

Walsh, J. B., 1966, Seismic wave attenuation in rock due to friction: J. Geophys. Res., v. 71, p. 2591–2599.

—— 1968, Attenuation in partially melted material: J. Geophys. Res., v. 73, p. 2209–2216.

—— 1969, New analysis of attenuation in partially melted rock: J. Geophys. Res., v. 74, p. 4333–4337.

Walsh, J. B., Brace, W. F., and Wawersik, W. R., 1970, Attenuation of stress waves in Ceder City quartz diorite: Air Force Weapons Lab. Tech. rep. AFWL-TR-70-8.

Wang, C., and Meltzen, M., 1972, Propagation of elastic waves in a rock undergoing phase transitions: Cordilleran Section, 68th Annual Meeting, GSA Abst., v. 4, p. 256–257.

Warren, N., Trice, R., and Stephens, J., 1974, Ultrasonic attenuation: Q measurements on 70215, 29: Geochim. Cosmochim. Acta, Supp. 5, p. 2927–2938.

White, J. E., 1975, Computed seismic speeds and attenuation in rocks with partial gas saturation: Geophysics, v. 40, p. 224–232.

Winkler, K., and Nur, A., 1978, Attenuation and velocity in dry and water-saturated Massilon sandstone: presented at the 48th Annual Int. SEG Meeting, October 31 in San Francisco.

Wyllie, M. R., Gregory, A. R., and Gardner, G. H. F., 1958, An experimental investigation of factors affecting elastic wave velocities in porous media: Geophysics, v. 23, p. 459–493.

Wyllie, M. R. J., Gardner, G. H. F., and Gregory, A. R., 1962, Studies of elastic wave attenuation in porous media: Geophysics, v. 27, p. 569–589.

Yamakawa, N., 1962, Scattering and attenuation of elastic waves: Geophysical Magazine, Tokyo, v. 31, p. 63–103.

Zoback, M. D., and Byerlee, J. D., 1976, Effect of high pressure deformation on permeability of Ottawa sand: AAPG Bull., v. 60, p. 1531–1542.

APPENDIX A
ELASTIC MODULI FOR CALCULATING ATTENUATION

The calculation of attenuation requires the knowledge of several elastic moduli and their pressure dependence. Given the matrix or grain moduli and density K, μ, and ρ, and the inclusion properties K', μ' and ρ', the effective properties of a composite medium may be found following the treatment of Kuster and Toksöz (1974). Cracks and large pores in the rock are represented by a discrete spectrum of various aspect ratio spheroids. Letting $c(\alpha_m)$ be the concentration of pores and cracks with aspect ratio $\alpha_m =$ thickness/diameter, the effective moduli are given by (Kuster and Toksöz, 1974)

$$\frac{K^* - K}{3K^* + 4\mu} = 1/3 \frac{K' - K}{3K + 4\mu} \cdot$$
$$\cdot \sum_{m=1}^{M} c(\alpha_m) T_{iijj}(\alpha_m), \quad (A-1)$$

and

$$\frac{\mu^* - \mu}{6\mu^*(K + 2\mu) + \mu(9K + 8\mu)}$$
$$= \frac{\mu' - \mu}{25\mu(3K + 4\mu)} \sum_{m=1}^{M} c(\alpha_m) \cdot$$
$$\cdot [T_{ijij}(\alpha_m) - 1/3\, T_{iijj}(\alpha_m)], \quad (A-2)$$

where the asterisk denotes effective properties, primed quantities refer to fluid properties, and unprimed quantities are matrix properties. K and μ represent bulk and shear moduli and T_{iijj} and T_{ijij} are

scalar quantities. The total porosity is

$$\phi = \sum_{m=1}^{M} c(\alpha_m),$$

and the density is

$$\rho^* = \rho(1 - \phi) + \rho'\phi.$$

The effect of pressure on the crack and pore distributions and, thus, the effective moduli and velocities of rocks, has been studied by Toksöz et al (1976). The strain field around an ellipsoidal cavity is calculated as a function of the elastic moduli of the matrix and an applied strain field at infinity. The dilatation of the applied field is $-P/K_A^*$ where P is the applied hydrostatic differential pressure and K_A^* is the effective static bulk modulus or frame bulk modulus. From this, the fractional change in pore volume (concentration) dc/c may be found. For the particular case of very thin cracks (i.e., $\alpha_m \to 0$),

$$\frac{dc}{c} = -\frac{P}{K_A^*} \left\{ \frac{4}{3\pi\alpha_m} \frac{(1 - \sigma^2)}{(1 - 2\sigma)} \right\}, \quad (A-3)$$

where $\sigma =$ matrix Poisson's ratio. This relationship also provides the basis for calculating the change in attenuation due to friction under increasing hydrostatic or differential pressure.

Anelasticity may be introduced into the effective moduli formulations by employing the concept of complex moduli (Anderson et al, 1965; Bland, 1960). This method is particularly useful in dealing with frequency independent Q mechanisms such as grain boundary and crack friction. Let the complex bulk and shear moduli be expressed as

$$K = K_R + iK_I,$$

and   (A-4)

$$\mu = \mu_R + i\mu_I,$$

where subscripts R and I refer to real and imaginary parts. If the attenuation is small, then the velocities and attenuation coefficients can be expressed conveniently. For compressional waves,

$$V_p = \left[\frac{K_R + 4/3\,\mu_R}{\rho} \right]^{1/2},$$

and   (A-5)

$$Q_p^{-1} = \frac{K_I + 4/3\,\mu_I}{K_R + 4/3\,\mu_R}.$$

For shear waves,

$$V_s = \left[\frac{\mu_R}{\rho} \right]^{1/2},$$

and   (A-6)

$$Q_s^{-1} = \frac{\mu_I}{\mu_R}.$$

To determine the imaginary part of the moduli, it is necessary to rely on observation and to follow an empirical approach. The magnitudes of K_I and μ_I should be chosen in each case to match observed Q values at appropriate conditions.

APPENDIX B
FLUID FLOW FROM CRACKS
FORMULATION AND ESTIMATION OF THE RELAXATION TIME

Flow will take place between thin cracks with aspect ratio $\alpha_m \approx 0$ and pores with $\alpha_m \approx 1$ due to a differential volume change induced by the stress wave. The fluid pressures and volume changes are given by

$$P_0 = -K'\theta_0, \qquad\qquad P_1 = -K'\theta_1,$$

and   (B-1)

$$dC_0 = C_0\theta_0,\ \alpha_m \approx 0, \qquad dC_1 = C_1\theta_1,\ \alpha_m \approx 1,$$

where C is the volume concentration of cracks (subscript 0) or pores (subscript 1) and θ is the dilatation. The pressure difference is $\Delta P = P_0 - P_1$. Letting the equalized pressure after flow be $\overline{P}$, then the corresponding dilatation in both the crack and pore is $\overline{\theta} = -\overline{P}/K'$. The total liquid volume displaced in order to equalize the pressure is given by

$$q_T = d\overline{C}_0 - dC_0 = dC_1 - d\overline{C}_1, \quad (B-2)$$

where $d\overline{C}_0 = C_0\overline{\theta}$ and $d\overline{C}_1 = C_1\overline{\theta}.$ Solving for $\overline{\theta}$, we obtain

$$\overline{\theta} = \frac{\varepsilon\theta_0 + \theta_1}{1 + \varepsilon}, \qquad (B-3)$$

where $\varepsilon = C_0/C_1$, or the volumetric ratio of connected cracks to pores. Furthermore,

$$q_T = C_0 \left[\frac{\theta_1 - \theta_0}{1 + \varepsilon} \right]. \qquad (B-4)$$

The instantaneous flow between two parallel plates (crack surfaces) separated by distance h is given by

$$q = \frac{h^2 A}{3\eta} \frac{dP}{dx}, \qquad (B-5)$$

where A now becomes the cross-sectional area of the crack and is equal to $\pi h^2/\alpha_m$ or πhl. If we let $dx = 2l$ (crack length), then from equations (B-1) and (B-5)

$$q = \frac{\pi h^3}{6\eta} K'(\theta_1 - \theta_0). \qquad (B-6)$$

Assuming a relaxation of the form

$$q_T = q \int_0^\infty e^{-t/\tau} dt = q\tau \qquad (B-7)$$

where τ is the relaxation time, we obtain

$$\tau = \frac{C_0(\theta_1 - \theta_0)/(1 + \varepsilon)}{\pi h^3 K'(\theta_1 - \theta_0)/6\eta}. \qquad (B-8)$$

Since the volume of the crack, $C_0 = 4\pi h^3/3\alpha_m^2$,

$$\tau = 8\eta/\alpha_m^2 K'(1 + \varepsilon). \qquad (B-9)$$

Viscoelastic formulation

We will now show that by using the correspondence principle for both the shear and bulk moduli of the fluid phase, the equations for the effective moduli may be written in terms of real and imaginary parts and two characteristic frequencies. Rewriting equation (A–1) for the effective bulk modulus by letting $\delta'' = (K'/K - 1) \, T_{iijj}/3$ (dropping the summation over aspect ratios), we obtain:

$$K^* = \frac{K + \dfrac{4CK\mu\delta''}{3K + 4\mu}}{1 - \dfrac{3CK\delta}{3K + 4\mu}}$$

$$= K\left[\frac{1 + 4\mu C\delta'}{1 - 3KC\delta'}\right], \qquad (B-10)$$

where $\delta' = \delta''/(3K + 4\mu)$. Letting δ' be complex, that is, $\delta' = a + ib$, then

$$K^* = K\left[\frac{1 + 4\mu C(a + ib)}{1 - 3KC(a + ib)}\right] = K_R^* + i K_I^*,$$

where

$$K_R^*$$
$$= K\left[\frac{(1 + 4\mu Ca)(1 - 3KCa) - 12K\mu C^2 b^2}{(1 - 3KCa)^2 + (3KCb)^2}\right],$$

$$(B-11)$$

and

$$K_I^*$$
$$= bK\left[\frac{4\mu C(1 - 3KCa) + 3KC(1 + 4\mu Ca)}{(1 - 3KCa)^2 + (3KCb)^2}\right].$$

Applying the correspondence principle, we let $K' = K_R' + i\omega g$ and $\mu' = i\omega\eta$, where η is the viscosity and g is considered an unknown to be determined from the relaxation time for flow. We now show that the equations for the effective moduli can be written in terms of two characteristic frequencies and that the real and imaginary parts of δ' are uniquely determined. For small aspect ratios

$$\delta' = \frac{1}{3K + 4\mu}\left(\frac{K'}{K} - 1\right)\frac{3K + 4\mu'}{3K' + 4\mu' + K_1} \qquad (B-12)$$

[Toksöz et al, 1976, equation (C–4)], where $K_1 = 3\pi\alpha_m\mu(3K + 4\mu)/(3K + 4\mu)$. Substituting the complex K' and μ', we obtain, after some algebra

$$\delta' = \frac{1}{3K + 4\mu}\left(K'' - 1 + \frac{i\omega}{\omega_c}\right) \cdot$$

$$\cdot \frac{1 + i\omega/\omega_d}{(K'' + K_2) + i\omega(1/\omega_c + 1/\omega_d)}, \qquad (B-13)$$

where $K'' = K_R'/K$, $K_2 = K_1/3K$ with $\omega_c = K/g$ and $\omega_d = 3K/4\eta$. ω_d is recognized as the characteristic frequency for viscous relaxation in isolated cracks (Walsh, 1969), and ω_c is the characteristic frequency for fluid flow from cracks. Finally, it can be shown that the real and imaginary parts of $\delta' = a + ib$ may be written as:

$$a = \frac{1}{A}\left[(K'' - 1)\left(K'' + K_2 + \frac{\omega^2}{\omega_d} \cdot\right.\right.$$

$$\cdot\left.\left(\frac{1}{\omega_c} + \frac{1}{\omega_d}\right)\right) - \frac{\omega^2}{\omega_c}\left(\frac{K'' + K_2}{\omega_d}\right.$$

$$\left.\left. - \frac{1}{\omega_c} - \frac{1}{\omega_d}\right)\right],$$

$$(B-14)$$

and

$$b = \frac{1}{A}\left[\omega\left(\frac{K'' + K_2}{\omega_d} - \frac{1}{\omega_c} - \frac{1}{\omega_d}\right) \cdot\right.$$

$$\cdot (K'' - 1) + \frac{\omega}{\omega_c}(K'' + K_2)$$

$$\left. + \frac{\omega^3}{\omega_c\omega_d}\left(\frac{1}{\omega_c} + \frac{1}{\omega_d}\right)\right],$$

where $A = [(K'' + K_2)^2 + \omega^2(1/\omega_c + 1/\omega_d)^2]/(3K + 4\mu)$. The equivalent result is obtained for the effective shear modulus.

From equation (B–9), we have

$$\frac{1}{\omega_c} = \frac{8\eta}{\alpha_m^2 K_R'(1 + \varepsilon)}, \qquad (B-15)$$

so that

$$g = \frac{8\eta}{\alpha_m^2(1 + \varepsilon)}\frac{K}{K_R'}. \qquad (B-16)$$

Chapter 4
Field measurements of attenuation

Measurement of seismic attenuation in the field is an important but difficult task. Amplitudes of seismic waves decay with distance from the source because of geometric spreading of the wavefront, reflection and scattering of waves from boundaries, and anelastic attenuation. With few exceptions, the effects of spreading and reflection on amplitudes are larger than are those of intrinsic damping. Thus, the successful field strategy for attenuation measurements requires either correcting for these other effects or eliminating them by using an appropriate geometry.

An important factor in attenuation measurements is that for most earth materials, attenuation increases with frequency. Thus, high frequencies decay more rapidly than low frequencies owing to intrinsic damping, while the geometric spreading factor is the same for all frequencies. As a result, spectral ratios of seismic waves at different distances have been used most frequently for obtaining attenuation values. For this method to work successfully, however, it is desirable for the wave to travel one wavelength or longer in a formation. This requires either that the measurements be done with short wavelengths or that the sections be thick and homogeneous. In practice, ideal conditions are not always found and methods must be devised to circumvent problems. These methods are described in the individual papers in this chapter.

The in-situ attenuation properties of the sedimentary layers and the crystalline basement rocks have been measured using a variety of different techniques. For sedimentary sections, the most successful method has been direct transmission from a repeatable near-surface source to a series of receivers at different depths in boreholes (vertical seismic profiling, or VSP). Spectral ratios of direct source-to-receiver pulses are then usually used to obtain the attenuation. *McDonal et al* (*1958*) applied this method successfully to the Pierre shale in Colorado. As discussed in the first paper in this chapter, they found in a homogeneous section of the shale that for vertically traveling *P*-waves, Q_p was constant in the frequency range of 50 to 450 Hz, and the average value for attenuation was $0.12 \cdot f$ dB per 1000 ft ($Q_p = 32$). For horizontally traveling shear waves in the frequency range of 20 to 125 Hz, they found again a nearly constant Q, and the attenuation was $1.0 \cdot f$ dB per 1000 ft ($Q_s = 10$).

Tullos and Reid (*1969*) carried out similar measurements (as discussed in the second paper of this chapter) in the Gulf Coast sedimentary section near Houston. They also found Q_p to be constant with respect to frequency in clayey sands and sand/clay lithologies.

More recently, Hauge (1981) presented a detailed VSP attenuation study for several Gulf Coast wells. Ganley and Kanasewich (1980) have used synthetic seismograms based on sonic and density logs to correct for frequency-dependent losses caused by reflections and transmissions. And, in a similar manner, Spencer et al (1979) have utilized model studies to determine the effects of receiver spacing, signal-to-noise (S/N) ratios, interference of up- and down-going waves at the receivers, cyclic zones within rock layers, and intrinsic attenuation on measured VSP apparent attenuation obtained from spectral ratios.

In-situ measurement of shear-wave attenuation is becoming important with the rapidly expanding shear-wave exploration program. These measurements can be done, in a manner similar to *P*-wave determinations, using a surface shear-wave source and three-component borehole seismometers.

The use of full waveform sonic logging tools will likely provide a useful method for obtaining in-situ attenuation values. Possible techniques may involve spectral ratios of direct *P*-wave arrivals, some measure of waveform change, and the use of borehole surface waves to obtain an indirect estimate of *S*-wave attenuation. Although little has been published on these topics, extensive research efforts are now underway in industry and universities.

Compressional wave attenuation in shallow marine sediments is described by *Hamilton* (*1972*) in the third paper. The sediment types ranged from coarse sand to clayey silts, and the frequencies ranged from 3.5 to 14 kHz. To these, data from other measurements at higher frequencies were added (Hamilton et al, 1970). It was found that the attenuation coefficient was approximately dependent upon the first power of frequency, suggesting a constant Q_p in the 3.5 to 100 kHz range. These measurements, although taken at higher frequencies than those normally used in seismic exploration, are especially important for offshore high-resolution seismic studies.

Theoretically, attenuation can be measured from the spectra of reflected waves, and "interval attenuation" can be obtained in addition to the interval velocities. In practice, however, this is complicated by the small amplitude of reflections and the interference from multiples and deeper reflections. In the fourth paper of this chapter, *Schoenberger and Levin* (*1974*) discuss this problem using synthetic data. Their results are encouraging, and multiples do not appear to mask the effects of intrinsic attenuation significantly.

With the increased use of surface sources (vibrators, weightdrop, etc.) for seismic wave generation, understanding wave attenuation in the soil and weathered layer has become important. Most measurements in soils have been carried out for engineering purposes with the emphasis on shear waves. In the fifth paper (*Kudo and Shima, 1970*), one set of such measurements is discussed. In soils and unconsolidated materials, the strain-amplitude dependence of attenuation is strong (see chapter 2 for strain dependence). Thus, for estimating losses in unconsolidated weathered layer materials, the strain amplitudes and saturation conditions must be considered.

For measurements of attenuation in the deeper crust of the earth, both seismic surface waves and body waves are used. Velocity gradients as well as intrinsic attenuation affect the amplitudes of body waves. In the last paper of this chapter, *Braile* (*1977*) discusses this problem and presents a method of obtaining the attenuation. When one deals with individual pulses propagating through the crust, it is found that pulse width as well as absolute amplitude depend upon Q. The pulse broadening caused by intrinsic attenuation is the topic of chapter 5, where this subject is discussed in several papers. Pulse width broadening has been used to determine Q in massive rock studies (Gladwin and Stacey, 1974).

The measurement of attenuation by surface waves, although widely used for studies of the crust and deeper interior of the earth, has limited application to exploration problems. For this reason, papers addressing this topic are not included in this compilation, although some key references are Knopoff et al (1964), Ben-Menahem (1965), Anderson and Hart (1978), and Lee and Solomon (1979).

REFERENCES

Anderson, D.L., and Hart, R.S., 1978, *Q* of the earth: J. Geophys. Res., v. 83, p. 5869–5882.

Ben-Menahem, A., 1965, Observed attenuation and *Q* values of seismic surface waves in the upper mantle: J. Geophys. Res., v. 70, p. 4641–4651.

Braile, L.W., 1977, Interpretation of crustal velocity gradients and *Q* structure using amplitude-corrected refraction profiles, *in* The earth's crust: AGU Geophys. monogr. 20, p. 427–439.

Ganley, D.C., and Kanasewich, E.R., 1980, Measurement of absorption and dispersion from check shot surveys: J. Geophys. Res., v. 85, p. 5219–5226.

Gladwin, M.T., and Stacey, F.D., 1974, Anelastic degradation of acoustic pulses in rock: Phys. Earth Planet. Int., v. 8, p. 332–336.

Hamilton, E., 1972, Compressional wave attenuation in marine sediments: Geophysics, v. 37, p. 620–646.

Hamilton, E.L., Bucker, H.P., Keir, D.L., and Whitney, J.A., 1970, Velocities of compressional and shear waves in marine sediments determined from a research submersible: J. Geophys. Res., v. 75, p. 4039–4049.

Hauge, P.S., 1981, Measurements of attenuation from vertical seismic profiles: Geophysics, in press.

Knopoff, L., Aki, K., Archambeau, C., Ben-Menahem, A., and Hudson, J.A., 1964, Attenuation of dispersed waves: J. Geophys. Res., v. 69, p. 1655–1657.

Kudo, K., and Shima, E., 1970, Attenuation of shear waves in soil: Bull. Earth Res. Inst., v. 48, p. 145–158.

Lee, W.B., and Solomon, S.C., 1979, Simultaneous inversion of surface wave phase velocity and attenuation: Rayleigh and Love waves over continental and oceanic paths: SSA Bull., v. 69, p. 65–95.

McDonal, F.J., Angona, F.A., Mills, R.L., Sengbush, R.L., Van Nostrand, R.G., and White, J.E., 1958, Attenuation of shear and compressional waves in Pierre shale: Geophysics, v. 23, p. 421–439.

Schoenberger, M., and Levin, F.K., 1974, Apparent attenuation due to intrabed multiples: Geophysics, v. 39, p. 278–291.

Spencer, T.W., Sonad, J.R., and Butler, T.M., 1979, Seismic *Q*-stratigraphy or dissipation: Presented at the 49th Annual International SEG Meeting, November 5, in New Orleans.

Tullos, F.N., and Reid, A.C., 1969, Seismic attenuation of Gulf Coast sediments: Geophysics, v. 34, p. 516–528.

Reprinted from Geophysics, v. 23, p. 421–439.

ATTENUATION OF SHEAR AND COMPRESSIONAL
WAVES IN PIERRE SHALE*

F. J. McDONAL,† F. A. ANGONA,† R. L. MILLS,‡ R. L. SENGBUSH,†
R. G. VAN NOSTRAND,§ AND J. E. WHITE¶

ABSTRACT

Attenuation measurements were made near Limon, Colorado, where the Pierre shale is unusually uniform from depths of less than 100 ft to approximately 4,000 ft. Particle velocity wave forms were measured at distances up to 750 ft from explosive and mechanical sources. Explosives gave a well-defined compressional pulse which was observed along vertical and horizontal travel paths. A weight dropped on the bottom of a borehole gave a horizontally-traveling shear wave with vertical particle motion. In each case, signals from three-component clusters of geophones rigidly clamped in boreholes were amplified by a calibrated, wide-band system and recorded oscillographically. The frequency content of each wave form was obtained by Fourier analysis, and attenuation as a function of frequency was computed from these spectra.

For vertically-traveling compressional waves, an average of 6 determinations over the frequency range of 50–450 cps gives $\alpha = 0.12$ f. For horizontally-traveling shear waves with vertical motion in the frequency range 20–125 cps, the results are expressed by $\alpha = 1.0$ f. In each case attenuation is expressed in decibels per 1,000 ft of travel and f is frequency in cps. These measurements indicate, therefore, that the Pierre shale does not behave as a visco-elastic material.

INTRODUCTION

The introduction of high-resolution seismic equipment, acoustic logging, seismic models, and the study of cores by high frequency acoustical methods has emphasized the need for a basic understanding of the propagation of sound through earth and seismic model materials. Even for the normal seismic band (20–100 cps) attenuation, and possibly dispersion, are important in unconsolidated materials. Attenuation of the higher frequencies is very evident in high-resolution recording and interval-velocity measurements. Furthermore, the common practice of applying high-frequency velocity determinations to the seismic band below 100 cps is inaccurate if dispersion is present.

Ricker (1953) has developed a theory of wavelet propagation based on the assumption of (1) an impulse as the initial displacement at the source and (2) a

* Presented at the 25th Annual Meeting of the Society at Denver on October 6, 1955. Manuscript received by Editor October 3, 1957.

† Magnolia Petroleum Company, Field Research Laboratory, Dallas, Texas.

‡ Summers and Mills, Inc., Dallas, Texas.

§ Société de Prospection et Exploitations Pétrolières en Alsace, Paris, France.

¶ Ohio Oil Company, Littleton, Colorado.

visco-elastic medium. This theory states that the breadth of a wavelet is proportional to the square root of the travel time of its center and that the amplitude of the particle velocity of the wavelet decreases as the five-halves power of travel time. This theory further states that, for the low frequency region of normal seismic work, attenuation is proportional to the square of frequency and velocity is constant. Above a specified transition frequency both the velocity of propagation and attenuation increase as the square root of frequency. Ricker (1953) also made an experimental study of the propagation of seismic wavelets in the Pierre shale near Limon, Colorado. No direct study of attenuation was made, but his measurements of wavelet breadth and wavelet amplitude are in reasonable agreement with his theoretical work.

Born (1941) conducted laboratory experiments with bars of shale, limestone, sandstone, and caprock, and he found that attenuation was proportional to the first power of frequency. The addition of small quantities of water, however, led to attenuation intermediate between the first and second powers of frequency. More recently, Collins and Lee (1956) have reported experiments in a quarry sandstone which indicate attenuation proportional to the first power of frequency, and they present theoretical considerations which indicate that the attenuation exponent may be arbitrarily close to unity without assuming a non-linear system. Knopoff (1956) has extended Zener's (1938, 1948) work on metals to earth materials, and suggested solid friction as a mechanism of attenuation. This leads to attenuation proportional to the first power of frequency and velocity independent of frequency.

ATTENUATION MEASUREMENTS

The measurements reported in this paper were undertaken to provide accurate data on attenuation of earth materials in place. Since the magnitude of attenuation in the normal seismic band is small, a thick section of uniform material is desirable. This requirement is met by the Pierre shale (Upper Cretaceous) in Eastern Colorado. In this area the Pierre shale is approximately 4,000 ft thick and exceptionally uniform. An acoustic velocity log of one of the test holes (Figure 1), however, does show minor variations in the interval between 250 and 800 ft. The largest break at 740 ft represents a reflection coefficient of only 0.04. Other variations do not exceed 0.03. On this basis multiple reflections should not produce significant interference. The observed velocity variations, however, are important for horizontal paths. Adjacent layers of slightly different velocities lead to the interference of overlapping signals. The result is an apparent increase in wavelet breadth and, in some cases, severely distorted wave forms.

Attenuation of vertically traveling compressional waves was determined from signals recorded by 5 detectors positioned in a cluster of boreholes (Figure 2) such that the detectors and source formed a straight line approximately vertical. Booster and one-pound charges were shot at depths of 250 to 300 ft. Detectors were placed at depths of 350, 450, 650, and 750 ft.

Since velocity measurements in these and adjacent boreholes indicated a uniform material, a correction for spherical spreading was made and attenuation was computed directly from a Fourier analysis of these signals. Typical records are shown in Figures 3 and 6. The corresponding Fourier analyses and attenuation values are shown in Figures 4, 5, 7, and 8. The average attenuation for 6

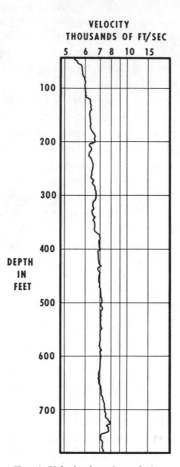

FIG. 1. Velocity log of test hole.

separate shots is given by Figure 9. The relation $\alpha = 0.12$ f fits the data in an acceptable manner, but the best fit of the data (by the least squares method) yields the expression, $\alpha = 0.065$ $f^{1.1}$ where α is the attenuation in db per 1,000 ft, and f is the frequency in cps.

Unfortunately we were not able to generate satisfactory vertically traveling shear waves, but we did produce acceptable horizontally traveling shear waves by dropping a 200-lb weight on the bottom of a 500-ft borehole. Shear signals were recorded by three component geophone clamps placed at this same depth in a line of boreholes at horizontal distances of 79, 142, 273, 321, 370, and 630 ft. Typical records for shear waves are shown in Figure 10. It is obvious from inspection that wavelet broadening and attenuation of the higher frequencies is much more pronounced than for compressional waves. Attenuation of shear waves is approximated by the relation $\alpha = 1.0$ f db per 1,000 ft (Figure 12).

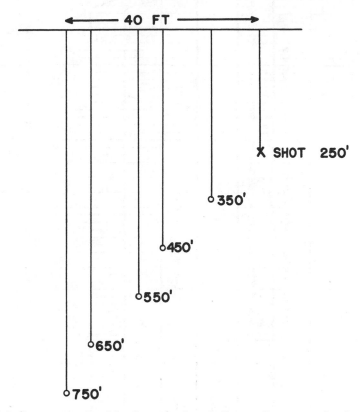

FIG. 2. Geometry employed for the study of vertically traveling compressional waves.

For vertical travel the average shear and compressional velocities were 2,630 and 7,100 ft/sec, respectively, for the depth interval of 450 to 550 ft. For horizontal travel at a depth of 500 ft the corresponding velocities were 2,680 and 7,360 ft/sec.

DISCUSSION OF RESULTS

The attenuation measurements reported in this paper disagree with the previous work of Ricker (1953) in several respects. Our measurements indicate that attenuation of vertically traveling compressional waves is proportional to the first power of frequency, the decay of wavelet amplitude as a function of travel

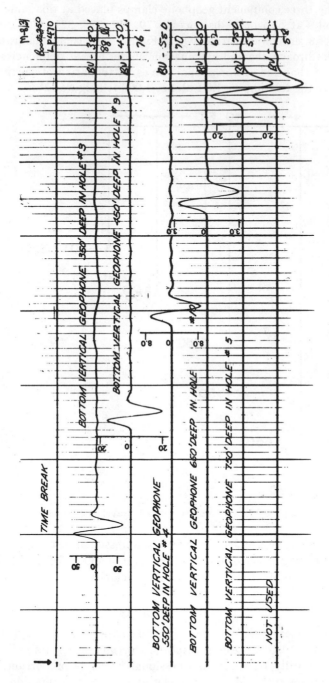

FIG. 3. Vertically traveling compressional waves generated by a booster charge at a depth of 250 ft. Record M. (Amplitudes are expressed in units of cm/sec$\times 10^{-3}$.)

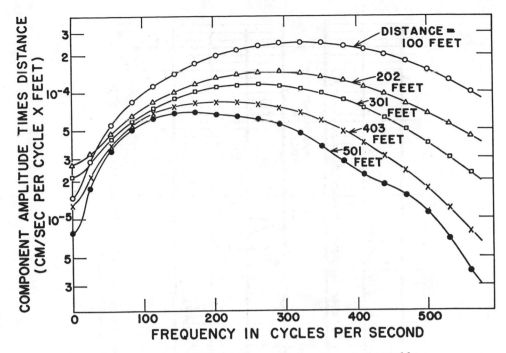

Fig. 4. Spectra of vertically traveling compressional waves generated by a booster charge at a depth of 250 ft. Record M.

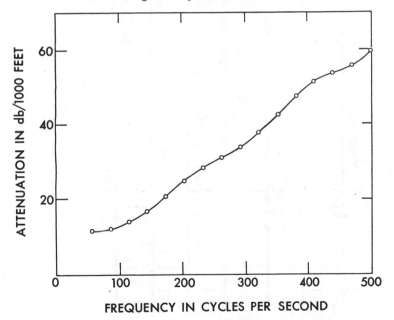

Fig. 5. Attenuation of vertically traveling compressional waves generated by a booster charge at a depth of 250 ft. Record M.

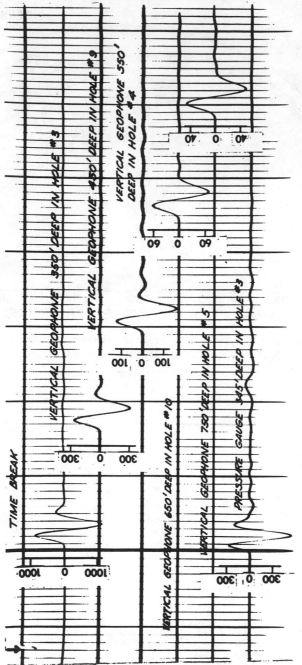

Fig. 6. Vertically traveling compressional waves generated by a charge of one pound of dynamite at a depth of 260 ft. Record T. (Amplitudes are expressed in units of $cm/sec \times 10^{-3}$ and $dynes/cm^2 \times 10^3$, respectively.)

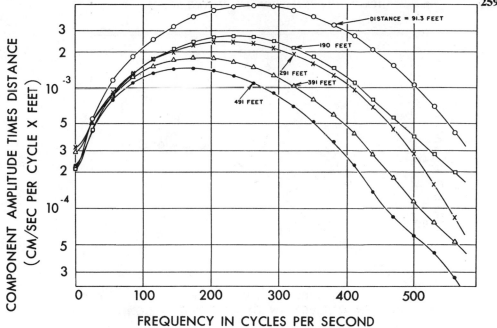

FIG. 7. Spectra of vertically traveling compressional waves generated by a charge of one pound of dynamite at a depth of 260 ft. Record T.

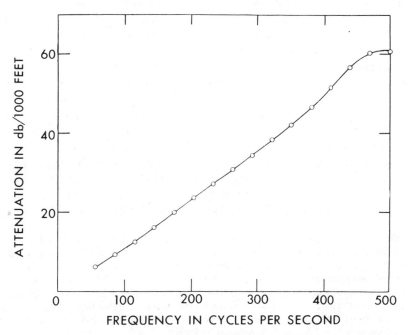

FIG. 8. Attenuation of vertically traveling compressional waves generated by a charge of one pound of dynamite at a depth of 260 ft. Record T.

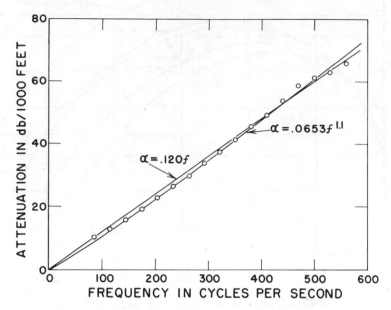

Fig. 9. Attenuation of vertically traveling compressional waves. This is the average attenuation for 6 records, including those of Figs. 5 and 8.

time is less rapid than Ricker's work indicates, and wavelet broadening is less.[1] The observed wave forms more nearly resemble a single cycle of a sine wave than they do Ricker wavelets. In summary, our measurements indicate that the Pierre shale does NOT behave as a visco-elastic material.

While unconsolidated material of the weathered zone may exhibit visco-elastic properties, the absence of dispersion further indicates that consolidated formations do not. Velocity determinations by acoustic velocity logs operating in the frequency range of 10 to 20 kc agree with geophone surveys to within a few percent. Since the two-receiver acoustic velocity log provides an absolute measurement of velocity determined by the distance and travel time between receivers, these measurements clearly indicate that no significant dispersion exists below 10 kc. For example, thick sections of Pennsylvanian shales and sands in Oklahoma (Magnolia No. 1 Bessie Weaver, McClain Co., Oklahoma), Gulf Coast sands and shales (Magnolia No. A-1X Louisiana State Lease 2922), and

---

[1] The authors consider a measurement of the attenuation more fundamental than a study of wavelet amplitude and breadth. We, therefore, have not emphasized wavelet amplitude and breadth. It is interesting to note, however, that for 6 records (same data used to compute results of Figure 9) the wavelet amplitude decreased as the -1.98, -1.99, -2.04, -2.15, -2.16, and -2.26 powers of travel time. These same records indicate that wavelet broadening can be satisfactorily represented by a relation of the form $b = a^{nt}$, where a and n are constants, t is travel time, and b is wavelet breadth. If one attempts to fit the data to the expression $b = t^n$, the exponent is approximately 0.25. These figures compare with -2.50 and 0.50 reported by Ricker.

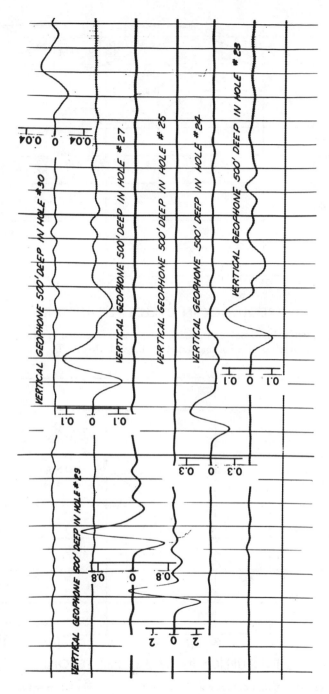

FIG. 10. Horizontally traveling shear waves generated by dropping a weight on the bottom of a 500-ft borehole. Record D. (Amplitudes are expressed in units of cm/sec$\times 10^{-3}$.)

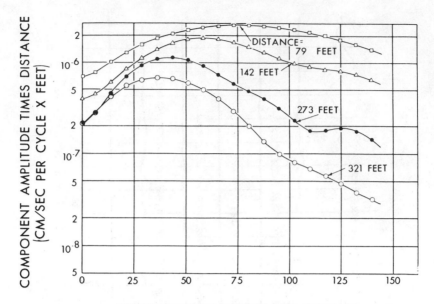

FrequEncy in Cycles Per Second

Fig. 11. Spectra of horizontally traveling shear waves generated by dropping
a weight on the bottom of a 500-ft borehole. Record D.

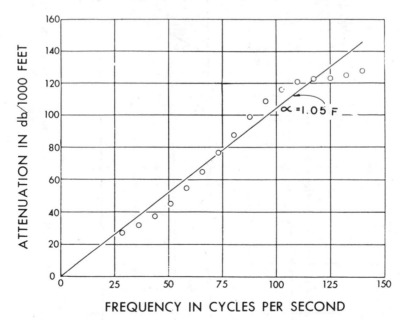

FrequEncy in Cycles Per Second

Fig. 12. Attenuation of horizontally traveling shear waves generated by
dropping a weight on the bottom of a 500-ft borehole. Record D.

thick carbonate and evaporite sections of West Texas (Magnolia No. 1 Hefflefinger, Roosevelt County, New Mexico) have been logged using Magnolia's two-receiver velocity logger. These wells check velocities obtained from conventional geophone surveys within one percent. (A single receiver type logging instrument was used to obtain the velocity log of Figure 1. Since this is not an absolute measurement the instrument is usually calibrated by a conventional geophone survey.)

An analysis of velocity measurements on cores under appropriate conditions of temperature and pressure is beyond the scope of this paper. It can be stated, however, that no significant dispersion has been reported for frequencies as high as one megacycle (Wyllie, Gregory, and Gardner, 1956).

The absence of dispersion is significant. In the broad field of physics it is indeed rare to find a transmission system which exhibits amplitude discrimination without the corresponding phase distortion. We believe that consolidated earth materials do just this, at least for the low frequency range employed in geophysical prospecting. In the preparation of synthetic seismograms and in the study of reflection wave shapes, the effects of attenuation can be represented by a filter which attenuates high frequencies and produces a constant time delay (no phase distortion) at all frequencies.

The absolute value of attenuation quoted in this paper applies to the Pierre shale to a depth of 1,000 ft. By estimating the reflection coefficient of the base of the Pierre shale on the basis of the velocity logs in the area, one may compute from reflection data the attenuation of the entire 4,000-ft section of Pierre shale. In this manner we have estimated the attenuation of the entire section of 4,000 ft of the Pierre shale as 2.0 ± 1.0 db per 1,000 ft at 50 cycles. This compares with 6.0 db per 1,000 ft for the depth interval of 250 to 750 ft (Figure 9). Both sets of measurements may be valid if we assume that attenuation decreases with depth.

APPENDIX

Requirements for Accurate Attenuation Measurements

A study of the changes of wave shape of a seismic wavelet as it propagates through the earth does require extreme care. This is equally true for measurements of peak amplitude, wavelet breadth, and attenuation. Filtering by the recording instruments, poor coupling of the detector, and inhomogeneities of the earth contribute to wavelet broadening. We believe the following requirements are necessary for an accurate study of attenuation and wavelet propagation.

1. A thick section of uniform material is needed. In the case of the Pierre shale a thickness of approximately 500 ft is required.
2. Attenuation determinations should be made only for signals produced by the *SAME SHOT*. "One cannot shoot a second time in the same hole because the same hole is not there anymore."[2] Even small charges and caps

---

[2] Quotation attributed to Dr. J. P. Woods.

do not accurately reproduce pulse amplitude and wave form on successive shots. The practice (Ricker, 1953) of increasing the shot-detector distance by moving the shot upward to shallower depths is very misleading. In general, shallow shots produce lower frequency signals, and therefore give the false impression of excessive high frequency attenuation and wavelet broadening.

3. Two or more detectors should be positioned in a straight line path from the shot to minimize the effect of inhomogeneities. A vertical travel path minimizes the effects of anisotropy.

4. The detectors should have uniform response to earth motion over the entire frequency range of the attenuation study. If geophone clamps are employed, one must avoid mechanical resonance. The mass of the clamp and stiffness of the coupling to the borehole wall determine the primary resonant frequency of the system. In general, it is necessary to keep the mass of the clamp small (less than 10 lb) and the pressure on the borehole wall high (greater than 200 psi). No geophone clamp can be considered reliable until its response, clamped in place, has been determined. The use of pressure gauges is questionable. Coupling of an acoustic signal into the borehole fluid depends upon local variations of the rigidity of the borehole wall.

5. A wideband recording system with calibrated and stabilized gain is required. The impulse response or the amplitude and phase response should be measured.

Recording System

To a first approximation the earth and a geophone may be treated as a low-pass resonant system which should exhibit flat response in the frequency region below the resonant peak. In order to insure that the resonant peak occurs at a frequency well above that of the normal recording band, it is necessary to employ a geophone clamp of low mass and rigid clamping to the borehole wall. In general, geophone clamps which depend only upon gravity for the clamping force do not meet this requirement.

The geophone clamps employed in this work (Figure 13) consist of two cylindrical half shells which are expanded by a hydraulic mechanism to provide the necessary clamping action. In one half shell four miniature geophones are imbedded in plastic which is mechanically isolated from the second shell by teflon gaskets. The hydraulic clamping action is obtained by releasing nitrogen gas under 2,000 psi from the cylinder positioned above the clamp. This actuates the piston between the cylindrical half shells which then forces the half shells against the opposite sides of the borehole. The release of the system is obtained by breaking one of the copper tubing pressure lines which is attached to the strain member of electrical cable. Of the four geophones mounted in the clamp, three serve to measure the vertical and horizontal components of motion while the fourth is

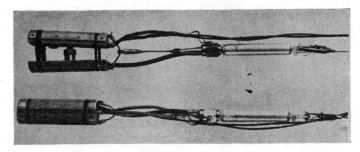

FIG. 13. Geophone clamp.

employed as a driver unit to test the effectiveness of the geophone clamp. This test was made by observing the output of one vertical geophone while a second geophone was being driven at a constant current over the frequency range of interest.

For the Pierre shale, the major resonant peak varies from 1,200 to 2,000 cps, but for any one geophone clamp position the response characteristic was constant. Furthermore, by comparing this response with that of the geophone clamp assembly hanging freely on long rubber bands (Figure 14), one may estimate the response of the system in a manner similar to that employed by Washburn and Wyllie (1941). On this basis, the high frequency response of the clamp was down approximately 10 percent at 450 cps. Since the response at low frequencies is limited only by the response of the geophones themselves, the overall band of the geophone clamp is then accurate to within 10 percent from 20 cycles to 450 cycles.

The pressure gauges used in this work show excellent response to pressure variations over the frequency range from 5 cps to 20 kc. However, it should be recognized that the acoustic coupling from the formation to the borehole fluid depends upon local variations of elastic properties in a borehole wall. The wave forms observed with pressure gauges (Figure 6) do frequently show additional legs of appreciable amplitude which are not observed on the corresponding signals measured with geophone clamps. We, therefore, believe that geophone clamps are more reliable, and we have used them in most of our work.

The 6-channel, high frequency recording truck employed combines broadband linear amplifiers and a high speed, drum-type oscillographic recorder to provide a band width of 5 to 2,000 cps. The amplifiers consist of an input cathode follower, three voltage amplifier plug-in units, and a power output stage. Stable gain and flat frequency response is achieved by employing 20 db negative feedback for each amplifier unit. This unit type construction also permits the use of three independent interstage plug-in filters which are unaffected by the amplifier feedback networks. With the filters not used, the amplifier response is flat within three percent from 3 to 20,000 cps. Gain calibration is achieved by employing precision attenuators in conjunction with a calibration oscillator, the output of which is monitored at a high output level by a diode-type AC voltmeter.

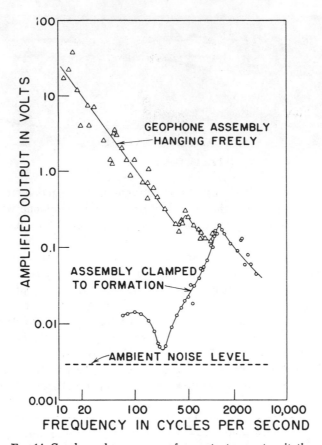

FIG. 14. Geophone clamp response for constant current excitation.

A modified Hathaway S14-C oscillographic recorder with a drum-type film magazine provides recording speeds up to 400 inches per second. Timing lines are reproduced at millisecond intervals. The galvanometers have an undamped natural frequency of 5,500 cps, but fluid damping employed by the manufacturer gives a response which is down 5 percent at 2,000 cps.

The geophone clamps described above limit the frequency response of the complete recording system to the band of 20 to 450 cps. However, with the barium titanate pressure gauges the system is limited by the amplifier-recorder combination to the band of 5 to 2,000 cps.

Fourier analyses were carried out by a digital computer from the digitized records. The field records were enlarged to eight times normal, and amplitudes were measured at 0.25 millisecond intervals.

A test of the uniformity of the 5 channels was conducted by placing 5 geophone clamps at the same depth, 310 ft, in 5 boreholes. The signal from a shot at a depth of 700 ft was recorded and Fourier analyses were made for each channel (Figures 15 and 16). Since all boreholes were grouped within a circle of

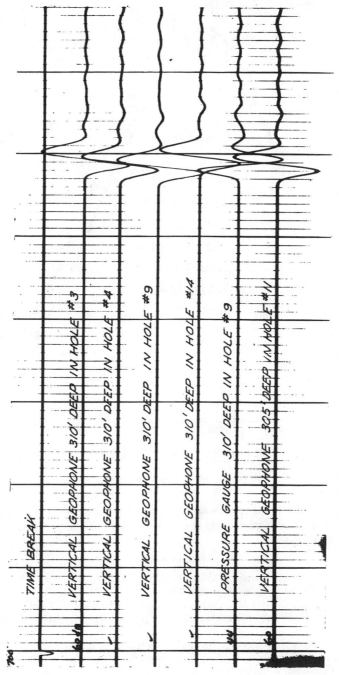

Fig. 15. Geophone huddle test. Source—Booster charge at a depth of 700 ft.

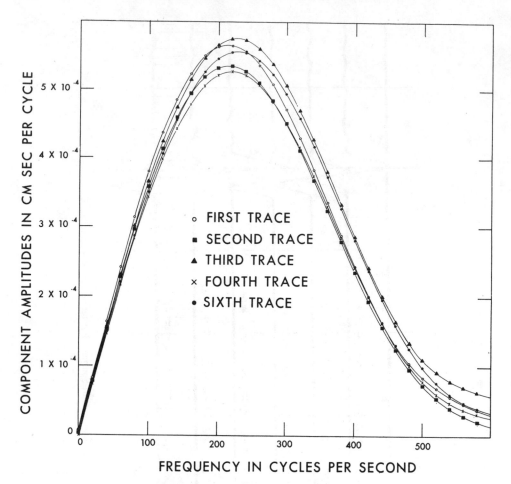

FIG. 16. Spectra of geophone huddle test.

a radius of 15 ft, the travel path was approximately vertical. The degree of similarity of the spectra of Figure 16 is the best measure of the precision of attenuation measurements.

REFERENCES

Birch, Francis, and Bancroft, Dennison, 1938a, Elasticity and internal friction in a long column of granite: Bull. Seis. Soc. Am., v. 28, p. 243–254.
———, ———, 1938b, The effect of pressure on the rigidity of rocks: Jour. Geol., v. 46, p. 59–141.
Birch, Francis, 1942, Handbook of physical constants: Geol. Soc. Amer., Special Papers Number, v. 36, p. 88–92.
Born, W. T., 1941, The attenuation constant of earth materials: Geophysics, v. 6, p. 132–148.
Bruckshaw, J. McG., and Mahanta, P. C., 1954, The variation of the elastic constants of rocks with frequency: Petroleum, v. 17, p. 14–18.

Collins, Francis, and Lee, C. C., 1956, Seismic wave attenuation characteristics: Geophysics, v. 21, p. 16–40.

Duvall, Wilbur I., 1953, Strain-wave shapes in rock near explosions: Geophysics, v. 18, p. 310–323.

Gross, B., 1947, On creep and relaxation: Jour. Appl. Phys., v. 18, p. 213–221.

———, 1948, On creep and relaxation II: Jour. Appl. Phys., v. 19, p. 257–264.

———, 1953, Mathematical structure of the theories of visco-elasticity: Paris, Hermann.

———, 1950, Frictional loss in visco-elastic substances: Jour. Appl. Phys., v. 21, p. 185.

Jeffreys, H., 1931, Damping in bodily seismic waves: Monthly Notices Royal Astronomical Society, Geophys. Suppl., v. 2, p. 318–323.

Knopoff, L., 1956, The attenuation of compression waves in lossy media: Bull. Seis. Soc. Am., v. 46, p. 47–56.

———, 1956, The seismic pulse in materials possessing solid friction, I: plane waves: Bull. Seis. Soc. Am., v. 46, p. 175–183.

———, 1954, On the dissipative visco-elastic constants of higher order: Jour. Acoust. Soc. Am., v. 26, p. 183–186.

Kolsky, H., 1953, Stress waves in solids: Oxford, Oxford University Press.

Lee, E. H., and Kanter, I., 1953, Wave propagation in finite rods of visco-elastic material: Jour. Appl. Phys., v. 24, p. 1115–1122.

Lomnitz, C., 1956, Linear dissipation in solids: Jour. Appl. Phys., v. 28, p. 201.

Mason, W. P., and McSkimmin, H. J., 1947, Attenuation and scattering of high frequency sound waves in metals and glasses: Jour. Acoust. Soc. Am., v. 19, p. 464–473.

Mattice, H. C., and Lieber, Paul, 1954, On attenuation of waves produced in visco-elastic materials: Trans. AGU, v. 35, p. 613–624.

McSkimmin, H. J., 1951, A method for determining the propagation constants of plastics at ultra-sonic frequencies: Jour. Acoust. Soc. Am., v. 23, p. 425–434.

Obert, Leonard, Windes, S. L., and Duvall, Wilbur I., 1946, Standardized tests for determining the physical properties of mine rock: U. S. Bur. Mines R.I. 3891.

Obert, Leonard, and Duvall, Wilbur I., 1949, A gage and recording equipment for measuring dynamic strain in rock: U. S. Bur. Mines R.I. 4581.

———, ———, 1950, Generation and propagation of strain waves in rock—part I: U. S. Bur. Mines R.I. 4683.

Parfitt, G. G., 1949, Energy dissipation in solids at sonic and ultra-sonic frequencies: Nature (London), v. 164, p. 489–490.

Quimby, S. L., 1925, On the experimental determination of the viscosity of vibrating solids: Phys. Rev., v. 25, p. 558–573.

Ricker, Norman, 1943, Further developments in the wavelet theory of seismogram structure: Bull. Seis. Soc. Am., v. 33, p. 197–228.

———, 1944, Wavelet functions and their polynomials: Geophysics, v. 9, p. 314–323.

———, 1951, The form and laws of propagation of seismic wavelets: Proc. Third World Petroleum Congress sec. I, p. 514–536. (Same as Richter (1953).)

———, 1953, The form and laws of propagation of seismic wavelets: Geophysics, v. 18, p. 10–40.

———, and Sorge, W. A., 1951, The primary seismic disturbance in shale: Bull. Seis. Soc. Amer., v. 41, p. 191–204.

Schreuer, E., 1952, The thermal damping of elastic vibrations: Z. Physik, v. 131, p. 619–628.

Sezawa, K., 1927, On the decay of waves in visco-elastic solid bodies: Bull. Earthq. Res. Inst., Tokyo Univ., v. 3, p. 43–53.

Thompson, W. T., 1950, Transmission of elastic waves through a stratified sodium medium: Jour. Appl. Phys., v. 21, p. 89–93.

Van Melle, F. A., 1954, Note on "The primary seismic disturbance in shale," by N. Ricker and W. A. Sorge: Bull., Seis. Soc. Am., v. 44, p. 123–125.

Washburn, Harold, and Wiley, Harold, 1941, The effect of the placement of a seismometer on its response characteristics: Geophysics, v. 6, p. 116–131.

Wegel, R. L., and Walther H., 1935, Internal dissipation in solids for small cyclic strains: Physics, v. 6, p. 141–157.

Wyllie, M. R. J., Gregory, A. R., and Gardner, L. W., 1956, Elastic wave velocities in heterogeneous and porous media: Geophysics, v. 21, p. 41–70.

Zener, Clarence, 1948, Elasticity and anelasticity of metals: Chicago, University of Chicago Press.

——, 1938, Theory of internal friction in metals, II: Phys. Rev., v. 53, p. 90-99.

Reprinted from Geophysics, v. 34, p. 516–528.

SEISMIC ATTENUATION OF GULF COAST SEDIMENTS†

FRANK N. TULLOS* AND ALTON C. REID*

Previous efforts to measure the attenuation of seismic energy as it propagates vertically through the earth have been restricted to thick, isotropic, homogeneous sections of material.

Values have now been obtained from normally layered Gulf Coast sediments. After normalizing with respect to the shot, the spectra of many traces over a small interval in depth can be averaged to minimize reflection interference and to yield a frequency response function associated with a given depth.

Except for the measurements in the low velocity layer (LVL), a cemented array of geophones 1000 ft deep was used. The LVL measurements were made using individually cemented geophones. Data were taken on a digital recorder using a 0.5 ms sampling rate, and the processing was performed on a digital computer.

Attenuation was found to vary exponentially with frequency to the first power in the range from 50 to 400 Hz.

The 1000-ft section of earth, made up largely of clay and sand layers, was separated into four sections at depths of 10, 100, and 500 ft. Attenuation in db/wavelength was found to be 13.1, 0.1508, 0.3641, and 0.2004 in the respective sections.

INTRODUCTION

To the authors' knowledge, the only in-situ measurements of the attenuation of seismic pulses that have been published are those of Ricker (1953) and McDonal et al (1958). They made their measurements in the "homogeneous, isotropic" Pierre shale. The work of McDonal et al, and the laboratory experiments of Birch and Bancroft (1938), Born (1941), Collins and Lee (1956), Peselnick and Zietz (1959), and Auberger and Rinehart (1960) indicated that the attenuation constant for a homogeneous, isotropic material is a linear function of frequency.

The presence of dispersion in the data recorded by McDonal et al has been pointed out by Horton (1959), Futterman (1962), Wuenschel (1965), and Strick (1967). O'Brien (1961) predicted attenuation of 0.1 to 1.0 db/wavelength in sedimentary rocks with dispersion less than one percent between 20 Hz and 20kHz, and hypothesized a nonlinear attenuation mechanism.

Born (1941) hypothesized a solid friction mechanism, and Collins and Lee (1956), attenuation proportional to frequency, with a probably nonlinear loss mechanism.

Since the near-surface earth is usually layered, methods of measuring the seismic wave transmission characteristics of such a medium are badly needed. The experiments described below were designed to measure attenuation and will do little to answer the dispersion question. However, the approach used may suggest an experimental method of measuring dispersion.

In order to make a reliable estimate of the relationship between absorption and frequency, one must first be able to find reliable estimates of the seismic amplitude spectra for points or regions in space. For example, if two inline observations of an outgoing pulse in a homogeneous isotropic medium were corrected for geometric spreading, the ratio of the spectrum observed at the second point to the spectrum observed at the first point

† Presented at the 37th Annual International SEG Meeting in Houston, Texas, October 31, 1967. Manuscript received by the Editor June 14, 1968; revised manuscript received March 14, 1969.
* Esso Production Research Company, Houston, Texas 77001.

271

would describe the decay of the frequency components between the two points.

Certainly, it is true for a layered medium that the spectrum observed at any one point will usually not be the true spectrum for the outgoing pulse. But we assumed that the *average* of spectra for a number of observations will tend toward a mean that will approach the true spectrum, provided individual observations are free of systematic errors and provided statistical independence is achieved. In this work, the average spectral behavior of the seismic pulses was estimated from a large number of observations.

DATA GATHERING

The site for these experiments is located approximately 20 miles south of Houston, in Gulf Coast sediments of Pleistocene age. Blasting caps in shotholes cased with thin-wall tubing were used for seismic sources, and geophones coupled to the earth with cement were used for seismic receivers.

This cement was a light-aggregate type with a density of approximately 0.75 times the density of the surrounding earth and a velocity of approximately 4000 ft/sec. For measurements between the LVL and 1000 ft, the geophones were set in a 6-inch diameter column of this special cement. Near the surface, where the velocity in earth is very low, the geophones were cemented in individual holes with 2-ft lateral and 1-ft vertical spacing. All geophones were carefully calibrated and oriented so that their axes were vertical.

Horizontal and vertical distances between sources and receivers were measured to within 0.5 ft, and the geophone and amplifier calibration errors were less than 0.1 percent. All data were recorded and processed digitally to preserve accuracy.

Figure 1 shows the relative location of the source and receivers as viewed from above. Figure 2 shows source and geophone locations used to measure attenuation between 10 and 1000 ft. The velocity curve on the right side of Figure 2 shows that velocity varied from 1000 ft/sec in the weathered layer to more than 7000 ft/sec at depth.

From the surface to 1000 ft, the section was artificially divided into four subsections, as indicated on Figure 2. Subsection 1, from 500 to 1000 ft, with geophones spaced at 20 ft, was used for exhaustive tests to prove the methods, to evaluate procedures, and to check repeatability

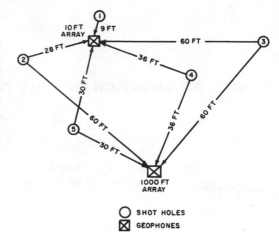

FIG. 1. Relative horizontal location of shotholes and geophones for the experiment.

of the experiments. This section was shot repeatedly over one year with 42 to 50-ft shot depths in two shotholes offset 30 and 36 ft from the array. Four sets of attenuation measurements were made with shots between 48 and 50 ft; the normalization process decribed below was evaluated for shots between 42 and 50 ft.

The interval from 100 to 500 ft was designated subsection 2. Subsection 3 was from 7.5 to 110 ft, and subsection 4 was the top (weathered) layer from one to 10 ft.

METHOD OF ANALYSIS

The concept of transfer functions as used in system analysis was used freely as an analysis tool in this work. Here the attenuating part of the transfer function of an earth section, neglecting phase, is found by introducing a test signal $A_1(f)$ at the input and measuring the output $A_2(f)$. The transfer function is then found by taking the ratio of the output to the input, or $Y_2 = A_2(f)/A_1(f)$. The seismic attenuation problem can be represented by a series of these transfer functions as shown in Figure 3. Here,

$$Y_1(f) = A_1(f)/A_0(f),$$
$$Y_2(f) = A_2(f)/A_1(f),$$
$$Y_3(f) = A_3(f)/A_2(f),$$

and

$$Y_{13}(f) = Y_1(f) \cdot Y_2(f) \cdot Y_3(f) = A_3(f)/A_0(f).$$

This type of analysis would be simple and

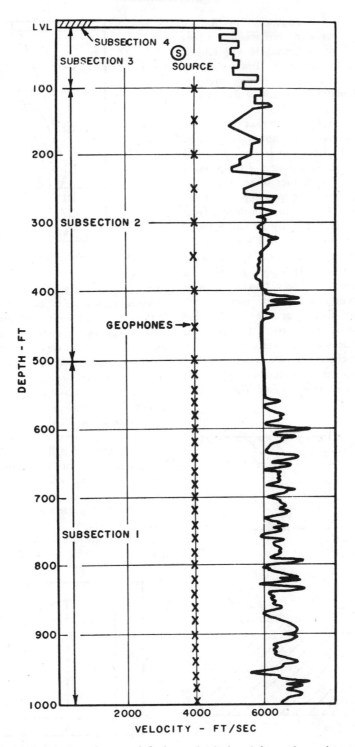

FIG. 2. Vertical location of geophones and the interval velocity of the section as determined from a CVL.

straightforward for a homogeneous medium of infinite extent, or for a layered medium where the layers are thick compared to the longest wavelength of the seismic signal. However, this is seldom the case for the normal seismic prospect which, in general, may have a number of layers per wavelength.

In the normal case, similar to that represented by Figure 4, a receiver does not sense a single pulse moving away from the source—rather it senses a series of pulses composed of the primary or direct pulse and a number of secondary reflected and refracted pulses overlapped in time. One would assume that the amplitude spectrum of a seismic pulse generated by an underground shot in a homogeneous, isotropic, elastic medium would normally be a smooth function because of the mode of generation. However, examination of $A(f)$ for a series of observations of a pulse propagating through a layered medium will show $A(f)$ to be highly variable for different points in space because of constructive and destructive interferences of the overlapping pulses. The amplitude of these spectra can vary between zero and two times the true value of the spectrum amplitude of the outgoing pulse. The extreme

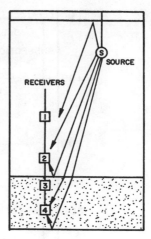

FIG. 4. Raypaths from source to receivers.

case would be for a reflecting interface, near the receiver, with a reflection coefficient of 1.0, that is the condition for total reflection.

In processing the data, the primary (outgoing) pulse of each trace was separated from the outstanding secondary (reflected or refracted) pulses by gating. These gated pulses were corrected with a scale factor made up of calibration constants, geometric corrections, and estimated acoustic impedance values. This was done so the Fourier analysis of the pulses would yield energy-density spectra in units of ft-lb of energy with spreading losses removed. The square root of this energy-density spectrum will be called the amplitude spectrum, or $A(f)$.

The primary pulse was gated from the seismic trace with the graded data window $D(t)$ shown in Figure 5. Here the window length is from t_0 to t_2 with $t_1 - t_0 = 0.25 \ (t_2 - t_0)$. And:

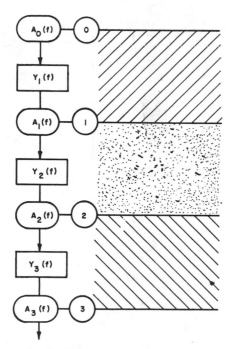

FIG. 3. The transfer function analogy.

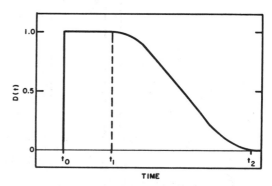

FIG. 5. Gate $D(t)$.

$$D(t) = \begin{cases} 0, & \text{for } t < t_0 \\ 1, & \text{for } t_0 \leq t < t_1 \\ \dfrac{1}{2}\left[1 + \cos\dfrac{(t - t_1)\pi}{t_2 - t_1} \right] \\ & \text{for } t_1 \leq t < t_2, \\ 0, & \text{for } t \geq t_2. \end{cases}$$

Here t is time along the seismic trace, and $D(t)$ is a weighting function applied to the data values of the seismic trace. This weighting function causes the trace to be zero until t_0, the beginning of the primary pulse. Then the trace has normal amplitude from t_0 to t_1. Between t_1 and t_2 the trace is squeezed to zero (and remains at zero for the rest of the trace). This method of gating avoids abruptly terminating the seismic trace at some value away from zero, which could cause fictitious frequency components to appear in the spectral analysis.

Straight-line raypaths and spherical divergence were used to compute the geometric corrections. The amplitudes of the pulses were corrected for the angles between the rays and geophone axis and for spreading. In the worst case for each subsection, the corrections found by this method were compared with corrections found by using Snell's law to compute the raypaths and spreading. Only in the LVL were the correction factors found by the two methods significantly different. Here too the less complicated method was used because the quality of the velocity data did not warrant the extra effort.

If the seismic trace is represented by $f(t)$, this operation is described by

$$f'(t) = D(t) \cdot f(t) \cdot (\text{constant})$$

with $f'(t)$ containing the gated and scaled primary pulse. Although the amplitude spectrum of $f'(t)$ is much smoother than the amplitude spectrum of $f(t)$, usually $f'(t)$ contains enough reflected and refracted energy to create interference patterns in the spectra of the gated pulses. This makes it necessary to smooth the spectra and spectral ratios further before the attenuation can be evaluated. Smoothing is accomplished by a running average of 25 samples spaced at 5 Hz intervals along the spectrum and by averaging the spectra and spectral ratios for a number of geophone stations. The running average operation

replaces the value of the 13th frequency sample with the average value of samples 1 through 25, replaces the value of the 14th sample with the average value of samples 2 through 26, and similarly across the spectrum. The last two operations were different for each subsection, and they are discussed separately below.

RESULTS OF ATTENUATION MEASUREMENTS

Subsection 1(500–1000 ft)

The continuous velocity log in Figure 2 showed an interval from 450 to 550 ft deep with almost constant sonic velocity of about 5800 ft/sec. For this reason, and because of the smaller offset angle, the interval from 500 to 1000 ft was used as the primary test section. This subsection, having geophones with 20-ft spacing, and subsection 2, from 100 to 500 ft, having geophones with 50-ft spacing, were used to gain experience with geophone spacing and to determine the total distance to be covered for reliable attenuation measurements.

The original set of data for Subsection 1 January 10, 1966 was recorded with shotpoint 4. When it became evident that the data gave consistent results, the experiment was repeated on April 11, 1966 from shotpoint 5, and on November 4, 1966 from both shotpoints. The 300-ft geophone was used as a reference for this section, and its output was recorded on channel one. The other five channels of the recorder were used for the outputs of the 20-ft spaced geophones of the subsection. Each succeeding record then was influenced by the attenuation of an additional 100 ft of coverage. (See Figure 6 for sketch of shotpoint and geophone locations.)

Outstanding events on the seismic traces were the primary pulses, followed 16 ms later by pulses reflected from the base of the low velocity layer. In each trace, the strong reflected pulse arising in the LVL was eliminated by the programmed graded data gate which accepted data beginning at the first data sample greater than a preset level and continued to accept data for the next 32 samples (or 16 ms). Here $D(t)$ extends 16 ms between t_0 and t_2. The corrected primary pulses from the six-channel records were then Fourier analyzed and normalized to yield a $Y_i(f)$ for the record. Here i is an index relating the records to the segments covered by the analysis. Normalization is necessary to remove the effects of shot parameter changes when multiple shots are used to cover a subsection.

The function $Y_i(f)$ in reality is an unsmoothed transfer function for the section between the reference geophone and the center of the other five

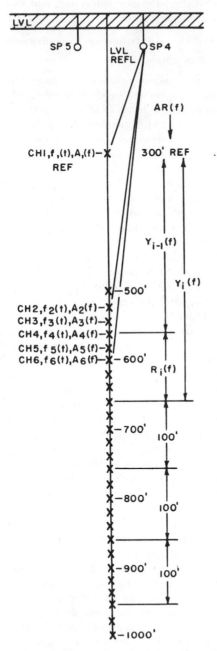

FIG. 6. Schematic diagram of shotpoint and geophone locations for subsection 1.

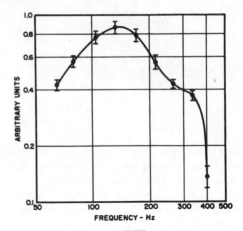

FIG. 7. Averaged spectra $\overline{A_2(f)}$, before normalization, for depth interval from 620 to 700 ft. Plotted are average value and one standard deviation above and below the average.

geophones whose signals are recorded on the record. $Y_i(f)$ is found by dividing $\overline{A_i(f)}$ by $A_{\text{ref}}(f)$ where $\overline{A_i(f)}$ is average of the spectra of the gated traces two through six and $A_{\text{ref}}(f)$ is the spectrum of gated trace one. Figure 7 shows $A_2(f)$ where i equals two and where channels two through six cover the segment from 620 ft to 700 ft. The normalization operation (see Figure 6) for the segments can be represented by $Y_i(f) = \overline{A_i(f)}/A_{\text{ref}}(f)$ with i progressing from one to five to cover the segments recorded on five records.

The next step in the analysis is to estimate the transfer functions for the 100-ft sections covered by the five records involved and to evaluate the results. Referring to Figure 8 and using the normalized, averaged spectra, we compute spectral ratios $Y_2(f)/Y_1(f)$, $Y_3(f)/Y_2(f)$, $Y_4(f)/Y_3(f)$ and $Y_5(f)/Y_4(f)$. These ratios are rough transfer functions that we expect to vary about some simple analytic function, $R(f)$, where

$$R(f) = \exp(-\alpha f^x \Delta t),$$

and

$$\text{Ln}\{\text{Ln}[1/R(f)]\} = \text{Ln}(\alpha \Delta t) + x \, \text{Ln} \, f.$$

Here $R(f)$ is an estimated transfer function, f is frequency, Δt is the traveltime between samples, and α (the attenuation constant) and x are unknown constants. Next we find, by the least-squares-fit method, the α and x for $R(f)$ that will cause the curve $\text{Ln}\{\text{Ln}[1/R(f)]\} = \text{Ln}(\alpha \Delta t)$

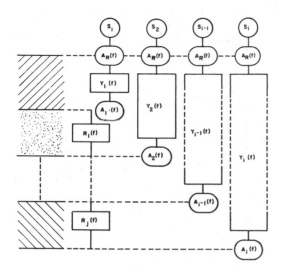

FIG. 8. Method of determining Incremental Transfer Function $R_j(f)$.

$+xLnf$ to be a best fit Ln Ln of the data points representing the spectral ratio, and call the resulting curve Ln Ln of $R_j(f)$, with j being an index associated with a given segment. This form of the attenuation law was chosen so that the attenuation constant α would have units of loss per wavelength when $x=1$.

When the values of α_j and x_j are found, a new function $\sigma_j(f)$ is generated to represent the error of Ln Ln of the data points about the theoretical curve Ln Ln of $R_j(f)$ representing the transfer

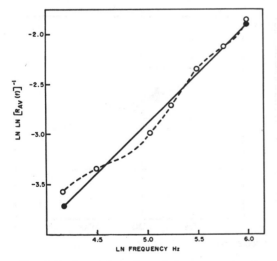

FIG. 9. Ln Ln of the reciprocal of average spectral ratios about Ln Ln of estimated $[R_{av}(f)]^{-1}$.

function. If we assume that the theoretical curve Ln Ln of $R_j(f)$ is the best estimate of Ln Ln of the transfer function for the given data, σ, the root-mean-square value of $\sigma_j(f)$, will represent the overall quality of the corresponding spectral ratio (σ is the standard deviation). The parameter α is the coefficient of attenuation, and x shows how attenuation varies with frequency. By averaging the four spectral ratios for the five 100-ft segments and solving for α and x for the average, we find Ln Ln of $R_{av}(f)$, or the best estimate of Ln Ln of the 100-ft transfer functions using all of the data. This curve and Ln Ln of the average spectral ratios are shown in Figure 9.

Values for σ and α, and x were used to determine the distance over which averages are needed for good estimates. In Table 1, σ for $R_{av}(f)$ is much smaller than it is for the parts making up $R_{av}(f)$. If $\sigma=0$ indicates a perfect curve fit to the data, the tabulated values of σ show that the transfer function found from the averaged data is much more dependable than the transfer functions found for the individual segments. If we compare σ, x, and α of $R_{av}(f)$ with σ, x, and α of $R_1(f)$ and $R_2(f)$, we find that the x's and α's compare favorably even though σ's for $R_1(f)$ and $R_2(f)$ are much greater than σ for $R_{av}(f)$. Figure 10 shows the behavior of the data points about $R_j(f)$ and $R_{av}(f)$. The data points for $R_3(f)$ and $R_4(f)$ actually exceed unity for part of the frequency band because of in-phase addition of reflected energy. Also in Table 1, $R_3(f)$ and $R_4(f)$ disagree with R_{av} in all aspects. In general, these data indicate that reliable estimates of x and α require a large number of statistically independent data samples for a layered medium.

Repeats of this experiment on April 11, 1966 and November 4, 1966 showed little change for the 10-month period beginning January 10, 1966. For a final analysis, the results from the first experiment are compared to the average of the

Table 1. Standard deviations, frequency exponents, and attenuation coefficients (subsection 1, January 10, 1966)

Function	σ	x	α
$R_{av}(f)$	0.0650	1.0021	3.65×10^{-4}
$R_1(f)$	0.2092	1.0040	3.54×10^{-4}
$R_2(f)$	0.3712	0.9920	4.38×10^{-4}
$R_3(f)$	1.4505	1.6550	9.00×10^{-6}
$R_4(f)$	1.1938	0.1178	46.30×10^{-4}

three in Figure 11a, b. Figure 11a is $R_{av}(f)$ for data of January 10, 1966, with vertical bars to show standard deviation among samples, and Figure 11b is $R_{av}(f)$ for the three sets of data, with bars to show the deviation among sets.

Student's t test (Neville and Kennedy, 1964, p. 178–183) was used to set confidence limits for the curve fit to the data. Since x was so near unity and since the t test showed there was no significant difference between x and one, we assumed x was one within the limits of our measurements. A scale factor was introduced to force x_{av} to one, and these scale factors were evaluated in terms of measurement error. The spectral ratio $[Y_i(f)/Y_{i-1}(f)]_{av}$ for January 10, 1966 was multiplied by 0.9999 for this adjustment, and the ratio averaged for three experiments was multiplied by 1.0034. If our assumption that $x_{av}=1$ is valid, the scale factors indicate measurement errors of 0.01 percent and 0.34 percent. Errors of this size could come from a number of sources. The curve-fit standard deviation σ_{av} and the attenuation constant α_{av} after these adjustments were 0.0650 and 3.69×10^{-4}, respectively, for the first set of data, and 0.0352 and 3.64×10^{-4} for the average over 10 months. The scale factor of 0.9999 caused no change in σ_{av} for the January data, but the scale factor of 1.0034 for all the data reduced σ_{av} from 0.0385 to 0.0352.

The value $x=1$ was accepted with a high confidence level and the data were used to obtain a least-squares fit of the line Ln $R(f)=-\alpha f \Delta t$; the

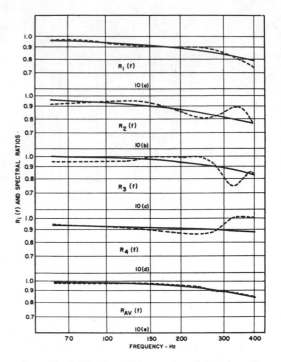

FIG. 10. $R_i(f)$ (the solid curve) and corresponding spectral ratios (dashed curve) compared to $R_{av}(f)$ and the average of 4 spectral ratios.

loss was computed in db per wavelength. Student's t test was again used to set a confidence level for α; α was found to be 0.2004 ± 0.0091 db/λ at a confidence level of 99.9 percent.

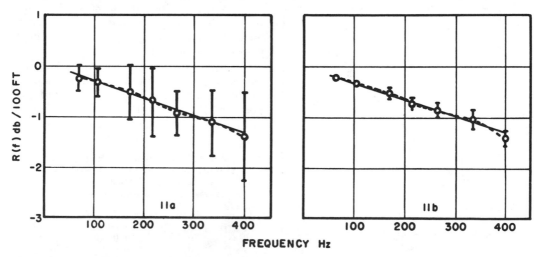

FIG. 11. Average spectral ratios about estimated $R_{av}(f)$—First set of subsection 1 data (Figure 11a) compared with average of three sets (Figure 11b).

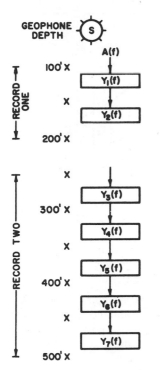

FIG. 12. Source-geophone setup for subsection 2.

Subsection 2 (100–500 ft)

Geophones for subsection 2 were spaced at 50-ft intervals from 100 to 500 ft. The wider spacing seemed to yield good results since σ for the curve fit was again small. Figure 12 shows the geophone and shotpoint setup; two six-channel records covered the subsection.

Here as for subsection 1, the primary pulse was gated, corrected, and Fourier-analyzed. The spectrum for each station was smoothed by averaging over frequency, and a transfer function for each 50 ft was obtained by dividing each smoothed spectrum by the preceding one. Only spectra of a common shot were used.

These operations gave very rough transfer functions for 50-ft intervals; seven of these rough transfer functions were averaged to give a smoother function for curve-fitting. The estimated transfer function, the mean data points, and the standard deviation of individual points about the mean are shown in Figure 13. The numerical results for this section are $\sigma = 0.03148$, $\alpha = 3.55 \times 10^{-4}$, and $x = 1.00682$. Here again, x is so near unity we felt justified in multiplying the

average of the original spectral ratios by 1.0004 to force x to unity; this caused no change in σ and gave a value of 0.3641 ± 0.0150 db/λ at the 99.9 percent confidence level. The σ for subsection 2 differed only slightly from σ for the average of all the data for subsection 1; however, Figure 13 shows that the standard deviation of the samples about their mean is appreciably larger for subsection 2.

Subsection 3 (7.5–110 ft)

In subsections 1 and 2 we achieved statistical diversity by moving the receiver location, or in reality by substituting receivers to sample the outgoing wave at different points in space. As the measurements approached the surface, it was no longer possible to have the source above the receivers, nor was it possible to get a large number of independent samples with the source below the receivers.

For the subsection 7.5 to 110 ft, diversity was realized by moving the source both vertically and horizontally and observing the direct outgoing pulse and the pulse reflected from 7.5 ft, at stations from 500 to 600 ft. Caps were fired in shotholes 2 and 3 at five-ft intervals between 80 ft and 140 ft, as shown in Figure 14. The data were processed in four groups, two groups for each shothole. There were 13 records for each hole; these 13 records were processed in groups of nine with a five-record overlap at the center. The nine six-trace records gave 54 direct pulses and 54

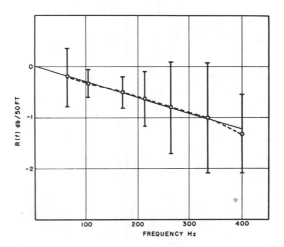

FIG. 13. Spread in average spectral ratios about $R(f)$ for subsection 2.

reflected pulses that were separately gated with an 8-ms graded data gate, corrected for spreading, and analyzed.

Spherical spreading was assumed and the spreading for the reflections was computed for an image source located above the reflector. A reflection coefficient was estimated from the square root of the average energies of the gated direct and reflected pulses. We assumed that the reflection coefficient would be somewhat larger than the square root of the quotient of total reflected energy divided by the direct energy when each had been corrected for spreading according to their separate raypaths. The square root of the

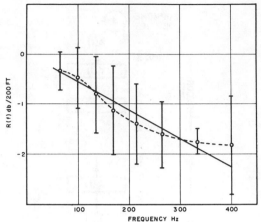

FIG. 15. Spread of average spectral ratios about $R(f)$ for subsection 3.

energy ratios was found to be 0.405; we assumed that there would be a 10 percent loss due to attenuation, so an estimated reflection coefficient of 0.446 was used for preliminary evaluation of the data.

The spectra for the 54 direct pulses were averaged, smoothed, and divided into the averaged and smoothed spectra of the corresponding 54 reflected pulses to get an estimated transfer function from the center of the shot group to 7.5 ft and back. The four transfer functions obtained in this manner (one for each data group mentioned above) were examined for smoothness and compared for consistency. Our conclusions were that none was a good estimate of the transfer function. However, the four, when averaged, gave a fair curve fit with $\sigma = 0.1775$ and $x = 1.325$. When the reflection coefficient was changed to 0.458, σ became 0.1291 (indicating a better fit), and x became unity. The difference of 0.053 between 0.458 and the energy relationship 0.405 was assumed to be the loss due to attenuation for the two-way travel between 110 ft and the reflector. The resulting $R(f)$ and the deviation of the average spectral ratios about $R(f)$ are shown in Figure 15. The numerical value for the attenuation coefficient is 0.1508 ± 0.0190 db/λ at the 99.9 percent confidence level.

Subsection 4 (1-10 ft)

The large reflection coefficient just discussed for subsection 3 is associated with the base of the

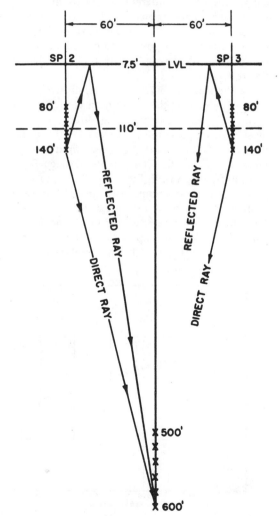

FIG. 14. Source-geophone setup for subsection 3; source moved horizontally and vertically.

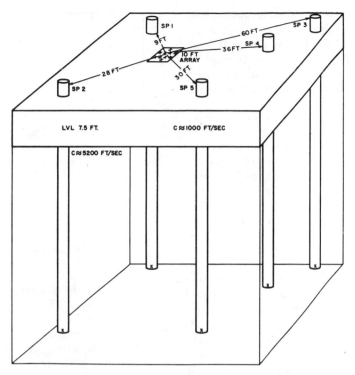

FIG. 16. Source-geophone setup for subsection 4.

LVL, where the sonic velocity changes from 5200 ft/sec at 8 ft to 1000 ft/sec at 7 ft. Pulses detected with a geophone located 1 ft below the surface in the 1000-ft cemented array showed very high energy losses due to absorption by the LVL.

To get an accurate value of the absorption, we

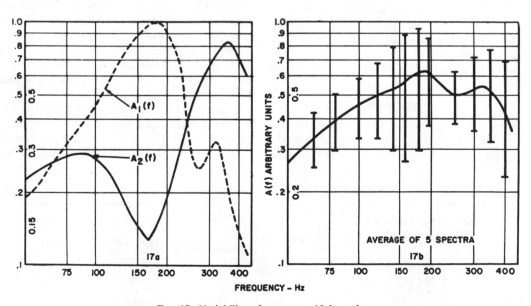

FIG. 17. Variability of spectra at 10-ft station.

cemented another set of 11 geophones in individual holes clustered near the 1000-ft array. These geophones were spaced at 1-ft-depth intervals from 1 ft to 10 ft, with a reference geophone placed at 20 ft; the individual holes had a 2-ft horizontal separation. Statistical diversity was achieved here by firing two shots at 50 ft in each of the five shotholes shown in Figure 16. The lateral offset for these shots ranged from nine to 60 ft. Figure 17 illustrates the effectiveness of averaging the spectra observed at the 10-ft station. Figure 17a shows the spectra for two of the shots, while Figure 17b is the average for five. The spectra for Figure 17 were partially normalized by dividing the spectra by the square root of the reference energy.

After correction for the energy loss due to reflection at the base of the LVL, the data for this section were processed in a manner similar to that described for subsection 1; that is, we averaged the spectra for channels 2–6 and normalized the result by dividing the average spectrum by the spectrum from channel 1. A spectral ratio for each shotpoint was then found by dividing the normalized average spectrum for shot number two by the normalized average spectrum from shot one. The spectral ratios for the five shotpoints were then averaged to get a set of data for the curve-fitting program.

Figure 18 shows the distribution of the spectral ratio averages about the estimated transfer function $R(f) = \exp(-\alpha f \Delta t)$. The attenuation rate for this section is approximately 100 times that of the section below the LVL. The attenuation constant is 13.1 ± 0.57 db/λ at the 99.9 percent confidence level.

CONCLUSIONS

Table 2 is a summary of the resulting attenuation estimates for this 1000 ft of layered earth. The experiments were designed to test a concept

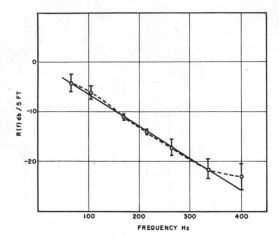

Fig. 18. Distribution of spectral averages about estimated transfer function $R(f)$ for subsection 4.

of making reliable absorption estimates using statistical methods. The operations began with a carefully recorded set of data and concluded with a number of averaging techniques. These techniques were chosen to preserve the attenuation characteristics of the traveling pulse while they destroyed the interferences of replica pulses that could not be separated from the primary pulse with a time gate. The degree of destruction of the interferences is indicated by the small value of σ for the curve-fit operation.

While these results are for this one location and would be expected to change with a change in the composition of the material, we believe that they are good estimates for this location within the limits shown in Table 2. The material classification in Table 2 is from observations of the drill mud returns.

Techniques used for subsections 1 and 2 should give good results in a large number of areas of seismic interest while those used for subsections 3 and 4 were special adaptations that gave fair re-

Table 2. Summary of attenuation measurements

Sub-section	Attenuation Coefficient with 99% Confidence Limits, db/λ	Number of Pulse or Spectra Samples	Approximate Depth Coverage Feet	Material in General	Date of Observation
4	13.1000 ± 0.5700	50	1–10	loam-sand-clay	8/5/66
3	0.1508 ± 0.0190	156	7.5–100	clay-sand	8/18/66
2	0.3641 ± 0.0150	9	100–500	sandy clay	1/10/66
1	0.2004 ± 0.0091	25	500–1000	clay-sand	1/10/66

sults here due to local conditions. These local conditions are the high reflection coefficient at the base of the LVL and the high attenuation coefficient within the LVL.

REFERENCES

Auberger, M. and Rinehart, J. S., 1960, Method for measuring attenuation of ultrasonic longitudinal waves in plastics and rocks: J. Acoust. Soc. A., v. 32, p. 1698–1699.
Birch, F. and Bancroft, D., 1938, Elasticity and internal friction in a long column of granite: Bull. Seis. Soc. Amer., v. 28, p. 243–255.
Born, W. T., 1941, The attenuation constants of earth materials: Geophysics, v. 6, p. 132–138.
Collins, F. and Lee, C. C., 1956, Seismic wave attenuation characteristics from pulse experiments: Geophysics, v. 21, p. 16–40.
Futterman, W., 1962, Dispersive body waves: J. Geoph. Res., v. 67, p. 5279–5291.

Horton, C. W., 1959, A loss mechanism for Pierre shale: Geophysics, v. 24, p. 667–680.
McDonal, F. J., Angona, F. A., Mills, R. L., Sengbush, R. L., Van Nostrand, R. G., and White, J. E., 1958, Attenuation of shear and compressional waves in Pierre shale: Geophysics, v. 23, p. 421–439.
Neville, A. M. and Kennedy, J. B., 1964, Basic statistical methods: Scranton, International Textbook Company.
O'Brien, P. N. S., 1961, A discussion on the nature and magnitude of elastic absorption in seismic prospecting: Geophys. Prosp., v. 9, p. 261–275.
Peselnick, L. and Zietz, I., 1959, Internal friction of fine grained limestones at ultrasonic frequencies: Geophysics, v. 24, p. 285–296.
Ricker, N., 1953, The form and laws of propagation of seismic wavelets: Geophysics, v. 18, p. 10–40.
Strick, E., 1967, The determination of Q, dynamic viscosity and transient creep from wave propagation measurements: Geophys. J. Roy. Astro. Soc., v. 13, p. 197–218.
Wuenschel, P. C., 1965, Dispersive body waves—an experimental study: Geophysics, v. 30, p. 539-551.

Reprinted from Geophysics, v. 37, p. 620–646.

COMPRESSIONAL-WAVE ATTENUATION IN MARINE SEDIMENTS†

EDWIN L. HAMILTON*

In-situ measurements of compressional (sound) velocity and attenuation were made in the sea floor off San Diego in water depths between 4 and 1100 m; frequencies were between 3.5 and 100 khz. Sediment types ranged from coarse sand to clayey silt. These measurements, and others from the literature, allowed analyses of the relationships between attenuation and frequency and other physical properties. This permitted the study of appropriate viscoelastic models which can be applied to saturated sediments. Some conclusions are: (1) attenuation in db/unit length is approximately dependent on the first power of frequency, (2) velocity dispersion is negligible, or absent, in water-saturated sediments, (3) intergrain friction appears to be, by far, the dominant cause of wave-energy damping in marine sediments; viscous losses due to relative movement of pore water and mineral structure are probably negligible, (4) a particular viscoelastic model (and concomitant equations) is recommended; the model appears to apply to both water-saturated rocks and sediments, and (5) a method is derived which allows prediction of compressional-wave attenuation, given sediment-mean-grain size or porosity.

INTRODUCTION

In the sea floor off San Diego, during the period 1966 to 1970, in-situ measurements were made of the attenuation of compressional (sound) waves from a deep-diving submersible in deep water, and by scuba diving in shallow water. In this program, probes were inserted into the sea floor, and compressional-wave velocity and attenuation (hereafter called "velocity" and "attenuation") were measured at three frequencies (14, 7, and 3.5 khz) without removing the probes from the sediment. This report also includes a few measurements at 25 khz from the bathyscaph *Trieste* program in 1962, and reevaluated measurements of attenuation at 100 khz from an earlier scuba diving program (Hamilton et al, 1956).

In-situ measurements from the above programs and other previously reported laboratory and in-situ measurements of attenuation allow study of the dependence of attenuation on frequency and dispersion (if any) of compressional-wave velocity.

The relationships between frequency and compressional-wave velocity and attenuation have important implications in forming parameters for permissible theoretical elastic or viscoelastic models for water-saturated, porous sediments.

The objectives of this study are 1) to report in-situ, measured values of velocity, attenuation, and associated physical properties (e.g., density, porosity, grain size) of marine sediments, 2) to assemble and analyze pertinent literature data on attenuation, 3) to discuss the relationships between frequency, velocity, attenuation, and other physical properties, 4) to discuss the causes of attenuation in saturated sediments, 5) to discuss elastic and viscoelastic models which can be applied to marine sediments, and 6) to suggest a method to predict attenuation in marine sediments given frequency and common physical properties such as porosity and grain size.

RESULTS

In this section, three sets of in-situ measurements of the attenuation of compressional waves

† Manuscript received by the Editor August 2, 1971; revised manuscript received January 12, 1972.
* Naval Undersea Research and Development Center, San Diego, California 92132.

Table 1. In-situ attenuation and velocity and other physical properties of sediments off San Diego; A. 1966–70 program; B. 1962 program*

Sediment type	No. stas.†	Water depth m	Mean grain diam. ϕ	Mean grain diam. mm	Density gm/cm³	Porosity percent	Velocity m/sec	Attenuation, db/m at f, khz 3.5	7.0	14.0	k‡		n‡	
A. 1966–1970 program														
Sand														
Coarse	1a	32	0.81	0.5704	2.060	38.0	1817	—	3.4	6.6	0.53	—	0.96	—
Medium	2a	20	1.02	0.4931	2.008	39.2	1798	1.5	3.8	6.8	0.41±0.12		1.09±0.14	
Fine	4a	8	2.55	0.1708	1.967	45.6	1686	1.7	3.2	7.2	0.45±0.07		1.04±0.07	
Very fine	1a	13	3.30	0.1015	1.933	47.0	1708	1.5	3.5	7.0	0.38±0.05		1.11±0.06	
Sand-silt-														
clay	1a	4	6.27	0.0130	1.512	72.3	1483	—	—	0.7	—	—	—	—
Clayey silt	1b	1030	7.52	0.0055	1.270	83.8	1453	—	—	1.0	—	—	—	—
Clayey silt	1b	1110	7.64	0.0050	1.300	82.5	1457	—	—	0.6	—	—	—	—
Clayey silt	1b	1087	7.42	0.0058	1.374	77.8	1459	—	1.2	3.8	0.19	—	0.94	—
Clayey silt	1b	1012	7.39	0.0060	1.349	79.1	1441	—	—	2.4	—	—	—	—
B. 1962 program								*f*=25 khz						
Fine sand	3a	9	2.08	0.2365	1.973	44.0	1645	9.4						
Sandy silt	1a	20	5.13	0.0286	1.702	60.9	1572	8.0						
Clayey silt	1c	951	7.50	0.0055	1.380	78.9	1449	4.0						

* Averaged values for number of stations indicated.
† Letter after number of stations indicates station occupied by: a—scuba divers, b—*Deepstar* 4000, or c—bathyscaph *Trieste*.
‡ In $\alpha = kf^n$, where, attenuation α db/m; frequency f, khz; k is a constant; and n, the exponent of frequency.

by the writer are listed in Tables 1 and 2 and plotted versus frequency in Figures 1 and 2. In addition, some in-situ and laboratory measurements from the literature are listed in Tables 3 and 4 and plotted in Figure 2. For comparison, and later discussions, all values are listed (many recomputed) in decibels/meter (db/m). The equations showing dependence of attenuation on frequency are listed in the form

$$\alpha = kf^n, \qquad (1)$$

where α is attenuation of compressional waves in db/m, k is a constant, f is frequency in khz, and n is the exponent of frequency.

In-situ measurements

1966–70 program.—The equipment and meth-

ods used to make in-situ measurements of velocity and attenuation from a submersible have been described and diagrammed in a previous report (Hamilton et al, 1970). In general, the equipment consisted of three stainless-steel probes, 7 cm in diameter, fastened to a 2-m-long rigid beam in such a manner that when the beam is on the sea floor, the probes are inserted a variable, preset depth into the sediment. This depth varied from 30 to 60 cm during the submersible dives. Three barium-titanate transducers were used as a sound source and two receivers. Velocity was determined by measuring traveltime over a 1-m path between the receivers. Attenuation was measured in decibels relative to that in the bottom water (assumed to be zero for 1 m). Both velocity and attenuation

Table 2. In-situ attenuation and velocity and other physical properties of sediments off San Diego; 1956 program*

Sediment type	No. stas.	Water depth m	Mean grain diam. ϕ	Mean grain diam. mm	Density gm/cm³	Porosity percent	Velocity m/sec	Attenuation db/m at 100 khz
Sand								
Coarse	1	20	0.79	0.5783	2.080	38.3	1752	53.1
Medium	1	19	1.99	0.2517	2.000	40.9	1630	47.3
Fine	17	10	2.46	0.1817	1.926	46.7	1684	52.1
Very fine	4	16	3.16	0.1119	1.938	47.4	1667	55.9
Sandy silt	2	13	4.23	0.0533	1.860	51.2	1619	74.3
Silty sand	2	17	5.07	0.0298	1.680	61.3	1537	60.9
Silt	1	15	5.53	0.0216	1.690	60.9	1465	15.9
Sand-silt-clay	1	17	5.05	0.0302	1.721	58.0	1490	59.7
Clayey silt	1	22	6.10	0.0146	1.600	65.6	1464	18.0

* Averaged values for number of stations indicated. All stations occupied by scuba divers.

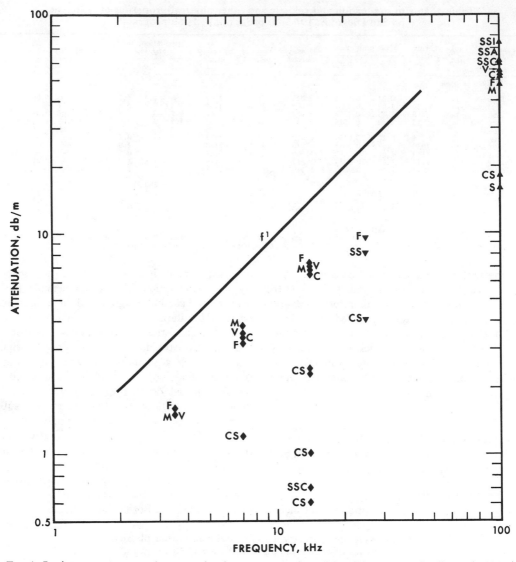

FIG. 1. In-situ measurements of compressional-wave attenuation off San Diego. Symbols: diamonds–1966–70, Table 1; inverted triangles–1962, Table 1; triangles–1956, Table 2. Letters indicate sediment type: C–coarse sand; M–medium sand; F–fine sand; V–very fine sand; SSI–sandy silt; SSA–silty sand; SSC–sand-silt-clay; S–silt; CS–clayey silt. Line labelled f^1 indicates slope of any line having a dependence of attenuation on the first power of frequency.

were measured at 3.5, 7, and 14 khz without disturbing the probes. Coring tubes attached to each end of the rigid beam obtained sediment samples for laboratory analyses. Sediment properties were obtained by standard laboratory procedures. Variations and interrelationships of these properties and sediment nomenclature were discussed by Hamilton (1970).

The submersible *Deepstar* 4000, from which some of the measurements were made, could not always insert the probes to maximum penetration required tò obtain measurements at 3.5 khz.

The same probes, described above, were used in shallow water from a small boat. Divers observed the sea floor, took samples, and helped insert the probes into the sediment.

The results of the above measurements and subsequent sediment laboratory analyses are listed in Table 1, and the attenuation-versus-frequency data are plotted in Figures 1 and 2.

1962 program.—In 1962, velocity and attenua- tion measurements were made at 25 khz from the bathyscaph *Trieste* and from a small boat in shallow water. The equipment, methods, and re- sults of the velocity measurements have been reported (Hamilton, 1963). The probes used for

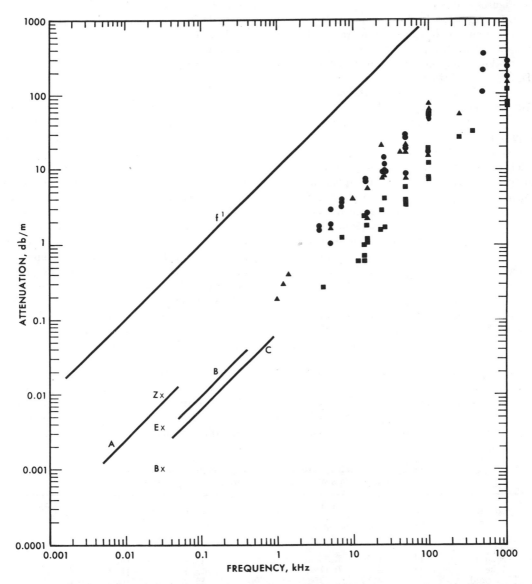

FIG. 2. Attenuation versus frequency in natural, saturated sediments and sedimentary strata; data from Tables 1–4b. Symbols: circles–sands (all grades); squares–clayey silt, silty clay; triangles–mixed sizes (e.g., silty sand, sandy silt, sand-silt-clay); sand data at 500 and 1000 khz from Busby and Richardson (1957). Low-frequency data: line A–Zemstov, 1969 (land, sedimentary strata); B–Tullos and Reid, 1969 (Gulf of Mexico coastal clay-sand); C–Bennett, 1967 (sea floor, reflection technique); Z–Zhadin (in Vasil'ev and Gurevich, 1962); E and B–Epinatyeva et al, and Berzon (in Zemstov, 1969). Line labelled f^1 indicates slope of any line having a dependence of attenuation on the first power of frequency.

these measurements were an earlier model of those used in the 1966–70 program; field methods were the same.

The few attenuation measurements made during this program (not previously reported) are listed in Table 1 and plotted versus frequency in Figures 1 and 2.

1956 program.—In 1956, sound velocity and attenuation measurements were reported for a number of in-situ and laboratory measurements in shallow water off San Diego (Hamilton, 1956; Hamilton et al, 1956).

In-situ velocity measurements were made at 100 khz. The equipment and methods have been reported (Hamilton et al, 1956). In general, the equipment consisted of two probes secured through a rigid base so that, when inserted 6 inches (15 cm) into the sea-floor sediment (by divers), velocity and wave-amplitude measurements could be made over a 1-ft (30-cm) path length. At the same stations, small containers (10-cm long and 5 cm in diameter) were inserted by hand into the sediment, for subsequent laboratory measurements of velocity and attenuation, using a resonant-chamber method (Shumway, 1956; Toulis, 1956).

In the 1956 report, velocities were reported for both the in-situ probe and laboratory resonant-chamber methods. Values of attenuation (at 23 to 40 khz) were reported for the resonant-chamber method.

During the in-situ probe measurements at 100 khz, the amplitude of the received wave was measured on the oscilloscope face in both bottom water and sediment; 3 to 5 separate measurements were made at each station. These amplitudes

were reported (Hamilton et al, 1956, Table II) as "20 log (amplitude water/amplitude sediment)," and not as linear attenuation, because of uncertainties concerning impedance loading in the transducers in the two media. Reevaluation of the method and results indicate these data can be reported as attenuation in the sediments. The basic data have been reanalyzed and extrapolated to 1 m and reported as db/m in Table 2; attenuation versus frequency is plotted in Figures 1 and 2.

Measurements from the literature (in situ)

In-situ measurements of attenuation in marine sediments are relatively rare compared with measurements of other properties. Those of the writer are noted above. In this section, brief resumés of other in-situ measurements in the literature will be noted. Included are in-situ measurements on land in some water-saturated sediments and sedimentary rocks which may be pertinent to lower layers of the sea floor, or to later discussions. These attenuation measurements are plotted versus frequency in Figure 2 and listed in Tables 3, 4a, and 4b. For those studies in which equations are given (e.g., McCann and McCann, 1969; Wood and Weston, 1964), points are plotted in Figure 2 for the lowest, highest, and an intermediate frequency.

Wood and Weston (1964) made in-situ measurements of attenuation (4 to 50 khz) in tidal mud flats in Emsworth Harbor, England. Maximum distance between probes was 180 ft (55 m); probe penetration was about 0.5 m.

Ulonska (1968) and Schirmer (1970) measured, in situ, compressional velocity and attenuation in

Table 3. Laboratory attenuation and other physical properties of sediments off San Diego from Shumway (1960)*

Sediment type	No. stas.	Mean grain diam. ϕ	Mean grain diam. mm	Density gm/cm³	Porosity percent	Velocity m/sec 23° C	Attenuation db/m	Attenuation khz
Sand								
Medium	13	1.44	0.3686	2.004	39.1	1737	9.4	26.6
Fine	13	2.49	0.1780	1.960	43.6	1693	9.6	26.0
Silty sand	1	2.99	0.1259	1.796	53.2	1551	20.6	23.9
Sandy silt	4	4.57	0.0421	1.737	57.4	1541	11.8	23.6
Silt	8	5.27	0.0259	1.653	61.5	1548	7.8	24.2
Clayey silt	4	6.31	0.0126	1.605	65.5	1508	2.8	23.2
Clayey silt	11	7.51	0.0055	1.336	80.0	1495	1.6	22.8

* Averaged values for number of stations indicated; velocity and attenuation for the second resonant mode (201) only.

shallow-water sediments in the Baltic Sea at sediment depths of about 2 m. Ulonska's measurements were over a 0.5-m path at 250 khz; Schirmer's were over a 12-m path at about 1 khz.

McCann and McCann (1969) reported in-situ measurements in water-saturated land and beach sediments at a mean sediment depth of 2 m. Frequencies were between 5 and 50 khz.

Bennett (1967), using reflection techniques from a surface vessel, computed values of attenuation at 12 khz, and between 40 and 900 hz, for sediments in the North Atlantic and Mediterranean.

Tullos and Reid (1969) measured attenuation in Gulf of Mexico coastal sediments (clay-sand) of Pleistocene age in three depth intervals to 1000 ft (305 m); frequencies were between 50 and 400 hz. Measurements in the depth interval of 7.5 to 110 ft (2 to 34 m) are listed in Table 4b.

Lewis (1971) measured attenuation, in situ,

at one station in shallow water off Puerto Rico by inserting probes into the sediment; frequencies were 5 to 50 khz. The data were scattered, but Lewis believed they represented a dependency of attenuation on frequency to the first power.

There are several references from Russian literature listed in Table 4b. Zhadin (in Vasil'ev and Gurevich, 1962) reported attenuation in "water-saturated clay" in the depth range of 0–20 m; frequency is not given but is presumed to be in the seismic range. Zemtsov (1969) reports his own studies of attenuation in the frequency range of 5 to 50 hz in "sedimentary strata," and notes attenuation-versus-frequency equations from Epinatyeva and Berzon in similar materials; values are shown at 30 hz in Figure 2.

Measurements from the literature (laboratory)

It is possible, but difficult, to make valid laboratory measurements of attenuation in natural or artificial, water-saturated sediments. Even

Table 4a. Attenuation and other properties of natural, saturated sediments (from the literature)*

Sediment type	Mean grain diam. ϕ	Mean grain diam. mm	Density gm/cm^3	Porosity percent	Velocity m/sec 23° C	k†	n†	Ref.
Fine sand	2.5	0.177	(1.99)	41.0	(1742)	0.13	1.26±0.13	1
Very fine sand	3.05	0.121	(1.95)	43.3	(1716)	0.27	1.17±0.13	1
Very fine sand	3.8	0.072	(1.97)	42.3	(1613)	0.56	1.00±0.01	1
Silt	5.32	0.025	(1.90)	46.6	(1622)	0.30	1.05±0.15	1
Medium sand	1.32	0.401	1.99	(40.8)	(1815)	0.164	1	2
Silty sand	—	—	1.32	(82.2)	(1540)	0.151	1	2
Clayey silt	—	—	1.20	(89.2)	(1530)	0.118	1	2
Clay-silt	—	—	1.15	(92.1)	(1525)	0.075	1	2
Silty clay	—	—	1.13	(93.3)	(1525)	0.072	1	2
"Mud"	(7.5)	(0.006)	1.32	76.0	(1480)	0.066	1	3
(Clayey silt)	(7.6)	(0.005)	—	(80.0)	—	0.049	1	4a
(Clayey silt)	—	—	—	—	—	0.066	1	4b
Sand-silt-clay	6.35	0.012	1.75	59.3	1527	(0.393)	(0.99)	5
Clay	9.93	0.001	(1.45)	74.8	—	31.9 db/m at 368 khz		6
—	—	—	—	—	(1615)	54 db/m at 250 khz		7a
(Clayey silt)	(7.5)	—	—	(75.0)	(1463)	26.2 db/m at 250 khz		7b
—	—	—	—	—	(1621)	0.2 db/m at 1 khz		8

* Values in parentheses computed or estimated by the writer of this report.

† In $\alpha = kf^n$, where: attenuation α, db/m; frequency f, khz; k is a constant; and n, the exponent of frequency.

1. McCann and McCann (1969): in situ, land and beach; velocity given at 20° C; 5 to 50 khz.

2. McLeroy and DeLoach (1968): lab, beach sand and St. Andrew Bay, Florida; 15 to 1500 khz; velocity from measured ratio.

3. Wood and Weston (1964): in situ, tidal mud flat; Emsworth, England; 4 to 50 khz; velocity from measured ratio.

4. Bennett (1967): in situ (reflection technique), N. Atlantic and Medit.; 4a: 12 khz; 4b: 40 to 900 hz.

5. Lewis (1971): in situ (best measurements), shallow water off Puerto Rico; 5 to 50 khz.

6. McCann and McCann (1969): lab, 8 core samples, N. Atlantic; carbonate <5 percent; 368 khz.

7a. Ulonska (1968): in situ, shallow Baltic Sea; velocity from measured ratio (Station 32).

7b. Ulonska (1968): in situ, shallow North and Baltic Seas; average for all stations with velocity ratios less than 1.00; 250 khz; velocity from measured ratio.

8. Schirmer (1970): in situ, shallow Baltic Sea (1 station); velocity from measured ratio.

Table 4b. Attenuation in sediments and
sedimentary strata (from the literature)*

Sediment type	k†	n†	Frequency hz	Ref.
Clay-sand	0.093	1	50 to 400	1
Water-saturated clay	0.326	1	(seismic)	2
Sedimentary strata	0.243	1	5 to 50	3
Sedimentary strata	0.119	1	(seismic)	4
Sedimentary strata	0.035	1	(seismic)	5

* (seismic) assumed.

† In $\alpha = kf^n$, where: attenuation α, db/m; frequency f, khz; k is a constant; and n, the exponent of frequency.

1. Tullos and Reid (1969): Gulf of Mexico coastal sediments; depth interval 2 to 34 m.
2. Zhadin (in Vasil'ev and Gurevich, 1962): depth interval 0 to 20 m.
3. Zemstov (1969).
4. Epinatyeva et al (in Zemstov, 1969).
5. Berzon (in Zemstov, 1969).

when valid measurements are made in some experimental materials, the results cannot always be extrapolated to natural marine sediments. Some of these difficulties are noted below.

One of the chief difficulties in measuring attenuation in "water-saturated" laboratory samples is air entrapped in the pore spaces of artificial sediments, or air or gas from organic materials in natural sediment samples transported to the laboratory. When air or gas is present in "water-saturated" sediments, the measured attenuation is apt to be too high and velocity too low (depending on frequency and bubble size). The experiences of Wood and Weston (1964) with sands and muds are instructive in this matter.

Some artificial sediments are not truly analogous to natural sediments. For example, kaolinite in distilled water containing a deflocculent acts as a Newtonian fluid without rigidity; whereas in a flocculated structure, rigidity is present (Cohen, 1968). As discussed in a later section, attenuation relations with frequency are different in the two cases. A favored laboratory material is round-grained quartz sand (e.g., St. Peters sand). This material has less rigidity and attenuation than the more common, angular-grained natural sands (discussed below).

Some laboratory measurements in natural marine sediments of particular significance to the present study are noted and briefly discussed in this section.

Shumway (1960) used a resonant-chamber technique to measure attenuation and velocity at various frequencies between 20 and 40 khz. The basic theory and equipment were reported by Toulis (1956) and Shumway (1956). These measurements have been widely referenced, and values have been extrapolated to other frequencies.

Unfortunately, Shumway's measurements, alone, cannot be used to determine frequency and attenuation relationships, although in a number of samples, measurements were made at two or three frequencies. A detailed examination of the published data and the unpublished records of measurements (in the writer's files) reveals numerous experimental errors. These errors are evident for the following reasons: 1) in individual samples where attenuation was measured at two or three frequencies there are 16 cases (Shumway, 1960) where attenuation *decreased* with increasing frequency, a result discordant with all theory and other experiments in similar materials; and 2) eliminating the 16 cases of 1), above, computations of the frequency dependence of attenuation (in the form: $\alpha = kf^n$) revealed that the exponent n varied in similar sediment types, with similar physical properties as follows: sand (0.9 to 3.1), sandy silt and silty sand (0.8 to 2.9), and clayey silt (0.6 to 3.4). Neither theory nor other experiments in similar materials support these variations.

The causes of the above-noted experimental errors are unknown. Shumway's samples were excellent: most were taken in situ by divers and never exposed to air before measurements. These errors may be related, in part, to the method. Shumway noted (1956, p. 318) that the resonant-chamber method assumes samples with no, or negligible, rigidity, which may be true in some shallow-water silt-clays, but not in sands and silty sands; see, for example, data in Hamilton (1971a). In addition, not all resonant modes were of equal reliability: Toulis (1956) suggested that the second resonant mode (201) was most reliable.

However, Shumway's (1960) attenuation data can probably be used, selectively, by averaging values for general sediment types; hopefully, experimental errors will cancel out. For comparison with the in-situ measurements off San Diego (this report), and with other values from the literature, the writer averaged attenuation values (and other properties) at the second resonant mode for various sediment types off San Diego; these are listed in Table 3, and plotted in Figure 2.

McLeroy and DeLoach (1968) measured (in

the laboratory) sound speed and attenuation, from 15 to 1500 khz, in natural sediments from sites in St. Andrew Bay, Panama City, Florida, and in sand from a Gulf of Mexico beach. They noted no indication of entrapped gases. These measurements are listed in Table 4a and plotted in Figure 2. The unusually low densities and computed (by the writer) high porosities of the Bay sediments are not typical of most open-ocean sediments.

The excellent study of attenuation in marine sediments by McCann and McCann (1969) has been previously noted in the section on in-situ measurements. These writers also measured attenuation at 368 khz in sediment cores from the North Atlantic. Most of these sediments contained appreciable amounts of calcium carbonate. There is some indication that attenuation in this type of sediment may be higher than in noncalcareous clay-silts with similar grain sizes and porosities. Because calcareous sediments are not included in this report, an average value of attenuation for 8 clay samples with less than 5 percent carbonate is listed in Table 4a and plotted in Figure 2.

DISCUSSION AND CONCLUSIONS

The study of elastic and viscoelastic models which can be applied to dry and saturated porous rock and sediments has concerned a large number of scientists and engineers. Many of these studies have considered the extent to which these media can be described by the equations of Hookean elasticity, or those of the Kelvin-Voigt, Maxwell or other viscoelastic models. It is somewhat surprising, therefore, to discover such a wide diversity of models, equations, and opinions on such an important subject. In the case of water-saturated, natural sediments, the reason for this diversity appears to be that, until recently, experimental evidence capable of restricting model parameters has been scarce. In the following discussion (unless otherwise noted) it is assumed that the medium is a porous, uncemented, mineral structure, fully saturated with water. The stress is that of a compressional or shear wave of low amplitude, and strains are of the order of 10^{-6} or less. Wavelengths are much greater than grain size; otherwise, Rayleigh scattering can occur, and attenuation is related to the fourth power of frequency (e.g., Busby and Richardson, 1957).

The equations of Hookean elasticity do not account for energy damping. Consequently, an adequate model must be anelastic if energy damping is considered. In the selection of an appropriate anelastic model, a critical factor is the extent of relative movement of pore water and mineral particles. If the pore water moves significantly relative to mineral structure, viscous damping and velocity dispersion must be considered. If the pore water does not move significantly with respect to the solids, there is little or no velocity dispersion, and energy damping is not dependent on viscosity of pore water and permeability of the mineral structure. The dependence of energy damping on frequency is different in the two cases. Consequently, two critical parameters for anelastic models are velocity dispersion, if any, and the dependence of energy damping on frequency. In the next section, the experimental evidence concerning these two factors will be reviewed. In following sections, a particular anelastic model is recommended, and the relationships between attenuation and other physical properties and causes of attenuation will be discussed. In the last section, a method for predicting attenuation (given grain size and sediment porosity) will be noted.

Some parameters of elastic and viscoelastic models for saturated sediments

Recent experimental studies in wave velocities and attenuation and dynamic rigidity have placed important restrictions on probable (and possible) elastic and viscoelastic models for water-saturated sediments. These parameters merit a more extended review because of their importance in geophysical studies. Additionally, the concepts and statements of this paper require documentation. These restrictions apply to porous sediments, saturated with water, and without a gas phase, when wavelengths are much greater than grain sizes. The frequency range is from a few hz to several hundred khz, or into the Mhz range. These restrictions are concerned with the question of the dependence, if any, of velocity on frequency ("velocity dispersion"), and the dependence of wave-energy damping, or attenuation, on frequency. In the following sections, these restrictions will be documented and discussed.

Velocity dispersion.—A number of investigators have measured compressional- and shear-wave velocities in rocks (laboratory and in situ). Most have concluded that there is no (or negligible) measurable velocity dispersion in the range from seismic frequencies into the Mhz range. Examples include work and reviews by Wyllie at al (1956), Birch (1961), Peselnick and Outerbridge (1961), White (1965), and Press (1966).

Most viscoelastic models requiring a dependence of velocity on frequency include movement of pore water relative to the mineral frame, which, in turn, causes water viscosity and sediment permeability to be important factors. Sands have relatively high permeability, large grain size, and interconnecting pores; consequently, these models indicate maximum velocity dispersion in sands. It is therefore instructive to examine the experimental evidence regarding velocity dispersion in sands across wide frequency ranges.

Because permeability is a factor in some models, it is pertinent that Wyllie et al (1956) found that velocity through water-saturated glass beads of various sizes was unaffected as permeability varied by a factor of 4.6×10^4.

No velocity dispersion was measured in clean, round-grained sands in the laboratory by Hunter et al (1961, 7 to 73 khz), Hardin and Richart (1963, 0.2 to 2.5 khz), Nolle et al (1963, 200 to 1000 khz), and Schön (1963, 20 to 64 khz).

In round-grained, pure quartz sands in distilled water at porosities of 36 percent, Shumway (1960, p. 463) and Nolle et al (1963) measured compressional velocities of 1744 m/sec (26 khz) and 1740 m/sec (400 to 1000 khz), respectively.

In soil-mechanics investigations in situ in sands, low-frequency vibrations were used in studies by Barkan (1962) and Jones (1958) to measure shear-wave velocities (Jones's measurements also included clay-slit). No velocity dispersion was measured in the frequency range from 10 to 400 hz.

Ideally, to test for velocity dispersion, measurements should be made on the same sample in the laboratory, or in situ in the same sediment, by merely changing the frequency. In natural sands, this has been done in the laboratory and in situ. In the laboratory, McLeroy and DeLoach (1968) measured a ratio, velocity in sediment/velocity in water, of 1.189 in medium sand over a frequency range of 15 to 1500 khz. In situ, McCann and McCann (1969) reported no velocity dispersion in fine or very fine sands between 5 and 50 khz. During the 1970 measurements off San Diego

(this report), special tests were made in fine sand at four stations to determine the presence or absence of velocity dispersion between 3.5 and 14 khz. The results indicated no measurable velocity dispersion between these frequencies.

In natural, saturated sands it is difficult to compare velocity measurements in different samples or from different stations because of the effects on velocity of grain size and shape, porosity, and other factors discussed in a previous report (Hamilton, 1970). However, no significant velocity dispersion is indicated in natural fine sand over a frequency range from 14 to 100 khz as the result of the following in-situ measurements (corrected to 23°C); Hamilton (this report, 6 stations)-1712 m/sec (porosity, 46.7 percent) at 14 khz; Hamilton et al (1956, 17 stations)-1704 m/sec (porosity, 46.7 percent) at 100 khz; McCann and McCann (1969)-1742 m/sec (porosity, 41.0 percent) at 5 to 50 khz. In very fine sand, McCann and McCann measured a velocity of 1716 (porosity, 43.3 percent) at frequencies between 5 and 50 khz.

In high-porosity silt-clays in the laboratory and in the field, compressional velocity is usually less in the sediment than it is in the water, and no velocity dispersion is indicated over very wide frequency ranges. The evidence is summarized in Table 5; velocity data are presented as the ratio: velocity in sediment/velocity in water. As discussed in a previous report (Hamilton, 1970), this ratio remains the same in the laboratory as in the surface sediments of the sea floor.

Some recent experiments in artificial clays in the laboratory have implications for studies of possible shear-wave velocity dispersion. Cohen (1968) measured complex rigidity $(\mu + i\mu')$ in flocculated kaolinite in distilled water. Both μ and $i\mu'$ were independent of frequency in the range 8.6 to 43.2 khz. Hardin and Black (1968), using a vibration technique to measure dynamic rigidity in kaolinite in distilled water, demonstrated that dynamic rigidity was independent of wave amplitude in their samples at amplitudes less than 10^{-4}, and independent of frequency between 200 and 300 hz. No dispersion in dynamic rigidity μ indicates no dispersion in shear-wave velocity.

Hampton (1967) measured compressional velocity and attenuation in artificial sediments and reported a marked velocity dispersion (4 to 6 percent) in silt-clays between frequencies of 3 to

200 khz. In his low-frequency measurements (3 to 30 khz) in flocculated kaolinite in distilled water, he reported (Hampton, 1967, Figure 11) anomalously low-velocity ratios (0.93 to 0.94), and unusually high-attenuation values (Hampton, 1967, p. 886). The experimental evidence of other investigators in the same (e.g., Urick, 1947; Cohen, 1968) or similar materials, and at the same frequencies, in both the laboratory and field (discussed above, Table 5), does not support Hampton's reported low-velocity ratios or velocity dispersion. There is a possibility that his clay-water, artificial sediments were not air or gas free. When air or gas bubbles from decaying organic material are trapped in pore spaces within any sediment they have a marked effect on both compressional velocity and attenuation (e.g., Meyer, 1957) depending on the concentration and size of bubbles, and the frequency (velocity is usually too low and attenuation is apt to be too high).

In summary, the following can be concluded in regard to velocity dispersion in saturated sediments. In sands, a number of studies (cited above) have reported no dispersion over restricted fre-

Table 5. Summary of ratios of compressional-wave velocities (V_p sediment/V_p water) in high-porosity silt-clays at various frequencies

Material	Velocity ratio	Frequency	Ref.
Kaolinite in distilled water	0.97 to 0.99	1 Mhz	1
Kaolinite in distilled water	0.97 to 0.99	9 to 43 khz	2
Deep-sea clay slurry	0.97	28 khz	3
Sand-silt-clay	0.98	14 khz	4
Clayey silt, San Diego Trough	0.98	23 to 40 khz	3
Clayey silt, San Diego Trough	0.98	25 khz	5
Clayey silt, San Diego Trough	0.98	14 khz	4
Clay-silt, St. Andrew Bay	0.997	15 to 1500 khz	6
Silt, clayey silt, Cont. shelf	0.992	2 Mhz	7
Silt-clays, Cont. shelf	0.994	2 Mhz	8
Silty clay, Cont. shelf	0.994	200 khz	9
Deep-sea silt-clay	0.977	30 to 200 hz	10
Deep-sea silt-clay	0.980	30 to 200 hz	11
Deep-sea silt-clay	0.985	200 khz	9
Deep-sea clay	0.975	200 khz	9
Deep-sea silt-clay (1 micron)	0.977	400 khz	12
Deep-sea silt-clay (2 microns)	0.987	400 khz	12
Deep-sea silt-clay	0.986	400 khz	13
"low-velocity layer," Cont. shelf	0.980	250 khz	14

1. Urick (1947): ratio dependent on concentration of solids.
2. Cohen (1968): ratio dependent on concentration of solids.
3. Shumway (1960): resonant chamber; laboratory.
4. Hamilton, this report: in-situ measurements.
5. Hamilton (1963): in-situ measurements.
6. McLeroy and DeLoach (1968): laboratory samples from St. Andrew Bay, Fla.
7. Bieda (1970): tops of 5 cores off Southern California.
8. Lasswell (1970): 9 cores; tidal mud flat; Southern California.
9. Hamilton (1970): laboratory; samples from North Pacific.
10. Fry and Raitt (1961): reflection technique; deep Pacific.
11. Houtz and Ewing (1964): reflection technique, deep Atlantic.
12. Horn et al (1968): laboratory, core samples from North Pacific.
13. Schreiber (1968): tops of 10 cores off Hawaii.
14. Ulonska (1968): in-situ measurements; shallow North and Baltic Seas.

quency ranges. These studies and comparison of the values of velocity in similar sands at different frequencies, as above, indicate that velocity dispersion, if present, is negligibly small from a few khz to the mhz range. The evidence (as in Table 5) indicates that velocity dispersion in higher-porosity silt-clays, if present, is negligible over a frequency range from less than one khz to two Mhz. However, it must be stated (as a reviewer pointed out) that most of the quoted tests of velocity dispersion were made over only an order of magnitude of frequencies or less, which is not necessarily enough to show dispersion. In other words, it cannot be stated on the basis of present experimental evidence that velocity dispersion is nonexistent; especially over very wide frequency ranges (from a few hz to several Mhz).

Energy damping.—The relationship between frequency and wave-energy loss, or damping, is a critical parameter in selection of an appropriate anelastic model for any medium. Recent reviews have summarized a large number of laboratory and field studies of wave-energy losses in rocks in which the specific attenuation factor $1/Q$ and the logarithmic decrement have been shown to be approximately independent of frequency over a range of at least 10^8 hz (Knopoff and Macdonald, 1958; Knopoff, 1965; White, 1965; Bradley and Fort, 1966; Attwell and Ramana, 1966). Attwell and Ramana included some sediment data.

Evidence that the specific attenuation factor $1/Q$ is independent of frequency implies that attenuation in db/unit length α increases linearly with frequency f (e.g., White, 1965, p. 98). Recent summaries of work in this field in the case of compressional waves (White, 1965; Knopoff, 1965; Attwell and Ramana, 1966) indicate that, for most rocks, there is a small variation around linearity in the range of frequencies of most interest in underwater acoustics and marine geophysics; that is, in the relationship, $\alpha = kf^n$, the exponent n is approximately one. These studies included dry rocks (usual in the laboratory) and in-situ measurements in rock strata which, below groundwater levels, are saturated.

Some recent studies of attenuation in saturated sediments which are pertinent to the present report have been noted in a previous section, with results in tables and figures. Of special interest in these studies, and in the measurements of this report, is the dependence of attenuation on fre-

Hamilton

quency. In this section the current experimental evidence on this subject will be reviewed.

In examining the dependence of attenuation on frequency, it is important to recognize that in natural marine sediments each set of measurements and the resulting equation (e.g., $\alpha = kf^n$) is apt to be unique. The reasons for this phenomenon will be discussed below, but in general, the important variables involve the varying sediment structures, porosity, grain size and shape, interparticle contacts and surface areas in sands and coarse silts, and physicochemical forces (cohesion) in the fine silts and clays. These variable factors result in much scatter of measured values of attenuation in similar sediments (Tables 1 to 4), and consequent scatter in computed values of the constant k. Also, because of the difficulties of making accurate measurements of attenuation in natural sediments (in situ or laboratory), it is to be expected that computed values of the exponent of frequency n will vary. As a result, generalized statements about the exact dependence of attenuation on frequency must be qualified, and, as far as justifiable, a statistical approach must be followed in studying data (which was the approach of Attwell and Ramana, 1966).

The important factor in predicting attenuation and extrapolating various frequencies is the exponent of frequency n (in the equation above). Also, this exponent is a critical parameter for selecting appropriate anelastic models which can be applied to saturated sediments.

In Tables 1 and 4a are values of the exponent of frequency for measurements made at two or more frequencies in a single sample or sediment type by an individual experimenter. In Table 4b, values are given for some low-frequency measurements in sediments or sedimentary rock sections (some of the reports did not specify the degree of lithification). Some of the meaurements in Table 4b may be pertinent to lithified or semilithified, deeper layers in the sea floor.

As can be seen in Tables 1 and 4, measured values of the exponent of frequency in these natural sediments from widely scattered geographic areas vary closely around one for a wide variety of deep- and shallow-water sediments, over a wide frequency range.

Computations were made (Table 6A) which interrelated in-situ measurements in similar sediments from various experiments by the writer off San Diego (Tables 1, 2). This allowed extension of the frequency range; frequencies usually included 3.5, 7, 14, 25, and 100 khz. Off Mission Beach, attenuation data for two stations came from Hamilton et al (1956, Table II). The in-situ data for "Same area off Mission Beach" and "San Diego Trough" involved only two or three stations in each area. In these locations the values of the exponent of frequency n (rounded to 2 decimals in the tables), are 1.007, 1.024, 0.947, and 0.969. The data for "General sediment type" for fine sand involved 24 stations; the value of n is 1.007.

In Table 6B, the writer's in-situ measurements (Tables 1, 2, 6A) were combined with averaged laboratory values for general sediment types off San Diego (Table 3) from Shumway (1960). The

Table 6. Attenuation and other physical properties of sediments off San Diego combined from various programs

Sediment type	Mean grain diam. ϕ	Mean grain diam. mm	Density gm/cm³	Porosity percent	Velocity* m/sec	k†	n†	Frequencies‡ (no. stas.)	Ref.
A. *In situ* (Tables 1 and 2)									
Sand									
Medium	1.51	0.351	2.00	40.1	1714	0.47±0.06	1.01±0.05	a, b, c (2); g (1)	1
Fine	2.40	0.190	1.94	46.4	1680	0.46±0.08	1.01±0.06	a, b, c (4); d (3); g (17)	2
Very fine	3.23	0.107	1.94	47.3	1725	0.45±0.04	1.02±0.03	a, b, c (1); g (4)	1
Clayey silt	7.44	0.006	1.37	78.6	1450	0.19±0.01	0.95±0.03	b, c (2); d (1)	3
Clayey silt	7.41	0.006	1.36	78.5	1450	0.18±0.03	0.97±0.05	b (1); c (2)	3
B. *In situ* (A, above) combined with laboratory (Shumway, 1960), Table 3									
Sand									
Medium	1.45	0.366	2.00	39.2	1740	0.48±0.10	0.98±0.07	a, b, c (2); f (13); g (1)	2
Fine	2.42	0.187	1.93	46.5	1703	0.46±0.09	0.99±0.07	a, b, c (4); d (3); e (13); g (17)	2

* A. Velocities in situ; B. Velocities at 23° C.
† In $\alpha = kf^n$, where, attenuation α, dp/m; frequency f, khz; k is a constant; and n, the exponent of frequency.
‡ Frequencies (khz): a—3.5; b—7; c—14; d—25; e—26; f—26.6; and g—100.
1. Same area off Mission Beach.
2. General sediment type.
3. San Diego Trough.

value of n for 37 stations in fine sand is 0.992; for medium sand at 16 stations, n is 0.982.

Because of the larger number of stations, the data for "General sediment type" in Table 6 are considered more reliable: 1.007, 0.992, 0.982. As in previous tables, the values of n vary closely around 1, with the more reliable data at, or slightly below, 1.

In reconciling acoustic theory with experimental acoustic energy losses of sound incident on the sea floor, it is usually necessary to assume values of attenuation at frequencies of interest for various sediment types and layers. Such reconciliation has been successful in several studies (frequencies from 0.1 to 4 khz) in which a first-power dependency of attenuation on frequency was assumed (e.g., Cole, 1965; Bucker et al, 1965; Hastrup, 1970; Morris, 1970). Cole concluded that the first-power dependency can be extended to 100 to 900 hz.

Compressional velocity and attenuation data from Ulonska (1968) and Schirmer (1970) are possibly significant in determining the relationship between attenuation and frequency. Schirmer, at a water depth of 20 m in the Baltic Sea, measured attenuations of 0.2 to 0.4 db/m at 1.0 to 1.4 khz; an average sediment/water velocity ratio of 1.086 (1.05 to 1.14) was measured in a core from the site. Ulonska (1968), at a water depth of 22 m in the Baltic Sea, measured (Sta. 32) a velocity ratio of 1.082, and attenuations of 50 to 58 db/m at 250 khz. Both sets of measurements were at about 2-m depth in the sea floor. The velocity ratios indicate the sediment types were about the same. If a first-power dependence of attenuation on frequency is assumed, extrapolation of Ulonska's average data (54 db/m at 250 khz) to 1 khz yields an attenuation of 0.22 db/m (Schirmer measured 0.2 db/m at 1 khz).

In a recent study of attenuation in quartz sand in distilled water (which has been widely referenced), Hampton (1967) reported attenuation dependent on the square root of frequency. The measurements do not support this conclusion. The data (Hampton, 1967, Figure 9) are more in accord with a first-power dependence (as noted, also, by Mizikos, 1971).

The values of the exponent of frequency which were measured, computed, and assumed above, are close to that statistically computed by Attwell and Ramana (1966) for frequencies between one and 10^8 hz: 0.911. Strick (1970) made a case that

the dependence of attenuation on frequency should be close to, but less than one to satisfy causality. The writer believes the best of his experimental data are in fine sand off San Diego ($n = 1.007 \pm 0.060$). When these data are combined with Shumway's (1960) averaged data for fine sand, $n = 0.992 \pm 0.065$.

Data listed in Tables 1 to 4, and 6, are plotted in Figure 2 (frequency versus attenuation). It can be seen that most of the data are consistent with an approximate first-power dependency of attenuation on frequency over a wide frequency range. The upper and lower bounds of the data-plot probably define the area in which most natural marine sediments will lie. With regard to sediment type, the silt-clays lie in a narrow band at the lower side of the data-plot, and very fine sands, silty sands, and sandy silts at the top. Extrapolation of the silt-clay data to frequencies below 1 khz, using a first-power dependency, results in attenuation values in accord with the data of Bennett (1967, 40 to 900 hz) and Tullos and Reid (1969, 50 to 400 hz).

Two recent studies of complex rigidity and energy damping in artificial clays have important implications in determining parameters for elastic and viscoelastic models in saturated sediments. Cohen (1968) measured both μ and $i\mu'$ in complex rigidity ($\mu + i\mu'$) in artificial laboratory sediments composed of kaolinite and bentonite in distilled water with and without a deflocculating agent. Cohen demonstrated that both μ and $i\mu'$ were independent of frequency in the range 8.6 to 43.2 khz in flocculated clay, but when a deflocculating agent was added, the flocculated structure of the clay sediment dispersed, the material lost all rigidity μ and behaved as a Newtonian fluid in which there was viscous damping of wave energy which was linearly dependent on frequency. The addition of 35.5 ppt of NaCl caused reflocculation, and complex rigidity was the same as before. Krizek and Franklin (1968), in studies of shear-wave energy damping in flocculated kaolinite in distilled water, demonstrated that $1/Q$ was independent of frequency in the range 0.1 to 30 hz, and that the stress-strain hysteresis loop for a given cycle was that of a linear viscoelastic medium.

The two studies noted above have several important implications for saturated clay sediments, at least in the frequency range covered (0.1 hz to 43 khz), and probably at much higher frequencies:

1) saturated, flocculated clay sediments respond to shear-wave energy as linearly viscoelastic media, 2) the independence of energy damping $(1/Q_s = \mu'/\mu)$ from frequency implies that linear attenuation of shear waves should be proportional to the first power of frequency, and 3) suspensions (without flocculated structures) do not respond to wave energy as do flocculated clay structures, and almost all natural, high-porosity silt-clays have this general type of structure.

Recent measurements of attenuation in water-saturated, natural sediments over a frequency range from 3.5 to 1500 khz can be summarized as follows. Tables 1, 4, and 6 list 25 values of the exponent of frequency n between 0.94 and 1.26; however, all but two of the values fall between 0.94 and 1.11. The experimental evidence indicates that the dependence of attenuation on frequency is close to f^1, and does not support any theory calling for a dependence of attenuation on $f^{1/2}$ or f^2. However, as in the discussion of velocity dispersion, the case should not be overstated. As a reviewer pointed out, no single data set covers more than two orders of magnitude in frequency. While these data are enough to show that the dependence of attenuation on frequency is more nearly f^1 than $f^{1/2}$ or f^2, it is not enough to verify an exact dependence.

Review of elastic and viscoelastic models

In the field of soil mechanics, large static or dynamic stresses have to be considered, and over the full range of stresses, sediments may be elastic, viscoelastic, or plastic. Yong and Warkentin (1966, p. 80–94) have a good discussion of the various models and elements within the models which describe this behavior.

In the fields of soil mechanics and foundation engineering, the Hookean model and equations are commonly used for derivations of dynamic elastic constants and studies of vibrating loads (e.g., Barkan, 1962; Heukelom, 1961; Jones, 1958; Evison, 1956; Hardin and Richart, 1963; Hall and Richart, 1963; Richart and Whitman, 1967). However, the dynamic moduli from most velocity data are for very small strains on the order of about 10^{-6}, and corrections to moduli should be made for greater strains (Whitman et al, 1969, have a correction curve).

In the fields of physics and geophysics, studies of the elasticity of minerals and rocks have demonstrated that the elastic equations of the Hookean system adequately define the velocities of compressional and shear waves; these equations are conveniently interrelated in a table by Birch (1961, p. 2206). This field has been summarized by Birch (1966) and by Anderson and Lieberman (1968); papers of special interest are by Christensen (1966a, b), Brace (1965a, b), and Simmons and Brace (1965).

The question of water movement relative to mineral frame is a critical key to whether or not the equations of elasticity can be used in studies of wave velocities in rock and sediments. If the pore water does not move significantly with respect to the solids, then the effective density of the medium is the sum of the mass of the water and the solids in a unit volume. There is negligible or no velocity dispersion; energy damping is negligibly dependent, or independent of frequency; and the equations of Hookean elasticity can be used to study wave velocities and elastic constants within the frequency range in which these parameters are effective, unless attenuation is involved in the study. This is the "closed system" of Gassmann (1951). The closed system, as a special case in studies of the elasticity or viscoelasticity of saturated porous media, has been noted in many experimental and theoretical studies (Biot, 1941, 1956, 1962; Biot and Willis, 1957; Gassmann, 1951; Morse, 1952; White and Sengbush, 1953; Zwikker and Kosten, 1949; Brandt, 1955; Hamilton, 1971a; Paterson, 1956; Wyllie, Gregory, and Gardner, 1956; Laughton, 1957; Jones, 1958; Knopoff and Macdonald, 1958; McDonal et al, 1958; Birch, 1960, 1961, 1966; Barkan, 1962; Geertsma and Smit, 1961; Nafe and Drake, 1963; Brutsaert, 1954; Simmons and Brace, 1965; White, 1965).

Although the elastic equations of the Hookean model adequately account for wave velocities in most earth materials, they do not provide for wave-energy losses in these media. To account for energy losses, various anelastic (viscoelastic and "near-elastic") models and equations have been proposed. Viscoelastic models frequently favored are the Kelvin-Voigt, Maxwell, or some other combination of Hookean elastic springs and Newtonian dashpots (see Yong and Warkentin, 1966, for a concise resumé), or some variation of Biot's models (1956) in which a basic assumption involves movement of pore water of the Poiseuille type (at lower frequencies).

In his various papers, Biot (e.g., 1956, 1962)

discussed the full range of systems in which water within pore spaces does or does not move with the solids upon imposition of a small stress such as that of a sound wave. In some of these acoustic models this movement or flow of water through the sediment mineral structure was considered to be of the Poiseuille type. In the last several decades it has been determined that the simple flow equations of the Poiseuille type (derived from flow of water through tubes) do not hold for real, in-situ sediments. These equations have to be considerably altered, even for clean sands, and are not applicable to relatively impermeable clays (e.g., Yong and Warkentin, 1966). In other words, models based on Poiseuille-type flow of pore water are probably not applicable to natural sediments.

One model which has been especially studied in connection with rocks and sediments, is the Kelvin-Voigt model, in which, as originally defined, compressional-wave velocity varies with frequency; and attenuation, at frequencies of most interest in underwater acoustics and geophysics, increases with the square of frequency. White (1965, p. 110–112) has a thorough discussion of theory and experimental evidence on this subject, and concludes (p. 112) that neither velocity nor attenuation shows this frequency dependence, and the Voigt solid cannot be considered an adequate model of earth materials. The evidence of this report and earlier ones (Hamilton et al, 1970; Hamilton, 1971a) are in accord with this conclusion.

A viscoelastic model for water-saturated sediments

In the absence of sufficient experimental evidence, it has been possible to construct rather elegant theoretical approaches, altered if necessary by constants to fit available data. To derive such theoretical models, one must start with assumptions. In the case of water-saturated sediments, some of the less tenable of these assumptions have been that, 1) all water-saturated sediments are analogous to suspensions of mineral particles in fluids, 2) all the mineral particles are spheres, 3) Poiseuille flow operates in natural sediments, 4) pore water necessarily moves relative to the mineral frame or sediment structure, and 5) sediments lack rigidity, in which case the shear modulus is zero, and Poisson's ratio is 0.5.

All of the above assumptions are invalid in part or in whole. In recent papers, Hamilton

(1970, 1971a) discussed several aspects of sediment structure and elasticity. Some conclusions are pertinent: almost all saturated sediments have mineral particles which are not spheres, near suspensions are unusual, and almost all sediments have sufficient rigidity to allow transmission of shear waves. As noted previously, Yong and Warkentin (1966) indicate Poiseuille flow (through small tubes) does not hold for natural sediments.

Given macroscopic isotropy, small, sinusoidal stresses, wavelengths much greater than grain size, and frequencies from a few Hertz to at least several hundred khz (and probably into the mhz range for most natural sediments), some parameters in addition to those in the preceding paragraph are as noted in previous sections: attenuation in db/linear measure is approximately dependent on the first power of frequency, and velocity dispersion, if present, is small. Some relative movement of pore water and mineral frame cannot be excluded on the basis of present evidence, although the above parameters indicate that, if present, it should be small.

The model proposed below is intended as a tentative, working model. It should be emphasized that other models are not excluded if they are within the stated parameters. The whole subject merits much more experimental and theoretical study.

A model and concomitant equations within the parameters noted above is a case of linear viscoelasticity. The basic equations of linear viscoelasticity have been summarized in an excellent treatise by Ferry (1961). For the model recommended in this paper, the basic equations (Adler, Sawyer, and Ferry, 1949) have been discussed in different form, including neglect of negligible factors, by Nolle and Sieck (1952), Ferry (1961, p. 93–94), Krizek (1964), White (1965), Krizek and Franklin (1968), Hamilton et al (1970), and others.

In the above model, the Lamé elastic moduli μ and λ are replaced by complex moduli, $(\mu+i\mu')$ and $(\lambda+i\lambda')$, in which μ, λ, and density govern wave velocity, and the imaginary moduli $i\mu'$ and $i\lambda'$ govern energy damping. The following (Ferry, 1961, p. 11–13) illustrate the stress-strain relations in this model. For a sinusoidal wave, if the viscoelastic behavior is linear, the strain will be out-of-phase with stress. The stress can be vectorially decomposed into two components: one in phase with strain and one 90 degrees out of phase. For a shear wave, the complex stress/strain ratio is $\mu^* = \mu + i\mu'$. The phase angle ϕ which expresses energy damping is, in this case: $\tan\phi = \mu'/\mu$.

The basic derivations of the above model are in Ferry (1961) and White (1965) and will not be repeated here. Without assumptions as to negligible factors, the equations of the model in the form of Bucker (in Hamilton et al, 1970, p. 4046), or in Ferry (1961, p. 94, 419), reduce to the following for both compressional and shear waves (with some changes in notation):

$$\frac{1}{Q} = \frac{aV}{\pi f - \dfrac{a^2 V^2}{4\pi f}}, \tag{2}$$

where $1/Q$ is the specific attenuation factor, or specific dissipation function, a is the attenuation coefficient, V is wave velocity, and f is frequency (circular frequency, $\omega = 2\pi f$).

Subscripts (p or s) can be inserted into equation (2) when referring to compressional or shear waves.

When energy damping is small (i.e., $\lambda' \ll \lambda$ and $\mu' \ll \mu$, White, 1965, p. 95; Ferry, 1961, p. 123: $r \ll 1$, where $r = aV/2\pi f$), the term in the denominator of equation (2), $a^2 V^2/4\pi f$, is negligible and can be dropped. This leaves the more familiar expression (e.g., Knopoff and Macdonald, 1958; White, 1965; Bradley and Fort, 1966; Attwell and Ramana, 1966):

$$\frac{1}{Q} = \frac{aV}{\pi f} \quad \text{and} \tag{3}$$

$$\frac{1}{Q} = \frac{2aV}{\omega} = \frac{\Delta}{\pi} = \tan\phi. \tag{4}$$

Additionally,

$$\frac{1}{Q_p} = \tan\phi_p = \frac{\lambda' + 2\mu'}{\lambda + 2\mu}, \tag{5}$$

$$\frac{1}{Q_s} = \tan\phi_s = \frac{\mu'}{\mu}, \tag{6}$$

$$\frac{\Delta E}{E} = \frac{2\pi}{Q}, \quad \text{and} \tag{7}$$

$$\alpha = 8.686\, a, \tag{8}$$

where (in addition to those symbols already defined) Δ is the logarithmic decrement (log of the ratio of two successive amplitudes in an exponentially decaying sinuosidal wave), $\tan\phi$ is the loss angle, $\Delta E/E$ is fraction of strain energy lost per stress cycle, and α is attenuation in db/linear measure (e.g., db/cm).

Equations involving compressional- and shear-wave velocities in Hamilton et al (1970), or in Ferry (1961), are (in Ferry's notation)

$$(\lambda + 2\mu) = \rho V_p^2 (1 - r^2)/(1 + r^2)^2 \quad \text{and} \tag{9}$$

$$\mu = \rho V_s^2 (1 - r^2)/(1 + r^2)^2, \tag{10}$$

where $r = aV/2\pi f$, $\lambda =$ Lamé's constant, $\mu =$ rigidity, and $\rho =$ density.

In equations (9) and (10), the term $(1 - r^2)/(1 + r^2)^2$ indicates the degree of velocity dispersion for linear viscoelastic media. When damping is small (defined above), this term is negligible, and can be dropped, as implied by Ferry (1961, p. 94). This leaves the more familiar Hookean equations

$$(\lambda + 2\mu) = \rho V_p^2 \quad \text{and} \tag{11}$$

$$\mu = \rho V_s^2. \tag{12}$$

This means that if the factor $(1 - r^2)/(1 + r^2)^2$ in equations (9) and (10) is considered negligible and is dropped, that wave velocity $1/Q$ and the log decrement are independent of frequency, and linear attenuation is proportional to the first power of frequency.

Computations with the data of this report and from the literature indicate that most water-saturated rocks and sediments qualify under the above definitions as media with "small damping." Therefore, equations (3), (4), (11), and (12) should apply to both water-saturated sediments and rocks. However, those investigators who wish to include velocity dispersion and $1/Q$ or a log decrement dependent on frequency can consider equations (2), (9), and (10).

In the sediments discussed in this report, the quality factor, or specific dissipation function Q, $1/Q$, and the logarithmic decrement Δ are approximately independent of frequency. This follows from the dependence of attenuation (approximately) on the first power of frequency. Table 7 lists these properties for the in-situ measurements in Tables 1 and 2. A conclusion from the data of this report is that the approximate frequency independence of $1/Q$, Q, and Δ, can be extended from rock (e.g., Knopoff, 1965) through most natural, water-saturated sediments.

Relationships between attenuation and other properties

The relationships between attenuation and other physical properties in saturated sediments are of considerable importance in determining the causes of attenuation and in selecting appropriate anelastic models. In this section, various relationships will be briefly noted and illustrated prior to discussions of the causes of attenuation.

As discussed in a previous section, attenuation in db/linear measure (e.g., db/m) is approximately dependent on the first power of frequency; that is, in the equation $\alpha = kf^n$, n is close to one. If n is taken as one, the only parameter in the equations for various sediments is the constant k. This constant is particularly useful in relating attenuation to other sediment properties such as grain size and porosity. Relations between k and other physical properties give an insight into

Table 7. The quality factor Q_p, the specific attenuation factor $1/Q_p$, and the logarithmic decrement Δ_p, from in-situ measurements of compressional-wave velocity V_p and attenuation α_p in sediments off San Diego*

Sediment type	Q_p	$1/Q_p$	Δ_p	Ref.
Sand				
Coarse	32	0.031	0.099	1
	29	0.034	0.107	2
Medium	31	0.032	0.101	1
	35	0.028	0.089	2
Fine	31	0.032	0.100	1
	31	0.032	0.101	2
	44	0.023	0.071	3
Very fine	32	0.031	0.093	1
	29	0.034	0.107	2
Sandy silt	23	0.044	0.138	2
	54	0.018	0.058	3
Sand-silt-clay	31	0.033	0.102	2
Clayey silt	104	0.010	0.030	2
	111	0.009	0.028	1
	114	0.009	0.028	1
	118	0.009	0.027	3
	263	0.004	0.012	1
	437	0.002	0.007	1
Sand-silt-clay	368	0.003	0.009	1

* Computed from equations

$1/Q_p = a_p V_p / \pi f = \Delta_p / \pi$; attenuation $\alpha_p = 8.686 a_p$.
1. Table 1(A): frequency, 14 khz.
2. Table 2: frequency, 100 khz.
3. Table 1(B): frequency, 25 khz.

causes of attenuation, and allow prediction of attenuation (as discussed in the last section) after deriving a value of k from its relationships with mean grain size and porosity.

Assuming that linear attenuation is dependent on the first power of frequency, values of k can be easily computed by dividing attenuation by frequency. This was done for all data in Tables 1 to 4 except as follows: 1) the values of k from Table 1a were determined from the 14 khz measurements (considered most reliable), and 2) for three measurements in Table 4a which did not show a first-power dependence, k was determined from an attenuation computed at a frequency near the mid-range of the measurements. These values of k were then plotted versus mean grain size and porosity in Figures 3, 4, and 5.

Relationships between k (or attenuation) and mean grain size, porosity, and dynamic rigidity are illustrated and discussed in the following sections. The data are for natural sediments; both in-situ and laboratory measurements are included. Averaged values from Tables 1 to 3 were used as primary data to establish regression equations (in the figure captions), but the individual measurements which were averaged are shown to better illustrate the trends and scatter of the data. These regression equations are included for use in predicting attenuation when grain size and porosity are known. These equations are strictly empirical and are recommended only within the limiting values indicated. The values of k so obtained are approximations, but it is predicted that most future measurements of attenuation will result in k values which fall within the indicated "envelopes."

Attenuation and grain size.—When wavelengths approach grain size, Rayleigh scattering takes place, and attenuation is proportional to the 4th power of frequency. This effect has been discussed by Knopoff and Porter (1963) in the case of granite. Rayleigh scattering appears to be a factor in Busby's and Richardson's (1957) measurements of attenuation in the mhz range in sands. For most sediments, Rayleigh scattering is not a factor to at least several hundred khz, and probably into the mhz range.

Figure 3 illustrates the relationships between the constant k (in $\alpha = kf'$) and mean grain size for most of the sediments listed in Tables 1 to 4. Mean grain size is plotted in logarithmic phi

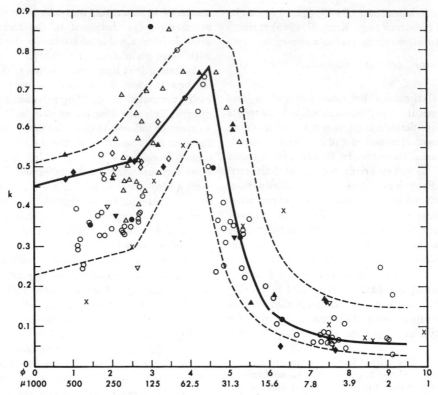

FIG. 3. Mean grain size M_z in ϕ units and microns μ versus k (in $\alpha = kf^1$) in natural, saturated sediments (Tables 1–4a); see text. Solid symbols are averaged values for data off San Diego: diamonds–1966–70; triangles–1956; inverted triangles–1962; circles–Shumway (1960), Table 3; open symbols–data in averages; X–literature value from Table 4a; area between dashed lines–predicted area within which most data should fall. Regression equations for solid lines (for data off San Diego and selected literature values recommended for use in similar sediments): Coarse, medium, and fine sand, in part (0 to 2.6ϕ): $k = 0.4556 + 0.0245\ (M_z)$. Fine sand (in part), very fine sand, and mixed sizes (2.6 to 4.5ϕ): $k = 0.1978 + 0.1245\ (M_z)$. Mixed sizes (4.5 to 6.0ϕ): $k = 8.0399 - 2.5228\ (M_z)$ $+ 0.20098\ (M_z)^2$. Silt-clays (6.0 to 9.5ϕ): $k = 0.9431 - 0.2041\ (M_z) + 0.0117\ (M_z)^2$.

units ($\phi = -\log_2$ of grain diameter in mm; see Inman, 1952, for discussion). In studying Figure 3, one can easily translate k to linear attenuation in db/m, because at any given frequency the only variable is k.

Figure 3 illustrates the distinct difference between mean grain size M_z and the constant k (or attenuation) in sands ($M_z = 0$ to 4ϕ, or 1 to 0.0625 mm) and in the silt-clays ($\phi > 4$). In sands (Figures 3, 4), k increases gradually with ϕ (or decreasing grain size) from coarse into fine sands (to about 2.6 ϕ), and then k increases rapidly into the finer sand sizes. Other investigators have measured increased attenuation in sands with decreasing grain size (e.g., Shumway, 1960; Hampton, 1967, Figure 9; McCann and McCann, 1969). The maximum values of k are in very fine sand, and mixtures of sand, coarse silt, and clay

(e.g., silty sand, sandy silt) in the grain-size range of 3.5 to 4.5ϕ (0.09 to 0.04 mm). Equivalent lower values of k are in coarser-grained sand and in silt, and finer-sized sandy silt. Attenuation decreases with decreasing grain size (increasing ϕ) from about 4.5ϕ to about 6ϕ, and then gradually declines with decreasing grain size into fine silts and clays.

A semilog plot (using phi units: Figure 3) tends to clarify relationships between grain size and other properties in the finer sizes (silt-clays), but obscures relationships in sands. Therefore, mean grain size in mm is plotted versus k in Figure 4. Figure 4 illustrates better than Figure 3 the gradual increase of k from coarse and medium sand sizes into fine sands, and the marked increase in k from about 0.17 mm (2.6ϕ) into the finer sand sizes.

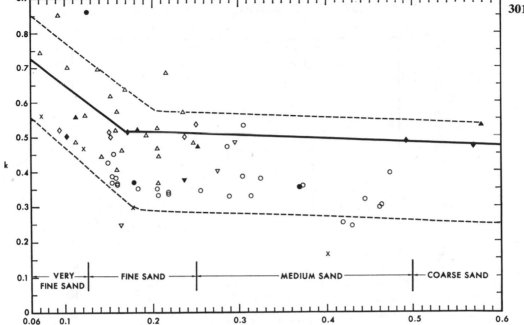

FIG. 4. Mean grain size mm versus k (in $\alpha=kf^1$). Data, symbols, and remarks: same as in caption for Figure 3. Plot of mean grain size in mm emphasizes relationships in sands (see text). Regression equations for solid lines: Coarse, medium, and fine sand, in part (0.6 to 0.167 mm): $k=0.5374-0.1113\ (M_z)$. Fine sand (in part), and very fine sand (0.167 to 0.063 mm): $k=0.8439-1.9431\ (M_z)$.

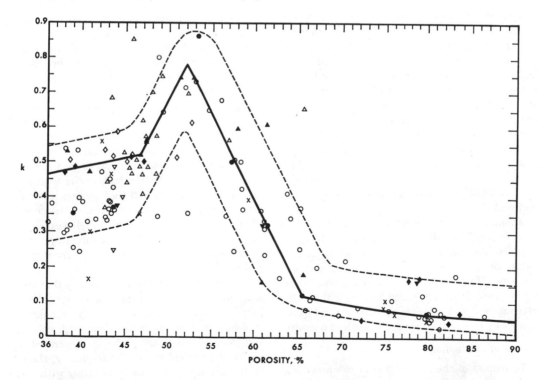

FIG. 5. Porosity n, percent, versus k (in $\alpha=kf^1$). Data, symbols, and remarks same as in caption for Figure 3. Regression equations for solid lines: Coarse, medium, and fine sand (36 to 46.7 percent): $k=0.2747+0.00527\ (n)$. Very fine sand and lower-porosity mixed sizes (46.7 to 52 percent): $k=0.04903\ (n)-1.7688$. Mixed sizes (52 to 65 percent): $k=3.3232-0.0489\ (n)$. Silt-clays (65 to 90 percent): $k=0.7602-0.01487\ (n)+0.000078\ (n)^2$.

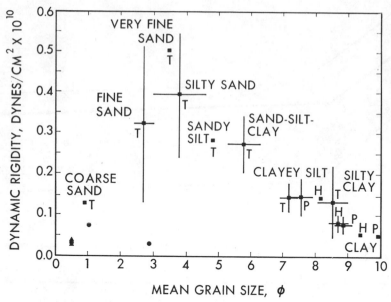

FIG. 6. Mean grain size versus computed values of the shear modulus (dynamic rigidity); from Hamilton (1971). Squares–mean values; crosses–3 times the standard error of the mean. T–samples from the continental terrace (shelf and slope); P–abyssal plain; H–abyssal hill. Lower left: St. Peter sand (circles) and Ottawa sand (triangles). See Hamilton (1971) for discussion.

Attenuation and porosity.—Porosity varies with k (Figure 5) in the same way as mean grain size (Figures 3 and 4). Equivalent values of k are apt to be found in the coarser sizes of sand and in higher-porosity silts and sandy silts. The highest values of k are in very fine sands, silty sands, and sandy silts in the porosity range of 50 to 54 percent. In silt-clays, k decreases with increasing porosity. Shumway (1960, Figure 6) indicates a similar variation between attenuation and porosity at frequencies between 20 and 40 khz. Relationships between grain size and porosity as they affect attenuation will be discussed in a section below.

Attenuation and dynamic rigidity.—Figures 6 and 7 (reproduced from Hamilton, 1971a) illustrate the probable variations of computed values of dynamic rigidity μ with grain size and porosity (Hamilton, 1970a). Comparison of these figures with Figures 3, 4, and 5 indicate that k (or attenuation) and values of rigidity respond in the same way to variations in grain size and porosity.

Causes of attenuation

In studies of the attenuation of compressional and shear waves in saturated porous media, there are usually two (or a combination of two) common viewpoints: 1) a viscoelastic model is used

in which pore water moves relative to mineral grains, and pore-water viscosity and media permeability are dominant factors in viscous sound absorption, or 2) a linear viscoelastic, or "nearly elastic," model is favored in which attenuation is mostly, or entirely, due to internal friction (i.e., energy is lost in intercrystalline or intergrain movements).

The case has been made that in crustal rocks, both dry and saturated, internal friction is by far the most probable, dominant process in wave-energy damping at seismic frequencies to at least several hundred khz, and that viscous losses are probably negligible (e.g., Knopoff and Macdonald, 1960; White, 1965). The writer believes, as discussed below, that internal friction is by far the dominant dissipative process in water-saturated sediments.

It is apparent from the present study that sands with grain-to-grain contacts and no cohesion (physicochemical net attraction) should be studied separately from silt-clays with cohesion and mineral particles probably separated by layers of adsorbed water. Consequently, the following discussion will deal separately with these two structural types. The structures of these and other sediment types were reviewed in earlier papers (Hamilton, 1970, 1971a).

Attenuation in sand.—Study of attenuation in sands is of particular interest in deciding on causes of attenuation because these sediments are the most permeable, and viscous losses, if present, should be higher than in other sediment types. On the other hand, if frictional energy losses are dominant, the intergrain reactions to stress-causing attenuation should be related to the same complex factors which affect dynamic rigidity and static shear strength. These factors will be reviewed briefly; a more extensive review (Hamilton, 1971a) contains numerous references to items outlined below.

Sands have grain-to-grain contacts between mineral particles, and resistance to shear stress is related to sliding and rolling friction between grains, to the number of intergrain contacts, and interlocking between grains [see Yong and Warkentin (1966) for a good resumé].

The number of interparticle contacts in sands depends on grain size and density of packing (i.e., loose, dense, or, more normal—an intermediate packing). At any given particle size, porosity is a measure of packing.

Interlocking of grains increases with density of packing and angularity of grains.

Laboratory experiments with clean quartz sand grains (St. Peters or Ottawa sand) have indicated the following (Hardin and Richart, 1963):

1) at the same grain size, dynamic rigidity increases with decreasing porosity in sands with round grains because of more interparticle contacts with denser packing, 2) at the same grain size and porosity, sands with angular grains have higher dynamic rigidities and shear-wave velocities than sands with round grains because of interlocking between grains, and 3) at the same grain size, increased angularity of grains (causing greater rigidity) can be more important than increases in porosity (causing lesser rigidity).

The grains of natural marine sands are much more angular than those in St. Peters or Ottawa sands at any given grain size. In natural sands, angularity increases and sphericity decreases with decreasing grain size (Pettijohn, 1957). Interlocking of grains should be greater when coarse silt particles are present.

The discussion above, and the results of Hamilton (1971a), imply and predict, for natural marine sediments: 1) coarse sand has fewer intergrain contacts and the grains are rounder than in the finer sizes; thus, dynamic rigidity should be a minimum (for sands) in coarse sand because of lesser intergrain friction, 2) dynamic rigidity should be less in sands composed of the highly rounded grains of St. Peters or Ottawa sand than in natural sands of the same density, porosity, and grain size; thus tests with St. Peters or

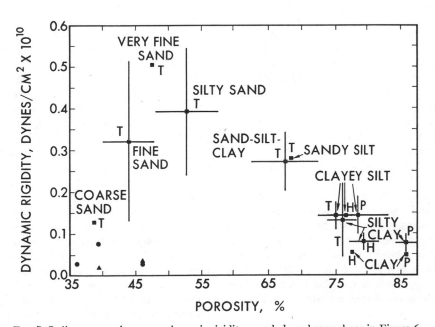

FIG. 7. Sediment porosity versus dynamic rigidity; symbols and remarks as in Figure 6.

Ottawa sand cannot always be directly related to natural sands; 2) Hardin and Richart (1963) observed that an increase in porosity will cause a decrease in rigidity—this is only true in sands of the same grain size and angularity; 3) in natural sands, porosity increases with finer sizes which have more numerous intergrain contacts and more angular grains which may cause rigidity to increase with increasing porosity; 4) at some relatively fine sand-silt size, intergrain friction and interlocking will reach a maximum and rigidity will be at a high point for natural, uncemented sediments; 5) when sand grains are no longer in contact because of increased amounts of fine silt and clay, intergrain friction and interlocking are no longer relatively effective, and rigidity is mostly due to cohesion between finer particles; 6) cohesion between fine particles varies with the numerous factors discussed below, but, in general, cohesion and rigidity decrease with increasing porosity and decreasing grain size in higher-porosity silt-clays.

Figures 6 and 7 illustrate the above phenomena. In the sands, coarse sand has the least rigidity, and natural sands have higher rigidities than round-grained St. Peters sand. Rigidity increases sharply with decreasing grain size and increasing porosity to maximum values in very fine sands, silty sands, and coarse silts. These maximum values of dynamic rigidity occur between mean grain sizes of 3.5 to 4.5ϕ (0.09 to 0.04 mm), and porosities between 55 and 60 percent. In a density-versus-dynamic-rigidity plot (not shown), maximum rigidities occur between densities of 1.7 and 1.8 g/cm^3. As grains become finer, rigidity decreases with increasing porosity and decreasing density (Figure 7), and decreasing grain size (Figure 6). Scatter diagrams of percent sand size by weight versus rigidity (not shown) indicate that maximum rigidity normally occurs when a sediment is composed of about 60 to 65 percent sand by weight.

Dynamic rigidity is a measure of resistance (friction and interlocking between grains) to shearing forces which tend to move grains. If attenuation is due to energy lost by friction between grains, then rigidity and attenuation should vary because of the same factors. However, as with rigidity, there is complex interaction between grain sizes, shapes, density of packing, or porosity.

Mean grain size is an important factor in both rigidity and attenuation. In coarse sands, the grains are larger and more round, interparticle contacts fewer, and surface areas smaller, than in the finer sands. Consequently, rigidity is relatively low in coarser-grained sands. When grains are moved, fewer interparticle contacts result, also, in low attenuation due to friction (Figures 3, 4). As grain size decreases, the grains become more numerous in a unit volume, they are more angular, and there are greater surface areas and more interparticle contacts; consequently, rigidity increases. When grains are relatively moved, however, the more numerous interparticle contacts and greater surface areas result in greater energy losses. This has been shown in the laboratory (e.g., Hampton, 1967, Figure 9) and in situ (this report; McCann and McCann, 1969).

The change in rate of increase of k from coarser sands at about 2.6ϕ, or 0.17 mm, into the finer sand sizes (Figures 3, 4) requires further discussion. The change is apparently related to the relationships between grain size, porosity, and surface area, or to the number of interparticle contacts.

A semilog plot of mean grain size (in phi units) versus porosity (Hamilton, 1970, Figure 2) indicates an almost linear relationship. This type of plot obscures the real relationship. Figure 8, an arithmetic plot of grain size in mm versus porosity, illustrates the very gradual increase of porosity with decreasing grain size in the coarse and medium sand sizes, and the marked increase in porosity which occurs as grain sizes decrease from fine sands into the silt-clay sizes (porosities greater than about 60 percent). There is much scatter in the relationships between grain size and porosity, because at any given grain size, porosity varies with packing (i.e., dense, loose).

If attenuation is related to energy lost in intergrain friction, a critical factor involves the number of intergrain contacts, or surface areas involved, in a unit volume. If grain sizes decreased without a change in porosity, the more numerous contacts should lead to an increase in attenuation. If porosity increased without a change in grain size, the fewer intergrain contacts should lead to decreased attenuation. In coarse and medium sands there is a relatively small increase in porosity with decrease in grain size (Figure 8). As grain sizes decrease from 0.500 to 0.180 (1 to

2.5ϕ), there is an increase in porosity of about 5 to 6 percent, and, according to Shumway and Igelman (1960), grain-surface areas increase by about 240 percent. From 0.180 to 0.063 mm (2.5 to 4ϕ), porosity increases about 9 to 10 percent, but grain surface area increases by about 320 percent. In summary, as grain sizes decrease from 0.500 to 0.063 mm (1 to 4ϕ) there is an increase of about 15 percent porosity, but grain surface areas increase on the order of 600 to 700 percent. If attenuation is due to energy lost between mineral particles in contact, then the increase in surface area (more mineral particles in contact in a unit volume), should cause increased attenuation (k higher) which is not offset by greater porosity. In addition, the rate of increase in attenuation should be less from 0.500 to 0.180 mm (coarse-medium sand into fine sand) than the rate from about 0.180 mm into finer sizes (the increased rate actually occurs at about 0.17 mm: Figures 3, 4).

Figure 5 (porosity versus k) illustrates these same relationships: k (or attenuation) increases gradually as porosity increases from 38 to about 47 percent, and then increases sharply from this porosity range to porosities of about 52 percent. These increases in attenuation with increasing porosity in sands, are related through grain size, and the more numerous particles and greater grain surface areas, as discussed above.

Attenuation in fine silts and clays.—In fine silts and clays, shear strength, dynamic rigidity, and attenuation are apparently related to cohesion between fine particles.

Cohesion is the resistance to shear stress which can be mobilized between adjacent, fine particles which stick, or cohere, to each other. Cohesion is considered to be an inherent property of fine-grained, clayey sediments which is independent of stress; it is caused by physicochemical forces of an interparticle, intermolecular, and intergranular nature. This subject was reviewed in an earlier paper (Hamilton, 1971a) which contains numerous references to the following items. Clay particles may not be in contact; they are apparently surrounded by layers of adsorbed water through which they interact with other particles; the amount of pore water, the distance between

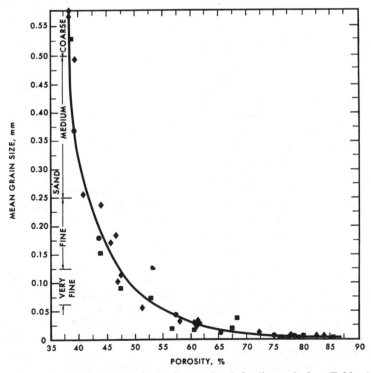

Fig. 8. Sediment porosity versus mean grain size in mm. Symbols: diamonds–from Tables 1 and 2; circles–Table 3, from Shumway (1960); squares–from Hamilton (1970).

particles, and the number of interparticle contacts are important influences on cohesion. At points of near contact between clay particles there is often bonding of the nature of cementation, especially in the presence of iron oxides, calcium, silica, and other minerals in solution in interstitial waters; where sediments have been exposed to overburden pressures there is apt to be pressure-point solution and redeposition. The structure of the mass of clay particles is important; for example, it has been demonstrated that the flocculated, or "cardhouse" structure (Hamilton, 1970, Figure 1d) is the strongest; these structures are largely determined by interparticle forces and the number of interparticle contacts. Differing clay minerals affect cohesion because of particle size and differing interparticle forces; for example, Na-montmorillonite has stronger cohesive bonds than kaolinite. Shear stress in clayey sediments occurs between particles and not through them; near-contact points will deform elastically, or plastically (depending on stress), by an amount to sustain the effective stress. Homogeneous clays are practically impervious; for example, the coefficient of permeability (cm/sec) in clean sand is of the order of 10^{-2}, whereas, this coefficient in homogeneous clay is of the order of 10^{-8} (Leonards, 1962).

The highest attenuation in marine sediments is in very fine sand and coarse silt, and mixtures of sand and silt. As grain sizes become smaller, the sand and coarser silt grains become fewer and separated, and relative movement between grains is controlled by complex interparticle forces (cohesion). Cohesion, dynamic rigidity, and attenuation become less as grain sizes decrease and porosities increase into high-porosity silt-clays. These effects are apparently due to the same causes: the strength of interparticle, net attractive forces, and the number of interparticle contacts. These forces become weaker, and contacts fewer with higher porosities, given the same mineral and structural type.

The sharp inflection in the curves relating k to porosity at about 65 percent (Figure 5), and at a mean grain size of about 6ϕ (0.016 mm; Figures 3, 4), indicates the probable range of porosity and mean grain size where the separation of larger grains (sand and silt) occurs. McCann and McCann (1969) also selected 6ϕ as the grain size

where interparticle forces become dominant.

It is important in studies of attenuation in clay-water systems to understand the different mineral structures. Clay, in the presence of an electrolyte (e.g., sea water), forms flocculated structures which have a finite rigidity and transmit shear waves (e.g., Hamilton, 1971a). When this flocculated structure is dispersed and the physico-chemical bonds are broken, the mixture is an actual, or near, suspension. Attenuation in dispersed or flocculated clay-water systems is distinctly different. Cohen (1968) demonstrated that both μ and $i\mu'$ were essentially independent of frequency in flocculated kaolinite in distilled water, but when a deflocculating agent was added, the mixture lost rigidity μ and behaved as a Newtonian fluid. This means that $1/Q$ for shear stresses (μ'/μ) was independent of frequency and attenuation dependent on the first power of frequency in the flocculated structure, but in the dispersed mixture, attenuation was dependent on f^2. When Cohen added NaCl, the mixture reflocculated and complex rigidity was the same as before. De Graft-Johnson (1968) experimentally showed that dispersed clay structures have higher log decrements than flocculated structures at the same water content; McCann (1969) explained this effect theoretically. Except as an approach to understanding theory, experimental data on attenuation in dispersed clay-water systems are not applicable to almost all marine sediments.

The data listed in the tables and plotted in the figures do not include attenuation in calcareous sediments. There is a small amount of evidence which indicates that high-porosity calcareous "oozes" may have higher attenuation values than sediments formed by clay minerals of the same general porosity and grain size. This might be due to the nature of interparticle bonds and the presence of foraminiferal tests in the calcareous oozes.

Effects of wetting.—When a dry sand is wetted, shear velocity and rigidity are both reduced (e.g., Hardin and Richart), 1963; Hamilton, 1971a: Figures 8, 9), and the log decrement increases (Hall and Richart, 1963). Pilbeam and Vaisnys (1971) also noted that lubrication of grains in a granular aggregate generally increased relative energy losses. The role of rigidity μ can be seen by substituting $(\mu/\rho)^{1/2}$ for V_s in equation (4).

$$a_s = \Delta_s(\rho)^{\frac{1}{2}} f / \mu^{\frac{1}{2}}. \qquad (13)$$

Dynamic rigidity μ is less in wet than in dry sand, apparently because lubrication allows grains to move under less shear stress (Hardin and Richart, 1963); rigidity is also less in saturated porous rock than in dry rock (Wyllie et al, 1962, p. 578). Hardin and Richart noted that an increase of only 1.4 percent in water content can reduce rigidity by as much as 15 percent. Experimentally, the log decrement of shear waves in sand and rock increases upon wetting (Hall and Richart, 1963; Wyllie et al, 1962), but shear-wave velocity decreases because of the decrease in rigidity; density would be greater wet than dry. Thus, all factors in equation (13) result in increase in a_s (after wetting) at any given frequency. A variation of equation (4), $\Delta_s = a_s V_s / f$, indicates that the increase in the log decrement upon wetting is due entirely to increase in a_s because V_s decreases. The apparent reason for the increase in attenuation of shear waves upon wetting or lubrication is that, given the same grain size and porosity (or the same number of interparticle contacts), the reduced rigidity allows greater slippage of grains under any given shear stress, resulting in greater energy losses because of greater intergrain friction.

Effects of pressure.—Although attenuation decreases with increasing effective pressure (e.g., overburden pressure), as discussed below, it is believed that the laboratory and in-situ measurements of this report are comparable because of the low pressures involved. This was also a conclusion of McCann and McCann (1969) after comparing their in-situ measurements, at depths of 2 m in sediments, with the laboratory measurements of Shumway (1960). The in-situ measurements using probes (listed in Tables 1–4a) were at sediment depths from about 0.3 to 2 m; thus, effective pressures were very small. Laboratory samples had negligible effective pressures. Thus, the data and conclusions of this report are concerned with surface sediments under very low, or negligible effective pressures. Some of the measurements at seismic frequencies in Table 4b involved greater effective (overburden) pressures, but these were not involved in calculations of equations relating attenuation and frequency.

Increase of effective pressure increases rigidity, and shear- and compressional-wave velocity in rocks (e.g., Birch, 1966; Nafe and Drake, 1967) and sand (e.g., Hardin and Richart, 1963; Hamilton, 1971a), but in both rocks and sand, attenuation and log decrement decrease (e.g., for rocks: Birch and Bancroft, 1938; Levykin, 1965; Balakrishna and Ramana, 1968; for sand: Hunter et al, 1961; Hall and Richart, 1963). As various authors have noted, these effects are apparently due to pressure effects on grain elastic moduli and grains in harder contact, which offers greater resistance to shear stress (shear modulus, or rigidity, greater), but allows less intergrain movement which reduces energy loss through intergrain friction (attenuation less).

Prediction of compressional-wave attenuation

Estimation or prediction of compressional-wave attenuation in saturated sediments is of considerable importance in geophysics and underwater acoustics. The relationships between attenuation (as expressed by the constant k) and grain size and porosity, previously discussed and illustrated (Figures 3, 4, 5) afford a very simple method for prediction of approximate values of attenuation at most frequencies of interest in geophysics and underwater acoustics in all major sediment types (with the possible exception of calcareous sediments).

In this report, attenuation is related to frequency in the form [equation (1)]: $\alpha = k f^n$, where attenuation α is in db/m and frequency f is in khz. The case has been made that the exponent of frequency n is close to one. If n is taken as one, the only variable in the equation is the constant k. This constant varies with mean grain size (Figures 3, 4) or porosity (Figure 5).

Mean grain size of a sediment can be determined easily in both wet and dry sediment; porosity can be determined in saturated materials. In the absence of measurements, both properties can be predicted, given general environment and predicted sediment type. Hamilton (1970, 1971b) discussed and listed averages of these properties for various sediment types in the major environments, and discussed methods of prediction. Technically, porosity should be a better index to attenuation because it is a better measure of the number of interparticle contacts (at the same grain size, porosity and the number of inter-particle contacts depends on packing).

The method for estimating or predicting approximate values of attenuation (given mean grain size, or porosity) follows (a numerical example is included):

1) Determine or predict mean grain size or porosity of the sediment.

Example: mean grain size $M_z = 7.5\phi$.

2) Enter the mean grain size or porosity-versus-k diagram (Figures 3, 4, and 5), and determine the value of k. Regression equations for these data are listed in the figure captions.

Example: entry into the mean-grain-size-versus-k diagram (Figure 3) yields: $k = 0.07$.

3) Insert k into equation (1), reproduced above.

Example: insertion of the value of k yields: α, db/m $= 0.07f^1$, where frequency f is in khz.

4) Compute attenuation in db/m at the desired frequency (khz).

The experimental evidence indicates that the exponent of frequency probably varies in saturated sediments between 0.9 and 1.1. Theoretically, or statistically, the value might be slightly less than one (Strick, 1970; Attwell and Ramana, 1966). Use of a value of k, assuming n is one (rather than 0.9 or 1.1), results in small differences in computed attenuation at lower frequencies which are of most interest in geophysics and underwater acoustics. For example, at 3 khz, if 0.07 is taken as k, and n is varied between 0.9, 1.0, and 1.1, the computed attenuations (db/m) are 0.19, 0.21, and 0.23; at $f = 50$ hz, the values of attenuation (db/m) are 0.005, 0.004, and 0.003. In a fine sand, if k is 0.5, the values of attenuation (db/m) at 3 khz (and $n = 0.9$, 1.0, and 1.1) are 1.34, 1.50, and 1.67; at 50 hz, attenuations are 0.034, 0.025, and 0.019.

ACKNOWLEDGMENTS

The considerable efforts of others in the laboratory and field have been acknowledged in previous reports (Hamilton, 1963, 1970, 1971; Hamilton et al, 1956, 1970). In connection with this report, aid in computer analyses were by Susan B. Dascomb, Jackie J. Phillips, and P. L. Sherrer. The writer benefitted from critical reading and discussions with H. P. Bucker, Halcyon E. Morris, D. G. Moore, E. C. Buffington, and official reviewers. This work was supported by the Naval Ship Systems Command (SF11-552-101; Task 00539) and the Ocean Science and Technology Division, Office of Naval Research.

REFERENCES

Adler, F. T., Sawyer, W. M., and Ferry, J. D., 1949, Propagation of transverse waves in viscoelastic media: J. Appl. Phys., v. 20, p. 1036–1041.

Anderson, O. L., Liebermann, R. C., 1968, Sound velocities in rocks and minerals: experimental methods, extrapolations to very high pressures and results *in* Physical acoustics, principles and methods: W. P. Mason, editor, New York, Academic Press, p. 330–466.

Attwell, P. B., and Ramana, Y. V., 1966, Wave attenuation and internal friction as functions of frequency in rocks: Geophysics, v. 31, p. 1049–1056.

Balakrishna, S., Ramana, Y. V., 1968, Integrated studies of the elastic properties of some Indian rocks, *in* The crust and upper mantle of the Pacific area: L. Knopoff, C. L. Drake, and P. J. Hart, editors, Geophys. Mon. No. 12, Washington, D. C., Am. Geophys. Union.

Barkan, D. D., 1962, Dynamics of bases and foundations: New York, McGraw-Hill Book Company, Inc.

Bennett, L. C., Jr., 1967, In situ measurements of acoustic absorption in unconsolidated sediments (abstract): Trans. Am. Geophys. Union, v. 48, p. 144.

Bieda, G. E., 1970, Measurement of the viscoelastic and related mass-physical properties of some continental terrace sediments, Monterey, Calif., U. S. Naval Postgraduate School, MS thesis.

Biot, M. A., 1941, General theory of three-dimensional consolidation: J. Appl. Phys., v. 12, p. 155–164.

———— 1956, Theory of propagation of elastic waves in a fluid-saturated porous solid. I. Low-frequency range: II. Higher-frequency range. J. Acoust. Soc. Am., v. 28, p. 168–191.

———— 1962, Mechanics of deformation and acoustic propagation in porous media: J. Appl. Phys., v. 33, p. 1482–1498.

Biot, M. A., and Willis, D. G., 1957, The elastic coefficients of the theory of consolidation: J. Appl. Mech., v. 24, p. 594–601.

Birch, F., 1960, The velocity of compressional waves in rocks to 10 kilobars, Part 1: J. Geophys. Res., v. 65, p. 1083–1102.

———— 1961, The velocity of compressional waves in rocks to 10 kilobars, Part 2: J. Geophys. Res., v. 66. p. 2199–2224.

———— 1966, Compressibility: Elastic constants *in* Handbook of physical constants: S. P. Clark, Jr., editor, Geol. Soc. Am. Memoir 97, p. 98–174.

Birch, F., and Bancroft, D., 1938, The effect of pressure on the rigidity of rocks: J. Geol., v. 46, p. 59–87, 113–141.

Brace, W. F., 1965a, Some new measurements of linear compressibility of rocks: J. Geophys. Res., v. 70, p. 391–398.

———— 1965b, Relation of elastic properties of rocks to fabric: J. Geophys. Res., v. 70, p. 5657–5667.

Bradley, J. J., and Fort, A. N., Jr., 1966, Internal friction in rocks, *in* Handbook of physical constants: S. P. Clark, Jr., editor, Geol. Soc. Am. Memoir 97, p. 175–194.

Brandt, H., 1955, A study of the speed of sound in porous granular media: J. Appl. Mech. paper no. 5-APM-37.

Brutsaert, W., 1964, The propagation of elastic waves in unconsolidated unsaturated granular mediums: J. Geophys. Res., v. 69, p. 243–257.

Bucker, H. P., Whitney, J. A., Yee, G. S., and Gardner, R. R., 1965, Reflection of low-frequency sonar signals from a smooth ocean bottom: J. Acoust. Soc. Am., v. 37, p. 1037–1051.

Busby, J., and Richardson, E. G., 1957, The absorption of sound in sediments: Geophysics, v. 22, p. 821–828.

Christensen, N. I., 1966a, Compressional wave velocities in single crystals of alkali feldspar at pressures to 10 kilobars: J. Geophys. Res., v. 71, p. 3113–3116.

—— 1966b, Shear wave velocities in metamorphic rocks at pressures to 10 kilobars: J. Geophys. Res., v. 71, p. 3549–3556.

Cohen, S. R., 1968, Measurement of the viscoelastic properties of water-saturated clay sediments: Monterey, Calif. U. S. Naval Postgraduate School, MS thesis.

Cole, B. F., 1965, Marine sediment attenuation and ocean-bottom-reflected sound: J. Acoust. Soc. Am., v. 38, p. 291–297.

de Graft-Johnson, J. W. S., 1968, The damping capacity of compacted kaolinite under low stresses, Proc. Int. Symp. Wave Propagation and Dynamic Properties of Earth Materials: Albuquerque, Univ. of New Mexico Press.

Evison, F. F., 1956, The seismic determination of Young's Modulus and Poisson's Ratio for rocks in situ: Geotechnique, v. 6, p. 118–123.

Ferry, J. D., 1961, Viscoelastic properties of polymers: New York, John Wiley and Sons, Inc.

Fry, J. C., and Raitt, R. W., 1961, Sound velocities at the surface of deep sea sediments: J. Geophys. Res., v. 66, p. 589–597.

Gassmann, F., 1951, Über die elastizität poröser medien: Vierteljahrsschrift Naturforschenden Gesellschaft in Zürich, v. 96, p. 1–23.

Geertsma, J., and Smit, D. C., 1961, Some aspects of elastic wave propagation in fluid-saturated porous solids: Geophysics, v. 26, p. 169–181.

Hall, J. R., Jr., and Richart, F. E., Jr., 1963, Dissipation of elastic wave energy in granular soils: J. Soil Mech. and Foundations Div., Am. Soc. Civil Engin., SM 6, p. 27–56.

Hamilton, E. L., 1956, Low sound velocities in high-porosity sediments: J. Acoust. Soc. Am., v. 28, p. 16–19.

—— 1963, Sediment sound velocity measurements made in situ from bathyscaph TRIESTE: J. Geophys. Res., v. 68, p. 5991–5998.

—— 1970, Sound velocity and related properties of marine sediments, North Pacific: J. Geophys. Res., v. 75, p. 4423–4446.

—— 1970a, Shear wave velocities in marine sediments (abstract): Trans. Am. Geophys. Union, v. 51, no. 4, p. 333.

—— 1971a, Elastic properties of marine sediments: J. Geophys. Res., v. 76, p. 579–604.

—— 1971b, Prediction of in-situ acoustic and elastic properties of marine sediments: Geophysics, v. 36, p. 266–284.

Hamilton, E. L., Shumway, G., Menard, H. W., and Shipek, C. J., 1956, Acoustic and other physical properties of shallow-water sediments off San Diego: J. Acoust. Soc. Am., v. 28, p. 1–15.

Hamilton, E. L., Bucker, H. P., Keir, D. L., and Whitney, J. A., 1970, Velocities of compressional and shear waves in marine sediments determined from a research submersible: J. Geophys. Res., v. 75, p. 4039–4049.

Hampton, L. D., 1967, Acoustical properties of sediments: J. Acoust. Soc. Am., v. 42, p. 882–890.

Hardin, B. O., and Richart, F. E., Jr., 1963, Elastic wave velocities in granular soils: J. Soil Mech and Foundations Div., Am. Soc. Civil Engin., SM1, p. 33–65; and discussions, SM 5, p. 103–118.

Hardin, B. O., and Black, W. L., 1968, Vibration modulus of normally consolidated clay: J. Soil Mech. and Foundation Div., Proc. Am. Soc. Civil Engin., v. 94, SM 2, p. 353–369.

Hastrup, O. F., 1970. Digital analysis of acoustic reflectivity in the Tyrrhenian Abyssal Plain: J. Acoust. Soc. Am., v. 70, p. 181–190.

Heukelom, W., 1961, Analysis of dynamic deflections on soils and pavements: Geotechnique, v. 11, p. 224–243.

Horn, D. R., Horn, B. M., and Delach, M. N., 1968, Sonic properties of deep-sea cores from the North Pacific Basin and their bearing on the acoustic provinces of the North Pacific: Lamont Geol. Obs. Tech. Rep. no. 10.

Houtz, R. E., and Ewing, J. I., 1964, Sedimentary velocities of the western North Atlantic margin: Seism. Soc. Am. Bull., v. 54, p. 867–895.

Hunter, A. N., Legge, R., and Matsukawa, E., 1961, Measurements of acoustic attenuation and velocity in sand: Acustica, v. 11, p. 26–31.

Inman, D. L., 1952, Measures for describing the size distribution of sediments: J. Sed. Petrol., v. 22, p. 125–145.

Jones, R., 1958, In situ measurement of the dynamic properties of soil by vibration methods: Geotechnique v. 8, p. 1–21.

Knopoff, L., 1965, Attenuation of elastic waves in the earth, *in* Physical acoustics: W. P. Mason, editor, New York, Academic Press, III-B, p. 287–324.

Knopoff, L., and MacDonald, G. J. F., 1958, Attenuation of small amplitude stress waves in solids: Rev. Modern Phys., v. 30, p. 1178–1192.

—— 1960, Models for acoustic loss in solids: J. Geophys. Res., v. 65, p. 2191–2197.

Knopoff, L., and Porter, L. D., 1963, Attenuation of surface waves in a granular material: J. Geophys. Res., v. 68, p. 6317–6321.

Krizek, R. J., 1964, Application of the one-sided Fourier transform to determine soil storage and dissipation characteristics: Proc. Symp. on Soil-Structure Interaction, Tucson, Univ. of Arizona Press.

Krizek, R. J., and Franklin, A. G., 1968, Energy dissipation in a soft clay: Proc. Int. Symp. Wave Propagation and Dynamic Properties of Earth Materials, Albuquerque, Univ. of New Mexico Press, p. 797–807.

Lasswell, J. B., 1970, A comparison of two methods for measuring rigidity of saturated marine sediments: Monterey, Calif., U. S. Naval Postgraduate School, MS thesis.

Laughton, A. S., 1957, Sound propagation in compacted ocean sediments: Geophysics, v. 22, p. 233–260.

Leonards, G. A., 1962, Engineering properties of soils, *in* Foundation engineering: New York, McGraw-Hill Book Co., Inc., p. 66–240.

Levykin, A. I., 1965, Longitudinal and transverse wave absorption and velocity in rock specimens at multilateral pressures up to 4000 kg/cm^2: Izv. Acad. Sci., USSR, no. 2, p. 94–98.

Lewis, L. F., 1971, An investigation of ocean sediments using the deep ocean sediment probe: Univ. of Rhode Island, Ph.D. thesis, p. 76.

McCann, C., 1969, Compressional wave attenuation in concentrated clay suspensions: Acustica, v. 22, p. 352–356.

McCann, C., and McCann, D. M., 1969, The attenuation of compressional waves in marine sediments: Geophysics, v. 34, p. 882–892.

McDonal, F. J., Angona, F. A., Mills, R. L., Sengbush, R. L., Van Nostrand, R. G., and White, J. E., 1958.

Attenuation of shear and compressional waves in Pierre shale: Geophysics, v. 23, p. 421–439.

McLeroy, E. G., and DeLoach, A., 1968, Sound speed and attenuation, from 15 to 1500 kHz, measured in natural sea-floor sediments: J. Acoust. Soc. Am., v. 44, p. 1148–1150.

Meyer, E., 1957, Air bubbles in water, chapt. 5, *in* v. II, Ultrasonic range, underwater acoustics, technical aspects of sound: E. G. Richardson, editor, London, Elsevier Publishing Co.

Mizikos, J. P., 1972, Attenuation of longitudinal sound waves in unconsolidated sediments with gravity dependent cohesion: Geophys. Prosp., in press.

Morris, H. E., 1970, Bottom-reflection-loss model with a velocity gradient: J. Acoust. Soc. Am., v. 48, p. 1198–1202.

Morse, R. W., 1952, Acoustic propagation in granular media: J. Acoust. Soc. Am., v. 24, p. 696–700.

Nafe, J. E., and Drake, C. L., 1963, Physical properties of marine sediments, *in* The sea, Vol. 3: M. N. Hill, editor, New York, Interscience Publishers, p. 794–815.

———— 1967, Physical properties of rocks of basaltic composition, *in* Basalts: H. Hess and A. Poldervaart, editors, New York, Interscience Publishers, p. 483–502.

Nolle, A. W., and Sieck, P. W., 1952, Longitudinal and transverse ultrasonic waves in synthetic rubber: J. Applied Phys., v. 23, p. 888–894.

Nolle, A. W., Hoyer, W. A., Mifsud, J. F., Runyan, W. R., and Ward, M. B., 1963, Acoustical properties of water-filled sands: J. Acoust. Soc. Am., v. 35, p. 1394–1408.

Paterson, N. R., 1956, Seismic wave propagation in porous granular media: Geophysics, v. 21, p. 691–714.

Peselnick, L., and Outerbridge, W. F., 1961, Internal friction in shear and shear modulus of Solenhofen limestone over a frequency range of 10^7 cycles per second: J. Geophys. Res., v. 66, p. 581–588.

Pettijohn, F. J., 1957, Sedimentary rocks, 2nd ed.: New York, Harper & Bros.

Pilbeam, C. C., and Vaisnys, J. R., 1971, Acoustic velocities and energy losses in granular aggregates (abstract): Trans. Am. Geophys. Union, v. 52, p. 288.

Press, F., 1966, Seismic velocities: Geol. Soc. Am. Memoir 97, p. 195–218.

Richart, F. E., Jr., and Whitman, R. V., 1967, Comparison of footing vibration tests with theory: J. Soil Mech. & Foundation Div., Am. Soc. Civil Engin., SM 6, pl 143–193.

Schirmer, F., 1970, Schallausbreitung im Schlick: Deut. Hydrogr. Z., Jahr. 23, Heft 1, p. 24–30.

Schön, J., 1963, Modellseismische Untersuchungen im Hinblick auf die Schallgeschwindigkeit in Lockergesteinen: Monatsberichte der Deut. A. K. Wissenschaften zu Berlin, Band 5, Heft 4, p. 262–273.

Schreiber, B. C., 1968, Marine geophysical program 65–67—Central North Pacific Ocean, Area V., v. 8, Cores: Alpine Geophys. Assoc., Inc., U. S. Naval Ocean. Off. Contract N62306-1688.

Shumway, G., 1956, A resonant chamber method for sound velocity and attenuation measurements in sediments: Geophysics, v. 21, p. 305–319.

———— 1960, Sound speed and absorption studies of marine sediments by a resonance method, Part I; Part II: Geophysics, v. 25, p. 451–467, 659–682.

Shumway, G. and Igelman, K., 1960, Computed sediment grain surface areas: J. Sed. Petrology, v. 30, p. 486–489.

Simmons, G., and Brace, W. F., 1965, Comparison of static and dynamic measurements of compressibility of rocks: J. Geophys. Res., v. 70, p. 5649–5656.

Strick, E., 1970, A predicted pedestal effect for pulse propagation in constant-Q solids: Geophysics, v. 35, p. 387–403.

Toulis, W. J., 1956, Theory of a resonance method to measure the acoustic properties of sediments: Geophysics, v. 21, p. 299–304.

Tullos, F. N., and Reid, C., 1969, Seismic attenuation of Gulf Coast sediments: Geophysics, v. 34, p. 516–528.

Ulonska, A., 1968, Versuche zur Messung der Schallgeschwindigkeit und Schalldampfung im Sediment in situ: Deut. Hydrogr. Z., Jahr. 21, Heft 2, p. 49–58.

Urick, R. J., 1947, A sound velocity method for determining compressibility of finely divided substances: J. Appl. Phys., v. 18, p. 983–987.

Vasil'ev, Y. I., and Gurevich, G. I., 1962, On the ratio between attenuation decrements and propagation velocities of longitudinal and transverse waves: English translation, Izv. Acad. Sci., USSR, Geophys. Ser., no. 12, p. 1061–1074.

White, J. E., 1965, Seismic waves: radiation, transmission, and attenuation: New York: McGraw-Hill Book Co., Inc.

White, J. E., and Sengbush, R. L., 1953, Velocity measurements in near-surface formations: Geophysics, v. 18, p. 54–69.

Whitman, R. V., Holt, R. J., and Murphy, V. J., 1969, Discussion of: vibration modulus of normally consolidated clay, by B. O. Hardin and W. L. Black: J. Soil Mech. & Foundation Div., Am. Soc. Civil Engin., SM 2, p. 656–659.

Wood, A. B., and Weston, D. E., 1964, The propagation of sound in mud: Acustica, v. 14, p. 156–162.

Wyllie, M. R. J., Gregory, A. R., and Gardiner, L. W., 1956, Elastic wave velocities in heterogeneous and porous media: Geophysics, v. 21, p. 41–70.

Wyllie, M. R. J., Gardner, G. H. F., and Gregory, A. R., 1962, Studies of elastic wave attenuation in porous media: Geophysics, v. 27, p. 569–589; with addendum, Geophysics. v. 28, p. 1074, 1963.

Yong, R. N., and Warkentin, B. P., 1966, Introduction to soil behavior: New York, The Macmillan Co.

Zemstov, E. E., 1969, Effect of oil and gas deposits on dynamic characteristics of reflected waves: Int. Geology Rev., v. 11, p. 504–509.

Zwikker, C., and Kosten, C. W., 1949, Sound absorbing materials: New York, Elsevier Publ. Co.

Reprinted from Geophysics, v. 39, p. 278–291.

APPARENT ATTENUATION DUE TO INTRABED MULTIPLES

M. SCHOENBERGER* AND F. K. LEVIN*

Experiments with synthetic seismograms from two wells let us confirm a conclusion of O'Doherty and Anstey that seismic attenuation due to layering alone arises from a combination of the transmission losses through the interfaces of a layered subsurface and the generation of intrabed multiple reflections. Transmission losses lower the amplitudes uniformly at all frequencies. Intrabed multiples tend to raise the amplitudes at the low-frequency end of the spectrum and lower those at the high-frequency end. For the wells studied, intrabed multiples higher than fifth order (10 msec delay) were unimportant. Attenuation due to layering accounted for 1/3 to 1/2 of the total frequency dependent attenuation estimated from field seismograms at the well locations.

INTRODUCTION

In a recent paper O'Doherty and Anstey (1971) showed that in many cases part or all of the attenuation apparent on seismograms can be attributed to intrabed multiple reflections rather than to the intrinsic properties of rock materials. Further, for a subsurface layered in a cyclic manner, they found intrabed multiples tend to compensate for simple reflection losses due to transmission across the numerous interfaces characteristic of such a subsurface.

Although O'Doherty and Anstey explained the major features of intrabed multiple contributions to a seismic trace, they left important questions unanswered. Does the restoration of amplitudes require many orders of intrabed multiples, or are a few sufficient? How nearly does spectral shaping of the transmitted pulse due to intrabed multiples simulate the shaping caused by intrinsic attenuation? What fraction of the losses computed for field records is due to each process? We have examined these questions and shall present tentative answers based on a limited investigation.

EXPERIMENTAL PROCEDURES AND RESULTS

Our study was based on synthetic seismograms for two wells, designated as well A and well B.

Logs and typical traces appear as Figures 1 and 2. For part of our work we modified the density and velocity logs of well A by adding a uniform section terminated with an isolated reflector at 15,000 ft. For computation of the seismograms, the logs were sliced to yield layers with 2 msec two-way traveltime (Wuenschel, 1960). Our computer program assumes plane waves generated by a surface source and normally incident on horizontal, plane interfaces; it produces four types of traces: input, total, direct, and multiple. Input traces include only primary reflections without transmission losses; total traces include all possible reflections and transmission losses.

In computing what we have labeled the direct trace, we considered transmission losses and as many intrabed reflections as a user desires. The division of multiple reflections into intrabed multiples and others is arbitrary. Following a discussion of Trorey (1962), we include in the M-order intrabed multiples all contributions which, generated by a time spike incident on each interface, trail that incident spike by M time steps ($2M$ msec in our case) or less. In practice, the spikes usually are convolved with an input pulse of, say, the Ricker type. Our method of computing intrabed multiples differs from that of Trorey only in our having eliminated the bot-

Manuscript received by the Editor November 29, 1973; revised manuscript received December 28, 1973.
* Esso Production Research Co., Houston, Tex. 77001.

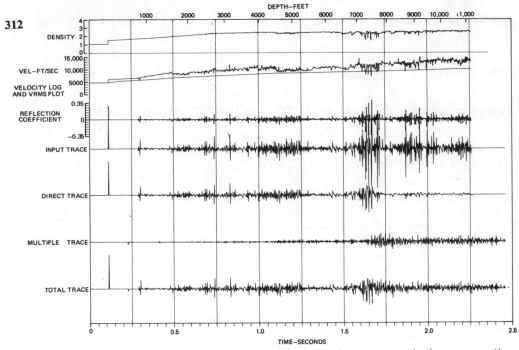

FIG. 1. Well A — velocity log, density log, and four synthetic seismograms — the input trace, direct trace, multiple trace, and total trace.

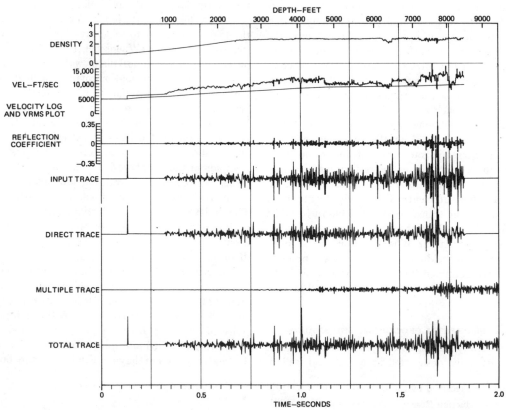

FIG. 2. Well B — velocity log, density log, and four synthetic seismograms — the input trace, direct trace, multiple trace, and total trace.

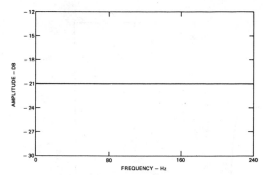

FIG. 3. Transmission loss encountered by a pulse which has passed through the media described by the logs of well A. This was obtained by placing an artificial interface 4000 ft below the bottom of well A.

tom triangle of Trorey's Figure A2. The elimination was a personal and arbitrary choice of one of the authors and has no significance. Shown also on Figures 1 and 2 are multiple traces found by subtracting a direct trace from the corresponding total trace. Multiple traces played no part in our study.

Most of the phenomena with which we will be concerned in this paper varied with frequency. To obtain input spectra independent of frequency we input spikes to the synthetic seismogram program. With the number of intrabed multiples M under user control, we were able to examine the effects of transmission losses and intrabed multiples separately. Thus, the ratio of spectra for direct traces with $M=M'$ and $M=0$ gave directly the contribution of M' intrabed multiples. The ratio of the direct spectrum for $M=0$ to the input spectrum revealed the transmission losses alone. Finally, the ratio of direct spectrum with $M=M'$ to input spectrum yielded the combined effects of transmission losses and intrabed multiples. By definition, the direct trace with $M=\infty$ and the total trace are identical.

We will look first at transmission losses, i.e., losses due to passage through interfaces. We examined time intervals selected along the direct and input traces; however, for clarity we will start with the reflection from the deep isolated interface added to well A (Figure 3). Since it consists only of a decrease in amplitude, the transmission loss for this reflection is independent of frequency. The transmission loss is $\prod_{i=1}^{n}(1-R_i^2)$ where R_i is the reflection coefficient at the ith interface

and n is the number of interfaces. The loss is large—21 db for energy that traversed the entire section of well A twice.

Twenty-one db represents a great deal of attenuation. Even more important, the transmission losses do not increase uniformly with depth; the rate of increase varies over the section (O'Doherty and Anstey, 1971). Since the transmission loss is independent of frequency, instead of forming the spectral ratio of a segment of direct trace with $M=0$ to the same segment of input trace, we can obtain an instantaneous transmission loss as the ratio of a point on the direct time trace to the same point of the input time trace. Figure 4 shows for well A the instantaneous transmission loss as a function of two-way traveltime. The sudden decrease between 1.6 and 1.8 sec is striking. Scrutiny of the sonic log (Figure 1) reveals the region of great amplitude loss is a cyclic interval, an interval where the velocity increases and decreases rapidly and repeatedly. O'Doherty and Anstey pointed out the fact that very large amplitude losses due to transmission should be expected for cyclic intervals. This results from the large values of R_i^2 in such intervals.

In addition to reducing the energy in the direct arrival, passage through a cyclic section appears to broaden the input pulse by producing a set of intrabed multiples which follow it. Figure 5 is the complex formed by the direct arrival and first 50 intrabed multiples for a spike reflected from the interface added to well A at 15,000 ft. The first eight intrabed multiples have the same polarity as the direct arrival, and the first five are larger than the direct arrival. Since the low-order multiples have the same polarity as the direct arrival, they add energy at low frequencies. In fact, if we assumed the important part of the reflection complex is composed of the direct arrival and first eight intrabed multiples, the spectrum of the complex (produced by a spike input) would peak at zero hertz and have a small high-frequency content. Each curve of Figure 6 shows for this reflection the spectral ratio of a direct trace with M intrabed multiples to the input trace. They illustrate for M between zero and 50 the anticipated change from frequency independence to a spectrum peaked at low frequencies.

As the number of intrabed multiples increased from zero, the low end of the spectrum was elevated. As larger and larger numbers of intrabed

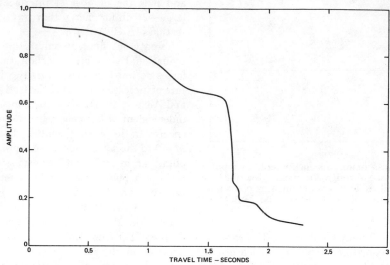

FIG. 4. Transmission loss encountered by a pulse that has traversed an earth
section corresponding to a given two-way traveltime, well A.

multiples were added, the tilt of the spectrum increased; but for the seismic frequency band of 0 hz to 100 hz, the change for M greater than 5 was negligible (Figures 6b and 6c). In a more restricted band of 0 hz to 70 hz, energy from intrabed multiples nearly compensated for transmission losses. Finally, when M became infinite (Figure 6d), the case for all multiples included, the ratios became jagged. Examination of the traces showed that for infinite M, the pulse from the deep interface was no longer isolated but was buried in multiples from shallow layers; hence,

the spectrum is deceptive. For modern records processed so as to minimize multiple reflections separated in time from the corresponding primary reflections, the direct trace with $M = 50$ probably represents data seen by an interpreter better than does the total trace ($M = \infty$).

Earlier we emphasized the importance of cyclic earth layering in causing transmission losses and in the restoration of those losses by intrabed multiples. To illustrate the phenomena, we first modified the logs from well A so that both velocity and density increased linearly from 8000 ft to

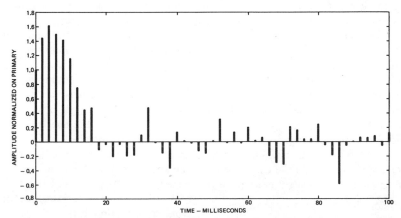

FIG. 5. The reflection from an isolated reflector 4000 ft below the bottom of well A when a spike was used as the input pulse. The amplitudes have been normalized on the primary (the value at time equal zero).

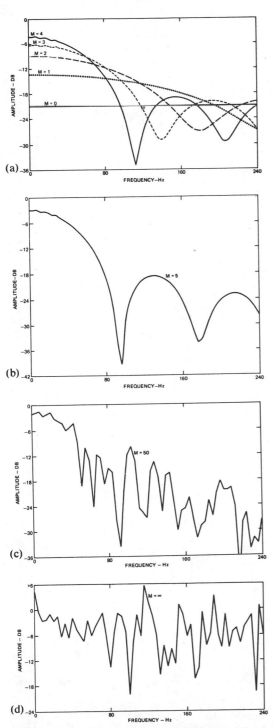

(a)

(b)

(c)

(d)

FIG. 6. The effective earth filtering of a seismic pulse. The synthetic pulses were from an isolated reflector 4000 ft below well A. (a) includes M intrabed multiples. (b) includes 5 intrabed multiples. (c) includes 50 intrabed multiples. (d) includes all multiples.

11,340 ft, then from 7000 ft to 11,340 ft. Thus, we eliminated in succession multiple contributions due to that portion of well A below the major cyclic section and contributions due to the cyclic section itself. Figure 7 shows reflections from the interface at 15,000 ft for the two cases.

For the complete section the low-order intrabed multiples were larger than the direct arrival (Figure 5). Removing variability from the lower part of the logs reduced the relative amplitudes of the intrabed multiples (Figure 7a); smoothing across the cyclic section also essentially eliminated the effect of the intrabed multiples (Figure 7b). The same information can be deduced from Figure 8, which contains plots of the spectral ratio of the direct trace with 5 intrabed multiples to the input trace for the three cases being considered. With no contributions from interfaces below 8000 ft, the transmission loss was 14.5 db instead of the 21 db for the entire section. With the cyclic section also gone, the transmission loss was only 4.1 db. Including the intrabed multiples restored the losses for the seismic band of frequencies in both cases.

Figures 7 and 8 illustrate significant features of the filtering of seismic energy due to effects of

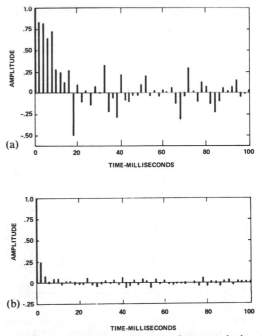

(a)

(b)

FIG. 7. The spike reflections from an isolated reflector at 15,000 ft in well A. Amplitudes have been normalized on the primary. (a) The interval below 8000 ft has been smoothed. (b) The interval below 7000 ft has been smoothed.

layering alone. From Figure 7b we see the transmitted pulse remains a spike in spite of having twice traversed approximately 7000 ft of earth. The corresponding spectral ratio of Figure 8 shows relatively slight attenuation of high frequencies. However, twice traversing an extra 1000 ft of earth section (Figure 7a) produces a transmitted pulse which differs considerably from a spike. The corresponding spectral ratio shows large attenuation of the high frequencies has occurred in this case. Finally, traversing the portion of the log from 8000 ft to 11,340 ft has a relatively small effect on the pulse shape (Figure 5) and results in little additional attenuation of high frequencies (Figure 8).

Let us next elucidate the effect of intrabed multiples on seismic data. In practice, we rarely have individual reflections available. Instead, we are faced with the multitude of overlapping reflections that comprise a seismic trace. To determine how the phenomena discussed in previous paragraphs show themselves on a seismic trace, we selected long and short intervals from the synthetic seismograms. We then formed spectral ratios just as we had done for the isolated reflection. For well A, the long interval included data from 1.6 to 2.2 sec (Figure 9), while the contiguous short intervals were 200 msec long beginning

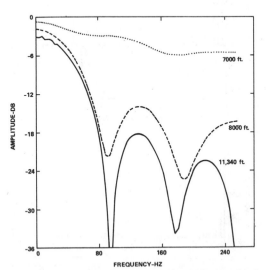

FIG. 8. The effective earth filtering of seismic pulses including 5 intrabed multiples. These pulses were from an isolated reflector at 15,000 ft in well A. Sections were smoothed below 7000 ft, 8000 ft, and 11,340 ft.

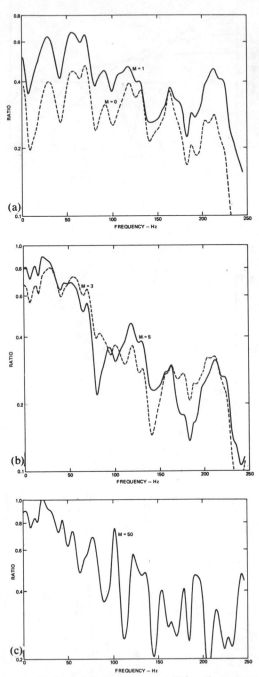

FIG. 9. Ratio of the spectrum of the direct trace with M intrabed multiples to that of the input trace for the long interval (1.6 to 2.2 sec), well A. These spectra represent the effects of transmission losses and intrabed multiple generation.

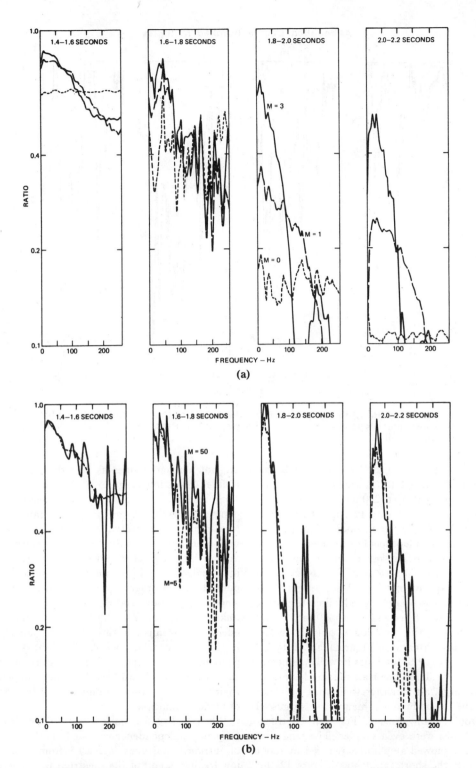

FIG. 10a, b. Ratio of the spectrum of the direct trace to that of the input trace for the short intervals (1.4 – 1.6 sec, 1.6 – 1.8 sec, 1.8 – 2.0 sec, 2.0 – 2.2 sec) with M intrabed multiples, well A.

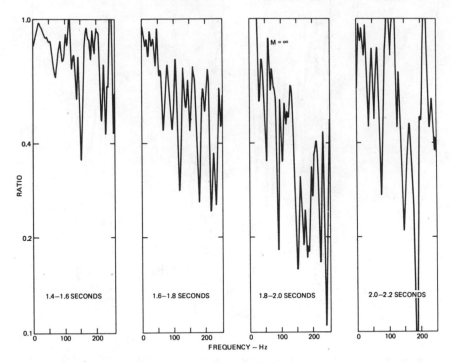

FIG. 10c. Ratio of the spectrum of the direct trace to that of the input trace for the short intervals (1.4 − 1.6 sec, 1.6 − 1.8 sec, 1.8 − 2.0 sec, 2.0 − 2.2 sec) with all multiples, well A.

with 1.4 to 1.6 sec (Figure 10). For well B, the long interval was 1.4 to 1.8 sec (Figure 11), and the 4 short intervals were 200 msec long beginning with 0.9 to 1.1 sec (Figure 12). In all cases, the losses for $M = 0$ (i.e., transmission losses only) were reasonably independent of frequency. Figure 9a shows transmission loss reducing the reflections in well A to 30 percent of their lossless amplitudes in approximately 1.9 sec. Figure 10a indicates the loss occurred predominantly in the intervals 1.6 to 1.8 sec and 1.8 to 2.0 sec, where, as noted before, the sonic log is notably cyclic. From the last interval of Figure 10a (2.0 to 2.2 sec), we see that the total loss in well A is nearly a factor of 10. Thus, the transmission losses can be fairly large, although for well A they did not attain the overwhelming magnitude discussed by O'Doherty and Anstey. For well B, transmission losses were even smaller. The long gate (Figure 11) showed amplitudes reduced to about 60 percent; the short gate losses (Figure 12) for well B were more regular than for well A.

The transmission loss plots of Figures 9 to 12

are not horizontal lines as was the plot for an isolated reflection (Figure 3). This may disturb some readers. The deviation from frequency independence is due to the increase of transmission losses within the interval being analyzed. If we had been able to analyze infinitely short time intervals, the curves would have been frequency independent. We could not. Also, we analyzed our data with a program which includes a spectral smoothing window that distorts the curves to an unknown degree. We showed earlier that the effect of transmission through many interfaces is simply a constant amplitude decrease independent of frequency. The transmission loss curves of Figures 9 to 12 are flat enough to permit their use in our subsequent analysis of the effects of intrabed multiples.

When we began to include intrabed multiples, frequency independence vanished. As the number of intrabed multiples increased from zero, the low-frequency end of the spectrum was elevated, just as for the isolated reflector.

Because the direct trace is a superposition of

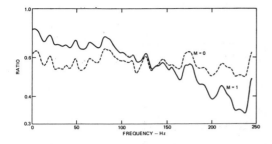

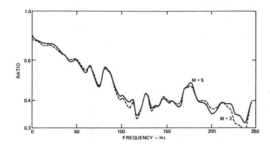

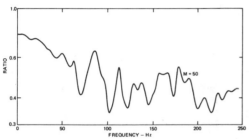

FIG. 11. Ratio of the spectrum of the direct trace with M intrabed multiples to that of the input trace for the long interval (1.4 – 1.8 sec), well B.

many pulses similar to that shown in Figure 5, we can explain the spectral behavior pictured on Figures 9 to 12. If those sections of the sonic log corresponding to the short intervals generated a large number of reflections, all with the same amplitude and randomly arranged, the smoothed spectra would approach the spectra of the isolated pulses (Robinson, 1957). An assumption of equal-amplitude, random arrivals is a good one for the first and fourth short intervals of well A but breaks down for the center two intervals where a few strong reflections dominate. Hence, spectra for the last interval should and do resemble the spectra of Figure 6. We anticipate that any interval free of strong reflections will produce a spectrum tilted to emphasize the low-frequency end.

The deeper the reflecting interface, the larger the number of overlying interfaces that can contribute intrabed multiples and, of course, the larger the transmission losses. We should expect the phenomena we have been discussing—transmission losses and tilting of the spectra to emphasize low frequencies—to become increasingly prominent with increasing record time.

Thus far we have discussed ratios of direct traces without intrabed multiples to input traces, which are ratios indicative of transmission losses, and ratios of direct traces with intrabed multiples to input traces, which give the combined effects of transmission losses and intrabed multiples. If we wish to examine the effects of intrabed multiples alone, we form ratios of direct traces with intrabed multiples to direct traces without intrabed multiples. Plots of this type in the form of spectra for the long intervals appear as Figures 13 and 14. Because the spectra without intrabed multiples ($M=0$, Figures 9 and 11) were nearly frequency independent, the curves of Figures 13 and 14 resemble those with $M>0$ in Figures 9 and 11. The unsystematic behavior of the curves for frequencies greater than 100 hz is obvious. At the higher frequencies, the spectra respond to detailed changes in the reflection train, which occur as intrabed multiples of increasing order are added. Low frequencies average over the individual arrivals. Since reflections from interfaces deep in a section rarely include frequencies as high as 100 hz, the systematic trend seen in the 10 to 100 hz band is really all that interests us.

To answer the third question in the introduction (whether the spectral tilting we have just discussed mimics the effects of intrinsic attenuation), we compared the frequency content of energy at large record times with the frequency content of an earlier section of the same record.

Data in early and late intervals of the direct trace for $M=50$ were frequency analyzed and the loss of energy at each frequency found. The normalized ratio of the spectra was assumed to have the form $e^{-\alpha f t}$, where f is frequency and t is elapsed time between intervals. α is the attenuation coefficient. For well A, we chose time intervals of 0.5 to 1.1 sec and 1.7 to 2.3 sec. For well B, the intervals were 0.4 to 1.1 sec and 1.1 to 1.8 sec. Taking the spectral ratios over the frequency range from 0 to 100 hz, we found α to be .0064 for well A and .0070 for well B. These values are

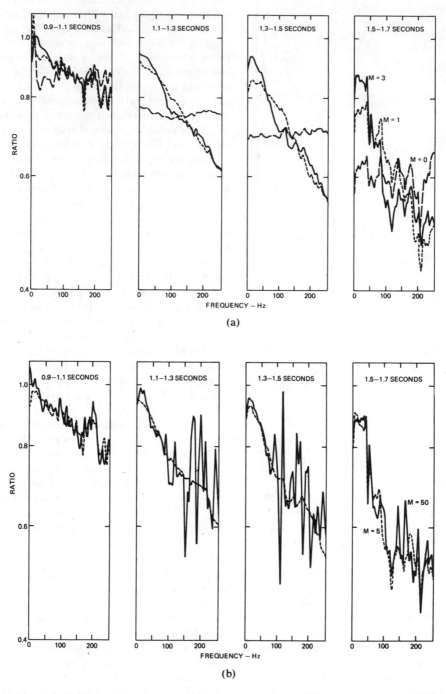

FIG. 12a, b. Ratio of the spectrum of the direct trace to that of the input trace for the short intervals (0.9 − 1.1 sec, 1.1 − 1.3 sec, 1.3 − 1.5 sec, 1.5 − 1.7 sec) with M intrabed multiples, well B.

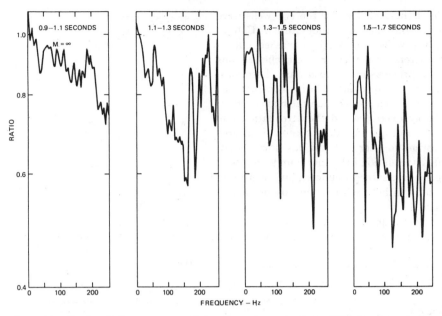

FIG. 12c. Ratio of the spectrum of the direct trace to that of the input trace for the short intervals (0.9 − 1.1 sec, 1.1 − 1.3 sec, 1.3 − 1.5 sec, 1.5 − 1.7 sec) with all multiples, well **B**.

equivalent to .056 db/wavelength and .061 db/wavelength, respectively, or 2.1 db/sec and 2.4 db/sec, respectively, at 40 hz.

Analysis of identical gates from seismic lines across the two wells yielded average values of α of .019 and .014 for wells A and B, respectively. Hence, for these two cases, apparent attenuation due to intrabed multiples accounts for 1/3 to 1/2 of the total attenuation observed on field data.

Once the attenuation constant for intrabed multiple losses was known, finding the intrinsic attenuation was—at least in principle—simple. If we let α_m be the constant for intrabed attenuation, α_i be the constant for intrinsic attenuation, and α be the constant from field data, we have

$$\alpha_i = \alpha - \alpha_m. \tag{1}$$

For a given interval of the subsurface, the part of the attenuation constant due to rock properties is, according to equation (1), the difference between the constant for field data and that computed for the corresponding interval of a synthetic seismogram from a well at that location. In writing equation (1), we have assumed the total loss for frequency f and elapsed time t is $e^{-(\alpha_i+\alpha_m)ft}$; i.e., that losses obey an exponential

law with a linear frequency dependence regardless of the loss mechanism involved. This assumption seems to provide a reasonable approximation to the data. In any case, in our present state of ignorance we do not have a more suitable approach.

We have now given answers to all the questions raised in the introduction. For the two wells examined, five intrabed multiples at 2 msec intervals were sufficient to restore, for the usual seismic frequency band, losses due to transmission through the interfaces of the subsurface. Spectral shaping caused by intrabed multiples did simulate, at least grossly, shaping caused by intrinsic attenuation. Finally, losses arising in the two processes were about the same size.

LIMITATIONS OF THIS INVESTIGATION

While confirming and, perhaps, extending in some directions the work of O'Doherty and Anstey, the investigation reported here suffers from serious limitations. Two wells are too few to allow us to draw general conclusions. Synthetic seismograms from many wells in different areas and field data from lines across the wells should be examined.

Although we have computed synthetic seismograms as if the sonic logs for the two wells were exactly accurate, of course they are not. Caliper logs for the wells show sections washed out to large diameters. In particular, the cyclic sections referred to previously were also washed sections. As a result, the apparent velocity contrast was likely to have been greater than the real contrast; we probably overestimated the attenuation due to intrabed multiples. We do not know the degree of overestimation nor is it obvious how we can find out. Attenuation constants found from field seismograms are subject to uncertainty; constants found from synthetic seismograms are not necessarily sacred either.

Two related questions await answers. First, how does the division of sonic logs into equal

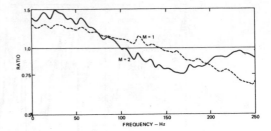

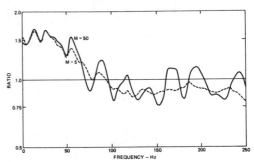

FIG. 14. Spectral ratios of direct traces with M intrabed multiples to direct traces without intrabed multiples, long interval (1.4 – 1.8 sec), well B. These spectra represent the effects of intrabed multiples alone.

traveltime slices affect the phenomena we have discussed? Second, are our results dependent on the duration of the time slices? There is no easy way of answering the first question, since all available one-dimensional synthetic seismograms divide a sonic log into equal time slices. No other procedure is practical for a very large number of layers. We partially answered the second question by exercising options in our synthetic seismogram program which allow one-msec, two-way travel-time slices, as well as the two-msec slices we used. For 1 msec sampling, the complex of Figure 7a changed slightly but not in a manner that would modify the conclusions of this paper (Figure 15). The primary ceased to be the largest arrival but the overall spike patterns (Figures 7a and 15a) are similar. Transmission losses increased from 14.5 db for 2 msec sampling to 25.5 db for 1 msec sampling. However, the recovery at seismic frequencies due to intrabed multiples was so complete that over the band of 0 hz to 75 hz the spectra for 1 and 2 msec sampling virtually coincided (Figure 15b). For the seismic frequency band, it is unlikely that other small interval sampling choices would have changed our results

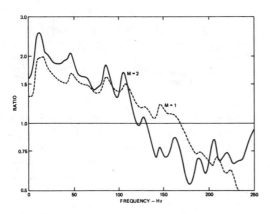

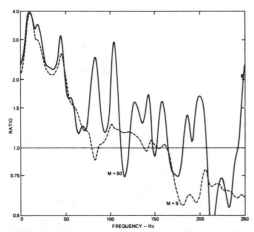

FIG. 13. Spectral ratios of direct traces with M intrabed multiples to direct traces without intrabed multiples, long interval (1.6 – 2.2 sec), well A. These spectra represent the effects of intrabed multiples alone.

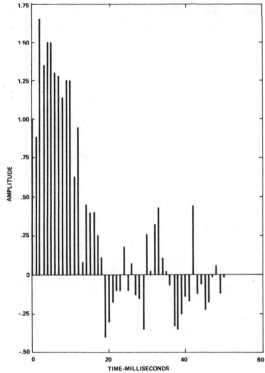

FIG. 15a. The spike reflection from an isolated reflector at 15,000 ft in well A. The interval below 8000 ft has been smoothed. Amplitudes have been normalized on the primary. The logs were sampled to give 1 msec two-way traveltimes.

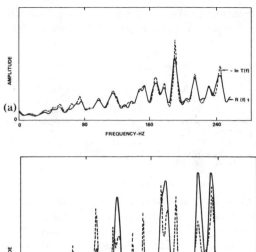

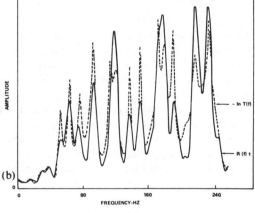

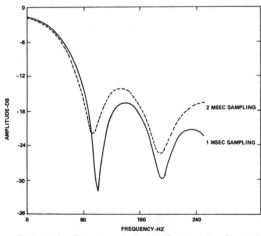

FIG. 15b. Spectra corresponding to the first 10 msec of intrabed multiples for 1 and 2 msec sampling for the reflection complexes of Figures 7a and 15a.

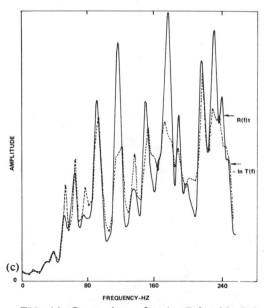

FIG. 16. Comparison of $-\ln T(f)$ with $R(f)t$. $T(f)$ is the amplitude spectrum of the transmitted pulse and $R(f)t$ is the power spectrum of the reflection coefficient train through which the pulse traveled in time t for the synthetic seismogram of well A. (a) Total interval, 7000 ft; (b) total interval, 8000 ft; (c) total interval, 11,340 ft.

significantly. We did not prove there would be no changes; the point awaits future investigations.

REFERENCES

O'Doherty, R. F., and Anstey, N. A., 1971, Reflections on amplitudes: Geophys. Prosp. v. 19, p. 430–458.
Robinson, E. A., 1957, Predictive decomposition of seismic traces: Geophysics, v. 22, p. 767–778.
Trorey, A. W., 1962, Theoretical seismograms with frequency and depth dependent absorption: Geophysics, v. 27, p. 766–785.
Wuenschel, P. C., 1960, Seismogram synthesis including multiples and transmission coefficients: Geophysics, v. 25, p. 106–129.

APPENDIX

O'Doherty and Anstey (1971) emphasized that energy reflected from an interface is not available for transmission to underlying interfaces. They summarized this fact with the phrase, "More up, less down."

The reflection process is not quite as simple as "More up, less down" would indicate, since energy traveling toward the surface can, of course, be reflected by overlying interfaces. Considering all the direct and multiple reflections, O'Doherty and Anstey derived a relation between the amplitude spectrum $T(f)$ of the transmitted pulse and the time-normalized power spectrum $R(f)t$ of the reflection coefficient series through which the pulse has traveled:

$$T(f) = \exp\left[-R(f)t\right], \qquad (A1)$$

where f is frequency and t is traveltime. An alternative way of writing equation (A1) is

$$\ln T(f) = -R(f)t. \qquad (A2)$$

In Figure 16 we have overlain plots of $-\ln T(f)$ and $R(f)t$ for the data from well A synthetic seismograms. $T(f)$ is taken from the isolated reflector on the direct trace with $M = 50$. The agreement is not perfect but it is good.

In view of Figure 16, we can hardly quarrel with the result of O'Doherty and Anstey's derivation. However, we could not check it either. We have not pursued the computation. Following a cowardly but prudent path, we leave the derivation to the original authors.

Reprinted from the Bulletin of the Earthquake Research Institute, v. 48, p. 145–158.

Attenuation of Shear Waves in Soil.

By Kazuyoshi KUDO and Etsuzo SHIMA,

Earthquake Research Institute.
(Read Nov. 22, 1969.—Received Jan. 24, 1970.)

1. Introduction

Since the early days of seismology, many seismologists and earth-quake engineers have paid much attention to the mechanism of intrinsic attenuation of waves in the earth materials. The problem is generally discussed in terms of attenuation coefficient or Q-value. As is well known, Q-value is proportional to frequency in the model suggested by Maxwell (1867). Such material is referred to as a Maxwell solid. Contrarily, it is inversely proportional to frequency in the model proposed by Meyer (1874a, 1874b), Kelvin (1878) and Voigt (1892). Such material is referred to as a Voigt solid. Many investigators attempted to explain the problem by means of the above mentioned simple models. Iida (1934, 1935, 1938a, 1938b, 1938c), Ishimoto and Iida (1936a, 1936b) investigated the properties of sand, clay and pitch-like materials by means of the vibration method and proposed that they behaved as the Voigt solids. Ricker (1953) studied the breadths and amplitudes of the seismic wavelets in Pierre shale, and explained the shale as the Voigt solid. While, Hirono (1935a, 1935b, 1941), Miyabe and Ooi (1949) attempted to explain the materials by the Maxwell solid. Adopting the Toda's model, Kubotera (1952) obtained the relaxation times, which are intimately related with the attenuation coefficient, associated with the crustal materials and the alluvial deposits using P waves.

On the other hand, from the laboratory experiment, Born (1941) showed that the attenuation of seismic pulses in the dry samples of earth materials was proportional to frequency, namely Q-values in such materials were constant. Similar results were obtained by Collins and Lee (1956) associated with the sandstone. McDonal et al (1958) studied the attenuation of compressional and shear waves in the Pierre shale. They found that the attenuation of seismic waves was proportional to frequency. Tullos and Reid (1969) acquired the same results as McDonal et al using P waves in the Gulf Coast sediments. Knopoff (1956), Knopoff and MacDonald (1958) emphasized that the attenuation is due to the solid friction. If this is the case, the attenuation is proportional to the frequency and no dispersion can exist in body waves.

Horton (1959), Futterman (1962), Wuenschel (1965) and Strick (1967), however, detected the dispersion of body waves from the seismograms obtained by McDonal et al (1958). Furthermore, as is represented by the papers of Lomnitz (1957, 1962), Futterman (1962) and others, there is a stream of thought that the attenuation is approximately proportional to frequency only in the limited frequency range, employing the theory of linear viscoelasticity. Thus, there are two schools of thought in the study of loss mechanism associated with the earth materials. This is mainly due to the fact that our observations are limited to a narrow range of frequency.

As mentioned above, the loss mechanism has not yet been investigated sufficiently, and it still remains as an outstanding problem to be clarified. Although there are many papers associated with the problem, almost all of them are related to the earth crust and mantle. Since the S wave measurement of soil layers in situ was believed to be difficult until quite recently, former investigators studied the attenuation of P waves and only a few papers associated with the S waves have hitherto been published. Nowadays, S wave measurement in situ has become an easy routine work. Therefore, persistent study of the attenuation of S waves measured in situ is necessary, since the study will supply a great deal of information clarifying the loss mechanism and the behavior of soil during earthquakes. In the following, an effort along the above mentioned lines is reported.

2. Experiments and Data

Experiments were carried out at four places in the Tokyo Metropolis. They are Adachi, Sunamachi, Yukigaya and Yayoi (Table 1). The natural period and the damping coefficient of the borehole seismometer used in these experiments were 0.2 sec and 0.6, respectively. The output voltage of the seismometer was amplified by means of the DC amplifier and then connected to the galvanometer having a natural frequency of 100 cps. Consequently, frequency characteristics of the total system were flat with respect to the ground velocity in the frequency range from 8 to 70 cps.

Table 1.

No.	Location	Address of the site
1	Adachi	Motogi-nishi-machi, Adachi-ku, Tokyo
2	Sunamachi	5, Kitasuna, Koto-ku, Tokyo
3	Yukigaya	Yukigaya, Ota-ku, Tokyo
4	Yayoi	1, Yayoi-cho, Bunkyo-ku, Tokyo

Fig. 1 shows the method of observation schematically. Seismic waves were generated at the position 1m from the borehole in each experiment by hitting horizontally, with a wooden hammer, the end of a slender weighted wooden plate laid down on the ground surface. Waves thus generated were observed by the borehole seismometer at different depths. The waves detected by the horizontal components, oriented in the same direction with that of the force, were confirmed as SH waves, since the inversion of corresponding phases was noticed when hitting the reverse end of the source plate.

Six vertical travel times for SH waves were obtained from independent trials at one depth. Travel times thus obtained were averaged and the amount of times, depth/200 m/s, were reduced. Such reduction is convenient for seeing roughly

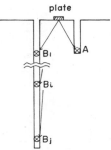

Fig. 1. Configuration of source plate and borehole seismometers. Seismometer at the position A was fixed all through the experiment and the seismometer at the position B was installed at various depths.

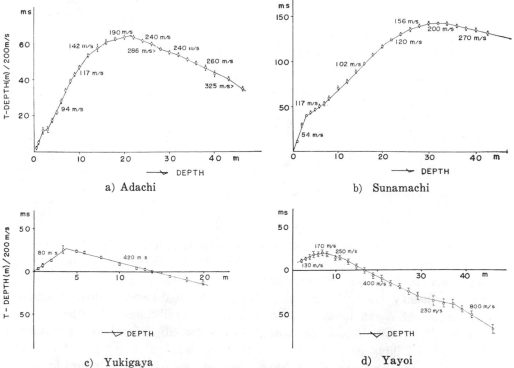

a) Adachi

b) Sunamachi

c) Yukigaya

d) Yayoi

Fig. 2. The reduced vertical travel time graphs for SH-waves. Bars show 95% confidence intervals.

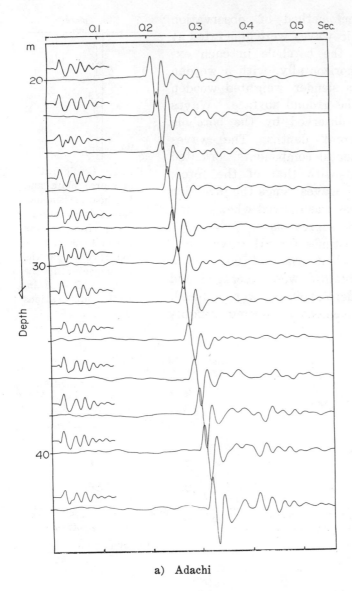

Fig. 3a. Sample seismo-
grams of vertically
travelling SH waves.
Shorter ones are refer-
ence seismograms ob-
tained at fixed position
A.

a) Adachi

if the soils are of diluvial or alluvial deposits, since the S wave velocities in alluvial soils are lower than 200 m/s while those in diluvial
soils are higher than 200 m/s (Shima et al, 1968).   Therefore, one may
easily classify the soils into the above mentioned two formations depending upon the gradients of the reduced travel times.   Reduced travel
times, shown by circles, are shown in Figs. 2a, 2b, 2c and 2d.   Confidence
intervals of 95% (shown by vertical bars) were also computed and are
shown in the figures.   From these figures, we found the layers in which
the SH wave velocities were fairly uniform.   The seismograms observed

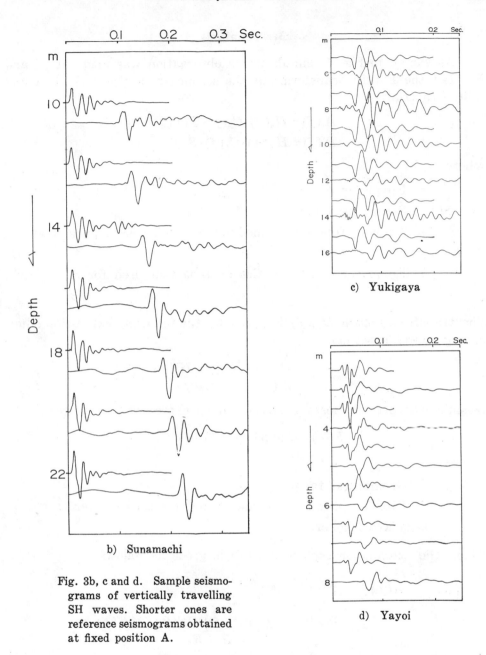

b) Sunamachi

c) Yukigaya

d) Yayoi

Fig. 3b, c and d. Sample seismograms of vertically travelling SH waves. Shorter ones are reference seismograms obtained at fixed position A.

in these strata were used in the present investigation. They are shown in Figs. 3a, 3b, 3c and 3d together with the reference seismograms (shorter ones in the figures) obtained at the position A fixed all through each experiment.

3. Analysis

As shown in Fig. 1, simultaneous observation was made at A and B. Let the Fourier transforms of the seismograms observed at A and B in i-th trial be

$$A_i(f) = H_A(f) \cdot C_A(f) \cdot S_i(f) \, ,$$
$$B_i(f) = H_{Bi}(f) \cdot C_B(f) \cdot S_i(f) \, ,$$

where

f Frequency
$S(f)$ Source spectrum
$H_A(f)$ Transfer function of medium between source point and A
$H_B(f)$ // B
$C_A(f)$ Frequency characteristics of apparatus used for A
$C_B(f)$ // B.

The transfer function $H_{Bi}(f)$ is given by the following expression eliminating source spectrum

$$H_{Bi}(f) = \frac{B_i(f)}{A_i(f)} \cdot \frac{C_A(f) \cdot H_A(f)}{C_B(f)} \, .$$

Seismic waves in the lossy medium can be expressed as follows ;

$$F(f) = G \exp \left[-\{\alpha(f) + ik\}R \right] \, ,$$

where

G Geometrical factor
R Distance between the source and the seismometer
k Wave number.

Then, the attenuation coefficient $\alpha(f)$ is given as follows

$$\alpha(f) = - \frac{\log \left[G_i^{-1} H_{Bi}(f) / G_j^{-1} H_{Bj}(f) \right]}{R_i - R_j}$$
$$= - \frac{\log \left[G_i^{-1} \{B_i(f)/A_i(f)\} / G_j^{-1} \{B_j(f)/A_j(f)\} \right]}{R_i - R_j} \, .$$

After Onda and Komaki (1965), waves observed in the direction bisecting the source plate are the waves of SH type alone. Since the observations were made at sufficiently far distances from the source point the effect of plate length can be safely ignored. Accordingly, the spherical spreading of waves was assumed in computing the geometrical factor.

The thickness of constant velocity layer used in the analysis was not so thick. In such a case, as pointed out by McDonal et al (1958), reflected and refracted waves at the interfaces at various depths may interfere with the useful signals. Fortunately, the velocity contrast in our experimental site was not so remarkable and the interference was negligible although the reflected wavelets were detected in the case of Adachi. As can be seen in Fig. 3a, the reflected wavelets from the deep interface were distinguishable from the down-going wavelets. So we could analyse the useful signals without the fear of interference.

A hanning window was applied in the spectral analysis. The duration of window was 0.08 second, and the sampling interval was 1 or 2 milisecond.

4. Results and Discussions

Fourier amplitudes of seismograms obtained at various depths were corrected by means of the geometrical factors. As an example, the case of Adachi is shown in Fig. 4. Figure 5 shows the relation between

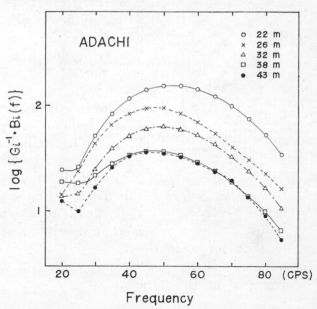

Fig. 4. An example of Fourier spectra corrected by the geometrical factor for spherical spreading (Adachi).

$\log \{G_i^{-1} B_i(f)\}$ and depth for various frequencies in the case of Adachi. The mean value and standard deviation for each frequency were calculated applying the regression line analysis. The relations between attenuation coefficients and frequencies are shown in Figs. 6a (Adachi),

6b (Sunamachi), 6c (Yukigaya) and 6d (Yayoi). The mean values and standard deviations are shown by circles and vertical bars, respectively. The attenuation coefficient in the Maxwell solid is almost independent of frequency and that in the Voigt solid is proportional to the square of frequency. Comparing this with our results, it was found that the attenuation coefficients agreed neither with the Maxwell model nor the Voigt model. If we assume that the attenuation of waves is due to the solid friction, attenuation coefficient is proportional to frequency. With this assumption in mind the straight lines were drawn as shown in Fig. 6a, 6b, 6c and 6d. These lines fitted better than those of the former cases. Figure 7 shows the phase velocity calculated from phase angles of Fourier transform, with respect to the frequency, in the case of Adachi. The regression line analysis was also applied to calculating the mean value (shown by circles) and standard deviation (shown by vertical bars) for each frequency. From the figure, it seems that the dispersion of SH waves is negligible, even if it does exist. As this result, it is concluded that it is hard to explain the loss mechanism by means of the Maxwell model or the Voigt model.

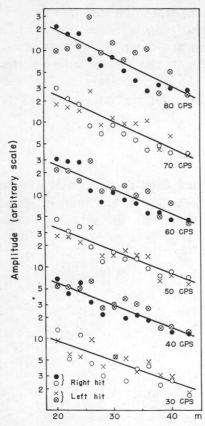

Fig. 5. Relation between $\log\{G_i^{-1}H_{Bi}(f)\}$ and depth for various frequencies in the case of Adachi.

Now we will discuss the loss mechanism in terms of Q-value. The Q-value can be expressed by the formula

$$Q = \frac{\pi f}{c \cdot \alpha(f)},$$

in which c is the phase velocity. The Q-values for various frequencies were obtained in the case of Adachi and are shown by circles in Fig. 8. As can be seen in the figure, Q-values scarcely depend on frequency. As a reference, Q-values were computed in the case of the Maxwell model and the Voigt model choosing the constants so that the values fit the observations at about 50 cps. They are also shown in Fig. 8.

Chae (1968) suggested the loss mechanism of snow and sand introducing the theory of linear viscoelasticity. The complex compliance is expressed as follows

$$J^*(i\omega) = J_\infty - \frac{i}{\omega\eta'} + \int_0^\infty \frac{D(\tau')}{1+i\omega\tau'} \, d\tau' \, ,$$

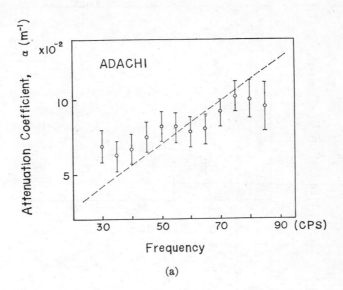

(a)

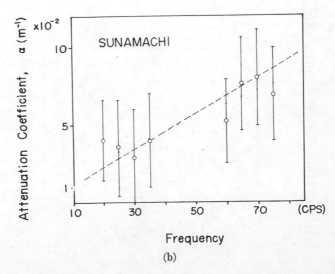

(b)

Fig. 6a, b. Attenuation coefficient vs. frequency. The mean values and the standard deviations are shown by circles and vertical bars, respectively.

in which J_∞ is compliance at infinitely large frequency, η' is viscosity, τ' is retardation time, $D(\tau')$ is the distribution function of retardation time. Chae assumed that

$$J_\infty = 0 , \qquad 1/\eta' = 0 ,$$
$$\left.\begin{array}{ll} D(\tau') = \beta'/\tau' & \tau_1' < \tau' < \infty \\ \quad\;\; = 0 & \tau' < \tau_1' . \end{array}\right\}$$

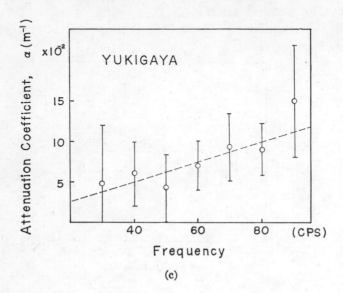

(c)

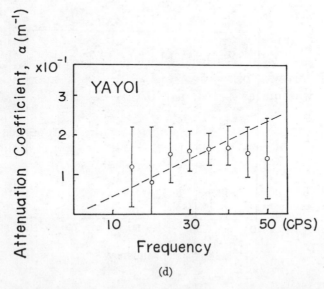

(d)

Fig. 6c, d. Attenuation coefficient vs. frequency. The mean values and the standard deviations are shown by circles and vertical bars, respectively.

Since the Q-value is expressed by the ratio of real to imaginary parts of complex compliance, Q-value vs. frequency for Chae's model is given as follows

$$Q = \frac{\log (1 + 1/\omega^2 \tau_1'^2)}{2 \arctan (1/\omega \tau_1')} .$$

The broken line in Fig. 8 shows the value computed when τ_1' equals to 10^{-7} sec. by the above equation.

Meanwhile, complex rigidity which is the reciprocal of complex compliance, is expressed as follows

$$G^*(i\omega) = G_0 + i\omega\eta + i\omega \int_0^\infty \frac{F(\tau)\tau}{1 + i\omega\tau} d\tau ,$$

in which G_0 is rigidity at zero frequency, η is viscosity, τ is relaxation time, $F(\tau)$ is the distribution function of relaxation time. Let us assume

$$\left. \begin{array}{ll} G_0 = 0 , & \eta = 0 , \\ F(\tau) = \beta/\tau & \tau_1 < \tau < \tau_2 \\ \quad\;\; = 0 & \text{otherwise} . \end{array} \right\}$$

Under the above assumption

$$Q = \frac{1}{2} \cdot \frac{\log (1 + \omega^2 \tau_2^2) - \log (1 + \omega^2 \tau_1^2)}{\arctan (\omega\tau_2) - \arctan (\omega\tau_1)} .$$

An example of Q-value vs. frequency, when $\tau_1 = 0$ and $\tau_2 = 10^3$ sec., is shown by the thick solid line in Fig. 8. As can be seen in the figure, this model agreed well with the observed value. However, considering the accuracy of our observation, and since the frequency range in our observations was quite limited, no further detailed discussion will be made.

The Q-values, SH waves velocities and other properties of soils obtained at four sites are tabulated in Table 2. The Q-values given in the Table are the averaged values over the frequency measured.

Table 2.

No.	Location	S-wave velocity	Q	Geology
1	Adachi	260m/s	8	Diluvial sand
2	Sunamachi	102m/s	20	Alluvial silt
3	Yukiga	420m/s	6.5	Tertiary mudstone
4	Yayoi	150m/s (mean)	5	Kwanto loam

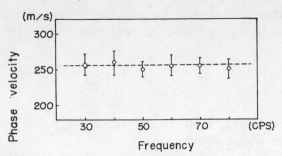

Fig. 7. Phase velocity vs. frequency (Adachi). The mean values and the
standard deviations are shown by circles and vertical bars, respectively.

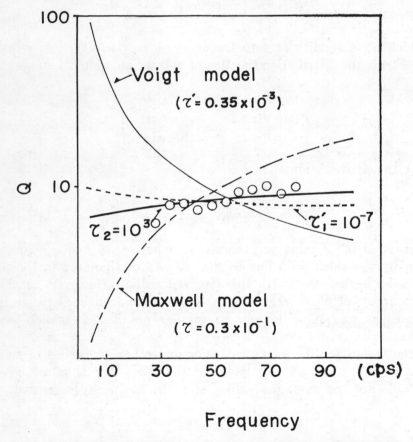

Frequency

Fig. 8. Relation between Q and frequency. The observed values are shown
by circles. Thin solid and part of a dotted line show the case of the
Voigt model ($\tau'=0.35\times10^{-3}$ sec.) and the Maxwell model ($\tau=0.3\times10^{-1}$ sec.),
respectively. The broken and thick solid lines show the Chae's model
($\tau_1'=10^{-7}$ sec.) and the model assumed in this paper ($\tau_1=0$, $\tau_2=10^3$ sec.),
respectively.

Acknowledgement

The authors express their hearty thanks to Professor H. Kawasumi for his valuable suggestion extended to them. Their thanks are also due to Dr. I. Onda and Dr. Y. Ohta for their discussions and to Mr. M. Yanagisawa and Mrs. S. Miyagi for their help in conducting the experiments.

References

BORN, W.T., 1941, The attenuation constant of earth materials, *Geophysics*, **6**, 132-148.

CHAE, Y.S., 1968, Viscoelastic properties of snow and sand, *J. Engineering Mechanics Division*, ASCE, **94**, 1379-1394.

COLLINS, F. and C.C. LEE, 1956, Seismic wave attenuation characteristics from pulse experiments, *Geophysics*, **21**, 16-39.

FUTTERMAN, W.I., 1962, Dispersive body waves, *J. Geophys. Res.*, **67**, 5279-5291.

HIRONO, T., 1935a, On the viscoelastic waves (I), *J. Met. Soc. Japan*, [ii], **13**, 413-424, (*in Japanese*).

HIRONO, T., 1935b, On the viscoelastic waves (II), *J. Met. Soc. Japan*, [ii], **13**, 512-520, (*in Japanese*).

HIRONO, T., 1941, On the viscoelastic waves (III), *Quart. J. Seismol.*, **11**, 337-348, (*in Japanese*).

HORTON, C.W., 1959, A loss mechanism for Pierre shale, *Geophysics*, **24**, 667-680.

IIDA, K., 1934, Experiments on the visco-elastic properties of pitch-like materials. (I), *Bull. Earthq. Res. Inst.*, **13**, 198-212.

IIDA, K., 1935, Experiments on the visco-elastic properties of pitch-like materials. (II), *Bull. Earthq. Res. Inst.*, **13**, 433-156.

IIDA, K., 1938a, Relation between the normal-tangential viscosity ratio and Poisson's elasticity ratio in certain soils, *Bull. Earthq. Res. Inst.*, **16**, 391-406.

IIDA, K., 1938b, Elastic and viscous properties of a certain kind of rock, *Bull. Earthq. Res. Inst.*, **17**, 59-78.

IIDA, K., 1938c, Determination of Young's modulus and the solid viscosity coefficients of rocks by the vibration method, *Bull. Earthq. Res. Inst.*, **17**, 79-92.

ISHIMOTO, M. and K. IIDA, 1936a, Determination of elastic constants of soil by means of vibration method, Part I, Young's modulus, *Bull. Earthq. Res. Inst.*, **14**, 632-657.

ISHIMOTO, M. and K. IIDA, 1936b, Determination of elastic constants of soil by means of vibration method, Part II, Modulus of rigidity and Poisson's ratio, *Bull. Earthq. Res. Inst.*, **15**, 67-88.

KELVIN, LORD, 1878, Elasticity, in *Encyclopedia Brittanica*, ninth edition.

KNOPOFF, I., 1956, The seismic pulse in materials possessing soild friction, 1, Plane waves, *Bull. Seismol. Soc. Am.*, **46**, 175-183.

KNOPOFF, I. and G.J.F. MACDONALD, 1958, Attenuation of small amplitude stress waves in solids, *Rev. Mod. Phys.*, **30**, 1178-1192.

KUBOTERA, A., 1952, Rheological properties of earth's crust and alluvial layers in relation to propagation of seismic waves, *J. Phys. Earth*, **1**, 25-34.

LOMNITZ, C., 1957, Linear dissipation in solids, *J. Appl. Phys.*, **28**, 201-205.

LOMNITZ, C., 1962, Application of the logarithmic creep law to stress wave attenuation in the solid earth, *J. Geophys. Res.*, **67**, 365-367.

MAXWELL, J.C., 1867, On the dynamical theory of gases, *Phil. Trans. Roy. Soc. London*, **157**, 49-88.

McDONAL, F. J., F. A. ANGONA, R. L. MILLS, R. L. SENGBUSH, R. G. VAN NOSTRAND and J. E. WHITE, 1958, Attenuation of shear and compressional waves in Pierre shale, *Geophysics*, **23**, 421-439.

MEYER, O. E., 1874a, Zur Theorie der inner Reibung, *J. Reine Angew, Math.*, **58**, 130-135.

MEYER, O. E., 1874b, Theorie der elastischen Nachwirkung, *Ann. Phys.*, **227**, 108-119.

MIYABE, N. and T. OOI, 1948, On longitudinal vibration tests, *Zisin (Bull. Seismol. Soc. Japan)*, [ii], **1**, 2-3, (*in Japanese*).

ONDA, I. and S. Komaki, 1965, Waves generated from a linear horizontal traction with finite source length on the surface of a semi-infinite elastic medium, with special remarks on the theory of shear wave generator, *Bull. Earthq. Res. Inst.*, **46**, 1-23.

RICKER, N., 1941, A note on the determination of the viscosity of shale from the measurement of wavelength breadth, *Geophysics*, **6**, 254-258.

SHIMA, E., Y. OHTA, M. YANAGISAWA, K. KUDO and H. KAWASUMI, 1968, *S* wave velocities of subsoil layers in Tokyo. 3, *Bull. Earthq. Res. Inst.*, **46**, 1301-1312, (in Japanese).

STRICK, E., 1967, The determination of *Q*, dynamic viscosity and transient creep from wave propagation measurements, *Geophys. J. Roy. Astro. Soc.*, **13**, 197-218.

TULLOS, F. and A. REID, 1969, Seismic attenuation of Gulf Coast sediments, *Geophysics*, **34**, 516-528.

VOIGT, W., 1892, Über Innere Reibung fester Körper, Insbesondere der Metalle, *Ann. Phys.*, **47**, 671-693.

WUENSCHEL, P. C., 1965, Dispersive body waves—An experimental study, *Geophysics*, **30**, 539-551.

---

10. 軟弱な地層における S 波の減衰

地震研究所 { 工　藤　一　嘉
　　　　　　 嶋　　悦　　三 }

　ごく最近に到るまで，現場における S 波の発生は困難であるとされていたため，S 波の減衰に関する研究は極めて少ない．地震波の減衰機構は，動的な応力・歪の関係と対応するから，物質の物理的性質を知る上で，詳しく調べておく必要がある．また応用の見地から，構築物が軟弱な地盤の上に建てられる場合が多いので，地震時における地盤の挙動を熟知する必要があるが，そのためにも軟弱な地層での減衰機構に関する知識が不可欠である．

　今回，板叩き法により S 波を発生させ，抗井法で得られた記録をフーリエ解析し，S 波の減衰を調べた．その結果，地質により多少異なるが，調べた周波数範囲内では，減衰係数が周波数に比例する傾向が強く，また位相速度の分散性も，非常に小さい．したがつて，*Q* の周波数依存性は，あるとしても非常に小さいことが解つた．この結果を単純な Maxwell 模型や Voigt 模型で説明することは難しい．しかしながら，必ずしもこの結果を固体摩擦で説明しなくとも，適当な仮定を設けることにより，線型粘弾性理論で説明出来る可能性のあることを示しておいた．

　このような点をさらに解明するためには，測定周波数範囲を広げなければならない．測定の精度の向上とあわせ，上記の努力を続けるつもりである．

Reprinted from The Earth's Crust, AGU monograph 20, p. 427–439.

INTERPRETATION OF CRUSTAL VELOCITY GRADIENTS AND Q STRUCTURE USING
AMPLITUDE-CORRECTED SEISMIC REFRACTION PROFILES

Lawrence W. Braile

Department of Geosciences, Purdue University
West Lafayette, Indiana 47907

   Abstract. Amplitudes of refracted and reflected seismic waves are
greatly affected by velocity gradients and anelasticity (Q^{-1}). Combined
interpretation of the amplitudes of upper crustal phases, the head wave
in the upper crust (P_g), and the reflection from the top of the lower
crustal layer (P_{cr}) allows inference of both velocity and Q structure
from amplitude-corrected seismic refraction data. Synthetic seismograms
were computed by using the reflectivity method modified to include effects of anelasticity for several upper crustal models. Modeling of the
amplitude-distance variations of the P_g and P_{cr} phases allows simultaneous determination of velocity gradients and Q structure. Interpretation of an amplitude-corrected seismic refraction profile from the eastern Basin and Range Province of the western United States suggests anomalously low compressional wave Q for the upper crust. Rapid attenuation with distance of the P_{cr} phase for this refraction profile suggests
that seismic wave energy absorption may occur primarily in the upper
crustal low-velocity high Poisson ratio layer, where a Q of ~ 50 is inferred.

Introduction

   While the general velocity structure of the earth's crust can be inferred from studies of the travel times of seismic waves, detailed velocity structure, including identification of velocity gradients within
layers and recognition of low-velocity layers, can best be determined by
a combined interpretation of the travel times and amplitudes of refracted and reflected waves. Seismic wave amplitudes are sensitive to the
presence of velocity gradients and anelasticity (Q^{-1}). Cerveny [1966]
has shown that even small velocity gradients (± 0.01 km s^{-1} km^{-1}) are
sufficient to affect amplitudes of head waves by a factor of 1 or 2
orders of magnitude. Such velocity gradients, however, result in only
slight curvature of the travel time curve. Hill [1971] has shown that
the effects of velocity gradients and anelasticity cannot be separated
by using the amplitude decay with distance of a head wave phase.
   The purpose of this paper is to present a method for determining both
velocity and Q structure in the earth by using combined interpretation
of amplitudes of refracted and reflected phases and to apply the method
to determination of the compressional wave Q in the upper crust in the
eastern Basin and Range Province of the western United States. The

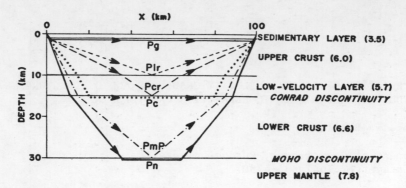

Fig. 1. Schematic diagram of prominent crustal and uppermost mantle
compressional wave phases. Characteristic velocities for Basin and
Range Province crustal structure are given in kilometers per second for
each layer.

method employs modeling of observed amplitudes from seismic refraction
profiles with theoretical amplitudes from synthetic seismograms. Synthetic seismograms are computed by using a modification of the reflectivity method [Fuchs and Müller, 1971] in which the effect of attenuation on body waves is included by introducing complex velocities to calculate the plane wave reflection coefficients. By following Schwab and
Knopoff [1972], if c and Q are the phase velocity and the quality factor, respectively, for either the compressional or the shear waves, then
the complex velocity is

$$V = V_1 + iV_2$$

where

$$V_1 = c[4Q^2/(4Q^2 + 1)]$$

$$V_2 = c[4Q^2/(4Q^2 + 1)]/2Q$$

Kennett [1975] has described a similar modification of the reflectivity
method, which he uses to investigate attenuation in the upper mantle.
In an alternative approach, Helmberger [1973] has employed an empirical
attenuation filter in order to introduce anelasticity into the calculation of synthetic seismograms by the Caignard-deHoop technique.

Model Studies

   Synthetic seismograms are calculated for several upper crustal velocity models both with and without the effects of attenuation. A 3-Hz
dominant frequency wavelet is used. Shear wave velocities are computed
from compressional wave velocities, a Poisson ratio of 0.25 being
assumed. Densities are computed from compressional wave velocities by
following Birch [1964]. Shear wave Q was specified to be 4/9 of compressional wave Q by following the hypothesis of Anderson and Archambeau
[1964] of no attenuation in pure compression.

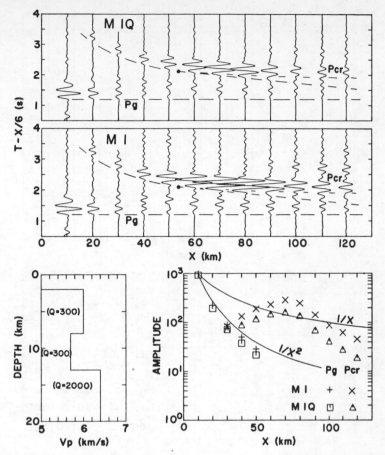

Fig. 2. Synthetic seismograms and amplitude-distance curves for models
M1 and M1Q. Velocity model is shown in lower left. Q values are infinite for model M1 and have values shown in the figure for model M1Q.
Synthetic seismograms are for the vertical component. Amplitudes are
multiplied by distance for convenient plotting. Solid circles indicate
the ray-theoretical critical point for the head wave from the lower
crustal layer. Phases are identified according to the notation in Figure 1. The solid lines in the relative amplitude versus distance plot
are reference lines showing $1/x$ and $1/x^2$ amplitude decay referenced to
10^3 at a distance of 10 km.

The crustal and uppermost mantle phases which are considered in this
study are shown schematically in Figure 1. The crustal model shown is
similar to models which have been proposed for the Basin and Range
Province and includes an upper crustal low-velocity layer (LVL).
Six upper crustal velocity models, synthetic seismograms calculated
for both the anelastic and the perfectly elastic case, and amplitude-
distance curves are shown in Figures 2-7. It will be seen that modeling
of the amplitudes of the head wave phase (P_g) and the reflection from
the top of the lower crustal layer (P_{cr}) allows separation of the ef-

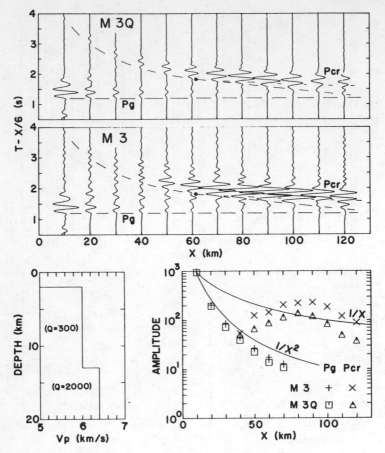

Fig. 3.   Same as Figure 2 for models M3 and M3Q.

fects of velocity gradients and Q structure.   However, inference of both
velocity and Q structure from P_g and P_{cr} amplitude decay requires accurate amplitude data and recognition of distinct phases over a considerable distance range.

The amplitude-distance curves of Figures 2-7 illustrate several features which bear on the problem of recognition of velocity gradients and
Q structure of the upper crust.

P_g amplitudes.   A positive velocity gradient in the upper crustal
layer, as is seen in models M8 and M11 (Figures 4 and 7), produces a
much less rapid amplitude decay than is produced in the homogeneous
layer case, model M3 (Figure 3).   Both a negative velocity gradient, M9
(Figure 5), and the presence of anelasticity (Q models) produce more
rapid amplitude decay.

P_{cr} amplitudes.   The distance at which the P_{cr} amplitude-distance
curve peaks (about 60-100 km for the models shown) is affected by the
velocity contrast at the upper crust-lower crust boundary.   Thus the P_{cr}
amplitude peak shifts from near 90 km for the homogeneous upper crustal
model (Figure 3) to about 70 km (model M1, Figure 2) owing to the

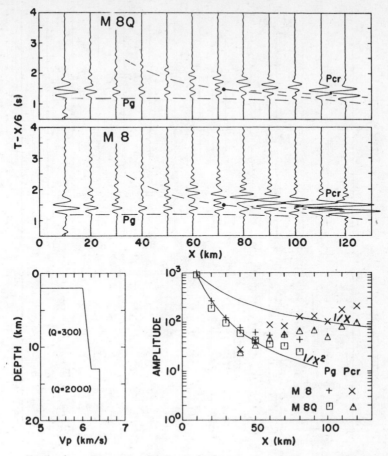

Fig. 4.   Same as Figure 2 for models M8 and M8Q.

presence of an LVL. The P_{cr} amplitude curve peaks sharply near 100 km
owing to a velocity transition zone at the base of the upper crust
(model M10, Figure 6). A positive velocity gradient in the upper crust
(Figures 4 and 7) produces large amplitudes at distances greater than
100 km, well beyond the critical distance.
    Separation of the effects of negative velocity gradients and anelas-
ticity is illustrated by models M3 and M9 (Figures 3 and 5, respectively). The negative velocity gradient infinite Q model (M9) can be
seen to have a P_g amplitude decay which is indistinguishable from the
homogeneous layer anelastic model M3Q (Figure 3) However, the P_{cr} amplitudes for M9 (Figure 5) and M3Q (Figure 3) differ by approximately a
factor of 2. Similarly, a positive velocity gradient low Q model
(similar to M11Q, Figure 7) may have a P_g amplitude decay equivalent to
the homogeneous perfectly elastic upper crustal model (M3, Figure 3).
However, the two models may be differentiated by their P_{cr} amplitudes.
    Application of this modeling method to determination of crustal vel-
ocity and Q structure is limited by several factors. Accurate amplitude
corrections and close station spacing are necessary so that local site

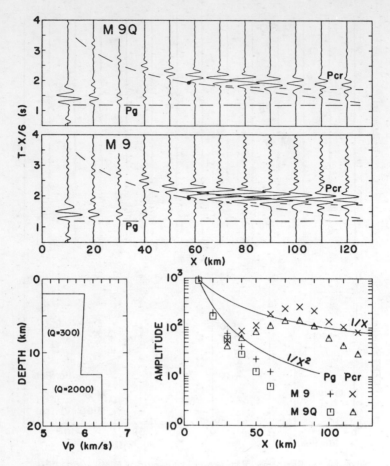

Fig. 5.   Same as Figure 2 for models M9 and M9Q.

responses may be 'averaged out.' Lateral homogeneity in both velocity
and anelasticity is assumed. Uniqueness of the derived model is not
assured by the modeling procedure, and since the interpretation is not
based on an inversion method, estimates of resolution are qualitative
and may only be determined by a lengthy perturbation analysis of the
derived crustal model. Despite these limitations, modeling of amplitude
data for crustal phases can yield important information on crustal velocity and Q structure. Significant improvements in interpretation
should be possible with high-quality amplitude data.

Application to Eastern Basin and Range Province

A detailed seismic refraction profile from the eastern Basin and Range
Province was presented by Keller et al. [1975]. These data, herein
called the 1972 Utah refraction profile (Figure 8), were recorded along
a line due south of the Bingham copper mine southwest of Salt Lake City,
Utah. This record section is conducive to amplitude interpretation because of high signal to noise ratio, the even distribution of stations

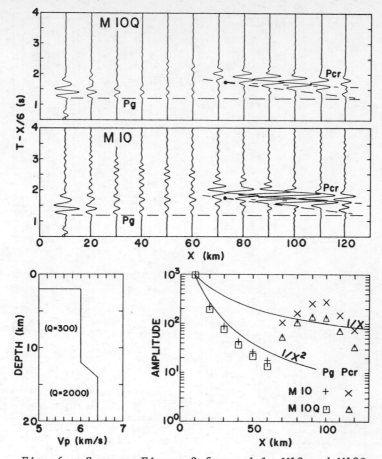

Fig. 6.   Same as Figure 2 for models M10 and M10Q.

over a distance range of 240 km, and the existence of distinct phases.
The record section was previously interpreted by Keller et al. [1975]
primarily on the basis of travel times.

   The data in Figure 8 were amplitude-corrected by removing the effect
of the instrument and accounting for differences in source sizes. This
correction was accomplished by deconvolving the seismograms within the
frequency range of 1–15 Hz by using

$$A_c(f) = k \frac{A_f(f)I_f(f)}{S(f)I_b(f)}$$

where

$A_c(f)$ amplitude spectrum of the corrected field seismogram;
$A_f(f)$ amplitude spectrum of vertical ground motion at the field
             station;

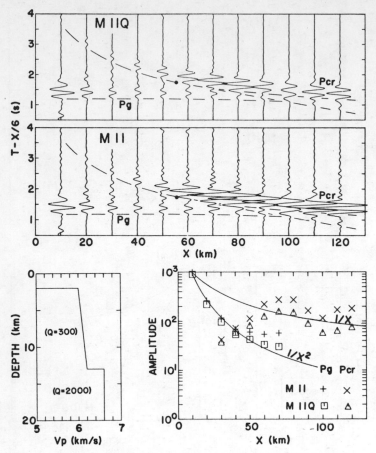

Fig. 7.   Same as Figure 2 for models M11 and M11Q.

$I_f(f)$ instrument response of the field seismograph;
$S(f)$ amplitude spectrum of the vertical component of the source
function at the base station;
$I_b(f)$ instrument response of the base station seismograph;
k   constant relating the relative amplifications of the base
and field station seismographs.

The base station for all sources was located approximately 12 km NNE of
the sources. The product $A_f(f) \cdot I_f(f)$ is the Fourier transform of the
field seismogram, and $S(f) \cdot I_b(f)$ is the Fourier transform of the first
0.3 s of the base station seismogram. Inspection of the base seismograms suggested that the source function was best represented by the
first 0.3 s of the seismogram. This length was equal to the first 1½–2
cycles for most sources. Since the seismographs were matched instruments, it is assumed that $I_f(f) = I_b(f)$ and that the effect of the instrument response is given by the constant k. This assumption was sub-
stantiated by daily instrument calibrations. The amplitude-corrected
seismograms are shown in record section form in Figure 9, and amplitudes

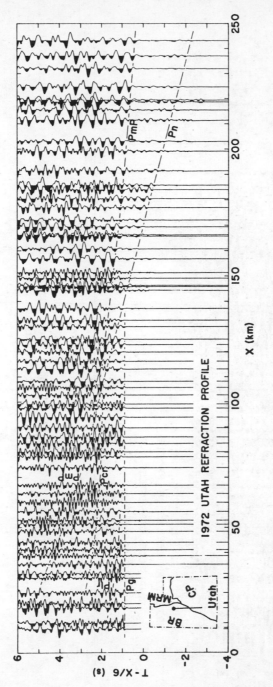

Fig. 8. Record section of vertical component seismograms for the 1972 Utah refraction profile. Phases are identified according to the notation in Figure 1. Inset map shows location of profile: BR, Basin and Range Province; CP, Colorado Plateau Province; MRM, Middle Rocky Mountain Province.

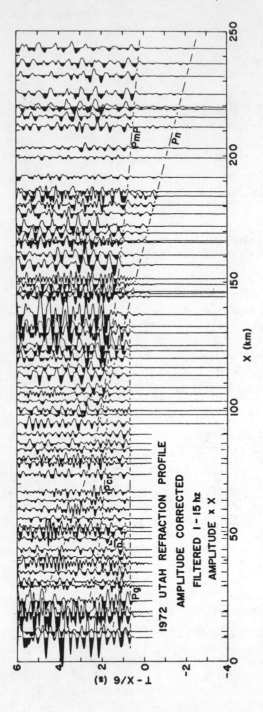

Fig. 9. 1972 Utah refraction profile after amplitude correction. Amplitudes have been multiplied by distance for convenient plotting.

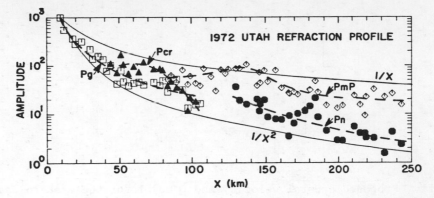

Fig. 10. Relative amplitude versus distance plot for prominent phases in Figure 9. Solid lines show $1/x$ and $1/x^2$ amplitude decay referenced to 10^3 at a distance of 10 km. Heavy dashed lines are inferred amplitude-distance curves fit through the data points. Theoretical amplitudes for the model shown in Figure 11 fall on the amplitude-distance curves for the P_g and P_{cr} phases, except near 10 km.

of prominent phases are plotted versus distance in Figure 10. Some scatter exists in the amplitude-distance curves owing to local site responses and possible inaccuracy in the amplitude correction. However, because of the close station spacing and the large number of amplitude observations the trend of the amplitude decay for each phase is reasonably well determined.

Two important features of the amplitude-distance (Figure 10) plot of upper crustal phases (P_g and P_{cr}) are that the amplitudes decay less rapidly than those of the homogeneous layer case (Figure 3) and that amplitude curve peaks at small distances ($\sim$60 km) and decays rapidly with distance at greater distances. The P_g decay suggests the presence of a positive velocity gradient in the upper crustal layer. However, the P_{cr} amplitude peak at short distances and the rapid amplitude decay at greater distances preclude the presence of a significant positive gradient. In fact, in order to explain the P_{cr} amplitudes an LVL, as previously inferred for this area by Keller et al. [1975], and a low compressional wave Q of about 100 are suggested for the upper crust. However, a Q of 100 produces a rapid decay of the P_g phase, which is not observed in Figure 10. A possible explanation of the P_g amplitude curve is that the P_{1r} phase, which is produced by reflection from the top of the LVL and which has arrival times only about 0.2 s after P_g at distances greater than about 40 km, interferes with P_g and causes a smaller apparent decay of P_g with distance.

In order to explain both the P_g and the P_{cr} amplitude-distance curves the crustal model in Figure 11 is proposed. The velocity structure is identical to that given by Keller et al. [1975]. The model produces calculated amplitudes of P_g and P_{cr} phases which match the observed amplitudes except in the short distance range (near 10 km). It is possible that the P_g and P_{cr} amplitudes observed in the 1972 Utah refraction profile could be explained by a model having a constant Q of about 100 in the upper crust instead of the variable Q model shown in Figure 11.

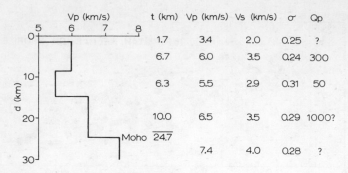

Fig. 11. Proposed crustal velocity and Q model for the eastern Basin and Range Province based on travel time and amplitude interpretation of the 1972 Utah refraction profile.

However, such a model would have to include a velocity structure more complex than the structures in the models shown in Figures 2-7, and the existing data are not sufficient to deduce such a feature. Furthermore, an anomalous Poisson ratio has previously been inferred for the LVL, and thus very low Q would not be unexpected. Two mechanisms which have been shown, under 'crustal' conditions, to produce an increase in Poisson ratio and a decrease in Q are increased temperature and increased pore fluid pressure [Spencer and Nur, 1976; Kissell, 1972; Volarovich and Gurvich, 1957; Grosenbaugh and Nur, 1976].

A complete analysis of the P_mP and P_n amplitude curves for the 1972 Utah refraction profile (Figure 10) will require considerable modeling and will be deferred for a future study. However, preliminary modeling using synthetic seismograms indicates that Q in the lower crust in the eastern Basin and Range Province must be at least 1000 to explain the P_mP amplitude curve.

Discussion

Few determinations of Q have been made in the upper crust. Press [1964] measured average values of Q for the upper crust in the Basin and Range Province from seismic profiles from the Nevada nuclear test site. He reported Q values of 450 and 260, inferred from L_g and P_g phases, respectively. However, the lower value (for P_g) was attributed to partial loss of P wave energy by mode conversion. Regional values of shear wave Q for the upper crust have been determined by Mitchell [1975] and by Lee and Solomon [1975] using Rayleigh wave attenuation. Mitchell reports shear wave Q values of ~125 and ~250 for the upper crust in the western and eastern United States, respectively. Lee and Solomon suggest crustal shear wave Q values of ~200 for a path in the western United States and ~300 for paths in the eastern United States. Compressional wave Q values for the upper crust consistent with averages of Mitchell's and Lee and Solomon's results are about 300 for the western United States and 750 for the eastern United States. Thus the proposed upper crustal Q values from this study (Figure 11) are exceptionally low and provide further evidence for an anomalous crust in the eastern Basin and Range Province [Smith et al., 1975].

Acknowledgements. I am grateful to Karl Fuchs and Gerhard Muller for providing me with a copy of the reflectivity computer program, which was modified for use in this research. I thank Robert B. Smith for his assistance in obtaining the information necessary to make amplitude corrections on the 1972 Utah refraction profile. This research was supported by the Office of Naval Research, Earth Physics Program, contract N00014-75-C-0972, and was presented at the Office of Naval Research symposium, 'The Nature and Physical Properties of the Earth's Crust,' held at Vail, Colorado, August 2-5, 1976.

References

Anderson, D. L., and C. B. Archambeau, The anelasticity of the earth, J. Geophys. Res., 69, 2071-2084, 1964.

Birch, F., Density and composition of the mantle and core, J. Geophys. Res., 69, 4377-4387, 1964.

Cerveny, V., On dynamic properties of reflected and head waves in the n-layered earth's crust, Geophys. J. Roy. Astron. Soc., 11, 139-147, 1966.

Fuchs, K., and G. Müller, Computation of synthetic seismograms with the reflectivity method and comparison with observations, Geophys. J. Roy. Astron. Soc., 23, 417-433, 1971.

Grosenbaugh, M., and A. Nur, Crustal low-velocity models (abstract), Eos Trans. A.G.U., 57, 961, 1976.

Helmberger, D. V., On the structure of the low-velocity zone, Geophys. J. Roy. Astron. Soc., 34, 251-263, 1973.

Hill, D. P., Velocity gradients and anelasticity from crustal body wave amplitudes, J. Geophys. Res., 76, 3309-3325, 1971.

Keller, G. R., R. B. Smith, and L. W. Braile, Crustal structure along the Great Basin-Colorado Plateau transition from seismic refraction studies, J. Geophys. Res., 80, 1093-1098, 1975.

Kennett, B. L. N., The effects of attenuation on seismograms, Bull. Seismol. Soc. Amer., 65, 1643-1651, 1975.

Kissell, F. N., Effect of temperature variation on internal friction in rocks, J. Geophys. Res., 77, 1420-1423, 1972.

Lee, W. B., and S. C. Solomon, Inversion schemes for surface wave attenuation and Q in the crust and the mantle, Geophys. J. Roy. Astron. Soc., 43, 47-71, 1975.

Mitchell, B. J., Regional Rayleigh wave attenuation in North America, J. Geophys. Res., 80, 4904-4916, 1975.

Press, F., Seismic wave attenuation in the crust, J. Geophys. Res., 69, 4417-4418, 1964.

Schwab, F. A., and L. Knopoff, Fast surface wave and free mode computations, Methods Comput. Phys., 11, 87-180, 1972.

Smith, R. B., L. W. Braile, and G. R. Keller, Upper crustal low-velocity layers: Possible effect of high temperatures over a mantle upwarp at the Basin Range-Colorado Plateau transition, Earth Planet. Sci. Lett., 28, 197-204, 1975.

Spencer, J. W., Jr., and A. M. Nur, The effects of pressure, temperature, and pore water on velocities in Westerly granite, J. Geophys. Res., 81, 899-904, 1976.

Volarovich, M. P., and A. S. Gurvich, Investigation of dynamic moduli of elasticity for rocks in relation to pressure, Bull. Acad. Sci. USSR, Geophys. Ser., Engl. Transl., no. 4, 1-9, 1957.

Chapter 5
The effect of attenuation on the seismic pulse

Seismic waves propagating in an infinite homogeneous medium are dispersed as a consequence of intrinsic attenuation. The shape of a pulse, propagating in a medium of finite Q, will change with distance or traveltime. When considered in the frequency domain, different frequency components will propagate with different phase and group velocities. Given the attenuation properties as a function of frequency, the dispersion characteristics of the pulse and the frequency dependence of phase and group velocities can be computed by invoking causality (i.e., the signal is not detected until sufficient time for travel has elapsed) and some assumptions easily justifiable for the earth. For typical earth materials where attenuation is not extremely high (e.g., $Q \geq 10$), the linear theory of wave propagation can be used to include the attenuation properties.

There are two separate approaches to seismic pulse propagation and dispersion in attenuating solids. The first assumes a viscoelastic rheology and evaluates its effect on the propagation of a seismic pulse. The work of *Kolsky* (1953, *1956*) and *Ricker* (*1953*, 1977) are examples of this method. The second approach assumes a constant or nearly constant Q for a medium. The Q is then used to calculate pulse broadening and dispersion (*Futterman, 1962; Strick*, 1967, *1970*, 1971; *Carpenter, 1966;* Fuchs and Müller, 1971; Liu et al, 1976; and *Kjartansson, 1979*). As was discussed in chapters 2 and 4, the constant-Q model is probably a better representation for most crustal rocks than is a Voigt solid. As a result, the second approach has gained greater acceptability in the past ten years.

There are six papers in this chapter. Two (*Kolsky, 1956; Ricker, 1953*) are outstanding classical examples of seismic-pulse propagation in solids of a given rheology. The Ricker wavelet concept has been used extensively in exploration seismology. In spite of the fact that the rheological model used may not be totally realistic, the *Ricker* (*1953*) paper demonstrates the application of pulse broadening, as observed in Pierre shale, to seismic exploration. *Kolsky's* (*1956*) paper, which uses laboratory measurement of pulse shapes propagating in polymer rods, complements Ricker's by providing a broader physical basis and a calculation method.

The four papers that follow (*Futterman, 1962; Strick, 1970; Carpenter, 1966;* and *Kjartansson, 1979*) provide a mathematical and physical treatment for cases with constant or nearly constant Q. Futterman demonstrates how dispersive properties can be calculated given a Q model and shows examples of pulse broadening. Strick carries this one step further by examining the effect of propagation in a constant-Q medium on both the onset of the pulse and the overall broadening of the pulse. Carpenter and Kjartansson develop practical techniques for incorporating the effect of attenuation on wave propagation in media where Q varies with depth but not with frequency. Convolution operators are developed that make it easy to incorporate the effects of attenuation in synthetic seismograms.

All papers in this chapter predict dispersion (i.e., velocity variation with frequency) caused by attenuation. For typical earth cases ($Q \geq 20$), dispersion is small in the frequency range encountered in exploration seismology. Dispersion has, however, been observed and measured in carefully planned laboratory experiments (Wuenschel, 1965; *Kolsky, 1956*).

REFERENCES

Carpenter, E.W., 1966, Absorption of elastic waves—An operator for a constant Q mechanism: UK Atom. Ener. Auth. AWRE Rep. 0-43/66.
Fuchs, K., and Müller, G., 1971, Computation of synthetic seismograms with the reflectivity method and comparison with observations: Geophys. J. Roy. Astr. Soc., v. 23, p. 417–433.

Futterman, W.I., 1962, Dispersive body waves: J. Geophys. Res., v. 67, p. 5257–5291.

Kjartansson, E., 1979, Constant Q—Wave propagation and attenuation: J. Geophys. Res., v. 84, p. 4737–4748.

Kolsky, H., 1953, Stress waves in solids: London, Oxford University Press; also, 1963, New York, Dover Publications.

———— 1956, The propagation of stress pulses in visco-elastic solid: Phyl. Mag., v. 1, p. 693–710.

Liu, H.P., Anderson, D.L., and Kanamori, H., 1976, Velocity dispersion due to anelasticity: Implications for seismology and mantle composition: Geophys. J. Roy. Astr. Soc., v. 47, p. 41–58.

Ricker, N.H., 1953, The form and laws of propagation of seismic wavelets: Geophysics, v. 18, p. 10–40.

———— 1977, Transient waves in visco-elastic media: New York, Elsevier.

Strick, E., 1967, The determination of Q, dynamic viscosity, and transient creep curves from wave propagation measurements: Geophys. J. Roy. Astr. Soc., v. 13, p. 197–218.

———— 1970, A predicted pedestal effect for pulse propagation in constant-Q solids: Geophysics, v. 35, p. 387–403.

———— 1971, An explanation of observed time discrepancies between continuous and conventional well velocity surveys: Geophysics, v. 35, p. 285–295.

Wuenschel, P.C., 1965, Dispersive body waves—An experimental study: Geophysics, v. 30, p. 539–557.

Reprinted from Geophysics, v. 18, p. 10–40.

THE FORM AND LAWS OF PROPAGATION OF
SEISMIC WAVELETS*

NORMAN RICKER†

ABSTRACT

An attack has been made upon the problem of the form of the seismic disturbance which proceeds outward from the explosion of a charge of dynamite and the laws of propagation of this disturbance. A mathematical theory was developed about ten years ago and a small amount of experimental work carried out at that time encouraged further efforts. Recently an extensive series of experimental researches have been carried out in a sensibly homogeneous shale section, and as a result of these studies a rather clear understanding of the form and nature of the primary seismic disturbance, and its laws of propagation, has been obtained for shale. The results are in good agreement with the theory.

INTRODUCTION

Among the various geophysical prospecting methods used in the search for petroleum, the seismic methods stand far ahead of the others in their ability to delineate buried geologic formation. These seismic methods have seen wide application to petroleum exploration and have been improved up to a certain stage during the period in which they have been in use. Attempts to obtain further improvement in seismic methods have met with quite serious obstacles some of which have not yielded to the customary experimental procedures of development.

About twelve years ago it was realized that some of the reasons for these difficulties lay in the very poor state of our knowledge of the basic principles of seismic wave propagation. In other words, the principal obstacle which lay in the path of further development appeared to be the fact that some of the basic problems of seismological physics had not been solved. One of these problems is concerned with answering the question: "What is the form of the seismic disturbance which proceeds outward from the explosion of a charge of dynamite in the earth and what are the laws of propagation of this disturbance?" This problem has lain unsolved during the history of seismology in spite of its basic importance. Questions relative to seismic resolution and the precise delineation of reflecting beds cannot be attacked properly until this question is answered.

A serious attack upon this problem was started about twelve years ago, and through mathematical researches a solution was obtained which seemed to be in accord with the small amount of experimental work carried out at that time. These mathematical studies resulted in what is now known as *The Wavelet Theory of Seismogram Structure*. Parts of this theory have been published.

---

* Presented before the Third World Petroleum Congress, The Hague, 1951. Manuscript received by the Editor October 1, 1952.

† Senior Research Physicist, Research Laboratories, The Carter Oil Company, Tulsa, Oklahoma.

To summarize the results of the theory, it has been found that the classical wave equation

$$\nabla^2 \Phi = \frac{1}{c^2} \frac{\partial^2 \Phi}{\partial t^2} \tag{1}$$

is not sufficient properly to explain seismic phenomena, and that it is necessary to make use of a modified wave equation

$$\nabla^2 \left[\Phi + \frac{1}{\omega_0} \frac{\partial \Phi}{\partial t} \right] = \frac{1}{c^2} \frac{\partial^2 \Phi}{\partial t^2} \tag{2}$$

which, in its plane wave form, was first given by Professor George Gabriel Stokes in 1845. Solutions of equation (2) have been given in the author's previous papers to describe the earth-displacement, the earth-particle velocity, and the earth-particle acceleration as a function of time and distance.

At the Rice Institute meeting of The American Physical Society in November, 1947, the following integral representations of the displacement and earth-particle velocity were given and are published herewith for the first time.

The displacement function is

$$\dot{\Phi} = \frac{2}{R} \int_0^\infty \beta e^{-R\beta(1+\beta^2)^{-1/4} \sin(\tan^{-1}\beta/2)}$$

$$\cdot \sin\left[R\beta(1+\beta^2)^{-1/4} \cos\left(\frac{\tan^{-1}\beta}{2}\right) - \beta T \right] d\beta \tag{3}$$

and the velocity function is

$$\ddot{\Phi} = -\frac{2}{R} \int_0^\infty \beta^2 e^{-R\beta(1+\beta^2)^{-1/4} \sin(\tan^{-1}\beta/2)}$$

$$\cdot \cos\left[R\beta(1+\beta^2)^{-1/4} \cos\left(\frac{\tan^{-1}\beta}{2}\right) - \beta T \right] d\beta. \tag{4}$$

The quantities R and T are dimensionless and are called the numerical distance and numerical time respectively. They are defined by

$$R = \frac{\omega_0 r}{c} \tag{5}$$

$$T = \omega_0 t \tag{6}$$

wherein r is the true radial distance, c the classical velocity of propagation, given by the square root of the elasticity over the density, and t is the time. ω_0 is the constant occurring in equation (2).

Since

$$\ddot{\Phi} = \frac{\partial \dot{\Phi}}{\partial T} \tag{7}$$

if we represent the actual displacement by Φ it follows that the actual earth-particle velocity is

$$\frac{\partial\dot{\Phi}}{\partial t} \quad \text{or} \quad \omega_0\ddot{\Phi}. \tag{8}$$

Equations (3) and (4) represent the seismic behavior of the earth when the displacement at the shot point is a simple impulse, that is to say, at the position of the explosion the walls of the cavity move outward and back in a very short interval of time. This behavior is illustrated in Figure 1 wherein the various wavelet forms have been drawn as functions of the numerical time, T.

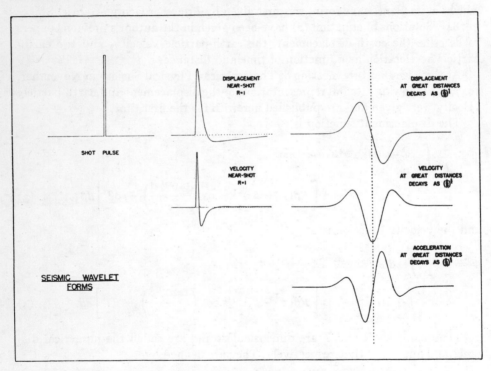

FIG. 1. Shot pulse and propagated wavelet forms.

We may also write the displacement, velocity and acceleration functions as

$$\dot{\Phi} = \left(\frac{2}{R}\right)^{4/2} \mathcal{U}(u \,|\, R) \tag{9}$$

$$\ddot{\Phi} = \left(\frac{2}{R}\right)^{5/2} \mathcal{V}(u \,|\, R) \tag{10}$$

$$\dddot{\Phi} = \left(\frac{2}{R}\right)^{6/2} \mathcal{W}(u \,|\, R) \tag{11}$$

wherein

$$u = \frac{T - R}{\left(\dfrac{R}{2}\right)^{1/2}}. \tag{12}$$

$\mathcal{U}(u\,|\,R), \mathcal{V}(u\,|\,R)$, and $\mathcal{W}(u\,|\,R)$ are called wavelet-form functions and are drawn as functions of u with R held fixed as a parameter. In Figure 2 a number of the wavelet forms

$$\mathcal{V}(u\,|\,R)$$

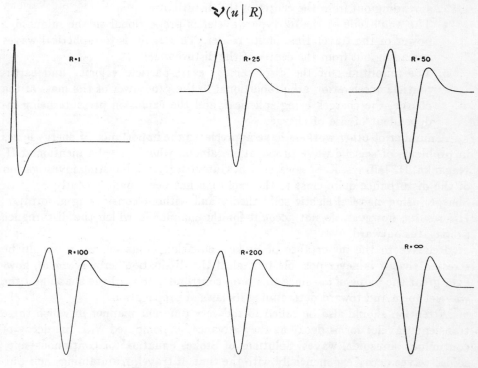

FIG. 2 Computed velocity-type wavelet forms.

are shown. These forms have been described in the published papers. They represent the motion of the trace of an electrical seismograph which is responsive to the velocity of earth-particle motion, and which is free from distortion. It is possible, then, to make a critical comparison of forms obtained experimentally with these forms computed from the theory.

The are also a number of laws of wavelet propagation derived from the theory and with which experimental observations may be compared:

1. It is the center of the wavelet which travels with the velocity, c, given by the square root of the elasticity over the density. This, in fact, defines the wavelet center.

2. The breadth of the wavelet, measured in seconds, say, is proportional to the square root of the propagation time of its center, the law holding for all three forms $\mathcal{U}, \mathcal{V}$, and $\mathcal{W}$.

3. The amplitude of the $\mathcal{U}$ type wavelet is proportional to the minus 4/2 power of the travel time of its center. This is for spherical waves spreading out from the center of the disturbance.

4. The amplitude of the $\mathcal{V}$ type wavelet is proportional to the minus 5/2 power of the travel time of its center. This also is for spherical waves spreading out from the center of the disturbance.

5. The amplitude of the $\mathcal{W}$ type wavelet is proportional to the minus 6/2 power of the travel time of its center. This again is for spherical waves spreading out from the center of the disturbance.

6. The amplitudes of the displacement earth-particle velocity and earth-particle acceleration are proportional to the 5/6 power of the mass of the charge, the charges being spherical, and the explosion pressure being independent of size of charge.

A number of other workers have appreciated the importance of energy losses in problems of seismic wave propagation, among whom may be mentioned H. Nagaoka, H. Jeffreys, K. Sezawa, and B. Gutenberg. An interesting investigation of the disturbance quite close to the explosion has been made recently by J. A. Sharpe, using classical elastic solid theory and without considering absorption. His results, however, do not account for the manner in which the disturbance propagates outward.

Relative to the importance of Stokes' equation (equation 2), it should be stressed that it is never possible to neglect the dissipation term $1/\omega_0 \partial/\partial t$, however great the value of ω_0, since this term contributes toward the shaping of the wavelet form and toward determining its laws of propagation.

Attention should also be called to the very different manner in which these transient wavelet forms decay as they advance, as compared with the decay of continuous sinusoidal waves. Solutions of Stokes equation for continuous sinusoidal waves decay exponentially with the time of travel, maintaining their uniform cycle breadth as they advance. On the other hand, the displacement, earth-particle velocity, and earth-particle acceleration forms, $\mathcal{U}(u|R), \mathcal{V}(u|R)$ and $\mathcal{W}(u|R)$ decay in accordance with the $-4/2, -5/2$, and $-6/2$ powers respectively of the travel time, and the three forms all broaden with the square root of the travel time.

EXPERIMENTAL STUDIES

Recently a rather extensive series of experimental studies have been carried out with the objective of determining the true form of the seismic disturbance which travels outward from the explosion of a charge of dynamite as a point in the earth. In order to carry out satisfactory experimental researches of this nature, it is of fundamental importance that the studies be conducted in a very simple earth, as nearly homogeneous as can readily be obtained. This is necessary in order that the seismograms may have as few disturbances as possible. Seismograms obtained from studies in the average severely bedded earth are far too complicated from the overlapping of successive disturbances to permit very much basic research information being obtained from them. The selection of such a simple homogenous and isotropic earth is not easy. The best choice to the knowledge of the author is the thick section of Pierre shale in the Denver basin in the eastern part of the State of Colorado. In this basin the shale, which is of Cretaceous age, ranges in thickness from o to around 8,000 feet. It lies exposed at the surface over a considerable area. At places it is covered by more recent Cretaceous beds of

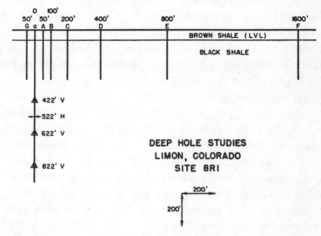

FIG. 3. Disposition of shot holes for the studies at Limon, Colorado.

Fox Hills sandstone or Laramie formation and at other places by a Tertiary gravel conglomerate known as the Ogalalla.

Two different sites were occupied in the Pierre shale, one about three miles north of Limon, Colorado, and the other about thirty-seven miles north of Limon at the crossroads village of Last Chance. At both of these sites the Pierre was exposed at the surface. At both sites the terrain was reasonably flat, the Limon site lying upon a ridge and the Last Chance site lying in a valley. In drilling into this shale, the bit first encounters a brown shale which is dry and which breaks easily. At a depth of about 55 feet (at Limon) the bit breaks suddenly into a black shale which is wet and somewhat soft at the top, but which firms rapidly so that it is quite hard at depths greater than, say, 100 feet. The upper crust of brown shale is lithologically the same as the underlying black shale and represents merely the dried and oxidized zone above the water table. The base of this brown shale was found to lie but very slightly above the base of the low velocity layer.

Figure 3 shows, in section, the disposition of holes drilled into the shale at Limon for the purpose of the experimental studies. A deep hole was drilled with a five inch drag bit and in this hole there were placed three vertical component geophones at depths of 822 feet, 622 feet, and 422 feet. These geophones had natural frequencies of $3\frac{1}{2}$ cycles per second, were damped to $\frac{1}{4}$ of critical damping, were clamped securely to the sidewalls of the bore hole, and each geophone had its separate cable to carry its signals to the surface. Before being placed in the hole, the geophones were carefully calibrated by dropping a small steel ball of known mass from a measured height and letting it fall into a small cup of grease placed on top of the geophone. By making a recording of the resulting disturbance for a known amplifier setting, the resulting trace amplitude could be compared with the velocity given the geophone case, calculated by the conservation of momentum, and so the sensitivity could be determined. The geophone was suspended from long rubber bands for this sensitivity determination.

The geophones formed part of a completely calibrated flat-response, distortion-free electrical seismograph, by means of which precise determinations of earth-particle velocities in centimeters per second could be obtained from measurements of trace amplitudes. A horizontal component geophone was also placed in the hole at a depth of 522 feet as shown in the figure.

The recording oscillograph was especially designed for the studies and contained a drum seventeen inches in diameter with a face wide enough to take a strip of eight inch photographic recording paper about fifty inches long wrapped once around the drum and held in place by clips. The drum rotated on ball bearings and was brought up to any desired speed by means of a hand-operated crank. When brought up to speed it was allowed to coast along on its own momentum. Since the friction was quite low, the paper speed was exceptionally uniform. In order to be able to place the seismogram on the strip of photographic paper in the proper position, the charge was fired in synchronism with the drum's rotation. This was accomplished by placing a relay in series with the shooting circuit, this relay being controlled by a micro-switch, cam-operated by the rotating drum. Thus the charge was fired, not at the time the shooter closed his switch, but at a short time thereafter when the camera drum was in the proper position to receive the seismogram. A shutter, similarly operated, remained open for but one revolution of the drum while the seismogram was being recorded.

The galvanometers in the camera were of the d'Arsonval type with natural frequencies around one thousand cycles per second and were damped to 7/10 of critical damping. Attention was also given to linearity of trace motion with respect to signal amplitude.

The frequency characteristic of the amplifiers was very flat, well beyond the range of frequencies of the seismic studies, and there was no measurable phase variation with respect to frequency throughout the range of interest. The amplifiers had calibrated input attenuators so that the over-all magnification of the seismograph from amplitude of earth-particle velocity to amplitude of trace motion was known for any setting of the attenuator.

Time lines were placed on the seismogram by flashes of light through a slotted wheel driven by a synchronous motor operating from a 1,000 cycle standard frequency generator. There were placed on the seismogram 1,000 timing

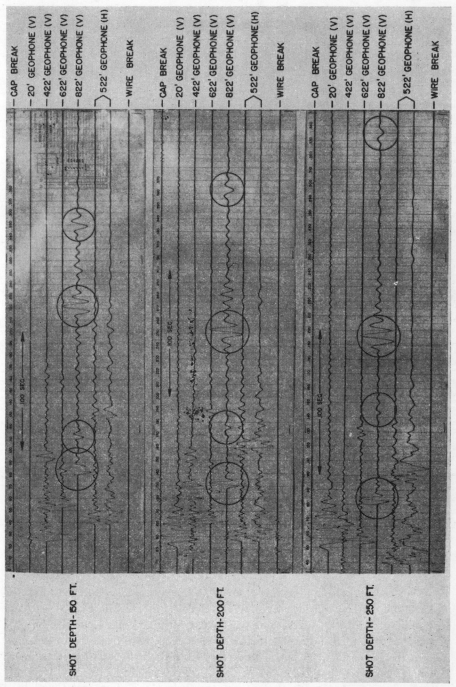

FIG. 4. Three typical seismograms made in Pierre shale.

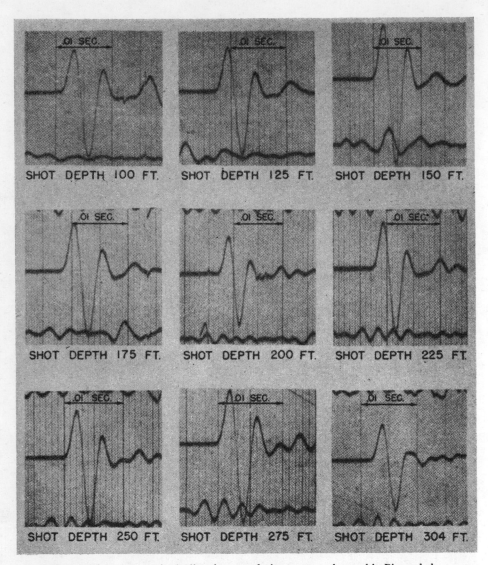

Fig. 5. The primary seismic disturbance, velocity type, as observed in Pierre shale.

lines per second and also a faint mid-line. Time lines were heavied each ten milli-seconds. Thus, a time line was available at half-millisecond intervals and time could be measured to the nearest one ten-thousandth part of a second.

A series of five inch shot holes, each 310 feet deep, were drilled into the shale at various distances from 50 feet to 1,600 feet from the instrument hole as shown in Figure 3. Uniform charges of one pound of 60 percent high-velocity dynamite were fired in these shot holes at depths varying in steps of 25 feet throughout the length of each shot hole. Thus, a large number of seismograms were obtained and these seismograms have enabled a large number of measurements to be made of the primary disturbance and other disturbances on the seismograms.

In Figure 4 there are shown three typical seismograms made at the Limon, Colorado, site. On each of these seismograms four prominent disturbances are shown circled. The first is the primary disturbance with which we are concerned in this paper. The second is the disturbance which started upward, was reflected from the base of the low velocity layer and then traveled downward to the geophone. The third disturbance contains both the shear wave and a tube wave which interfered with it (see Sharpe 2). The fourth disturbance is paired with the third through a reflection from the base of the low velocity layer.

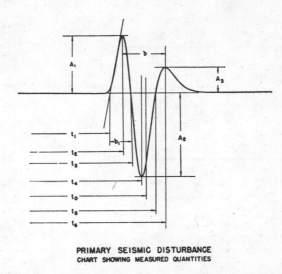

PRIMARY SEISMIC DISTURBANCE
CHART SHOWING MEASURED QUANTITIES

FIG. 6. Primary seismic disturbance—chart showing quantities measured.

In Figure 5 there are shown nine enlarged forms of the primary disturbance as delivered by the 822 foot geophone and initiated by shots at varying depths in the shot hole, A, placed 50 feet from the instrument hole. It will be noticed that the small disturbances which follow the form vary from seismogram to seismogram while the principal disturbance retains a fairly constant shape. The minor disturbances have been investigated and have been found to be caused by reflections from the minor stratifications in the shale. The shale, of course, is not perfectly uniform. In drilling the deep instrument hole, a careful log of the hole was kept and only twenty beds were found throughout the 1,005 feet to which the hole was drilled which could be called harder than the general body of the shale. These beds of harder calcareous shale ranged from one inch to ten inches in thickness.

Figure 6 shows the measurements which were made on the primary disturbance. It will be noticed that no attempt has been made to measure a *first kick* arrival time in the conventional manner. This is because there is no sudden *takeoff* of the trace when the disturbance arrives. The motion begins gradually and if a *first kick* arrival time is attempted, the time picked will depend upon the

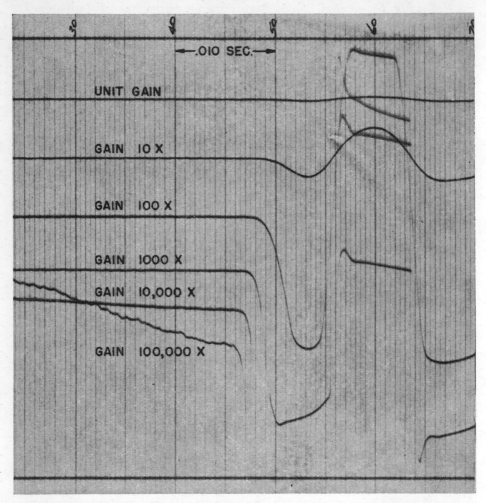

FIG. 7. Seismograms from six amplifiers, all on the same geophone but with gains increasing by factors of ten.

over-all magnification of the seismograph. This is demonstrated on Figure 7 which shows the traces of six amplifiers whose gains vary in steps of ten-fold increase in gain, all amplifiers being driven by the same geophone. It will be seen that the conventional *first kick* moves ahead about one millisecond for each tenfold increase in magnification. This is for a wavelet breadth of about ten milli-

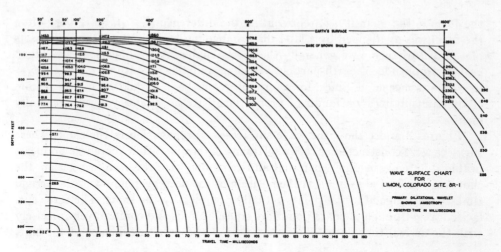

FIG. 8. Wave surfaces for propagation in the Pierre shale at Limon, Colorado.

seconds. In place of the *conventional first kick* the time t_1 is substituted. This time, called the *intercept first kick* is obtained by picking the time where a tangent drawn to the first motion through the point of inflection intersects the time axis. This point does not advance as the amplification of the seismograph is varied. The other time picks t_2, t_3, t_4, t_5, and t_6 are made on the successive bend points and axial crossings of the form. The time t_0 is the arrival time of the *wavelet center*. At great distances, after all dispersion has disappeared and the form has become symmetrical, t_4 and t_0 are coincident. In the plotted data, the arrival time of the disturbance is taken as t_4 because it is reasonably near the wavelet center. The amplitudes measured are indicated as A_1, A_2, and A_3 on Figure 6. The breadth of

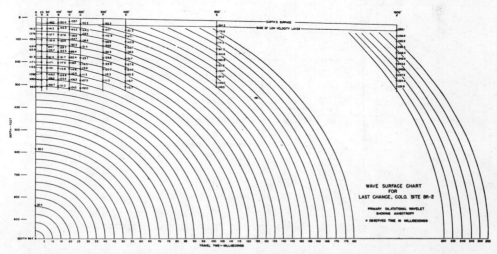

FIG. 9. Wave surfaces for propagation in the Pierre shale at Last Chance, Colorado.

the form is the quantity, b, representing the difference $(t_6 - t_2)$. Since very often the form was overlapped by a later disturbance, measurements of the early parts of the disturbance were generally the most reliable. Accordingly, A_1 was taken as the amplitude of earth-particle velocity, and the breadth b_1 given by $b_1 = (t_3 - t_1)$ was measured. Since, for a form of given shape b could be obtained from b_1 by multiplying by a factor of proportionality, the quantity b_1 was generally measured.

Figures 8 and 9 show the wave surfaces for propagation in the shale. These surfaces were constructed as follows. Having determined that there was no measurable difference in time when shot and geophone were reversed in position, the shot was considered as lying at the position of the deep geophone and the arrival times were assigned to the points actually occupied by the shots. The wave surfaces drawn through the data points are true ellipsoids and conform very precisely with the arrival time data. The smoothness of the data is evidence of the suit-

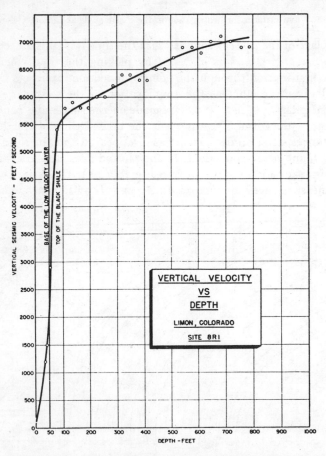

FIG. 10. Wavelet velocity as a function of depth—Pierre shale, Limon, Colorado.

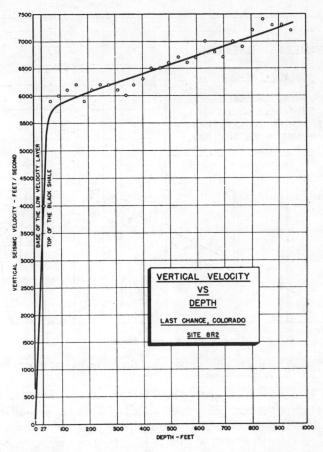

FIG. 11. Wavelet velocity as a function of depth—Pierre shale, Last Chance, Colorado.

ability of the shale as a medium for the studies. Such smooth surfaces cannot be drawn in a more severely bedded earth.

The anisotropy was such that horizontally traveling dilatational disturbances had a higher velocity of propagation than vertically traveling disturbances by about 18 percent at Limon and by about 14 percent at Last Chance.

The ray paths are but slightly curved, the centers of curvature lying in a horizontal plane about 3,000 feet above the earth's surface.

Figures 10 and 11 show the velocity of vertically traveling dilatational waves as a function of depth in the earth for the Limon and Last Chance areas respectively. These curves were prepared from interval times and interval distances taken from the wave surface charts. Very low values of the velocity are reached near the earth's free surface. Since the earth at the very surface should have the elasticity of air and the density of earth, it may readily be seen that these velocities may drop to as low a value as 100 feet per second.

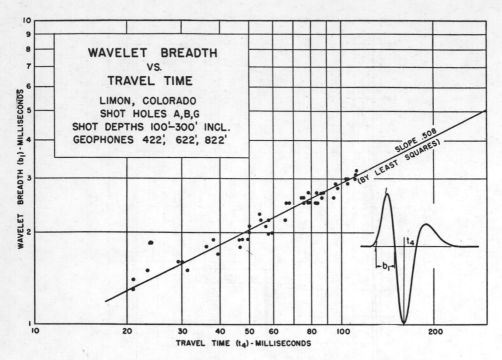

FIG. 12. Wavelet breadth as a function of travel time to reveal rate of broadening law.

Figure 12 shows the manner in which the breadth of the disturbance varies with the travel time, the plot being made on logarithmic paper, and drawn for the Limon area. The straight line drawn through the data points by a least square technique has a slope of 0.508. A similar plot for the Last Chance area had a slope of 0.498. Since, according to the wavelet theory, this slope should be $\frac{1}{2}$ it may be seen that the rate of broadening law is very closely satisfied.

In order to study the form of the disturbance properly, it is necessary to have some standard of comparison. The wavelet theory provides such a standard of comparison in the forms calculated from the theory. Some of these forms are given in Figure 2. Now the numerical distance, R, which is a parameter of the series of forms, may be determined for the observed forms from the approximate relation

$$ R = \sqrt{\dfrac{10}{\left(\dfrac{A_1}{A_2} - .446\right)^3}} . $$

It was found that, for the observed forms, R ranged from 5 to 32. Since the only computed form in this range is that for $R = 25$, the forms selected for comparison are those whose numerical distance fell reasonably close to 25. One comparison is shown in Figure 13. It will be seen that the fit between observed and

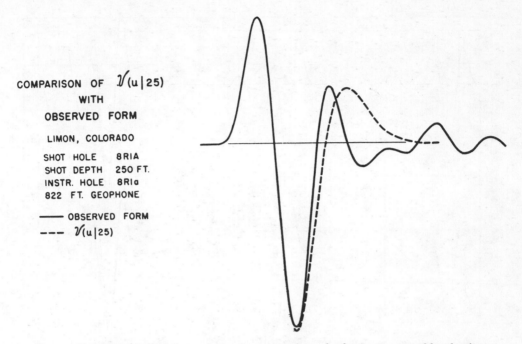

COMPARISON OF $\mathcal{V}(u\,|\,25)$
WITH
OBSERVED FORM

LIMON, COLORADO

SHOT HOLE   8RIA
SHOT DEPTH   250 FT.
INSTR. HOLE 8RIa
822 FT. GEOPHONE

——— OBSERVED FORM
– – – $\mathcal{V}(u\,|\,25)$

Fig. 13. Showing good fit of observed form to computed wavelet form up to second bend point.

theoretical forms is excellent up to the second bend point. On Figure 14 twelve additional comparisons are shown and in addition, the residual obtained by subtracting the ideal form from the observed form is drawn in. Very often this residual has a beginning very much like another wavelet form.

In Figure 15 a more complex disturbance, obtained in the Last Chance Area, has been broken down into three wavelet forms, each of numerical distance 25, each of a lesser amplitude than its predecessor and with the centers equally spaced. This suggests that the shock created by the explosion may not be a single impulse but a succession of impulses equally spaced in time, similar to the *bubble pulses* obtained by Ewing for shots made in open water. The interval of $3\frac{1}{2}$ milliseconds between shocks, however, is too small for gravity controlled bubble pulses and suggests that an elastic rebound of the walls of the shot cavity is causing a succession of slaps and so creating the multiple impulses.

Another important matter to investigate is the spectrum of the observed form and its comparison with the spectrum predicted by the wavelet theory. In order to make this comparison a number of the observed forms were analyzed by means of a Henrici rolling sphere harmonic analyzer and some of the analyses are shown on Figure 16.

The first form is the theoretical form $\mathcal{V}(u\,|\,25)$ and below it is the analysis made of it by the harmonic analyzer, the solid circles representing measured values of amplitude of the harmonic components. The dashed line is the wavelet

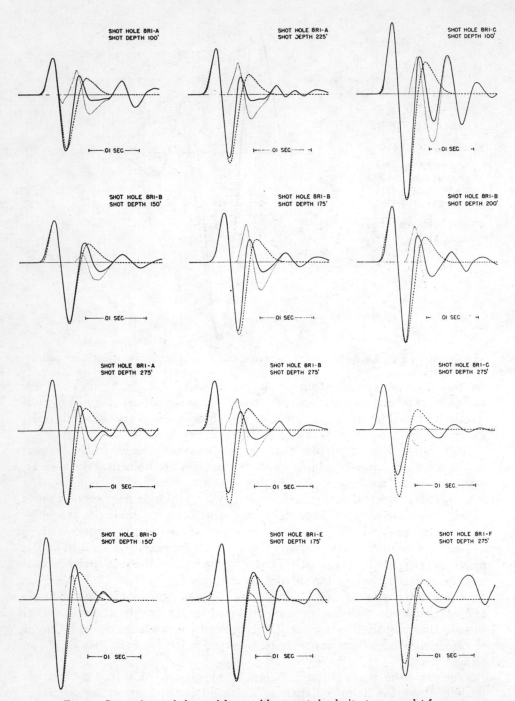

FIG. 14. Comparisons of observed forms with computed velocity-type wavelet form.

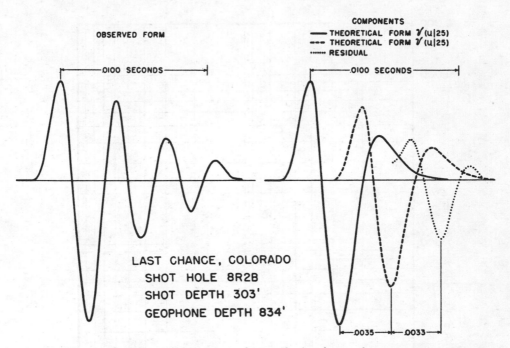

FIG. 15. Breakdown of observed wavelet-complex into its wavelet components.

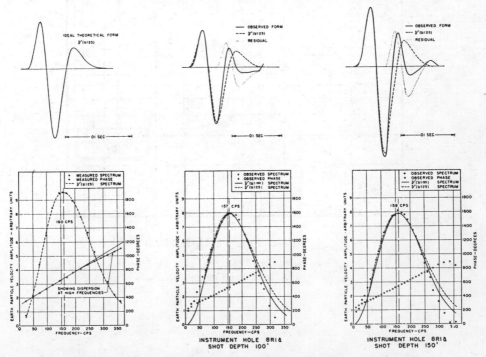

FIG. 16. Comparison of measured spectra with the theoretical spectral curve.

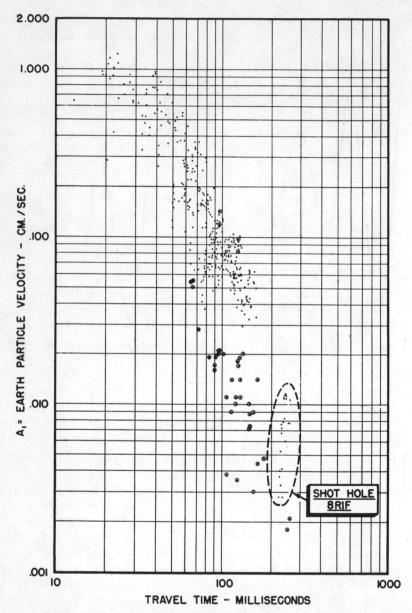

Fig. 17. Amplitude of earth-particle velocity plotted against the travel time for all data obtained at the Limon, Colorado site.

spectrum as plotted from its equation. The purpose of this comparison is to show first the precision with which the analyzer is able to measure the spectrum, and it may be seen that the points are well placed with respect to the theoretical curve. The open circles represent the measured phase while the solid line through these points was drawn from the theoretical expression for the phase. With the assurance, then, that the analyzer is able to make precision amplitude and phase measurements of the forms, the analyses of the other two forms are to be studied. In these other two graphs the measured spectra of the two observed forms are compared with the spectra of $\mathcal{V}(u|25)$ and $\mathcal{V}(u|\infty)$. It is apparent that the agreement is good, but that the precision is not sufficient to determine the numerical distance of the form from its spectrum.

An important part of the studies was the determination of the law of decay in amplitude of the earth-particle velocity with increasing travel time of the disturbance. On Figure 17 there are shown all the data points obtained in the Limon area, with the amplitude A_1 plotted against the travel time on logarithmic paper. There is a very considerable scatter but it must be remembered that although the charge is held constant at one pound there are three geophones at different depths, and shots at twelve different depths in six shot holes at varying distances from the instrument hole. Also, the circled points represent data from shots in the friable brown shale low velocity zone. A study of the data indicated that the greatest cause of this scatter was varying shot depth. Accordingly the data were replotted so that on each graph the shot depth was held fixed. The results are shown on Figure 18 for the Limon Area and on Figure 19 for the Last Chance Area. On each of the graphs a straight line has been drawn through the data points by least square adjustment. The slopes of these lines range from minus 2.19 to minus 2.90 for Limon and with an average value of minus 2.42. For the Last Chance area the slopes range from minus 2.18 to minus 2.96 with an average value of minus 2.43. The averages differ from the minus 2.50 given by the wavelet theory by only 3 percent. It may then be safely concluded that the experimental data agree quite satisfactorily with the theory.

In order to investigate the effect of varying shot depth the upper tier of graphs shown on Figure 20 were drawn. Here each graph is for a separate shot hole. In a given shot hole twelve shots have been fired at different depths. Consider a single shot. There are three data points associated with this shot from the observations made by the three geophones. Through these three data points the best straight line of slope minus 5/2 has been drawn by least square adjustment. These lines do not coincide but stand at different heights for a specified travel time. This is because the several shots are not equally effective as a seismic source. If we extrapolate along any line to a travel time of one second, the amplitude at that time is a measure of the strength of the shot. We call this the *seismic punch*. This seismic punch may be determined for any observation by forming the product $A_1 t_4^{5/2}$. In the second tier of graphs of Figure 20 this seismic punch has been plotted as a function of shot depth for each of the shot holes. It

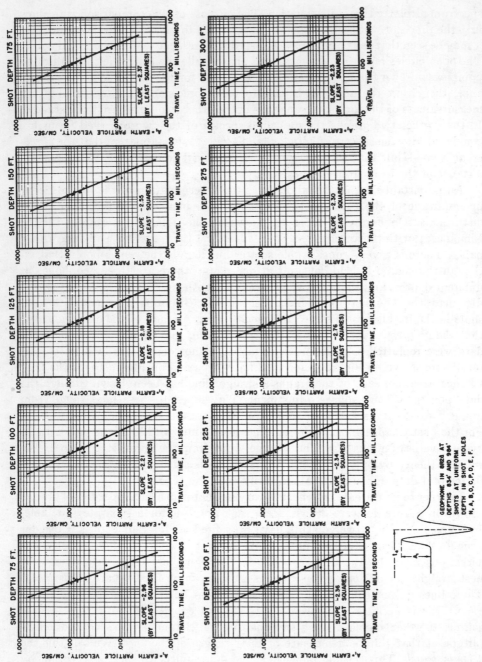

FIG. 18. Amplitude of earth-particle velocity plotted against travel time with shot depth held fixed—Limon, Colorado.

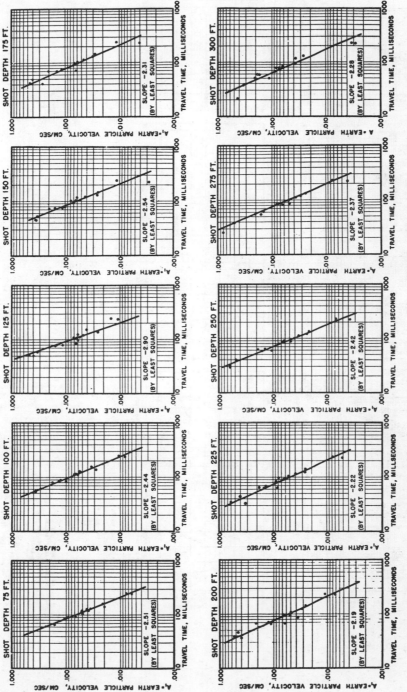

FIG. 19. Amplitude of earth-particle velocity plotted against travel time with shot depth held fixed—Last Chance, Colorado.

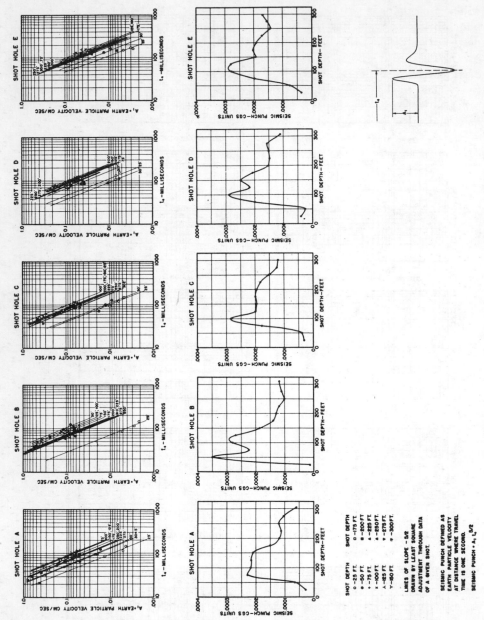

Fɪɢ. 2o. The determination of seismic punch and its variation with depth of shot, Limon, Colorado.

will be seen that the pattern is quite similar for each of the shot holes, the seismic punch being very low for shallow shots, increasing to a maximum as the shot is deepened and then falling for the deeper shots.

On Figures 21 and 22 the average seismic punch for holes *A, B,* and *C* has been plotted as a function of the shot depth for Limon and Last Chance respectively. The values for shots deeper than 300 feet were obtained at the close of the program when the geophones were withdrawn and placed in redrills of the shot holes and shots were made in the deep hole.

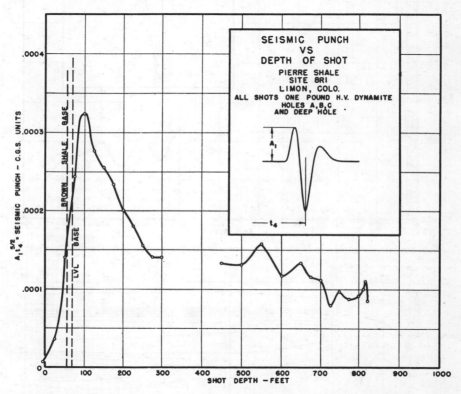

FIG. 21. Seismic punch vs. depth of shot, average for three holes, Limon, Colorado.

In Figure 23 the seismic punch has been contoured in section for the Limon area. There are a number of possible explanations for the variation of seismic punch with shot depth but a clear understanding of this effect will not be available without further studies. On Figure 24 there are drawn a set of graphs giving the earth-particle displacement, earth-particle velocity and earth-particle acceleration, the wavelet breadth and the distance traveled by the wavelet as functions of the travel time of the wavelet center. All are based on the predictions of the wavelet theory but are drawn for the propagation constants determined for the

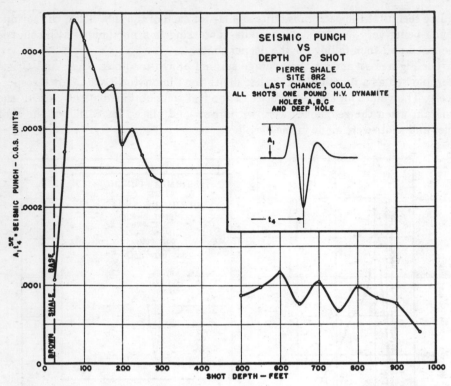

FIG. 22. Seismic punch vs. depth of shot, average for three holes, Last Chance, Colorado.

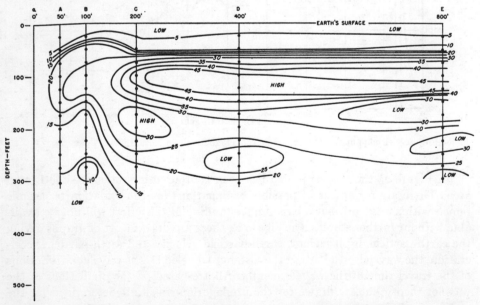

FIG. 23. Seismic punch contoured in section at Limon, Colorado.

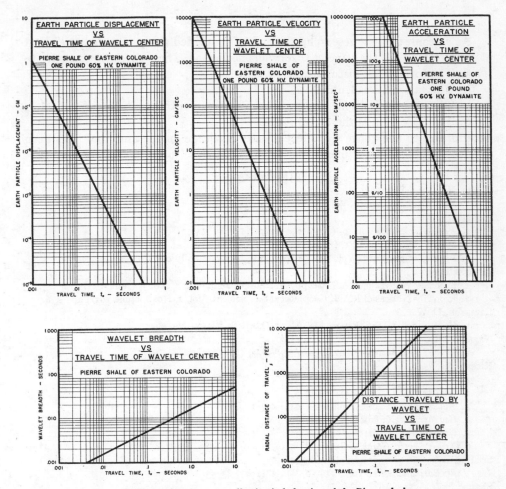

FIG. 24. Charts showing over-all seismic behavior of the Pierre shale.

Pierre shale by these experimental studies. The constants used in drawing these graphs are

$$c = 7,000 \text{ feet/second.}$$

$$\omega_0 = 50,000 \text{ radians/second.}$$

The quantity ω_0 is 2π times the transition frequency, f_0. Thus a representative value of the transition frequency for the shale at a depth of around 300 to 600 feet is 8,000 cycles per second. For a further discussion of this transition frequency reference may be made to the literature (Ricker 3, 4). This transition frequency marks a position in the frequency scale above which dispersion becomes important for sinusoidal waves and where the nature of the absorption changes.

SUMMARY AND CONCLUSION

As a result of these researches it may be concluded that, in shale, agreement
of experiment with the wavelet theory has been good. These checks are relative to

a. The form of the disturbance.

b. The law of broadening.

c. The law of amplitude decay with travel time for the earth particle velocity.

d. The spectrum of the disturbance.

In addition to the above the effect of size of charge on the amplitude has been
checked in special experiments and found to be close to the 5/6 power law.

BIBLIOGRAPHY

Ewing, Vine, and Worzel

    1. Bull. Geol. Soc. Amer., Vol. 57, (1946) 909–934.

Gutenberg, B.

    1. *Physik Zeitschr.*, Vol. 30, (1929) 230–231.

    2. *Physik Zeitschr.*, Vol. 31, (1930) 745–752.

    3. Physics of the Earth, Vol. VII, Chapter XV, McGraw-Hill Book Co., (1939) 361–384.

Jeffreys, H.

    1. Monthly Notices of R.A.S. Geophysical Supplement, Vol. 2, No. 7, (Jan. 1931) 318–323.

    2. Monthly Notices of R.A.S. Geophysical Supplement, Vol. 2, No. 8, (June 1931) 407–416.

Lampson, C. W.

    1. P. B. 50860 NDRC Report No. A-479, O.S.R.D. Report No. 6645.

Nagaoka, H.

    1. *Proc. Tokyo Math.—Phys. Soc. Ser. 2*, Vol. 3, (1906) 17–25.

Ricker, N.

    1. *Geophysics*, V, (1940) 348–366.

    2. *Geophysics*, VI, (1941) 254–258.

    3. *Bull. Seis. Soc. Amer.*, Vol. 33, (1943) 197–228.

    4. *Geophysics*, IX, (1944) 314–323.

    5. *Geophysics*, X, (1945) 207–220.

    6. *Trans. Amer. Geophys. Union.*, Vol. 30, (1949) 184–186, 457–458.

Sezawa, K.

    1. *Bull. Earthquake Res. Inst.*, Tokyo, Vol. 3, (1927) 43–53.

Sharpe, J. A.

    1. *Geophysics*, VII, (1942) 144–154.

    2. *Geophysics*, VII, (1942) 311–321.

Stokes, G. G.

    1. *Trans. Cambridge Phil. Soc.*, Vol. 8, Part III, (1845) 287–319.

DISCUSSION

MR. D. T. GERMAIN-JONES (Anglo Iranian Oil Co., London, U.K.) commented upon the vagaries of the first kick as demonstrated in Figure 7, and he
inquired whether the same effect is operating if the amplification is kept constant
while the size of the charge is increased. He had observed a similar effect in long
range refraction work, as sometimes the conventional first kick times recorded
by the near geophone stations, with larger trace amplitudes, appear to be relatively too short; consequently the apparent velocity indicated by these first
kicks would be too small. He felt that for the purpose of velocity determinations

first peaks give more reliable times than first kicks, provided of course that the first peaks are not complicated by later refractions.

In connection with the ellipsoidal form of the wave front shown in Figure 8, he asked whether the author had any comments on the question of a suitable overburden velocity to be used in refraction computations, as the horizontal velocity is too high whereas the vertical or cross-bedding velocity obtained from well shooting appears to be too low.

Referring to Figures 10 and 11 he wondered whether the velocity curves can actually be extrapolated down to values as low as 100 ft/sec. As a layer of such a low velocity at the earth's surface is probably immeasurably thin, he would have expected a cut-off of the curves at a velocity approximating that of sound in air.

Mr. H. Closs (Geological Survey, Hanover, Germany) commented on the increase of wavelet breadth with travel time, and the corresponding decrease of apparent frequency. In a shot hole in Germany he had observed increasing frequency with increasing travel distance, but with increasing depth of the charge.* He drew the conclusion that the increase of frequency must be ascribed to a change in the originated spectrum of frequencies, which is perhaps a function of the depth of the charge; and he inquired if the author could confirm this opinion.

In a written comment Mr. S. Kaufman (Shell Oil, Houston, Texas, U.S.A.) pointed to a discrepancy in the numerical values given by the author for the classical velocity of propagation ($c = 7,000$ ft/sec.), the transition frequency $f_0 = 8,000$ c.p.s.) and the numerical distance ($R = 25$). If these values are substituted in equation 5, the shooting distance r is found to be $3\frac{1}{2}$ feet, which is an unreasonably small value as the distances used in the experiment have ranged from 50 to 1,600 ft.

Mr. C. H. Dix (Calif. Inst. of Techn., Pasadena, Calif., U.S.A.) commented in writing on the importance of the subject treated by the author. Then he pointed out that in his opinion failure of the medium to be perfectly elastic is not the only cause of wavelet shape variation. Heterogeneity contributes to this effect through scattering and path disturbances; imperfectly plane reflectors also lengthen wavelets. Only specially selected transformations would shorten a wavelet, while random transformations generally lengthen it.

Mr. T. Krey (Seismos, Hannover, Germany) commented on the residual obtained by subtracting the ideal wavelet form from the observed one. If we are indeed dealing with bubble pulses, then the time difference of $3\frac{1}{2}$ milliseconds between the center of the ideal wavelet and the center of the succeeding residual wave must be independent of the travel time. However, for very large distances the two waves must practically coincide, because the interval of $3\frac{1}{2}$ milliseconds will be small compared with the breadth of the wavelets, on account of the law of broadening. Consequently, for large distances there should not be any meas-

---

* Vide Closs, Hans, "Applied Geophysics in West Germany during the Last Five Years," Proceedings Third World Petroleum Congress, Leiden 1951, Section I, p. 564.

urable difference between the observed motion and the ideal wavelet form. The speaker considered this point to be worth investigating.

Mr. N. D. Smith (Shell Oil, Houston, Texas, U.S.A.) pointed out that the importance of the paper lies in the demonstration that the Stokes' equation, and the assumption of a simple shot pulse containing all frequencies, predict the qualitative changes in the seismic signal as a function of the travel distance. He felt that the initial disturbance produced by an explosion is certainly more complicated than that assumed by the author, and that such differences in frequency distribution may account for the deviations observed between theory and experiment in the later stages of the wavelet, and for the rather wide variations of the cut-off frequency with the location of the shot.

The author's seismic punch and its variation as a function of the position of the shot are well known in a qualitative way to every field seismologist, but the quantitative approach and the detailed study of figure 23 are new and very interesting; the speaker considered that additional studies would be of great practical importance.

Speaking on behalf of the author, Mr. T. V. Moore (Standard Oil Development Co., New York, U.S.A.) agreed with Mr. Dix that there are many factors that can result in broadening of wavelets besides the theory of friction, which is taken into account by the Stokes' equation. However, the experiments have indicated that this theory goes a long way toward explaining many of the seismic phenomena.

He felt that an experimental study along the lines suggested by Mr. Krey would be handicapped by the complex nature of a seismogram which records energy that has travelled a long distance through the earth.

In a written reply to the various comments the author, Mr. Ricker, points out that according to theory the conventional first kick should advance through any agency that increses its trace amplitude, but he has not made an experimental study of the effect of charge size on the advance; he agrees with Mr. Germain-Jones' preference for first peak times.

Relative to the 100 ft/sec. velocity near the earth's surface, he revealed that in addition to the theoretical arguments presented in the paper this estimated value is based on special experiments. Geophones were set at intervals of one foot, and records were taken of the explosion of a blasting cap, which was covered with a sound-deadening cover to diminish the air wave. Velocities of 500 ft/sec. were observed quite frequently, and in one instance a velocity as low as 300 ft/sec. has been recorded. These very low velocities must lie within the first few inches of the earth's surface.

In reply to Mr. Closs the author reports that in observations with a deeply placed geophone he has indeed observed a very great change in wavelet breadth with charge depth. When the depth of the shot is varied from 300 to 100 ft., the wavelet breadth may change from .004 to .006 second. This effect is further enhanced as the shot is placed up in the highly absorbent low velocity zone, the breadth increasing to .009 sec. for a shot depth of 50 ft, and becoming much broader for still shallower shots. The author's measure of these changes in the absorbing properties of the rocks at various depths is through what he calls the transition frequency, which has been discussed in his earlier papers. For sinusoidal waves decaying exponentially as

$$A = A_0 \epsilon^{-\alpha x} \cos 2\pi f \left(t - \frac{x}{c} \right)$$

the absorption coefficient α is given by

$$x = \frac{\pi f^2}{c f_0}$$

for low frequencies, c being the wave velocity, f the wave frequency, and f_0 the transition frequency of the earth. This transition frequency may be determined experimentally by observations of wavelet breadths; the procedure is to plot b^2 against the travel time t_4, draw a smooth curve through the data points and then draw tangents to the resulting curve. One has then:

$$f_0 = \frac{6}{\pi} \cdot \frac{\Delta t_4}{\Delta(b^2)} \cdot$$

For depths of 300, 200, 100 and 50 feet the transition frequency is found to be 8,000, 5,000, 2,000 and 300 cycles/sec. respectively; very low values are reached at shallower depths.

The discrepancy mentioned by Mr. Kaufmann must according to the author be explained by the fact that the Pierre shale is not homogeneous. He points out that the wavelet is shaped in the near vicinity of the shot and moves downward into a region of higher and higher transition frequency. The peak frequency of the wavelet spectrum is around 200 cycles, which is low compared with the transition frequency of the earth below it. Since a change in form of the wavelet is to be effected only through dispersion (a change in breadth is not a change in form), it may be seen that this dispersion is too small to permit the parameter R of the form to increase appreciably. If v is the velocity of waves of frequency f, f_0 the transition frequency of the medium, and c the wave velocity approached as f becomes vanishingly small, there exists for low frequencies the approximate relation

$$\frac{v}{c} = 1 + \frac{3}{8} \cdot \frac{f^2}{f_0^2}$$

which for $f = 200$ and $f_0 = 8,000$ gives $v/c = 1.00024$. Thus the dispersion is quite small and evidently does not allow the parameter R of the wavelet form to keep pace with the numerical distance as determined for a homogeneous medium by equation 5 of the paper. This observation is one of the interesting things which point to the desirability of making further studies in the zone immediately surrounding the shot.

In commenting on the remarks made by Mr. Dix the author points out that although there are many matters yet to be considered it was his idea to first determine the basic form (or series of forms) assumed by a seismic disturbance in an ideal homogeneous absorbent earth. This form is offered as a better basis for further studies than continuous sinusoidal waves. The seismic wavelet, as the author defines it, is an ideal mathematical entity and not subject to the vagaries of passage through a heterogenous medium; in the author's opinion Dix is actually thinking of modifications in the wavelet complex due to changes in the spacings and relative amplitudes of the component wavelets.

Referring to some of the comments made by Messrs. Smith and Krey, the author expresses his agreement with Mr. Krey's conception of the processes involved in the propagation of a seismic wave. He adds that in addition to the bubble pulses also the effects of other phenomena, such as finite strains immediately surrounding the shot and the variation of pressure in the shot cavity as a function of time, are being ironed out at greater distances. Regardless of the shot characteristics the earth finally takes over and controls the shaping of the disturbance.

Reprinted from Philosophical Magazine, v. 1, p. 693–710.

LXXI. *The Propagation of Stress Pulses in Viscoelastic Solids*

By H. KOLSKY ‡

Imperial Chemical Industries Research Department, Welwyn †

[Received December 28, 1955]

ABSTRACT

The propagation of short mechanical pulses along rods of three polymers, polythene, polystyrene and polymethylmethacrylate, has been investigated experimentally. The pulses were produced by the detonation of small quantities of explosive at one end of a rod and a condenser microphone was employed to record the displacement of the opposite end. It is shown that if the response of the material to sinusoidal stresses over a wide frequency range is known, the pulse shapes can be predicted accurately by means of a numerical Fourier synthesis. Further, that where the damping loss is not too large and is constant over a wide frequency range, as it is for many polymers, a general solution of the problem can be obtained which gives the pulse shape for all such polymers and for all distances of travel. Some experiments on pulses through blocks of plastic are also described.

---

§ 1. INTRODUCTION

THE mechanical behaviour of high polymers, such as rubbers and plastics, depends very markedly on the rate at which they are deformed and, in recent years, the stress–strain relations of these materials under dynamic loading has received considerable attention both experimentally and theoretically. For small deformations the relation between stress and strain can generally be expressed as a linear differential equation, involving the stress, the strain and their derivatives with respect to time, so that a stress which varies sinusoidally with time produces a sinusoidally varying strain. The phase difference between stress and strain is a measure of the internal friction of the material. The rather involved theory of linear viscoelastic behaviour has been ably summarized in books by Leaderman (1943) and Gross (1953) and relations derived in these will be used below. It is the purpose of this paper to describe some experiments on the propagation of pulses of short duration along rods of high polymers and to discuss the interpretation of the results in terms of the known viscoelastic properties of these materials. Some preliminary experiments have been described earlier (Kolsky 1954 a).

---

† Communicated by the Author.

‡ Now at Brown University, Providence, Rhode Island, U.S.A.

For nearly all high polymers it is found that the elastic modulus, i.e. the ratio between the stress amplitude and the strain amplitude for sinusoidal vibrations, increases very markedly with increasing frequency whilst the internal friction, which may be defined by the fractional energy loss per cycle $\triangle W/W$ or alternatively by the tangent of the phase lag, tan δ, between stress and strain, changes comparatively little, except in the region when the polymer is at a temperature close to the transition from rubber-like to glass-like behaviour.

If a plane sinuosoidal stress wave travels through a viscoelastic material and its amplitude is $\sigma_0 \cos pt$ at the origin, its amplitude at a distance x from the origin is $\sigma_0 \exp(-\alpha x) \cos p(t-x/c)$, where p is 2π times the frequency, α is the attenuation coefficient and c is the velocity of propagation of the wave, which is given by $(E/\rho)^{1/2}$ where E is the elastic modulus and ρ is the density of the medium. α is related to the loss factor tan δ by the equation

$$\alpha = \frac{p}{4\pi c}(\varDelta W/W) = \frac{p}{2c} \tan \delta \text{ (see Kolsky 1953, p. 106).} \qquad (1)$$

Thus if tan δ is constant the attenuation α increases linearly with the frequency.

When a mechanical pulse is propagated through a viscoelastic solid it is dispersed, as the high frequency components travel faster and are attenuated more rapidly than those of lower frequency. A purely analytic approach to the problem of pulse propagation in a general viscoelastic solid leads to intractable mathematical expressions and several attempts have been made to treat the solid in terms of a simple model. This was first done by Thompson (1933) and subsequently by several other workers including Ricker (1943), Zverev (1950), Eubanks *et al.* (1952), Lee and Kanter (1953) and Glauz and Lee (1954). Whilst these treatments are undoubtedly of theoretical interest and may well have some practical application they suffer from the inherent limitation that the mechanical behaviour of most plastics and rubbers does not conform over more than a very limited frequency range to that of a simple mechanical model. This limitation is discussed fully by Leaderman (1943) and by the author (Kolsky 1953), and figs. 1, 2 and 3 are included here to illustrate the fact. Figure 1 shows the experimental results obtained in this laboratory by Hillier (1949) for the attenuation α and the phase velocity c of sinusoidal waves in a polythene filament. The dynamic measurements were made at frequencies between 1 kc/s and 16 kc/s. The point for 'zero frequency' was obtained for a stressing time of several seconds. Figure 2 shows the theoretical curves for two simple models, the Maxwell model which consists of a perfectly elastic spring in series with a dashpot which has Newtonian viscosity, and the Voigt model in which the spring is joined across the dashpot. The relations for velocity and attenuation in materials whose mechanical behaviour is given by such models have been discussed earlier (Kolsky 1953, Chap. V). The curves here are plotted non-dimensionally, τ being the characteristic time given by the ratio of the viscosity of the dashpot divided by the elasticity of the spring in each case. (For the Maxwell model τ is termed the *relaxation time* and for the Voigt model τ is the *retardation time*.) The scale for the vertical axis was fixed for the Maxwell curve by plotting

Fig. 1

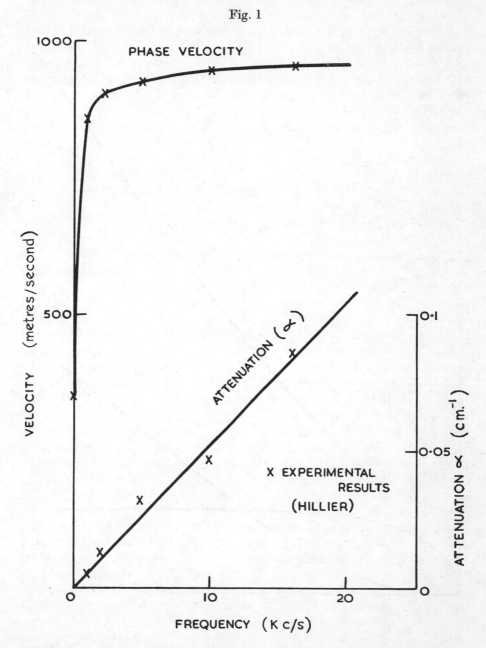

Experimental measurements of attenuation and velocity in polythene filaments at 10°c.

α/α_{max} and c/c_{max}, whilst for the Voigt solid c/c_{min} is plotted and α is in arbitrary units. It may be seen that even over a very limited frequency range it is impossible to fit the observed behaviour of a polymer such as polythene to the dispersion curves obtained from such two element

Fig. 2

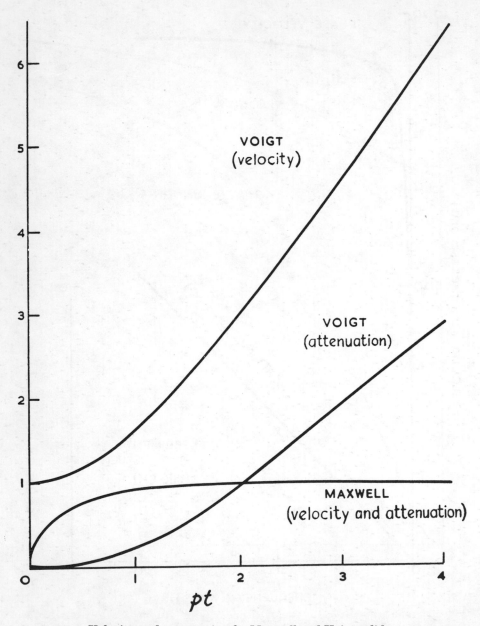

Velocity and attenuation for Maxwell and Voigt solids.

models. Figure 3 shows the corresponding curves for a simple three element model which is sometimes known as the *standard linear solid*. This model can be represented by a second spring in series with a Voigt model and its mechanical behaviour corresponds to the most general

Fig. 3

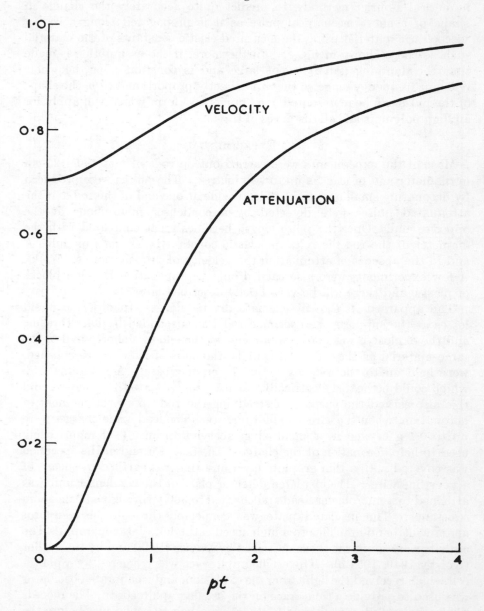

Velocity and attenuation in standard linear solid. Model corresponds to two equal springs in series, across one of which there is a dashpot.

linear equation in stress and strain involving only stress, strain and their first derivatives with respect to time.   It may be seen from fig. 3 that such a model gives a somewhat better approximation to the behaviour of a real viscoelastic solid such as shown in fig. 1 but still can only be fitted over a limited frequency range.

In the work which follows it will be shown that it is possible, by numerical Fourier methods, to predict quite accurately the change in shape of a short mechanical pulse as it is propagated along a rod of viscoelastic material, using the measured elastic constants of the material such as those shown in fig. 1.   Furthermore, if the assumption is made that the damping (tan δ) is not large and is constant over the whole relevant frequency range so that α is directly proportional to p, the shape of the pulse can be represented in a ' universal ' form which will apply for all high polymers and all distances of travel.

§ 2. Experimental

Most of the experiments were carried out on rods of material 1·25 cm in diameter and of lengths up to two metres.   The pulses were produced by detonating small quantities of lead azide at one end of the rod and the attenuated pulses were detected by a condenser microphone at the opposite end.   Once the pulse length becomes great compared with the diameter of the rod the relevant elastic constant is Young's modulus E and in the absence of attenuation the velocity of propagation is $(E/\rho)^{1/2}$. A few experiments were also carried out on propagation through blocks of plastic and these will be very briefly described later.

The apparatus is shown schematically in fig. 4.   In order to avoid losses at the supports the specimen rod was suspended by long threads, and the explosive was placed in a recess at the end of a short anvil of the same material as the rod.   The anvils, the ends of which were polished, were held on to the rods by a thin layer of grease.   A nichrome wire which could be heated electrically was used to detonate the explosive and this also served the purpose of steadying the rod up to the moment of detonation, when the wire was blown away.   The lead azide charges were up to 0·1 g in weight, a small silver acetylide primer ($\sim$1 mgm) being used to help detonation of the charge.   The detector end of the specimen was covered with a thin graphite layer and this was earthed by means of a very fine wire.   The insulated detector plate of the condenser unit was attached to a micrometer head and charged to 360 v through a 50 megohm resistance.   The insulated plate was connected through the feed unit and a wide band amplifier to a high-speed cathode-ray oscillograph.   The circuits were similar to those described by Davies (1948) and by the author (Kolsky 1949, 1954 b).   The oscillograph trace was triggered by a photo-cell which received the light from the explosion, and a calibrated oscillator was used to provide a time trace on the oscillograph record.   The records were obtained photographically, the camera being opened just before the charge was detonated.

Any movement of the end of the rod resulted in a change in the capacity of the condenser consisting of the detector plate and the earthed graphite layer at the end of the rod, and since the plate could only lose its charge through large resistances a sudden change in capacity resulted in a corresponding change in the potential of the plate. This change of potential was amplified and recorded on the oscillograph. When the displacement of the end of the rod is small compared with the separation between it and the detector plate, the change in voltage across the condenser becomes proportional to this displacement (cf. Davies 1948). Under these conditions the oscillograph gives a true record of the motion of the end of the rod, and by differentiating this displacement–time curve, the

Fig. 4

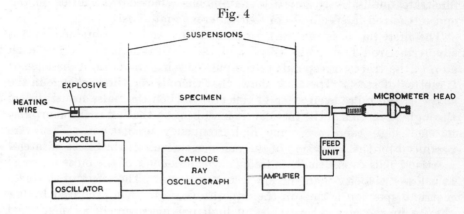

Experimental arrangement.

stress–time curve can be obtained. The relation used applies when the pulse length is great compared with the diameter of the rod and is

$$\sigma = \tfrac{1}{2}\rho c \frac{du}{dt} \qquad \qquad (2)$$

where σ is the stress and u is the displacement at time t. Integrating eqn. (2) it can be seen that the total displacement U produced by the reflection of a pulse at the end of the rod is given by

$$\int \sigma \, dt = \tfrac{1}{2}\rho c U. \qquad \qquad (3)$$

The left-hand side of the equation represents the momentum associated with the pulse. In order to measure the values of this momentum, explosive charges similar to those used in the later experiments were detonated against a small ballistic pendulum. The total movement U in each case could then be calculated from (3) and the condenser gap set in each experiment to be sufficiently wide for there to be virtual linearity between the voltage change and the displacement.

In the experiments, the compression pulse was reflected at the detector end of the specimen as a pulse of tension (since the end is free). This travels back to the firing end where it is once again reflected with a reversal of phase so that it again approaches the detector end as a pulse of compression.   A second reflection then takes place resulting in a further movement in this end of the specimen.   In between reflections the ends of the specimen are at rest and in the absence of attenuation of the pulse the ends would continue to move in jumps indefinitely (cf. Kolsky 1953, p. 46).   If an oscillograph record is taken of the motion of the detector end during several reflections the shape of the pulse after it has traversed the rod once, three times, five times, etc. can be found and such records were used to determine the change in pulse shape for long distances of travel in the material.   Figures 5, 6, 7 and 8 are oscillograph records which illustrate qualitatively how attenuation has affected the motion of the end of the rod in specimens of four different materials.

The first, fig. 5, is included for comparison and was obtained with a silver steel rod 1 metre in length and 1·25 cm in diameter.   Thus each step in the trace corresponds to the pulse having traversed a distance of 2 metres of steel.   The trace shows that there is very little change in the shape of the pulse even after 11 reflections when the pulse has travelled through 21 metres of the metal.   It may be seen that after the sharp rise in each case there are some high frequency oscillations.   These are produced by the dispersion of the pulse as a result of the effect of lateral inertia.   This causes the high frequency components of the pulse to travel at a lower velocity than the low frequency ones.   The effect of this geometric dispersion is thus in the opposite sense to that produced by the change in the elastic properties of high polymers which, as mentioned earlier, results in the velocity of propagation increasing with the frequency of the waves.   The pulse produced by the explosive is initially about 2 microseconds in duration and in the 1·25 cm diameter steel rod, because of the effect of lateral inertia, the pulse is extended to a length of about 15 microseconds ; this is followed by a train of high frequency oscillations. Davies (1948) has discussed the effect fully and has shown that a pulse must be at least ten times as long as the diameter of the rod for it to be propagated without appreciable change in shape.   In the present investigation this effect placed a lower limit on the length of pulse that could be studied, since it was necessary to ensure that the observed changes in the pulse shape were due to the properties of the material and not to inertia effects.

Figure 6 is a record of the movement of the end of a rod of polystyrene 1·25 cm in diameter and 50 cm in length.   Polystyrene is a rather brittle glass-like polymer which has comparatively low internal friction (see Lethersich 1950).   It may be seen that there is little change in the shape of the pulse as it travels back and forth along the rod although a slight rounding off of the steps occurs on the 6th and 7th reflections, showing a slight lengthening of the pulse.   It may be noted that the oscillations which in steel follow the main pulse do not appear in this record.   This is due to the attenuation by the plastic of the high frequency Fourier components which produce these oscillations.

Fig. 5

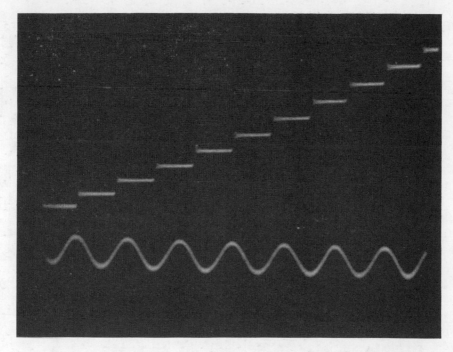

Oscillograph record of displacement of end of steel rod 1 metre long and 1·25 cm
in diameter when 0·01 grams of lead azide have been detonated at the
opposite end.   (Period of timing wave 500 microseconds.)

Fig. 6

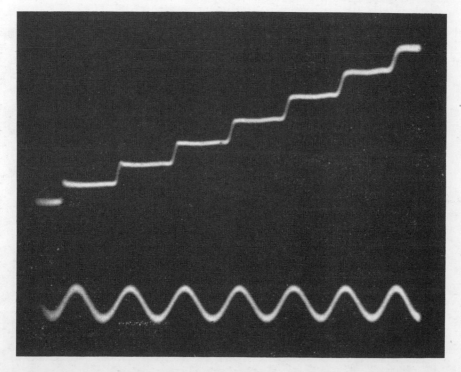

Oscillograph record of displacement of end of polystyrene rod 50 cm long and
1·25 cm in diameter when 0·005 grams of lead azide have been detonated
at the opposite end.   (Period of timing wave 500 microseconds.)

Fig. 7

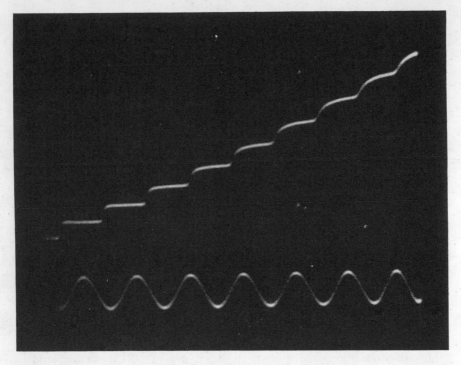

Oscillograph record of displacement of end of polymethylmethacrylate rod 46 cm long and 1·25 cm in diameter when 0·005 grams of lead azide have been detonated at the opposite end. (Period of timing wave 500 microseconds.)

Fig. 8

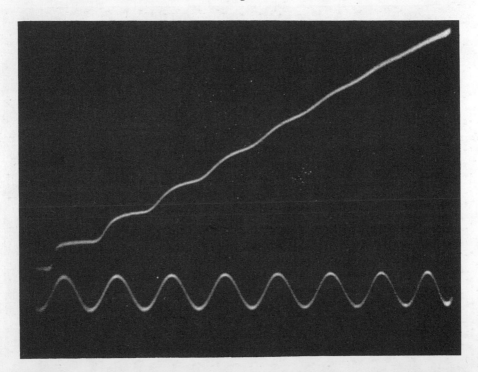

Oscillograph record of displacement of end of polythene rod 20 cm long and 1·25 cm in diameter when 0·005 grams of lead azide have been detonated at the opposite end. (Period of timing wave 500 microseconds.)

Figure 7 shows the record obtained from a specimen, 40 cm. long, of another glass-like polymer, 'Perspex' (a form of polymethyl-methacrylate). This material has a higher mechanical loss than polystyrene and it may be seen that as a result the steps in the record become progressively more gradual, the pulse length increasing considerably between each reflection. It may also be noted that the high frequency oscillations produced by lateral inertia effects are here completely absent.

Lastly fig. 8 was obtained with a specimen 20 cm long of the softer polymer polythene which has a high coefficient of internal friction and here the pulse is rapidly lengthened so that after a few reflections the motion of the end of the bar has become almost continuous.

In order to determine the shape of the pulse, after travelling through various distances in the material, rods of different lengths were used and the records for the first few reflections were measured. Wherever possible only the first and second reflections were used, since this enabled a larger and hence more accurate trace to be obtained and also made it possible to employ a condenser gap which was sufficiently large, compared with the total relevant movement of the end of the bar, for linearity to obtain. The measured displacement–time curves were differentiated numerically to give the shapes of the stress pulses. Figure 9 shows the pulse shapes for polythene after traversing 30, 60 and 90 cm and fig. 10 shows similar curves in 'Perspex' for travel distances of 120, 360, 600 and 840 cm. The attenuation in polystyrene was too low for reliable changes in pulse shape to be determined. The figures show that the pulses in both materials are asymmetrical in shape and that they continually lengthen and decrease in amplitude as they progress through the specimen. Since the area under each pulse is proportional to the momentum associated with it this area will remain constant as the pulse travels through the material. The dependence of the pulse shape on the viscoelastic properties of the medium is discussed in §3.

As mentioned earlier, a few experiments were also carried out on the propagation of pulses in blocks of polymer. There is no dispersion due to lateral inertia in this case, so that the propagation of pulses of very much shorter duration can be investigated. Since, however, the pulse under these conditions is spreading out radially its amplitude is inversely proportional to the distance travelled. Furthermore, at distances from the source comparable with the pulse length changes in shape occur even in perfectly elastic media as the pulse travels out. The form of spherically divergent stress pulses has been discussed by several authors (e.g. Selberg 1952) and they show that the displacement u_r at distance r is represented by

$$u_r = \frac{c}{r} f'(r-ct) - \frac{c^2}{r^2} f(r-ct). \qquad \cdots \cdots (4)$$

Thus for large r the first term on the right-hand side of the equation is predominant, whilst for small r the second term becomes important. The shape of the displacement–time curve at large distances from the source is thus the differential of the curve close to it. It is consequently

Fig. 9

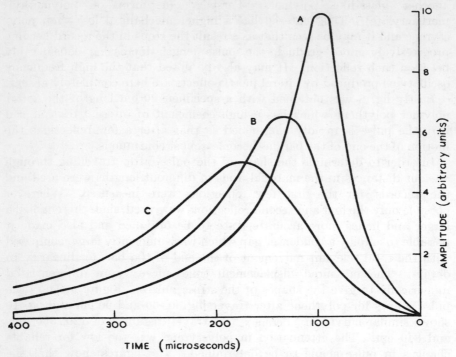

Experimental results for attenuation through different lengths of polythene rod
at 15°c.   A, 30 cm ; B, 60 cm ; C, 90 cm.

Fig. 10

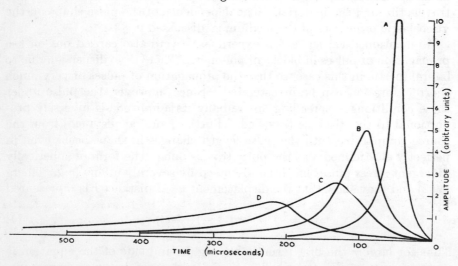

Experimental results for attenuation through different lengths of ' Perspex '
at 15°c.   A, 120 cm ; B, 360 cm ; C, 600 cm ; D, 840 cm.

Fig. 11

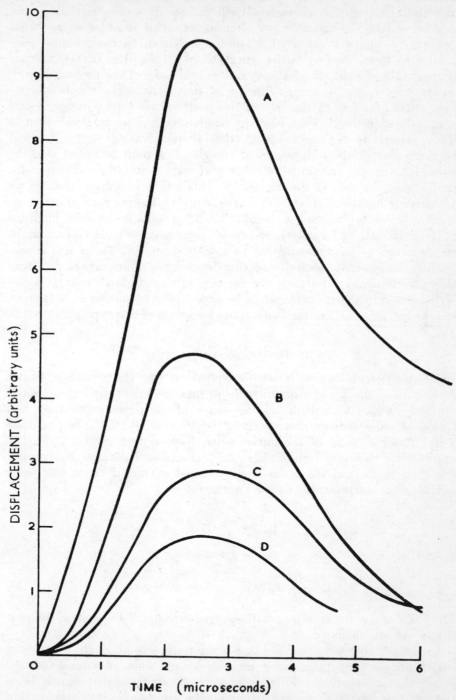

Displacement-time curves through blocks of polythene of different thicknesses
at 15°c.   A, 2·5 cm ; B, 5 cm ; C, 7·5 cm ; D, 10 cm.

difficult to separate the effects of attenuation by the medium from changes in pulse shape due to geometrical effects, but it was nevertheless considered worthwhile to carry out some experiments in order to see what qualitative changes occurred. The experimental arrangements were similar to those used in earlier work (Kolsky 1954 b) to investigate the propagation of pulses in elastic cylinders and most of the present experiments were carried out on cylinders of polythene. The displacement–time curves for four thicknesses of this material are shown in fig. 11 and it may be seen that in each case the displacement is tending to return to zero. This is to be expected from eqn. (4) since it is $f'(r-ct)$ which will be dominant at these distances of travel. It should be noted that the **pulse** lengths are here much shorter and increase more slowly than for propagation in rods of the material. This might be expected since we are dealing here with dilatation waves in which the relevant elastic constant is $(k+4/3\mu)$, where k is the bulk modulus and μ is the shear modulus of the material. k for polythene is considerably greater than μ and might be expected to be less sensitive to rate of loading. There will consequently be less dispersion due to the dependence of velocity of propagation on frequency. Further, the fraction of stored elastic energy which is dissipated in a stress cycle might be expected to be smaller so that there will be less attenuation due to damping for waves of this type.

§ 3. Discussion of Results

As mentioned earlier, it was intended to use the dynamic elastic constants of the plastics obtained from measurements with sinusoidally applied stresses for calculating the shape of the pulses after they had travelled some distance along rods of these materials. This can be done if a Fourier analysis of the initial pulse-shape is first carried out ; the appropriate changes in amplitude and phase of each Fourier component are then made and the pulse re-synthesised. Thus if the stress pulse at the origin is represented by the Fourier integral

$$\int_0^\infty A_p \cos pt \, dp$$

after travelling distance x it will be given by

$$\sigma(x, t) = \int_0^\infty A_p \exp(-\alpha x) \cos p(t - x/c) \, dp. \qquad . \quad . \quad . \quad (5)$$

Thus if α and c are known over the relevant range of frequency, the new shape can be calculated.

In the experiments described in § 2 the pulse was originally very short (ca. 2 microseconds) compared with its length when measured (several hundred microseconds). Further, in the early stages of its progress down the bar two separate types of dispersion were present, viz. : attenuation by the material of the bar and dispersion due to lateral inertia effects.

In order to treat the problem theoretically, it therefore seemed reasonable to investigate the propagation of a pulse which was infinitely sharp at the origin and hence can be represented by a δ-function. In this case A_p is a constant and all that is required is a Fourier synthesis for any value of x.

In order to carry this out numerically the Fourier integral must be replaced by a Fourier series and it is desirable to choose a frame of reference which is moving with the pulse. If we take v as the velocity of the pulse we may write

$$\sigma(x,t) = \sum_0^n A \exp(-\alpha x) \cos np\left(t - \frac{x}{c} + \frac{x}{v}\right) \quad \ldots \ldots \quad (6)$$

where t is now measured from the time at which the pulse arrives at the point at which it is observed, a distance x from the origin.

Equation (6) can be expanded in the form

$$\sigma(x,t) = \sum_0^n A_n \cos npt + \sum_1^n B_n \sin npt \quad \ldots \ldots \quad (7)$$

where $\qquad\qquad A_n = A \exp(-\alpha x) \cos npx(1/c - 1/v)$

and $\qquad\qquad B_n = A \exp(-\alpha x) \sin npx(1/c - 1/v)$.

A_n and B_n can be calculated for the terms in the series if the values of α and c are known at each frequency $np/2\pi$. v is a constant for the summation and must be chosen so that the pulse remains within the basic interval of the series. The value of v is not too critical in that a change in v results only in a shift of the pulse with respect to the chosen axes of reference ; in the present work v was taken as the phase velocity at the basic frequency $p/2\pi$. The basic interval of the series is $2\pi/p$ and for accuracy this must be chosen sufficiently narrow for the pulse to occupy a large part of the time covered. On the other hand it must be sufficiently wide to preclude any appreciable part of the pulse falling outside this interval, or the periodicity assumption inherent in the use of a Fourier series is no longer justified. The numerical method which was employed for the Fourier synthesis is that due to Danielson and Lanczos (1942) and by using 32 terms for both the cosine and sine summations 64 ordinates of the final pulse could be calculated.

Using the data shown in fig. 1 the pulse shapes to be expected in polythene rods were determined. Two difficulties arise in comparing these with those found experimentally. First, the initial pulse is not infinitely sharp and secondly, as mentioned earlier, the dispersion due to inertia effects in the bar is important until the length of the pulse becomes several times the diameter of the bar. The result of both these effects is to make the pulse longer than that predicted by the calculation. For 1·25 cm polythene rods it was found that at large distances of travel the agreement was best when the calculated distance was about 10 cm greater than the actual distance travelled. Figure 12 shows a comparison between the pulse observed experimentally after travelling through 60 cm of polythene and that calculated for 70 cm.

In comparing the pulse shapes shown in figs. 9 and 10 it was found empirically that they were all approximately similar in shape.   By this is meant that if the time scales are suitably extended or contracted and the amplitudes decreased or increased in the same ratio (so that the area under the curve remains constant) all the pulses approximate to the same shape.   This result was surprising in view of the very different properties of the two plastics and it therefore seemed worth seeing whether there was any underlying reason for this similarity.

Fig. 12

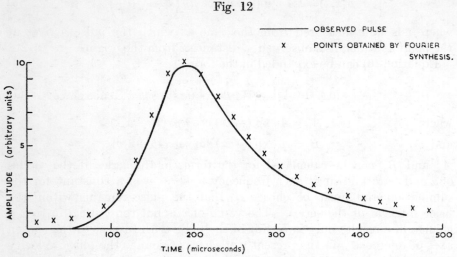

OBSERVED PULSE

x   POINTS OBTAINED BY FOURIER SYNTHESIS.

Comparison between observed pulse for 60 cm rod and curve calculated for 70 cm.

The observed similarity implies that if the stress pulse, after travelling a distance x_1 is expressed as

$$\sigma_1 = f(t) \qquad \qquad (8)$$

at distance x_2 the stress will be given by

$$\sigma_2 = \frac{1}{m} f\left(\frac{t}{m}\right). \qquad \qquad (9)$$

Choosing a frame of reference moving with velocity v as before, and using the expression for α from eqn. (1) we may rewrite eqn. (5)

$$\sigma_1 = A \int_0^\infty \exp(Kpx_1) \cos p\left(t - x_1\left[\frac{1}{c} - \frac{1}{v}\right]\right) dp \qquad (10)$$

where $K = -\tan \delta/2c$.

Now if we put $m = x_2/x_1$, the expression at x_2 may be written

$$\sigma_2 = \frac{A}{m} \int_0^\infty \exp(Kmpx_1) \cos mp\left(\frac{t}{m} - x_1\left[\frac{1}{c} - \frac{1}{v}\right]\right) d(mp). \qquad (11)$$

This is of the form (9) if K and $(1/c-1/v)$ are either independent of p or can be expressed as a function of the product px. As mentioned earlier, for most polymers tan δ does not vary markedly over limited ranges of frequency whilst c increases with frequency. The relevant frequency range of the Fourier integral which effectively determines the pulse shape is only of the order of a decade or two and within this region tan δ will in general vary very little. Further, if tan δ is independent of frequency and small, the theory of linear viscoelasticity shows that

$$\frac{dE}{d(\ln p)} = \frac{2 \tan \delta}{\pi}$$

$$E = E_0 \left(1 + \frac{2 \tan \delta}{\pi} \ln [p/p_0] \right)$$

where E_0 is the modulus at a standard frequency p_0. Hence since

$$c = (E/\rho)^{1/2}$$

$$c \simeq c_0 \left(1 + \frac{\tan \delta}{\pi} \ln [p/p_0] \right). \quad \ldots \ldots \quad (12)$$

Thus if tan δ is small c will not vary much over the relevant frequency range and K may be assumed to be constant. The same, however, does not apply to the expression $(1/c-1/v)$, since it is just the small variations of c which produce dispersion effects in the pulse. Since this expression is not constant the similarity condition will only hold if $(1/c-1/v)$ can be expressed as $f(px)$. This would, in general, not be possible, but from eqn. (12) we have

$$\frac{1}{c} \simeq \frac{1}{c_0} \left(1 - \frac{\tan \delta}{\pi} \ln [p/p_0] \right)$$

and so if we take

$$\frac{1}{v} = c_0 \left(1 + \frac{\tan \delta}{\pi} \ln \frac{x}{x_0} \right) \quad \ldots \ldots \ldots \quad (13)$$

where x_0 is a disposable constant, we have

$$(1/c-1/v) = \frac{\tan \delta}{c_0 \pi} \ln \frac{px}{p_0 x_0}. \quad \ldots \ldots \ldots \quad (14)$$

which is of the required form. Equation (13) implies that the ' pulse velocity ' v is not constant but decreases slowly as the pulse progresses.

It has thus been shown that if tan δ is constant and sufficiently small for its higher powers to be neglected, similarity will apply between the pulse shapes at different distances of travel through any one material. If we now use eqn. (14) to substitute for $(1/c-1/v)$ in eqn. (11) and take $K = -\tan \delta/2c_0$, we find that the properties of the material occur only in

Fig. 13

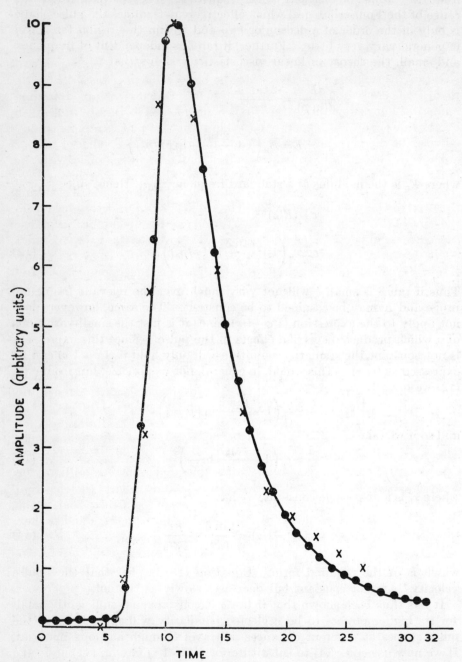

Experimental results for polymethylmethacrylate compared with universal
pulse calculated by Fourier synthesis.

the form of the parameter $(x \tan \delta/c_0)$. Thus all linear viscoelastic materials which fulfil the assumptions made will give the same pulse shapes after suitably scaled travel distances and if the integral is evaluated it will give the pulse shape for all such viscoelastic materials and all distances of travel when the time scale is suitably chosen.

In order to calculate the shape of the generalized pulse the integral was replaced by the sum of a sine and a cosine Fourier series, as given in eqn. (7). The basic interval of the series was taken to correspond to $(4\pi x \tan \delta)/c$ (writing c for c_0), so that the basic value of p was $c/(2x \tan \delta)$. Figure 13 shows the central 32 of the 64 ordinates obtained from a numerical Fourier synthesis, the x axis being plotted in units of time given by $(\pi x \tan \delta)/16c$. It should be possible to fit this plot to a pulse in any polymer of the type postulated at any distance x if the value of $\tan \delta/c$ is known. The figure shows the remarkably good fit obtained with the pulse shape observed in polymethylmethacrylate after a distance of travel of 6 metres. The time scale here was fitted empirically, taking $(x \tan \delta/c)$ to be 100 microseconds. The value of c can be determined from the time of transit of the pulse across the rod (see fig. 6) and for this plastic the value is about 2300 metres/second. This makes the value of $\tan \delta$ about 0·04, which is in reasonable agreement with that found by Lethersich (1950) for torsional oscillations of this material at comparable frequencies.

§ 4. CONCLUSION

When mechanical pulses are propagated along rods of high polymers they change in shape as a result of two separate causes. First, the lateral inertia of the rods results in dispersion, and secondly, the variation of mechanical properties of the material produces dispersion and attenuation of the pulse. A theoretical investigation of the combined effects of the two types of dispersion would be of interest but would certainly involve heavy mathematical analysis. When, however, the length of the pulse is large compared with the diameter of the rod the effects of lateral inertia can be neglected and it has been shown that under these conditions the pulse shape can be accurately predicted from a knowledge of the variation of elastic modulus and damping with frequency. Further, that if the damping $(\tan \delta)$ is constant and sufficiently small for powers higher than the first to be neglected, (i.e. $\tan \delta < 0·1$), the effects of damping and velocity change can be combined into a single parameter and the results of pulse attenuation expressed in a non-dimensional form which applies to all such viscoelastic solids.

ACKNOWLEDGMENTS

The author wishes to express his thanks to Dr. F. C. Roesler for many valuable discussions on this work and to Mrs. P. M. Weaver for her help with the experiments.

REFERENCES

DANIELSON, G. C., and LANCZOS, C., 1942, *J. Franklin Inst.*, **233**, 365.

DAVIES, R. M., 1948, *Phil. Trans.* A, **240**, 375.

EUBANKS, R. A., MUSTER, D., and VOLTERRA, E. G., 1952, *Illinois Inst. of Techn. Tech. Rep.* No. 1, Contract No. N7onr–32911, Project No. NR 064 369.

GLAUZ, R. D., and LEE, E. H., 1954, *J. Appl. Phys.*, **25**, 947.

GROSS, B., 1953, *Mathematical Structure of the Theories of Viscoelasticity* (Paris : Hermann).

HILLIER, K. W., 1949, *Proc. Phys. Soc.* B, **62**, 701.

KOLSKY, H., 1949, *Proc. Phys. Soc.* B, **62**, 676 ; 1953, *Stress Waves in Solids* (Oxford : Clarendon Press) ; 1954 a, *Proc. 2nd Int. Congr. on Rheology*, p. 79 (London : Butterworths) ; 1954 b, *Phil. Mag.*, **45**, 712.

LEADERMAN, H., 1943, *Elastic and Creep Properties of Filamentous Materials* (Washington : Textile Foundation).

LEE, E. H., and KANTER, I., 1953, *J. Appl. Phys.*, **24**, 1115.

LETHERSICH, W., 1950, *Brit. J. Appl. Phys.*, **1**, 294.

RICKER, N., 1943, *Bull. Seism. Soc. Amer.*, **33**, 197.

SELBERG, H. L., 1952, *Arkiv. fysik*, **5**, 97.

THOMPSON, J. H. C., 1933, *Phil. Trans.* A, **231**, 339.

ZVEREV, I. N., 1950, *Appl. Math. Mech., Leningr.*, **15**, 295.

Reprinted from the Journal of Geophysical Research, v. 67, p. 5279–5291.

Dispersive Body Waves

Walter I. Futterman[1]

Lawrence Radiation Laboratory
University of California, Livermore

Abstract. Although the attenuation of body waves in an infinite medium has often been considered, little emphasis is usually given to the attendant dispersion, which is a necessary consequence of the medium absorption and is determined unambiguously by it. Dispersion equations are obtained from the application of an integral transform in the frequency domain. Such relations are of the Kramers-Krönig type, relating the dispersive part of the index of refraction of the medium to the absorptive part by means of an integral over the entire frequency range. They are a consequence of the principle of causality and follow without recourse to a specific wave equation. In particular, our results are consistent with and presume linearity for the wave motion. Three forms of the dispersion equation are exhibited. In all cases, the absorption coefficient has a linear dependence upon frequency for frequencies above a certain characteristic value—chosen low enough to be outside the range of measurement. However, it is necessary in all cases to introduce a low-frequency, nonzero cutoff for the absorption, which can in principle be taken arbitrarily small. Group and phase velocities are determined as functions of the medium Q and a characteristic velocity, which for small absorption can be related to the measured signal velocity. Seismic phenomena occur in materials where the absorption is small, $Q \geq 30$, so that the corresponding dispersion presumably will be difficult to measure in the earth. For appropriate laboratory materials the effect can be significant.

Introduction

Continuing doubt is expressed in the literature of theoretical and experimental seismology about the existence of dispersion for body waves in the earth, although the attenuation of such waves due to absorption has often been observed [*McDonal et al.*, 1958; *Knopoff and MacDonald*, 1958]. Absence of accompanying dispersion implies a nonlinear wave equation, for in a linear theory of wave propagation the presence of absorption is a necessary and sufficient condition for the presence of dispersion.

These negative experiments and calculations have been supplemented by the measurement of the frequency dependence of the absorption coefficient, $\alpha(\omega)$. For many earth materials it varies linearly with the frequency, [*Knopoff and MacDonald*, 1958]. From theoretical studies, *Knopoff and MacDonald* [1958] and *Collins and Lee* [1956] conclude that a strictly linear dependence is incompatible with a linear wave theory, and hence this measurement is used as an additional argument for a nonlinear theory.

The obvious advantages of a linear theory are given up only for sufficient cause. There exists, moreover, a reasonably large body of theoretical seismology successfully founded on linear wave motion [*Bullen*, 1953; *Ewing et al.*, 1957; *Jeffreys*, 1959]. Thus, to reconcile or to disprove the linearity of the theory by confronting it with the preceding observations would resolve a matter of principle as well as of practice. Contrary to the conclusion of the theoretical studies cited in the preceding paragraph, we find that the experimental results are consistent with a linear theory. The origin of this discrepancy is discussed in the context of the paragraph on absorption-dispersion pairs. The demonstrations by means of a Kramers-Krönig dispersion relation will determine the nature of the pulse dispersion in terms of explicit expressions for the phase and group velocities and will make possible a direct examination of the pulse form itself after it has traveled through the absorbing medium.

We define the absorption coefficient $\alpha(\omega)$ of a one-dimensional plane wave displacement amplitude as

$$u(R, T) = u_0 e^{iKR} e^{-i\omega t} \qquad (1)$$

[1] Now at Physics Department, Missiles and Space Division, Lockheed Aircraft Corporation, Palo Alto, Calif.

where the propagation constant K can be expressed in terms of the phase coefficient k and the absorption coefficient α:

$$K(\omega) = k(\omega) + i\alpha(\omega) \qquad (2)$$

Here α is taken positive to assure us that energy is lost from the wave to the medium. The first expression can then be written as

$$u(R, t) = u_0 e^{-\alpha R} e^{i(kR-\omega t)} \qquad (3)$$

For our calculations we make two assumptions: (a) The absorption coefficient $\alpha(\omega)$ is strictly linear in the frequency, over the range of measurement. (b) The wave motion is linear; i.e., the principle of superposition is valid. The second assumption is of a more fundamental nature.

Thus we can express any pulse $u(R,t)$ as a superposition of plane waves. In particular, we choose a complex representation of the form

$$u(R, t) = \int_{-\infty}^{\infty} d\omega \hat{u}(0, \omega) e^{i[K(\omega)R-\omega t]} \qquad (4)$$

where $\hat{u}(R,\omega)$ is the Fourier transform in time of the displacement $u(R,t)$, and where $R = 0$ defines a plane boundary of the material for which $R \geq 0$. Incidental to the determination of $k(\omega)$ from $\alpha(\omega)$ we shall see that assumptions (a) and (b) are not incompatible.

THE DISPERSION RELATIONS

The method we use is model independent, with the disadvantage that it yields no information about the physical mechanism that accounts for the absorption or the corresponding dispersion. Conversely, its major advantage is that we need not appeal to the physical details that are characteristic of a particular theory. They are bypassed to yield the dispersion directly from the absorption coefficient, in this case taken from experiment.

Before proceeding to the expression of interest, it is convenient to express the complex wave number $K(\omega)$ in terms of the index of refraction of the medium. We make the following definition:

$$n(\omega) = K(\omega)/K_0(\omega) \qquad (5)$$

where K_0 defines the nondispersive behavior of K at the same frequency, and where, in its component form, the complex index of refraction

will be written as

$$n(\omega) = \text{Re } n(\omega) + i \text{ Im } n(\omega) \qquad (6)$$

In view of the experimental observation that the absorption coefficient decreases with decreasing frequency, it seems reasonable to assume that the absorption is negligible for sufficiently low frequencies. We hypothesize, therefore, that for the circular frequency ω less than some as yet unspecified cutoff frequency ω_0, characteristic of the material, no dispersion exists. It will later be shown that the theory requires ω_0 different from zero. From these remarks it follows that for $\omega < \omega_0$

$$K(\omega) = \omega/c \qquad (7)$$

c being the nondispersive limit of the phase velocity, which occurs at low frequencies. By definition then, $K_0(\omega)$ is given by the same expression at all frequencies.

Thus, we have

$$n(\omega) = K(\omega)/(\omega/c) \qquad (8)$$

$$\lim_{\omega \to 0} n(\omega) = n(0) = 1 \qquad (9)$$

Dispersion relations of the Kramers-Krönig (K-K) type [*Kramers*, 1927; *Krönig*, 1926], well known in electric circuit theory [*Bode*, 1945; *Guilleman*, 1949], determine, for example, the real part of the propagation constant from the values of the imaginary part summed over the entire range of frequencies for wave motions that are linear. The K-K relations have been derived many times in the literature of electromagnetic wave theory [*Toll*, 1952; *Jauch and Rohrlich*, 1955].

For wave fields whose phase velocity limit at high frequencies is less than the speed of light, a slightly modified proof of the classical K-K relations is needed. Essentially, the only principle required in order to obtain the expressions is that of causality: the signal is not detected at the apparatus until sufficient time has passed for it to arrive there. In the appendix we deduce a K-K relation which is expressed in terms of the low-frequency behavior of the index

$$\text{Re } [n(x) - n(0)]$$

$$= \frac{2x^2}{\pi} P \int_0^{\infty} dx' \frac{\text{Im } n(x')}{x'(x'^2 - x^2)} \qquad (10)$$

where (8) tells us that Re $n(0) = 1$. The P

indicates that Cauchy's principal value is to be taken where singularities of the integrand appear. The dimensionless variable x is defined by $x = \omega/\omega_0$, where ω_0 is as yet an arbitrary circular frequency to which we will later give the significance of a cutoff frequency. In deriving this K-K relation, use was made of a symmetry condition on the index of refraction as a function of the frequency. This condition is readily deduced from the superposition principle (4), using the fact that the displacement is a real function of position and time. It amounts to the 'crossing symmetry' relationship,

$$K(x) = K^*(-x) \qquad (11)$$

where the asterisk denotes the complex conjugate. In component form this result becomes

$$k(x) = -k(-x)$$
$$\alpha(x) = \alpha(-x) \qquad (12)$$

Noting that the components of the complex index of refraction are of opposite parity to the respective wave number counterparts, which contain an additional power of the frequency, we have the result

$$\text{Re } n(x) = \text{Re } n(-x)$$
$$\text{Im } n(x) = -\text{Im } n(-x) \qquad (13)$$

It should be emphasized that these conditions hold for all linear theories, independent of the specific form of K or of n. Thus, for example, the extension of α to negative frequencies is even, i.e. symmetric, in the frequency for whatever power law is used in the physical domain of positive frequencies.

Before proceeding to the actual calculation of the dispersion we need some additional expressions for physical quantities that characterize the absorptive medium.

PHASE AND GROUP VELOCITIES AND THE MEDIUM Q

Velocities. For absorptive media a little care is needed in defining the phase velocity v_p and the group velocity u_g. The notion of phase velocity relates to a single-frequency component of the wave packet of interest (4), which we rewrite in the form

$$u(R, t) = \int_{-\infty}^{\infty} d\omega \, u_\omega(R, t)$$

The amplitude and phase of the component u_ω are determined from the expression

$$u_\omega(R, t) \equiv A_\omega(R)e^{i\phi_\omega} \qquad A, \phi \text{ real}$$

where

$$A_\omega(R) = \hat{u}(0, \omega)e^{-\alpha(\omega)R}$$
$$\phi_\omega(R, t) = k(\omega)R - \omega t \qquad (14)$$

Considering the space-time dependence of ϕ, we define the phase velocity as that velocity which maintains ϕ constant with respect to variations in R and t. Thus we find at once that

$$v_p(\omega) = (dR/dt)_{\phi=\text{const}} = \omega/k(\omega) \qquad (15)$$

In terms of the index of refraction, therefore,

$$v_p(\omega) = c/\text{Re } n(\omega) \qquad (16)$$

The expression for the group velocity u_g is simply obtained by application of the method of stationary phase to the wave packet (14). This velocity corresponds to the requirement that the phase be stationary with respect to variations in the frequency, $\partial\phi/\partial\omega = 0$. Thus, the roots of this equation, ω_s, which are the stationary points of the phase, occur at

$$u_g = R/t = (\partial k/\partial\omega)^{-1} \qquad (17)$$

In terms of the index of refraction,

$$u_g = c\left\{\frac{\partial[\omega \text{ Re } n(\omega)]}{\partial\omega}\right\}^{-1} \qquad (18)$$

We note that

$$\lim_{\omega\to 0} v_p = \lim_{\omega\to 0} u_g = c$$

Comparison of these formulas for the velocities with those that derive from the usual dispersion relations for nonabsorptive media (for example, as the result of application of boundary conditions to a system of linear differential equations) shows that in both situations it is the real part of the propagation constant that determines the velocities (Im $K = 0$, of course, in the usual case).

For sufficiently small absorption the group velocity can be identified with that velocity with which a measured signal propagates, as well as with the velocity of energy propagation. In general, however, the concepts differ, and for large absorption these velocities are not equal [*Brillouin*, 1960].

Q. Among the methods of measuring the dissipative properties of a material are those of (1) free vibration and (2) wave propagation [*Kolsky*, 1953]. These methods suggest a simple way to relate attenuation in space to the damping in time for linear systems. (1) Consider a sample, say a rod or filament which, to avoid losses, is not coupled to the exterior medium. If the rod is struck impulsively, standing waves will be set up in the material which in time will decrease in amplitude owing to dissipation in the material. (2) If we consider a sample whose dimensions are large compared with the width of a pulse, boundary effects upon its propagation can be neglected. An impulse given to a boundary surface will propagate as a pulsed body wave through the material, the amplitudes of all its constituent frequencies attenuating in space and, hence, in time. Since the material behaves linearly, any (damped) pulse form can be constructed from (damped) standing waves. In particular, the damping of those standing waves of (1) which constitute the pulse in (2) determine the damping of the pulse.

(1) Consider the single-frequency component of the displacement,

$$u = Ae^{-\gamma t} \cos (\omega t + \beta)$$

a 'standing wave' with damping constant γ and phase constant β. Within one period, $t = 2\pi/\omega$, the amplitude will be reduced by a factor exp $(-2\pi\gamma/\omega) \equiv \exp(-\triangle)$, where $\triangle$ is the logarithmic decrement. Defining $\triangle W/W$ as the ratio of the energy loss per cycle to the maximum energy stored in the specimen, we easily find that, for a linear material where the energy stored is proportional to the square of the displacement,

$$\triangle W/W = 1 - e^{-2\triangle}$$

$$\approx 2\triangle \quad \triangle \ll \tfrac{1}{2}$$

This fractional energy loss per cycle is defined conveniently in terms of the dimensionless quality factor Q, which has the same significance here as the quantity used in electric circuit theory, $2\pi/Q = \triangle W/W$. Thus,

$$Q^{-1} = (2\pi)^{-1}(1 - e^{-2\triangle})$$

$$\approx \triangle/\pi \quad \triangle \ll \tfrac{1}{2}$$

It is a measure of the absorption strength of a material. The result, $Q_{max}^{-1} = (2\pi)^{-1}$, is in accord with the physically reasonable $(\triangle W/W)_{max} = 1$.

(2) Consider a propagating wave which is approximated by the infinitely long sinusoidal train,

$$u_\phi(x, t) = c\, e^{-\alpha x} \cos \phi(x, t)$$

$$\phi(x, t) = \omega(x/v_p - t)$$

The position of successive maximums of displacement occurs at $\phi = 0, 2\pi, 4\pi, \ldots, 2\pi n, \ldots$ ($\cos \phi = 1$). At $\phi = 0$, $x = v_p t$. At this same time t, the next crest occurs at $\phi = 2\pi = \omega\, [(x + \delta x)/v_p - t)]$, where $\delta x = 2\pi\, v_p/\omega$. Since $u_0(x, t) \approx \exp(-\alpha v_p t)$ and $u_{2\pi}\, (x + \delta x, t) \approx \exp[-\alpha v_p(t + 2\pi/\omega)]$,

$$\triangle = 2\pi\alpha v_p/\omega$$

and

$$Q^{-1}(\omega) = (2\pi)^{-1}[1 - \exp(-4\pi\alpha v_p/\omega)] \quad (19)$$

For $4\pi\,\alpha v_p/\omega \ll 1$,

$$Q(\omega) \approx \omega/2\alpha(\omega)v_p(\omega) \quad (20)$$

a result which agrees with *Mason* [1958, p. 214]. We note that (20) is, strictly, an approximate expression, good for small absorption. Equivalently, we have $Q(\omega) \approx k(\omega)/2\alpha(\omega)$.

It is convenient to define the reduced quality factor,[2]

$$Q_0(\omega) = \omega/2\alpha(\omega)c \quad (21)$$

where now the only intrinsic dependence on frequency occurs in α. Expressing W in terms of Q_0, we have

$$Q = cQ_0/v_p \quad (22)$$

In terms of the index of refraction we then have

$$\mathrm{Im}\, n(\omega) = 1/2Q_0(\omega) \quad (23)$$

and, combining the results of (22), (23), and (8), we find that

$$\lim_{\omega \to 0} Q = Q_0(0) = \infty$$

---

[2] This quantity is called Q by *Knopoff and MacDonald* [1958]. It is perhaps not out of place here to remark that our expression is strictly correct for an infinite homogeneous material. In the case of finite material, the boundary losses must also be considered, the velocity of energy transport replacing the phase velocity in this component of Q.

since α departs from assumption (a) at sufficiently low frequencies. In the region of interest where α satisfies assumption (a), Q_0 and Im n are frequency independent.

ABSORPTION-DISPERSION PAIRS

Let us now consider three different forms of absorption which satisfy assumption (a). By the crossing symmetry, the index need be defined only for the physical frequencies, $x \geq 0$,

$$\text{Im } n(x) = \frac{1}{2Q_0} \begin{cases} 0 & 0 \leq x \leq 1 \\ 1 & x > 1 \end{cases} \qquad \text{(A1)}$$

$$\text{Im } n(x) = \frac{1}{2Q_0} \begin{cases} x/x_c & 0 \leq x \leq x_c \equiv \omega/\omega_c \\ 1 & x > x_c \end{cases} \qquad \text{(A2)}$$

$$\text{Im } n(x) = \frac{1}{2Q_0} (1 - e^{-x}) \qquad 0 \leq x \qquad \text{(A3)}$$

where the last form satisfies assumption (a) asymptotically. These forms are illustrated schematically in Figure 1. Should we wish to write these expressions explicitly over the entire frequency range there is no difficulty involved. For example, (A3) becomes

$$\text{Im } n(x) = \frac{1}{2Q_0} (1 - e^{-|x|}) \text{ sgn } x$$

where the function sgn x is equal to -1 for $x < 0$ and to $+1$ for $x > 0$.

Rather straightforward integrations of each of these forms, according to the dispersion formula (10), yield the respective dispersive component of the index of refraction. The results are listed below.

$$\text{Re } n(x) - 1 = -\frac{1}{\pi Q_0}$$

$$\begin{cases} \frac{1}{2} \ln |x^2 - 1| & \text{(D1)} \\ \frac{1}{2}\left[\ln \left| 1 - \left(\frac{x}{x_c}\right)^2 \right| - \frac{x}{x_c} \ln \left| \frac{1 - x/x_c}{1 + x/x_c} \right| \right] & \text{(D2)} \\ \ln \gamma x - \frac{1}{2}[e^{-x}\overline{Ei(x)} + e^x Ei(-x)] & \text{(D3)} \end{cases}$$

where $\ln \gamma = 0.5772157 \ldots$, Euler's constant, and the Ei symbols define the exponential integral and its conjugate [*Janke and Emde*, 1945],

$$-Ei(-x) = \int_x^\infty \frac{e^{-t}}{t} dt \qquad x > 0$$

$$Ei(x) = -P \int_{-x}^\infty \frac{e^{-t}}{t} dt \qquad x > 0$$

These expressions satisfy the imposed boundary condition, $n(0) = 1$, and for large x[i.e., $x \gg 1$ in (D1), $x \gg x_c$ in (D2), and $x \gg \gamma$ in (D3)] the three functional forms coincide. The behavior of the absorption-dispersion pair (A3, D3) is illustrated in Figure 2. The asymptotic form common to all three functions is simply

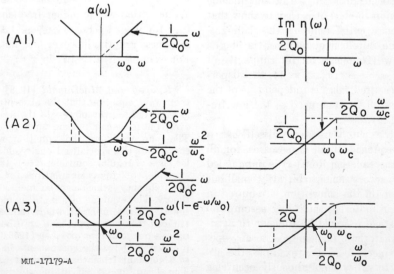

MUL-17179-A

Fig. 1. Schematic behavior of the absorption coefficients, $\alpha(\omega)$, and the corresponding Im n (ω).

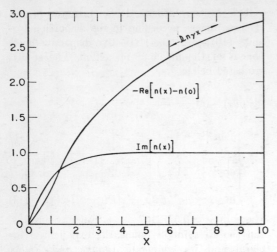

Fig. 2. The absorption-dispersion pair (A3, D3).

$$\operatorname{Re} n(x) = 1 - \frac{1}{\pi Q_0} \ln x \qquad (24)$$

This formula clearly fails for $x \geq e^{\pi Q_0}$. For $Q_0 \geq 30$, $x \gtrsim 10^{100}$, a very large number. The existence of a failure point, beyond which $\operatorname{Re} n < 0$, is not surprising since (24) treats assumption (a) as if it were valid out to infinite frequency. Deviations from this behavior must be expected for frequencies such that the wavelength is of the order of molecular dimensions, where resonance phenomena will become effective.. At infinite frequency no absorption is expected (see appendix). Indeed, although we do not include the calculation in the text, it is easy to show that a high-frequency cutoff removes this failing.

Thus, if the cutoff frequency, ω_0, is chosen sufficiently low (from now on for simplicity of discussion let $\omega_e = \omega_0$, i.e., $x_e = 1$), the dispersion in the measured range is independent of the form of the cutoff function used at the low frequencies, by (24).

However, the cutoff frequency itself has a physical consequence. In all three cases, for all values of ω, the function $\operatorname{Re} n(\omega)$ is unbounded for ω_0 equal to zero. A finite, arbitrarily small but nonzero, cutoff on the absorption is required in this theory for materials satisfying assumption (a). At this stage it would be difficult to attribute the characteristic frequency ω_0 to any specific physical mechanism. For example, the absorption due to other simultaneously occurring low-frequency effects, which we have left out of consideration, could produce this cutoff in earth materials.

We note that the application of the dispersion equation 10 is not limited to absorption coefficients satisfying assumption (a). For example, any power law, ω^β, where β is not necessarily an integer, can be assumed, subject of course to the crossing symmetry (12).[3] A discussion of the convergence problem for the integral (10) is relegated to the appendix. It can be resolved by a subtraction procedure.

RESULTS

Velocities. From the expressions D1, D2, and D3 we can immediately deduce the corresponding phase and group velocities. We restrict our attention to the most refined calculation, that of (D3). From a tabulation of values of the exponential integrals we find that, for $x \geq 6$ ($f = \omega/2\pi \gtrsim \omega_0$),

$$v_p(x) = c\left[1 - \frac{1}{\pi Q_0} \ln \gamma x \right]^{-1} \qquad (25)$$

$$u_g(x) = c\left[1 - \frac{1}{\pi Q_0} (1 + \ln \gamma x) \right]^{-1} \qquad (26)$$

and by (21) we have the high-frequency relation,

$$Q(x) = Q_0\left(1 - \frac{1}{\pi Q_0} \ln \gamma x \right)$$
$$Q_0 = \text{const} \qquad (27)$$

We note that at the higher frequencies, $x \gg \gamma^{-1}$ $e \simeq 1.5$, $u_g \simeq v_p$. For $x \gg \gamma \simeq 1.8$, $\ln \gamma x \simeq \ln x$. (These criteria, however, are not very useful. For example, if $x = 10^4$, $\ln x = 9.2$, whereas 1

---

[3] *Knopoff and MacDonald* [1958] deduce from their wave equation that the absorption coefficient is restricted to even powers of the frequency. However, linearity can put no such restriction on α. Their linear differential equation with constant coefficients is a specialized one, yielding a particular class of regular solutions $K(\omega)$. It does not, for example, encompass singularities for real values of the frequency, which can be included in a linear theory. The example of a metal in electromagnetic theory is a case in point, where the complex dielectric constant has a pole at $\omega = 0$, so that n has a branch point. *Collins and Lee* [1956] derive an expression (14) for the components of the propagation constant from their assumed wave equation. The absorption coefficient does not have the proper crossing symmetry.

$+ \ln \gamma = 1.577 > 0.1 \ln x$.) Asymptotically then,

$$v_p(x) = u_g(x) = c\left(1 - \frac{1}{\pi Q_0} \ln x\right)^{-1} \quad (28)$$

in agreement with (24).

This expression makes evident the difficulty in measuring the phase shift due to this kind of absorption. A direct measure of the observability of this phase change is given by the fractional change in the phase velocity $\Delta v_p/v_p$ for a given material, the velocity being measured at two different frequencies which are components of the pulse. For M discontinuity arrivals [*Werth et al.*, 1962], $Q_0 \simeq 300$ and the phase velocity for 1-cps waves is less than 10^{-3} greater than that at 1/20 cps. For Pierre shale [*McDonal et al.*, 1958] $Q_0 \simeq 30$, $\Delta v_p/v_p \simeq 0.04$, and on comparison with the case of no dispersion, the fractional change, $\Delta v/c$, for 1-cps waves is approximately 8 per cent. Thus, in the highest attenuating earth material available, one must be able to resolve phase changes which are probably beyond the present accuracy of measurement. From the approximate expression, for the increment in v_p with frequency due to dispersion,

$$\delta v_p = \frac{\delta x}{x\pi Q_0}\left(1 - \frac{1}{\pi Q_0} \ln x\right)^{-1} v_p \quad x \gg 1$$

which we obtain from (28), we see qualitatively that, given a large frequency difference δx, the effect is most enhanced for small $Q_0 x$. The dependence on Q_0 is not surprising, and that upon x follows from the fact that the largest percentage change in the logarithm function occurs at the smallest argument. It is in this range that experiments should be run. We note, however, that the effect of lowering Q_0 is no more effective than lowering the frequency in this range.

From measurements at a given frequency of v_p and u_g, we have, in principle, a means of determining the parameters c and ω_0, by solving simultaneously (25) and (26). However, because of the logarithmic dependence on frequency, the solutions are insensitive to variations in ω_0. Hence, in practice, we are free to choose ω_0 according to the phenomenological criterion that it be small compared with the lowest measured frequency ω. For the frequency range of interest, the magnitude $\omega_0 = 10^{-3}$ sec^{-1} is consistent with this constraint, and the pre-

ceding argument tells us that the value of c so obtained depends only weakly on this choice.

If we bear in mind the considerations of the preceding paragraph, there is a practical way to estimate the low-frequency limiting velocity c, based upon a single measurement at high frequency of either the signal or the phase velocity. In the study of the onset motion, for example, we look for the highest frequency, ω_M, which contributes significantly to the pulse, for by (24) or (28) it is ω_M which corresponds to the first arrival. There are two obvious methods for determining ω_M:

(1) Fourier analysis of the seismic pulse, using the noise level as a lower bound (where $\omega \simeq \omega_M$) on the amplitude $|A_\omega|$ of equation 14.

(2) Use the crude but direct estimate of ω_M, taken from the somewhat arbitrary assumption that it corresponds to a single e-folding of the amplitude. The uncertainty as to how many e-foldings, δn_M, constitute the proper choice yields a corresponding error, $\delta \omega_M = (\delta n_M/n_M)\omega_M$, in the maximum frequency, which is large compared with ω_M. For $\delta n_M \sim 4$, $n_M = 1$, $\delta \omega_M/\omega_M \sim 4$. But because of the logarithmic dependence on frequency, the uncertainty does not significantly alter the value of c which we deduce by the iteration procedure, defined according to the scheme,

$$c^{(n+1)} = u_g\left[1 - \frac{1}{\pi Q_0}\left(1 + \ln \frac{2Q_0\omega_0\gamma}{R} + c^{(n)}\right)\right]$$

where

$$c^{(0)} \equiv u_g$$

$$x_{max} \approx 2Q_0 c^{(0)} \omega_0/R$$

For example, for $u_g = 7.38$ km/sec, $\omega_0 = 2\pi \cdot 10^{-3}$ sec^{-1}, $f_{max} = 10$ cps, and $Q = 300$, we find $c \simeq 7.29$ km/sec.

Pulse shapes. In the time domain it is the pulse forms that are of interest. The nature of the absorption and dispersion of media satisfying assumption (a) is demonstrable in terms of the change of source functions caused by propagation through the medium. Figures 3 and 4 are illustrative. Equation 4 was integrated numerically, introducing as an input or boundary condition on the pulse, $u(0,t) = \delta(t)$, so that $\hat{u}(0,\omega) = \frac{1}{2}\pi$. We have chosen cases of extremely low Q, to make the effects clearly demonstrable.

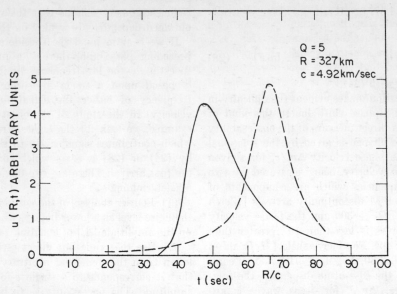

Fig. 3. Two pulses which originate as equal strength δ functions at $R = 0$ are compared at $R = 327$ km. For the causal pulse (straight line) the absorption-dispersion pair (A3,D3) is used. For the acausal pulse (dashed line) the presence of dispersion is neglected.

In Figure 3 we see two pulses which have traversed identical media, defined by the three parameters Q_0, c, and ω_0, where the last quantity will henceforth be taken as 10^{-8} sec^{-1}. Both pulses originate as δ functions at $t = 0$ and have traveled the same distance R. However, in the case of the right-hand pulse, we have re-moved from the phase its dispersive contribution, so that $k = \omega/c$, whereas the other wave form contains the complete contribution of the pair (A3, D3). The parameters have the following values: $Q_0 = 5$, $c = 4.92$ km/sec, $R = 327$ km. We see that the complete pulse has the lower maximum, a sharpened front (since v_p

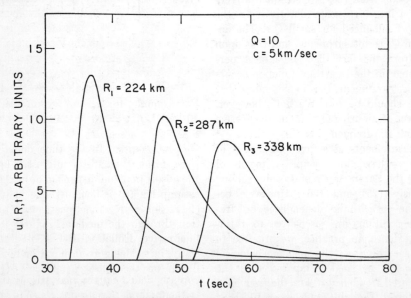

Fig. 4. A causal pulse (A3,D3) which originates as a δ function at $R = 0$ is plotted as a function of time at three successive distances from the source.

increases with frequency), and a definite arrival time. Without dispersion, the wave form is symmetrical about the maximum; this peak travels with the constant velocity c. Finally, it must be noted that there is some wave present at this distance even at $t = 0$. This acausality of the wave is the result of leaving out the dispersion; small as it is, its presence is required.

Figure 4 corresponds to a given medium, $Q_0 = 10$, $c = 5$ km/sec, traversed by three pulses which have moved through the respective distances of 224 km, 287 km, and 338 km. In this plot we see two effects working against each other. The phase velocity (or dispersion) sharpens up the wave front, for it rushes forward the highest frequency components present. However, as time passes, a given pulse will have traveled farther from its origin, so that the attenuation becomes increasingly effective—the higher frequencies are ironed out and the front falls as the effect of the absorption finally dominates. This last result is easily seen from the e-folding of the amplitude. For a given e-folding, n_M, corresponding to ω_M, $\omega_M = n_M Q_0/R$, goes to zero as R becomes infinite.

Linear viscoelasticity. Our expression (28) for the phase velocity resembles a result due to *Kolsky* [1956], $v_p(\omega) \simeq c_0 [1 + (\pi Q)^{-1} \ln \omega/\omega_0]$, where c_0 is the phase velocity at the reference frequency ω_0, and Q is defined by (20). The deduction of this approximate result [*Kolsky*, 1956; *Hunter*, 1960] depends on the model of a viscoelastic filament of constant Q. It should be realized that the similarity of the two expressions is only an apparent one, for in our results Q is frequency dependent, by (22), whereas the above expression implies $\alpha \simeq \omega/2Q$ $c_0 [1 + (\pi Q)^{-1} \ln \omega/\omega_0]$, contrary to assumption (a).

The same approximation is inappropriate to the case of constant Q_0 for this viscoelastic model, since we then obtain the result that the phase velocity is negative for all Q_0, implying that the direction of wave propagation is reversed even by a very small absorption.

CONCLUSION AND SUMMARY

It has been shown, independently of the details of a specific physical model, that an absorption coefficient strictly proportional to the frequency is compatible with a linear theory of wave propagation. For a linear theory, the existence of absorption implies dispersion and determines its form unambiguously up to certain constants characteristic of the material. It is the presence of this phase shift in the wave, due to the absorption, which guarantees a causal arrival of the signal. It was shown that neglecting dispersion would permit the arrival of a noncausal precursor. In the study of seismic arrivals from the M discontinuity *Werth et al.* [1962] found satisfactory agreement with the experimental wave forms of Blanca, Logan, and Tamalpais when the dispersion effect was included. No gradual onset of a nondispersive, attenuated wave is in evidence on the experimental records.

We saw that the effect of a small Q was not particularly significant in terms of the behavior of the single-frequency component of the wave, but that in terms of the shape of the pulse itself —in the time domain—a low Q clearly demonstrates the result of the presence of both the dispersion and attenuation of the wave-number components, for measurements made at not too large a distance from the source. Thus, low Q laboratory materials, which have linear attenuation, could test the theory.

The phase shift itself depends upon a nonzero low-frequency cutoff which would be difficult to measure but which must be finite in order that meaningful predictions may be made. The difficulty in directly measuring the dispersion, from a given frequency component of the wave, is evident from the logarithmic dependence of the phase velocity on frequency. On the other hand, it is this very dependence that is consistent with the unobserved dispersion in the field.

Finally, it should be emphasized that, while the preceding results, based upon a first power law for attenuation, are consistent with the linear theory of wave propagation and with recent observation, we have not actually proved the correct theory to be a linear one.

APPENDIX

The K-K dispersion relations will be deduced in a nonrigorous but, it is hoped, heuristic way. Analyticity, in a part of the complex frequency plane bounded by the real axis, is the key idea to the method. Specifically, we shall show that as a consequence of causality, the quantity $z/c[n(z)$

$-n(\infty)]$ is analytic in the upper half of the complex frequency plane (u.h.p.), where $n(\infty)$ denotes the value of the index of refraction at infinite frequency. Cauchy's theorem can then be applied to this quantity in the u.h.p. and the results follow easily. It may be remarked that the domain of discussion is off the real frequency axis, on which the function of interest can be outside the domain of analyticity. In the final section (3A) we deduce a crossing symmetry for the propagation constant analytically continued into the u.h.p.

(1A). Consider a real valued disturbance such as a displacement pulse, $u(R,t)$, which travels through a series of absorbing media. A detector is located on the boundary $R = 0$, of the semi-infinite medium of interest. Let the wave arrive at the detector at $t = 0$. By causality then

$$u(0, t) = 0 \qquad t < 0 \qquad \text{(A-1)}$$

Representing this pulse by means of a Fourier integral, we have

$$u(0, t) = \int_{-\infty}^{\infty} d\omega_1 \hat{u}(\omega_1) e^{-i\omega_1 t} \qquad \text{(A-2)}$$

where $\hat{u}(\omega_1)$ is identical with the function $\hat{u}(0, \omega_1)$ of equation 4.

Lemma: A necessary (N) and sufficient (S) condition for the existence of the causality condition (A-1) is that $\hat{u}(\omega_1)$ be analytic in the u.h.p. of the frequency.

(S) Assume the expression (A-1). Inversion of the integral (A-2) immediately gives

$$\hat{u}(\omega_1) = \frac{1}{2\pi} \int_{-\infty}^{\infty} dt\, u(0, t) e^{i\omega_1 t}$$
$$\qquad \text{(A-3)}$$
$$= \frac{1}{2\pi} \int_{0}^{\infty} dt\, u(0, t) e^{i\omega_1 t}$$

Define the expression

$$\phi(\omega) = \frac{1}{2\pi} \int_{0}^{\infty} dt\, u(0, t) e^{i\omega t}$$
$$\qquad \text{(A-4)}$$
$$\omega = \omega_1 + i\omega_2$$

Evidently $\phi(\omega)$ is of the same functional form as $\hat{u}(\omega_1)$, the complex variable ω replacing ω_1. No confusion will arise if we take account of this identity of form explicitly, writing $\phi(\omega) \equiv \hat{u}(\omega)$. The integrand occurring in (A-4) contains the factor $e^{-\omega_2 t}$, which for $\omega_2 > 0$ acts as a

convergence factor over the entire range of integration, $t > 0$. Thus, all derivatives with respect to ω exist, and therefore $\hat{u}(\omega)$ is analytic in the u.h.p. (One may draw the same conclusion from the example given by *Whittaker and Watson* [1950, p. 93].)

(N) Consider the integral

$$u_\epsilon(t) = \int_{-\infty+i\epsilon}^{\infty+i\epsilon} d\omega\, e^{-i\omega t} \hat{u}(\omega)$$
$$= \int_{-\infty+i\epsilon}^{\infty+i\epsilon} d\omega\, e^{-i\omega_1 t} e^{+\omega_2 t} \hat{u}(\omega)$$

We can close this contour at infinity in the u.h.p. By Jordan's lemma, this additional path, $|\omega| \mapsto \infty$, does not contribute to the integral for $t < 0$. But, by hypothesis, $\hat{u}(\omega)$ is without singularities in the u.h.p. Hence, by Cauchy's residue theorem, $u_\epsilon(t) = 0$, $t < 0$.

It is plausible that for a reasonably behaved function $\hat{u}(\omega)$ in the u.h.p.,

$$\lim_{\substack{\omega \to \omega_1 + 0 \\ \text{a.e}}} \hat{u}(\omega) = \hat{u}(\omega_1) \qquad \text{(A-5)}$$

That is, almost everywhere, $\hat{u}(\omega_1)$ is the boundary value of the analytic function $\hat{u}(\omega)$. Boundedness is a sufficient condition for (A-5) to be valid:

$$\int_{-\infty}^{\infty} |\hat{u}(\omega_1 + i\omega_2)|^2\, d\omega_1 < K \qquad \omega_2 > 0$$

[*Titchmarsh,* 1948, theorem 93]. We shall henceforth assume this property to be satisfied and that $\lim_{\epsilon \to 0} u_\epsilon(t) = u(0, t)$. Hence, the lemma is proved.

In passing we may remark that, for $t > 0$, closing of the contour for $u_\epsilon(t)$ in the lower half-plane shows, again using Jordan's lemma, that $u(0,t)$ will be determined by the sum of the residues of $u(\omega)$ occurring there.

Since our theory is a linear one, we have a superposition principle which we express in the form of (4), for a pulse in the medium $R > 0$,

$$u(R, t) = \int_{-\infty}^{\infty} d\omega_1\, u(\omega_1) \exp i[K(\omega_1)R - \omega_1 t]$$
$$\qquad \text{(A-6)}$$

where $K(\omega_1) = \omega\, u(\omega_1)/c$ is a complex function of a real variable.

As in the preceding lemma, define a function

$$V_\epsilon(t) = \int_{-\infty+i\epsilon}^{\infty+i\epsilon} d\omega\, \hat{u}(\omega) \exp\left\{i\omega\left[\frac{n(\omega)}{c} R - t\right]\right\}$$

By the same argument as that used earlier we have

$$V_\epsilon(t) = 0 \qquad \mathrm{Re}\left[t - \frac{n(\omega)}{c} R\right] < 0$$

or by extending the definition (16) of v_p to complex argument, for $t - R/v_p(\omega) < 0$, $v_p(\omega) = c/\mathrm{Re}\,\hat{n}(\omega)$, when ω is in the asymptotic part of the plane. The last qualification, $|\omega| \to \infty$, is needed to assure us that the contribution to the integrand from the path at infinity which is discarded actually vanishes. Thus, the requirement of causality implies that $v_p(\omega)$ has an asymptotic limit [and therefore that $k = \mathrm{Re}\, K \sim 0(\omega)$]. We will use the notation $v_p(\omega) \sim v_p(\infty)$ and $\mathrm{Re}\, n(\omega) \sim \mathrm{Re}\, n(\infty)$. Letting $\epsilon \to 0$, this condition becomes

$$u(R, t) = 0 \quad t < R/v_p(\infty) \equiv \tau \quad R \geq 0 \quad \text{(A-7)}$$

This result tells us that no signal can travel faster than the asymptotic limit of the phase velocity. To assure ourselves that there is no contribution from the integrand of $u(R, t)$ for $t < 0$, we require:

(1) $n(\omega)$ has no singularities in the u.h.p. [$\hat{u}(\omega)$ has already been shown to have none], and the condition on signals taken from special relativity,

(2) $v_p(\infty) \leq c_{\mathrm{light}}$ or $\mathrm{Re}\, n(\infty) \geq c/c_l > 0$.

We note that as long as (A-7) is satisfied we can allow $v_p(\omega) > c_l$ in the finite part of the frequency plane, since (A-7) is an asymptotic condition. Condition (1) is proved next.

Inversion of the integral (A-6) gives

$$\hat{u}(\omega_1)e^{iK(\omega_1)R} = \frac{1}{2\pi}\int_{-\infty}^{\infty} dt\, u(R, t)e^{i\omega_1 t}$$

Invoking the causality condition (A-7), we have

$$\hat{u}(\omega_1)e^{iK(\omega_1)R} = \frac{1}{2\pi}\int_{\tau}^{\infty} dt\, u(R, t)e^{i\omega_1 t} \quad \text{(A-8)}$$

We could use (A-8) to show that $K(\omega)$ is analytic in the u.h.p. except, perhaps, at the infinite point. [The analyticity of $e^{iK(\omega)r}$ everywhere in the u.h.p. follows readily from (A-8) owing to the factor $e^{-\omega_2 t}$.] However, it is more convenient at the outset to perform a displacement transformation on the time variable to obtain an exponent that is everywhere analytic in the u.h.p. Defining the new variable, $\rho = t - \tau$, we find that

$$\hat{u}(\omega_1)e^{iK(\omega_1)R} = \frac{1}{2\pi}\int_0^{\infty} d\rho\, u(R, \rho + \tau)e^{i\omega_1\rho}e^{i\omega_1\tau}$$

or that

$$\hat{u}(\omega_1)e^{if(\omega_1)R} = \frac{1}{2\pi}\int_0^{\infty} d\rho\, u(R, \rho+\tau)e^{i\omega_1\rho} \quad \text{(A-9)}$$

where

$$f(\omega_1) \equiv \omega_1/c[n(\omega_1) - \mathrm{Re}\, n(\infty)] \quad \text{(A-10)}$$

Now, since $\rho \geq 0$, we can define, as we did earlier in (A-4), a new function $\psi(\omega)$ identical in form with the left-hand side of (A-9), where ω_1 has been replaced by the complex variable ω. It is a continuous function of R, analytic everywhere in the u.h.p. for all $R \geq 0$. [For $R = 0$ we recover the expression (A-4).] This result was arrived at for an arbitrary Fourier transform $\hat{u}(\omega_1)$ of the pulse $u(0, t)$. It holds, in particular, for a $\hat{u}(\omega)$ having no roots in the u.h.p. [For example, for a δ function input, $u(0, t) = \delta(t)$, so that $\hat{u}(\omega_1) = \frac{1}{2}\pi = \hat{u}(\omega)$.] Thus, we have shown that $e^{if(\omega)R}$ is analytic in the entire u.h.p., $R \geq 0$.

It can be easily shown that the exponent f itself is an analytic function everywhere in the u.h.p. [*Toll*, 1952]. Write f in the form, $f = 1/iR \ln e^{ifR}$. The logarithm is analytic except at roots of e^{if}. The proof, by contradiction, then follows: Assume that such a root, ω_r, exists in the u.h.p. Then since e^{if} is regular we can expand about that root, $e^{if} = C(\omega - \omega_r)^P + \ldots$, where P is an integer ≥ 1, $C = $ constant, and $e^{ifR} = [C(\omega - \omega_r)^P + \ldots]^R$. Letting $R = R'/2P$, $e^{ifR} = [C'(\omega - \omega_r)^{R'/2} + \ldots]$. Now e^{ifR} is analytic in the u.h.p. for all $R > 0$, hence, for all $R' > 0$. In particular, let R' not be an even number, so that $R'/2$ is not an integer, contrary to the assumption of analyticity. Therefore, no roots exist in the u.h.p.

(2A). Having proved the analyticity of $f(\omega)$, $\omega = \omega_1 + i\omega_2$, $\omega_2 > 0$, we can in principle apply Cauchy's residue theorem directly to it. In practice, however, it proves convenient to close the contour, taken along the real frequency axis, in the u.h.p., with a semicircular contour Γ at infinity. In our case, we know of no general argument which gives information about the behavior of $f(\omega)$ asymptotically. However, we bypass this difficulty most easily by means of a physical assumption applied to the index of refraction: It is difficult to envision earth structures of such small dimension that they resonate to the infinite frequency component of the incident displacement wave. We must therefore expect that no absorption occurs at infinite frequency, so that $\mathrm{Im}\, n(\infty) = 0$, and $\mathrm{Re}\, n(\infty) = n(\infty)$. Hence $n(\infty)$ is bounded, and the expression for f becomes $f(\omega) = [\omega/c]\Delta n(\omega)$, where $\Delta n(\omega) = n(\omega) - n(\infty)$, an expression familiar in electromagnetic theory].

The theorem is now applied to $\Delta n(\omega)$, which is evidently analytic in the u.h.p. It is the ratio of two functions analytic in the u.h.p., where the denominator ω is without roots. We note the important property, $\Delta n(\infty) = 0$, so that the contribution $\Delta n/\omega$ to the residue integral from the contour Γ goes faster than $1/\omega$. Hence, Cauchy's residue theorem can be written as

$$\Delta n(\omega_1) = \frac{1}{i\pi} P \int_{-\infty}^{\infty} d\omega\, \frac{\Delta n(\omega)}{\omega - \omega_1} \qquad (A\text{-}11)$$

for a point ω_1 on the boundary of analyticity of $\Delta u(\omega)$.

Splitting (A-11) into its real and imaginary parts we obtain the Hilbert transforms [*Titchmarsh*, 1948],

$$\mathrm{Re}\, \Delta n(\omega_1) = \frac{1}{\pi} P \int_{-\infty}^{\infty} d\omega\, \frac{\mathrm{Im}\, \Delta n(\omega)}{\omega - \omega_1}$$
$$\qquad (A\text{-}12)$$
$$\mathrm{Im}\, \Delta n(\omega_1) = -\frac{1}{\pi} P \int_{-\infty}^{\infty} d\omega\, \frac{\mathrm{Re}\, \Delta n(\omega)}{\omega - \omega_1}$$

Thus, the presence of absorption, $\mathrm{Im}\, \Delta u(\omega_1)$, is both a necessary and sufficient condition for the presence of dispersion, $\mathrm{Re}\, \Delta u(\omega_1)$.

With the result $P \int_{-\infty}^{\infty} d\omega/(\omega - \omega_1) = 0$, the first of the transforms (A-12) becomes

$$\mathrm{Re}\, [n(\omega_1) - n(\infty)]$$
$$= \frac{1}{\pi} P \int_{-\infty}^{\infty} d\omega\, \frac{\mathrm{Im}\, n(\omega)}{\omega - \omega_1} \qquad (A\text{-}13)$$

Therefore,

$$\mathrm{Re}\, [n(0) - n(\infty)]$$
$$= \frac{1}{\pi} P \int_{-\infty}^{\infty} d\omega\, \frac{\mathrm{Im}\, n(\omega)}{\omega} \qquad (A\text{-}14)$$

Subtracting these two equations, we find

$$\mathrm{Re}\, [n(\omega_1) - n(0)]$$
$$= \frac{\omega_1}{\pi} P \int_{-\infty}^{\infty} d\omega\, \frac{\mathrm{Im}\, n(\omega)}{\omega(\omega - \omega_1)} \qquad (A\text{-}15)$$

The interval of integration can be reduced to the physical range of positive frequencies by introducing the crossing symmetry relation (13), $\mathrm{Im}\, n(\omega_1) = -\mathrm{Im}\, n(-\omega_1)$, so that

$$\mathrm{Re}\, [n(\omega_1) - n(0)]$$
$$= \frac{2\omega_1^2}{\pi} P \int_0^{\infty} d\omega\, \frac{\mathrm{Im}\, n(\omega)}{\omega(\omega^2 - \omega_1^2)} \qquad (A\text{-}16)$$

which is equation 10. Since the low-frequency behavior of $n(\omega_1)$ is known experimentally, $n(0) = 1$, this last form is particularly convenient for comparison with experiment.

We collect here (A-13) and (A-14), after application of the crossing symmetry relations,

$$\mathrm{Re}\, [n(\omega_1) - n(\infty)]$$
$$= \frac{2}{\pi} P \int_0^{\infty} d\omega\, \frac{\mathrm{Im}\, n(\omega)\omega}{\omega^2 - \omega_1^2} \qquad (A\text{-}13')$$
$$\mathrm{Re}\, [n(0) - n(\infty)]$$
$$= \frac{2}{\pi} P \int_0^{\infty} d\omega\, \frac{\mathrm{Im}\, n(\omega)}{\omega} \qquad (A\text{-}14')$$

In a similar way we can deduce expressions for the inverse transform,

$$\mathrm{Im}\, n(\omega_1) = -\frac{1}{\pi} P \int_{-\infty}^{\infty} d\omega\, \frac{\mathrm{Re}\, n(\omega)}{\omega - \omega_1} \qquad (A\text{-}17)$$
$$= -\frac{2\omega_1}{\pi} P \int_0^{\infty} d\omega\, \frac{\mathrm{Re}\, n(\omega)}{\omega^2 - \omega_1^2} \qquad (A\text{-}18)$$

where these results depend on the assumption that n has no pole with imaginary coefficient on the real axis. The result (A-17) can easily be modified in this case, and no change is necessary in the forms deduced from (A-13).

These relations may also provide general information about n from a study of their limiting values at particular frequencies. For example, from (A-17), putting ω_1 equal to zero, we find

$$P \int_{-\infty}^{\infty} d\omega\, \mathrm{Re}\, n(\omega)/\omega = 0$$

which is a simple consequence of the crossing symmetry of $n(\omega)$. More interesting, however, is the equation

$$\lim_{\omega_1 \to 0} \frac{\mathrm{Im}\, n(\omega_1)}{\omega_1} = \frac{d}{d\omega_1}\, \mathrm{Im}\, n(\omega_1)\Big|_{\omega_1 = 0}$$
$$= -\frac{2}{\pi} P \int_0^{\infty} d\omega\, \frac{\mathrm{Re}\, n(\omega)}{\omega^2} \qquad (A\text{-}19)$$

It tells us that the slope of the absorptive part of the index of refraction at zero frequency determines a weighted sum, taken over the entire frequency range, of the dispersive part of the index. In our examples, the left-hand side of (A-19) takes on values of zero (A-1) or $\frac{1}{2}Q_0\omega_0$ (A-2, A-3). It follows, therefore, that the phase velocity in (A-1) will take on negative values for some frequencies, in agreement with (24). For these examples, which satisfy condition (a), we see that without a cutoff, $\omega_0 \neq 0$, the integral would be unbounded, in agreement with our earlier results. The low frequencies are given the greatest weight in this expression, which explains how our three examples of dispersion (D1, D2, D3) can be identical asymptotically but nevertheless give different evaluations to the sum (A-19).

(*3A*). The representation (A-8) for the amplitude and phase of the Fourier transform of the pulse enables us to extend the domain of definition of the propagation constant into the u.h.p. If in this expression we let $u(0, t) = \delta(t)$, $\hat{u}(\omega) = \pi/2$, and (A-8) reduces to

$$e^{iK(\omega)R} = \int_\tau^\infty dt \, e^{i\omega t} u(R, t) \qquad \text{(A-20)}$$

which is analytic in the u.h.p. of ω.

The substitution, $\omega \rightarrow -\omega^*$, gives

$$e^{iK(-\omega^*)R} = \int_\tau^\infty dt \, e^{-i\omega^* t} u(R, t)$$

Taking the complex conjugate of both sides of this equation, we find

$$e^{-iK^*(-\omega^*)R} = \int_\tau^\infty dt \, e^{i\omega t} u(R, t)$$

which is just our original expression. Hence

$$K(\omega) = -K^*(-\omega^*)$$

or

$$-K(-\omega) = K^*(\omega^*) \qquad \omega_2 > 0 \qquad \text{(A-21)}$$

The expression (A-8) can also be used to extend $\hat{u}(\omega)$ into the u.h.p. by applying it at $R = 0$. In a similar way, we obtain

$$\hat{u}(-\omega) = u^*(\omega^*) \qquad \omega_2 > 0 \qquad \text{(A-22)}$$

Both expressions reduce to the original symmetry conditions on the real frequency axis.

Acknowledgments. I should like to thank my colleagues Roland Herbst and Glenn Werth for many interesting and useful discussions, and Warren Heckrotte for a critical reading of the manuscript. David Brooks was very helpful with his program for Fourier synthesis. I should like also to thank Professor Knopoff for an interesting conversation about his paper.

References

Bode, H. W., *Network Analysis and Feedback Amplifier Design,* Van Nostrand, Princeton, N. J., 1945.

Brillouin, L., *Wave Propagation and Group Velocity,* Academic Press, New York, 1960.

Bullen, K. E., *An Introduction to the Theory of Seismology,* Cambridge University Press, 1953.

Collins, F., and C. C. Lee, Seismic wave attenuation characteristics from pulse experiments, *Geophysics, 21,* 16–40, 1956.

Ewing, W. M., W. S. Jardetzky, and F. Press, *Elastic Waves in Layered Media,* McGraw-Hill Book Co., New York, 1957.

Gross, B., *Theories of Viscoelasticity,* Hermann et Cie, Paris, 1953.

Guilleman, E. A., *Mathematics of Circuit Analysis,* John Wiley & Sons, New York, 1949.

Hunter, S. C., Viscoelastic waves, chapter 1 in *Progress in Solid Mechanics,* vol. 1, edited by R. Hill and I. N. Sneddon, Interscience Publishers, New York, 1960.

Janke, E., and F. Emde, *Tables of Functions,* 4th ed., Dover, New York, 1945.

Jauch, J. M., and F. Rohrlich, *The Theory of Photons and Electrons,* appendix 7, Addison-Wesley, Cambridge, Mass., 1955.

Jeffreys, H., *The Earth,* 4th ed., Cambridge University Press, 1959.

Knopoff, L., and G. J. F. MacDonald, Attenuation of small amplitude stress waves in solids, *Revs. Modern Phys., 30,* 1178–1192, 1958. This reference contains a review of the relevant experiments yielding the first power law data. See especially the work of *McDonal et al.* [1958] and *Collins and Lee* [1956].

Kolsky, H., *Stress Waves in Solids,* Clarendon Press, Oxford, 1953.

Kolsky, H., The propagation of stress pulses in viscoelastic solids, *Phil. Mag.,* (8) *1,* 693–710 (p. 704), 1956.

Kramers, H. A., La diffusion de la lumiere par les atomes, *Atti congr. intern. fisica, Como 2,* 545–557, 1927.

Kronig, R., On the theory of the dispersion of X-rays, *J. Opt. Soc. Am., 12,* 547–557, 1926.

Mason, W. P., *Physical Acoustics and Properties of Solids,* Van Nostrand, Princeton, N. J., 1958.

McDonal, F. J., F. A. Angona, R. L. Mills, R. L. Sengbush, R. G. Van Nostrand, and J. E. White, Attenuation of shear and compressional waves in Pierre shale, *Geophysics, 23,* 421–439, 1958.

Titchmarsh, E. C., *Introduction to the Theory of Fourier Integrals,* 2nd ed., Clarendon Press, Oxford, 1948.

Toll, J. S., The dispersion relation for light, unpublished thesis, Princeton University, Princeton, N. J., 1952.

Werth, G. C., R. F. Herbst, and D. Springer, Amplitudes of seismic arrivals from the M discontinuity, *J. Geophys. Research, 67,* 1587–1610, 1962.

Whittaker, E. T., and G. N. Watson, *Modern Analysis,* Cambridge University Press, 1927.

(Manuscript received June 27, 1962.)

Reprinted from the United Kingdom Atomic Energy Authority AWRE Report, 1966.

Absorption of Elastic Waves - An Operator
for a Constant Q Mechanism

E. W. Carpenter

Note: This paper has previously been available only as a
       xerographic copy. It has been retyped and corrected
       for inclusion in this volume.

                                            The Editors

SUMMARY

Theoretical analysis is given of absorption of elastic waves in the frequency range of seismological interest. The impulse response of a system satisfying the constant Q hypothesis is calculated and the results discussed.

1.   INTRODUCTION

Over the past few years much has been written about the absorption of elastic waves in the frequency range of seismological interest. That there is attenuation of elastic waves is easy to observe, as is the increase in attenuation with increasing frequency. What is difficult to measure with enough precision to support any theoretical analysis is the amount of attenuation. Bearing in mind the variable nature of geologic materials, this is not too surprising.

In any theoretical work on the amplitude of seismic signals, the effect of absorption must be taken into account. Since the experimental data are not adequate to define the appropriate parameters uniquely, the natural approach is to postulate an analytical model which satisfies the data.

The literature contains several relevant papers, those by Kolsky [1] (see also Hunter [2]) and Futterman [3] being particularly explicit, while a review article by Knopoff [4] is also very valuable. The essential point brought out in these papers is that absorption must be accompanied by dispersion [5]. In using the dispersion relations, it has proved convenient to derive an operator which can be used for convolution in the time domain. Although Kolsky derived essentially the same operator in his early paper, there does seem to be some value in repeating the calculation with appropriate comment on the applicability of the results to seismic problems.

2.   ONE DIMENSIONAL PROPAGATION WITH NO ATTENUATION

In addition to absorption of energy by non-elastic processes, there is, in general, attenuation during propagation because of geometrical spreading. For simplicity we shall therefore consider one dimensional propagation so that any attenuation is entirely due to non-elastic effects. The co-ordinate system is chosen such that x represents distance and t, time.

Suppose we generate at $x = 0$, $t = 0$ a pulse having Fourier components $\overline{A}(\omega)$ then the waveform $A(0,t)$ is given by

$$A(0,t) = \frac{1}{2\pi} \int_{-\infty}^{+\infty} \overline{A}(\omega) \exp(i\omega t)d\omega, \qquad (1)$$

where ω is the angular frequency.

If the pulse now propagates in one dimension with constant velocity U the wave form at distance x and time t can be written

$$B(x,t) = \frac{1}{2\pi} \int_{-\infty}^{+\infty} \overline{A}(\omega) \exp i\omega(t - x/U)d\omega. \tag{2}$$

If we now refer time to a new origin moving with velocity V and call the new time t^1, then

$$B(x,t^1) = \frac{1}{2\pi} \int_{-\infty}^{+\infty} \overline{A}(\omega) \exp i\omega(t^1 - x/U + x/V)d\omega \tag{3}$$

Comparing (1) and (3) we note that if $V = U$, $B(x,t^1) = A(0,t)$; which is obvious from physical considerations.

3. OBSERVATION AND THE CONSTANT Q POSTULATE

In practice it is found that absorption of energy takes place during propagation and we may write quite generally that

$$|\overline{B}(x,\omega)| = |\overline{A}(0,\omega)| \exp (- \alpha\omega x). \tag{4}$$

Experience shows that high frequencies are preferentially absorbed and to the degree of experimental accuracy obtainable the absorption coefficient α is usually found to be independent of frequency over quite wide frequency bands.

If we <u>postulate</u> that as an identity α = constant, then certain complications arise. Both Futterman [3] and Kolsky [1] show that if we require a theory which is both linear (i.e., one which obeys the principle of superposition, and has the attendant mathematical advantages) and obeys the principle of causality, then:

(a) There must be a low frequency cut-off below which the absorption coefficient is not constant but decreases.

(b) There must be a phase shift, i.e., dispersion <u>must</u> occur.

The first condition is relatively unimportant. Since all observations refer to some finite bandwidth we simply take the low frequency cut-off, ω_0, well below the lowest frequency of interest, so that the fact that there is this finite cut-off frequency becomes of purely academic importance.

The second condition, that there is necessarily some phase shift, is of very considerable importance. From Section 2, we had, for no absorption

$$\overline{B} (x,\omega) = \overline{A}(0,\omega) \exp - (i\omega x/U), \tag{5}$$

where U is a constant. Now with absorption we must write

$$\overline{B} (x,\omega) = \overline{A}(0,\omega) \exp (i\omega Kx), \tag{6}$$

where $K = i\alpha - 1/C(\omega)$, in which α is, by definition, constant for $\omega>\omega_0$ and $C(\omega)$, the phase velocity at (angular) frequency ω, is a function of ω.

Although the variation of C with ω is probably too small to measure by direct means, the fact that there is a variation has fundamental implications to the subsequent analysis. Essentially it derives from the causality condition and is analogous to the similar theorem in electronic circuit theory whereby a filter cannot give an output before it receives an input. Thus, as is shown in many texts (e.g., Mason and Zimmermann [6]), the real and imaginary parts of the filter transfer function (corresponding to α and C respectively) are intimately related via the Hilbert transform.

Futterman's analysis is analytical and makes no appeal to physical processes, whereas Kolsky's analysis is based on the concept of linear elasto-viscosity [2]. Their results are for practical purposes identical, although in the following analysis Futterman's notation will be used.

We first define Q_o by the relation

$$\alpha = 1/2Q_o C_o,$$ (7)

where C_o, a constant, is the velocity of the very low frequency waves ($\omega < \omega_o$) which suffer no absorption. Since α and C_o are both constants, Q_o is a constant also. Then the phase velocity, $C(\omega)$, of the wave having frequency ω is given by

$$C = C_o \left[1 - \frac{\ln(\gamma\omega/\omega_o)}{\pi Q_o} \right]^{-1},$$ (8)

where γ is Euler's constant.

This expression is derived by Futterman for a particular model of the behaviour of α for $\omega < \omega_o$. The details need not concern us; it is sufficient to accept that C varies but slowly with ω according to the logarithmic law.

Note that for a given frequency, ω, the conventional definition of Q relates the absorption coefficient α (which is defined as being constant in the frequency range of interest) to the phase velocity C (which is a function of ω) by the equation

$$Q = 1/2\alpha C,$$ (9)

so that to say that we have a constant Q model is strictly incorrect. However to insist on the difference between constant Q and constant Q_o would be pedantic and henceforth the subscript is omitted.

From equations (6), (7) and (8) we can therefore write

$$B(x,t) = \frac{1}{2\pi} \int_{-\infty}^{+\infty} \overline{A}(0,\omega) \exp(-\omega x/2QC) \exp i\omega \left[t - \frac{x}{C_o} \left\{ 1 - \frac{\ln(\gamma\omega/\omega_o)}{\pi Q} \right\} \right] d\omega.$$ (10)

If instead of going to the frequency domain as in equation (10) we use convolution in the time domain, we can write

$$B(x,t) = \int_{-\infty}^{+\infty} A(0,\tau) I(x, t - \tau) d\tau,$$ (11)

where $I(x,t)$ is derived from equation (10) by putting $A(0,\omega) = 1$, i.e., $I(x,t)$ is the response to unit impulse, $\delta(t)$, applied at $x = 0$. Our aim now is to evaluate the impulse response.

4.   THE NORMALIZED IMPULSE RESPONSE

Consider the impulse response derived in the previous section

$$I(x,t) = \frac{1}{2\pi} \int_{-\infty}^{+\infty} \exp(-\omega x/2QC_0) \, \exp i\omega \left[t - \frac{x}{C_0} \left\{ 1 - \frac{\ln(\gamma\omega/\omega_0)}{\pi Q} \right\} \right] d\omega. \tag{12}$$

Since the function is a real function of space and time $K(\omega) = K^*(-\omega)$ and we can write

$$I(x,t) = \frac{1}{\pi} \int_{0}^{+\infty} \exp(-\omega x/2QC_0) \, \cos \omega \left[t - \frac{x}{C_0} \left\{ 1 - \frac{\ln(\gamma\omega/\omega_0)}{\pi Q} \right\} \right] d\omega. \tag{13}$$

As in Section 2 it would now be convenient to refer $I(x,t)$ to a moving frame of reference. Examination of equation (13) shows that choosing a frame of reference moving with constant velocity C_0 is one possibility. Kolsky [1] suggests using a frame of reference which travels with a velocity

$$V = C_0 \left[1 + \frac{\ln(x/x_0)}{\pi Q} \right], \tag{14}$$

where x_0 is an arbitrary constant. This has a very important result that it introduces similarity (see later) but the choice of x_0 has no physical basis.

The question which then arises is, can the advantages of similarity be preserved with a more physically satisfying choice of velocity?

In any particular case where the integral in equation (13) is evaluated numerically, some limit will necessarily be set on accuracy. For instance, we may choose to work to an accuracy of 1 part in 10^6. Then to this degree of approximation there will be for each case some value of frequency, ω^1 say, above which all contributions to the integral will be negligible, e.g., $\int_{\omega^1}^{\infty}$, of equation (13) is less than 10^{-6}. Examination of equation (13) shows that it is more convenient to work in terms of the dimensionless parameter

$$D = \frac{\omega^1 x}{QC_0}. \tag{15}$$

The accuracy of the solution is then directly related to D, a value of 10π being adequate for most calculations.

The velocity to be used as the velocity of the frame of reference should be related to ω^1. Neither the phase velocity

$$C(\omega^1) = C_0 \left[1 - \frac{\ln(\gamma\omega^1/\omega_0)}{Q} \right]^{-1} \tag{16}$$

nor the group velocity

$$U(\omega^1) = C_0 \left[1 - \frac{1}{\pi Q} \left\{ 1 + \ln(\gamma\omega^1/\omega_0) \right\} \right]^{-1} \tag{17}$$

are appropriate, the signal commencing before the origin in both cases. If, however, we heuristically define an average "signal velocity" $S(\omega^1)$ from the equation

$$\frac{X}{S(\omega^1)} = \int_0^X \frac{dx}{U(\omega^1)} , \tag{18}$$

where X is the travel distance, then it can readily be shown that

$$S(\omega^1) = C_0 \left[1 - \frac{1}{\pi Q}\left\{ 2 + \ln(\gamma\omega^1/\omega_0) \right\} \right]^{-1} \tag{19}$$

It is interesting to note that if $U(\omega^1)$ and $C(\omega^1)$ are substituted in equation (18) in place of $S(\omega^1)$ and $U(\omega^1)$ respectively then the solution for $U(\omega^1)$ is in fact equation (17), which can equally be derived from the more familiar equation

$$U(\omega) = C - \lambda dC/d\lambda. \tag{20}$$

This velocity, $S(\omega^1)$, is then the pulse velocity, and is a function of distance. Note that the pulse velocity is to some extent arbitrary because it depends on the highest frequency resolvable, but the resolution changes so rapidly with frequency while the velocity changes so slowly with frequency that no real problem arises.

Referred to an origin moving with velocity $S(\omega^1)$ the impulse response can be written

$$I_0(t^1) = \frac{1}{\pi} \int_0^{\omega^1} \exp\left(- \omega x/2QC_0\right) \cos \omega\left[t^1 + \frac{x}{\pi QC_0}\left\{ \ln(\omega/\omega^1) - 2 \right\} \right] d\omega, \tag{21}$$

where $\omega^1 = (QC_0D)/x$.

This equation is a function of $\omega x/QC_0$ only and further simplification is therefore possible.

5.   THE SIMILARITY SOLUTION, E(g)

The principle of similarity asserts that if $G_1(m\omega) = G_2(\omega)$ and $f(t)$ is the time transform of G_1, then the time transform of G_2 is

$$\frac{1}{m}f(t/m). \tag{22}$$

Applying this principle to equation (21) we can write, ignoring the superscript,

$$I_0(t) = \frac{QC_0}{x} E(g)$$

where

$$E(g) = \frac{1}{\pi} \int_0^D \exp(-h/2) \cos h \left[g + \frac{1}{\pi} \left\{ \ln(h/D) - 2 \right\} \right] dh \; . \tag{23}$$

g is the dimensionless time

$$g = t \div (x/QC_o)$$

and h is the dimensionless (angular) frequency, $h = \omega(x/QC_o)$.

This then is the impulse response which is to be calculated.

Note that to a good approximation x/C_o can be written as T the travel time. This is particularly useful when considering cases where Q and C_o vary with position for then we can replace, $\int \frac{dx}{QC_o}$ along the propagation path by T/Q_{av} where T is the travel time and Q_{av} an effective "average" value for the whole path travelled. Thus in practice we estimate T/Q and use the function $(Q/T)E(T/Q)$ as the convolution operator which allows for absorption.

6.   EVALUATION OF THE FUNCTION E(g)

The function

$$E(g) = \frac{1}{\pi} \int_0^D \exp(-h/2) \cos h \left[g + \left\{ \frac{\ln(h/D) - 2}{\pi} \right\} \right] dh \tag{24}$$

was evaluated by standard Fourier series techniques, replacing the integral by a summation.

Two parameters are required, the maximum and minimum (excluding zero) frequencies. The maximum, or Nyquist, frequency FNYQ cycles is selected directly and gives the maximum value of h according to the equation

$$HMAX = 2\pi FNYQ \equiv D.$$

This parameter defines the sampling interval in the g ($\equiv$time) domain, i.e.,

$$DELG = 1/(2 \cdot FNYQ).$$

The minimum frequency is defined indirectly as the reciprocal of the signal duration which is, in turn, derived as a number,

$$GMAX = Z \cdot DELG$$

$$FMIN = 1/GMAX = 1/(Z \cdot DELG)$$

and $HMIN = 2\pi \cdot FMIN = 2\pi/(Z \cdot DELG) = 4\pi \cdot FNYQ/Z.$

The integral is now replaced by a summation using increments of h, DELH, equal to the minimum frequency.

Thus, E'(R·DELG) =

$$\frac{DELH}{\pi}\left[0.5 + \sum_{n=1}^{n=z/2} \exp(-n\cdot DELH/2) \cos n\cdot DELH\left[R\cdot DELG + \frac{1}{\pi}\left\{\ln\frac{(n\cdot DELH)}{HMAX} -2\right\}\right]\right] \qquad (25)$$

where R takes integral values from 0 to Z, where the prime is used to denote a calculated value as opposed to an absolute value.

The first term inside the brackets, 0.5, takes account of the zero frequency component which is unaltered during the pulse transmission. This ensures that

$$\int_{R=0}^{Z} E'(R\cdot DELG)\cdot DELG = 1,$$

i.e., the area under the E'(g) curve is always unity, as required by the condition that very low frequencies are unattenuated. By using the series expansion we have also forced E'(0) = E'(GMAX). In general, the series expansion is an approximation to the integral, hence the use of the superscript to indicate an approximate evaluation.

7. RESULTS

The expression for E'(g) was evaluated for several combinations of the parameters FNYQ and Z. The results for FNYQ = 5, Z = 1000 are given in Table 1 and plotted in Figure 1 together with the curve for FNYQ = 2, Z = 80. Note that E'(g) is positive for all values of g. If, therefore, one imagines convolving E'(g) with any input waveform, it is clear that the model cannot provide a mechanism for the inversion of first motion. Despite the fact that phase changes of π are produced, the changes are virtually linearly dependent upon frequency (from equation (21)) and a time shift rather than inversion occurs.

Inaccuracies in E'(g) (i.e., differences from E(g)) arise in three ways. Firstly, there is an error due to replacing an integral over frequency from zero to infinity by an integral from zero to some finite maximum, FNYQ. Secondly, there are inaccuracies in evaluation of the integral by approximating it with a series of finite length. The third is the inaccuracy due to the requirement that the integral over the duration of the transient should be unity. The first type of error, which must be less than exp (- πFNYQ), can be reduced by increasing FNYQ, the second and third by increasing the duration of the transient, i.e., Z. In general, the third type of error is numerically the most significant, but appears simply as a base line shift. This effect is well demonstrated in Figure 2, where E'(g) is plotted on a logarithmic scale for different values of Z.

The significance of the second source of inaccuracy can be evaluated by examining how first differences vary with Z, thus removing the larger base line shift. For FNYQ = 5, Z = 1000 the errors are of the order of 10^{-7}, which are comparable with the first type of inaccuracy for FNYQ = 5. If, therefore, the value of E'(0) is subtracted from the E'(g) values given in Table 1, the results represent the values of E(g) to an accuracy of at least the last digit.

The problem of defining an arrival time is well illustrated in Figures 1 and 2. The plotting accuracy in Figure 1 is about 10^{-3}, and to this accuracy FNYQ = 2, Z = 80 is adequate. The time origin is chosen as the arrival of FNYQ travelling with the signal velocity as defined by equation (19), the corresponding group and phase arrival times being $1/\pi$ and $2/\pi$ units respectively. For FNYQ = 5 the origin is referred to the signal velocity for FNYQ = 5. In Figure 1 the plotting accuracy is not sufficient to show the early high frequency arrivals, but Figure 2 demonstrates how the apparent onset gets earlier with increasing resolution.

8.   CONCLUSIONS

The main purpose of this report was to calculate the impulse response of a system satisfying the constant Q hypothesis. There seems to be no practical justification for regarding the models of Futterman and Kolsky as different, and although Futterman's analysis has been more widely quoted, Kolsky's work contains several valuable features which can be applied directly to Futterman's model. In particular the principle of similarity means that a single operator in the time domain can be scaled for use with any values of the independent variable. In a homogeneous medium this variable is the distance divided by $C_o Q$, but for a inhomogeneous model the variable can be replaced by the parameter

$\int (C_o Q)^{-1} dx$, where the integration is taken over the whole path. Although for

the model the velocities are a continuous function of frequency, the changes are so slight that for practical purposes the parameter can be written as T/Q_{av}, where

T is the travel time. Nevertheless the change of velocity with frequency, though small, does present some interesting problems in rigorously defining the arrival time. The arrival time is a function of resolution, and for practical applications it corresponds to the "signal" velocity of a frequency of about (2Q/T) cycles per second. Perhaps the most important observation is that, since the operator is positive for all time, there can be no reversal of first motion.

The operator as calculated is suitable for taking account of absorption by convolution in the time domain. In many cases it is more convenient to work in the frequency domain throughout, but even then a knowledge of the impulse response of the absorptive part of the system will help assess its importance. In particular the above nomenclature may assist in avoiding that irritating feature of numerical calculation, the precursor.

APPENDIX A

Although as was stated in the text, pulse experiments cannot be expected to show the dispersion associated with absorption, there does appear to be possibility of an observable effect. Suppose the model is physically realized. The percentage difference in velocity between two periods T_1 and T_2 is approximately $30 \cdot Q^{-1} \ln(T_2/T_1)$. Now in current geophysical techniques applied to the earth, T varies between 1 s for body waves and 1 h for the earth's fundamental oscillations, giving the $\ln(T_2/T_1)$ term value of up to 10. In the "low velocity" or "low Q" layer beneath the Moho, Q may be as low as 40 for shear waves and 100 for P waves. Therefore, we might expect that, since velocity decreases with period, the low velocity layer becomes more pronounced for S waves than for P waves, and more pronounced for the longer periods. It is certainly true that over the past few years, as longer period waves have been observed, the evidence of the low velocity layer has increased substantially. Although the low velocity (or Q) layer is of relatively small thickness, the accuracy of 1 part in 10^4 or 10^5 with which fundamental oscillations are observed does raise the possibility that changes of velocity with period could be observed. In essence it means that we would need a velocity depth structure which is frequency dependent.

REFERENCES

1. H. Kolsky: (1956) "The Propagation of Stress Pulses in Viscoelastic Solids", Phil. Mag., 8, 693-710.

2. S.C. Hunter: (1960) "Viscoelastic Waves, Chap. 1: Progress in Solid Mechanics", Interscience Publishers, New York.

3. W.I. Futterman ,(1962) "Dispersive Body Waves", J. Geophys. Res., 67, 13, 5279-5291.

4. L. Knopoff: (1964) Quarterly Review of Geophysics, 2, 4, 625-660.

5. P.N.S. O'Brien: (1961) "A Discussion of the Nature and Magnitude of Elastic Absorption in Seismic Prospecting", Geophysical Prospecting, 9, 2, 261-275.

6. S.J. Mason and H.J. Zimmermann: (1960) "Electric Circuits, Signals and Systems", Wiley.

TABLE 1
E (g) for FNYQ = 5, Z = 1000

g	0	0.1	0.2	0.3	0.4
E'(g)	0.00005	0.00005	0.00006	0.00016	0.00158
g	0.5	0.6	0.7	0.8	0.9
E'(g)	0.01060	0.04149	0.10813	0.20906	0.32507
g	1.0	1.1	1.2	1.3	1.4
E'(g)	0.43128	0.50986	0.55443	0.56760	0.55647
g	1.5	1.6	1.7	1.8	1.9
E'(g)	0.52902	0.49219	0.45116	0.40948	0.36930
g	2.0	2.1	2.2	2.3	2.4
E'(g)	0.33186	0.29769	0.26695	0.23954	0.21524
g	2.5	2.6	2.7	2.8	2.9
E'(g)	0.19376	0.17481	0.15810	0.14334	0.13031
g	3.0	3.1	3.2	3.3	3.4
E'(g)	0.11876	0.10853	0.09943	0.09132	0.08408
g	3.5	3.6	3.7	3.8	3.9
E'(g)	0.07759	0.07177	0.06653	0.06180	0.05753
g	4.0	4.2	4.4	4.6	4.8
E'(g)	0.05366	0.04693	0.04153	0.03664	0.03266
g	5.0	5.2	5.4	5.6	5.8
E'(g)	0.02927	0.02636	0.02385	0.02167	0.01977
g	6.0	6.2	6.4	6.6	6.8
E'(g)	0.01810	0.01663	0.01532	0.01416	0.01312
g	7.0	7.2	7.4	7.6	7.8
E'(g)	0.01219	0.01136	0.01060	0.00992	0.00930
g	8.0	8.5	9.0	9.5	10.0
E'(g)	0.00873	0.00752	0.00654	0.00574	0.00508
g	10.5	11.0	11.5	12.0	12.5
E'(g)	0.00453	0.00406	0.00365	0.00331	0.00301
g	13.0	13.5	14.0	14.5	15.0
E'(g)	0.00275	0.00253	0.00233	0.00215	0.00199
g	16.0	17.0	18.0	19.0	20.0
E'(g)	0.00173	0.00151	0.00133	0.00119	0.00106
g	22.0	24.0	26.0	28.0	30.0
E'(g)	0.00087	0.00072	0.00061	0.00053	0.00046
g	32.0	34.0	36.0	38.0	40.0
E'(g)	0.00040	0.00036	0.00032	0.00029	0.00026
g	50.0	60.0	70.0	80.0	90.0
E'(g)	0.00017	0.00012	0.00010	0.00008	0.00006

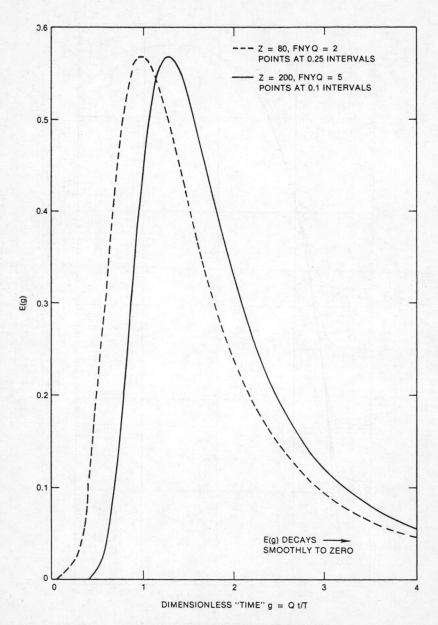

FIGURE 1. THE FUNCTION E(g) TO g = 4

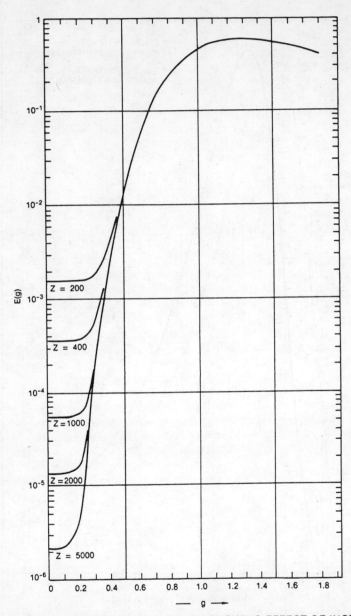

FIGURE 2. LOGARITHMIC PLOT OF E(g) SHOWING EFFECT OF INCREASING Z

Reprinted from Geophysics, v. 35, p. 387–403.

A PREDICTED PEDESTAL EFFECT FOR PULSE PROPAGATION IN CONSTANT-Q SOLIDS†

E. STRICK*

Under the assumption of almost constant-Q behavior of solids over a wide range of frequencies, together with some meaningful assumptions about the linear frequency behavior of the attenuation function, we have been able to obtain theoretical expressions describing the waveform distortion of an impulse-excited plane wave as it decays and spreads on passing through large distances of the solid. When typical values for the parameters (as obtained from laboratory model or short range field experiments) for a solid having a mechanical Q of about 50 are used, the resulting waveforms at first glance appear to have a simple and not unexpected behavior. The peak amplitude of the waveform in the time domain varies roughly with the inverse of the square of the travel distance (this includes an inverse first power due to geometrical spreading). Also, a spreading of the waveform occurs that varies roughly linearly with the travel distance. This spreading is such that a positive impulse simply broadens as it travels without developing any zero crossings in its wave shape. However, it turns out that the part of the waveform leading up to the visible onset does not admit of a purely elastic interpretation, so that one cannot relate conceptually the arrival time with a purely elastic-wave velocity. For our solid of $Q \approx 50$, about one-seventh of the arrival time duration is due to a purely inelastic behavior; during this period of inelastic behavior the amplitude is not zero but it is deceptively small. We have designated this portion as a "pedestal." On observing seismic records, we are never aware of this conceptual division and consequently attribute to the observed arrival an elastic wave velocity, which in our example happens to be about 15 percent less than the actual elastic wave velocity of the solid. Although the existence of the pedestal may not appear to be significant when we restrict our observations to seismic records limited in bandwidth to the low frequencies encountered in exploration or earthquake seismology, an upward curvature which is not associated with a discontinuity in the slope at the time of visual onset will remain and can be of great importance if very accurate arrival time measurements are made. This part of the pedestal can become even more important when we consider refraction arrivals, where the onset undergoes an additional time integration that further enhances the upward curvature of the pedestal.

INTRODUCTION

It is basic in seismology to be able to predict the behavior of a solid to an impulsive mechanical load. In particular, it is important to be able to predict the behavior of rock at the frequencies normally encountered in exploration and earthquake seismology. Because of the enormous expense required, very little controlled field experimentation of the type necessary to obtain this kind of knowledge at low frequencies has been

† Manuscript received by the Editor August 26, 1969; revised manuscript received February 2, 1970.
* University of Pittsburgh, Pittsburgh, Pennsylvania 15213.

published since the classic work of McDonal et al (1958) in the Pierre shale. Instead, investigators have invariably resorted to laboratory models, i.e. cores or thin sheet models of the rock. However, in such laboratory tests, all frequencies in the excitation pulse are well above those encountered in exploration or earthquake seismology. For our needs the most useful laboratory model data are those that came from a plexiglas sheet used as the model for sedimentary rock. In an extremely well planned and executed laboratory experiment, Wuenschel (1965) showed that a scaling factor close to 300 enabled him to model exploration seismology with plexiglas ($Q \approx 50$); the Q of sedimentary rock ranges from about 10 to about 100. Pierre shale with a Q of close to 30 lies near the midpoint of this range. Using a theoretical treatment of dispersive body waves developed by Futterman (1962), Wuenschel obtained a parametric fit which enabled him to predict with great accuracy the waveforms at large model distances from waveforms recorded at short distances. For our needs, however, his approach must be altered in one important detail: Wuenschel's final waveforms include the actual source excitation, which is frequency band-limited. A band-limited signal restricts our insight as to how the solid is really dispersing the waveform. In particular, we would like to obtain the response of the solid to an impulse excitation. There are two ways to obtain the desired impulse response. One way is to Fourier analyze the source excitation waveform, to divide the transform of the source waveform into the Fourier transforms of the response spectra, and finally to transform the quotients back to the time domain. The second method follows more closely that used by McDonal et al and Wuenschel. If we Fourier analyze the traveling waveforms

$$A_n(\omega) \exp\left[j\omega t - j\Phi_n(\omega)\right]$$

at selected points R_n in space, we can identify

$$\ln\left[A_n(\omega)R_n\right] = + a_0(\omega) - \alpha(\omega)R_n$$

and

$$\Phi_n(\omega) = -\theta_0(\omega) + \theta(\omega)R_n,$$

where the source excitation function

$$\exp\left[a_0(\omega) + j\theta_0(\omega)\right]$$

is, of course, independent of R_n. The impulse response is contained only in the source-free part of the transform,

$$\exp\left[j\omega t - \alpha(\omega)R_n - j\theta(\omega)R_n\right].$$

Thus if we use *only* the slopes, i.e. $\alpha(\omega)$ and $\theta(\omega)$, as obtained from the straight line plots of $\ln\left[A_n(\omega)R_n\right]$ and $\Phi_n(\omega)$ versus R_n in the above transform of the traveling wave, we obtain our desired impulse response.

In addition to modifying the procedure of McDonal and Wuenschel so as to get just the impulse response, we have also been able to obtain an analytical expression for the spreading and attenuation of the impulse response in the time domain. By working in the Laplace rather than the Fourier domain, we have been able to accomplish this for both a generalization of Futterman's truncated linear frequency approximation and our power-law approximation (Strick, 1967). Both of these parametric formulations are consistent with the essentially constant-Q condition discussed by Kolsky (1960) and Knopoff (1964). Although our power-law approximation to the attenuation,

$$\alpha(\omega) = k\,|\,\omega\,|^s, \tag{1a}$$

involves two inelastic parameters, s and k, rather than the single parameter b in Futterman's truncated linear frequency form,

$$\alpha(\omega) = \begin{cases} b\,|\,\omega\,| & \text{for } \omega_0 < \omega < \omega_H \\ 0 & \text{otherwise,} \end{cases} \tag{1b}$$

the added complexity is only apparent, since equation (1b) contains ambiguities which can only be partially resolved through the assignment of additional parameters. In this paper we shall compute the dispersion of waveforms according to both formulations and shall show that an interesting "pedestal" effect is predicted by both. Because our power-law formulation has less inherent ambiguity, we shall study it first.

THE POWER-LAW, CONSTANT-Q (PL) SOLID

Before entering into the development of the expression for the response function, we should discuss the construction and validity of our constant-Q, power-law attenuation assumptions and make a minor change in the definition of one of the parameters. Because we shall refer to this formulation many times, we shall call such a medium a PL solid. In an earlier paper, Strick (1967), we demonstrated that the introduction of numerical values for the parameters, as obtained from two-dimensional model experiments with a plexiglas sheet, into our PL formulation enabled us to predict for rocks independently measured physical quantities, such as dynamic viscosity and transient creep, with unexpected high accuracy over frequency ranges and time durations much greater than those of seismological interest. This success has given us great confidence that our PL formulation is not only meaningful but also an accurate approximation to the real world. We shall assume in this paper that in the frequency range of seismological interest plexiglas has the mechanical behavior typical of sedimentary rock.

We summarize here the development of our PL model as given in our earlier paper. We consider a propagating harmonic plane wave

$$P(R, t) = \exp\left[pt - \gamma(p)R\right], \qquad (2)$$

where $p = j\omega = j2\pi f$ and $\gamma(p)$ is the complex propagation function

$$\gamma(p) = \alpha(\omega) + j\theta(\omega). \qquad (3)$$

If we invoke the Paley-Weiner causality condition (Papoulis, 1962, p. 215) to relate the phase lag function $\theta(\omega)$ to the attenuation function $\alpha(\omega)$, we can write

$$\theta(\omega) = \omega\tau + \hat{\alpha}(\omega), \qquad (4)$$

where the first term on the right is of all-pass nature (Papoulis, 1962, p. 97). The traveltime is τR, so that τ is a purely elastic parameter. The second term on the right, i.e. $\hat{\alpha}(\omega)$, is the Hilbert transform of $\alpha(\omega)$. Thus $\hat{\alpha}(\omega)$ is the minimum phase contribution to the propagation function and is purely inelastic in nature. The Paley-Weiner condition restricts the permissible choices of $\alpha(\omega)$ to those that increase slower than the first power of the frequency as the frequency becomes infinite. At first glance, this condition appears to place very little restriction on the choice of $\alpha(\omega)$, but upon invoking the constant-Q requirement, where the spatial Q is defined by

$$Q(\omega) = \left|\frac{\theta(\omega)}{2\alpha(\omega)}\right| = \frac{\omega\tau + \hat{\alpha}(\omega)}{2\alpha(\omega)}, \qquad (5)$$

we see that the spatial Q cannot be completely independent of ω unless both $\alpha(\omega)$ and $\hat{\alpha}(\omega)$ vary linearly with ω. But for $\hat{\alpha}(\omega)$ to vary precisely with ω would require that it have all-pass rather than a minimum-phase behavior, a requirement which would violate the Paley-Weiner condition. Although a precisely constant Q is not possible, we can obtain an essentially constant Q if the dependence of $\alpha(\omega)$ and $\hat{\alpha}(\omega)$ on ω is not greatly different from linear. The PL choice (Strick, 1967) for the complex propagation function $\gamma(p)$ is one way to satisfy the near-linear requirement. Thus, if we form

$$\gamma(p) = \tau p + K(s)p^s, \qquad (6)$$

it is easily seen that

$$\left.\begin{aligned} \alpha(\omega) &= K(s)\cos\left(\frac{s\pi}{2}\right)|\omega|^s \\ \text{and} & \\ \theta(\omega) &= \omega\tau + K(s)\sin\left(\frac{s\pi}{2}\right)|\omega|^s\operatorname{sgn}\omega \end{aligned}\right\} . \quad (7a, b)$$

Causality is satisfied as long as $s < 1$ (Guillemin, 1963, p. 556); for constant-Q solids, s must be rather close to 1. For compressional waves in plexiglas, s is 0.9227. Upon comparing relations (6) and (7) with those in our 1967 paper, where $\alpha(\omega)$ has the notation of equation (1a), i.e.,

$$\alpha(\omega) = k|\omega|^s = k_0|f|^s, \qquad (8)$$

we find that we must have

$$K(s) \equiv \frac{k}{\cos\left(\dfrac{s\pi}{2}\right)} = \frac{k_0}{(2\pi)^s\cos\left(\dfrac{s\pi}{2}\right)}. \qquad (9)$$

We have introduced the dependence of K on s to bring out the fact that causality is satisfied for any dependence of K on s, and it was only for reason of simplicity in determining k and s from a log-log plot of experimentally determined $\alpha(f)$ that we assumed k not to depend on s. We now see that equations (7) allow $\alpha(\omega)$ to become zero and $\theta(\omega)$ to be finite as s approaches 1, whereas the earlier formulation yielded a finite $\alpha(\omega)$ from equation (8) but an infinite $\theta(\omega)$ from its Hilbert-derived phase lag function. Hence, we shall use equations (7) in this paper, with the reservation that they can be changed still further as experi-

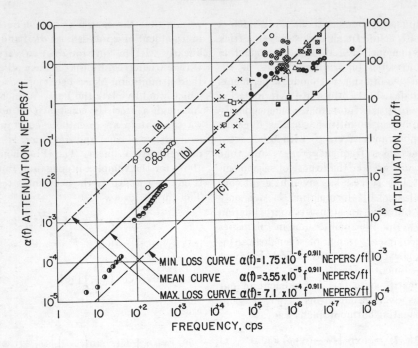

Fig. 1. Compilation of experimental values of attenuation as a function of frequency for compressional waves in sedimentary rock. (after Attewell and Ramana—1966)

mental data involving other PL-solid s-values become available.

Although we do not know much about the variation of s with different sedimentary rock, we note that Attewell and Ramana (1966) have summarized the data that exist for compressional waves in sedimentary rock. In Figure 1 we have reproduced their results after deleting the surface wave data. For the meaning of the various symbols, we should refer to Attewell and Ramana's paper. Though much of the data of Figure 1 are not very reliable, the average curves that indicate loss should be good, and it is this average behavior that is of interest to us. For the attenuation function $\alpha(f)$ expressed in the form of equation (8), involving the parameters k_0 and s, we compare Attewell and Ramana's results with those for plexiglas (see Strick, 1967) in Table 1. The difference in value for s between the mean value for sedimentary rock and our plexiglas value is not significant for problems of seismological interest. However a mean value of k_0 for sedimentary rocks nearly double that for plexiglas does drop the Q at 100 hz, as computed from equation (5), almost in half. (These Q values are based upon a common value of 120

$\times 10^{-6}$ sec/ft for the elastic parameter τ.) In addition to the mean value parameters, we have also listed extreme values for k_0 as crudely estimated from the dashed curves of Figure 1. We have not attempted to estimate similar bounds on the parameter s from the data of Figure 1. Since the mean value curve of Attewell and Ramana yields a Q at 100 hz which is too small to allow us to adopt their mean values of s and k_0 for our typical sedimentary rock, we have chosen to remain with the plexiglas parameters as those of our reference solid. The listed Q for plexiglas is less than the value of 50 stated earlier. The

Table 1. PL parameters for plexiglas and sedimentary rock as deduced from the compilation of Attewell and Ramana (1966)

	s	k_0 (fps units)	Q ($f = 100$ hz)
Plexiglas	0.9227	2.034×10^{-5}	40
Sed. Rock; Mean loss	0.911	3.55×10^{-5}	20
Sed. Rock; Max. loss	0.911	7.10×10^{-4}	4
Sed. Rock; Min. loss	0.911	1.75×10^{-6}	300

smaller value listed here has its origin in the fact that with $\alpha(f)$ from equation (8) inserted in equation (5) for the spatial Q, we obtain a slowly increasing function of frequency [i.e. to the power $(1-s)$], which adds about 10 to the Q when f is increased by a factor of 100 to reach the frequency range used in two-dimensional model work.

DISTORTION OF A PROPAGATING PULSE
BY A PL SOLID

We shall now investigate the behavior of an impulse-excited plane wave as it travels through large distances of a PL solid. We start by superposing the harmonic waveforms of equation (2) over all frequencies to obtain for a traveling pulse

$$P(R, t) = \frac{1}{R} \int_{-\infty}^{+\infty} d\omega \exp\left[j\omega t - \gamma(\omega)R\right]. \quad (10)$$

Note that in (10) we have introduced a geometrical spreading factor $1/R$ in order to obtain an expression that is valid for three-dimensional spherical symmetry. The complex propagation function $\gamma(\omega)$ must also correspond to propagation in three dimensions. The validity of this dimensional transition requires that R be large, but a large R is precisely what we shall require in order to carry out an integration of equation (10) by the method of steepest descent. In order to perform this integration, we find it convenient to transform equation (10) into the Laplace domain by the usual substitution $p = j\omega$. We have then to evaluate

$$P(R, t) = \frac{1}{2\pi jR} \int_{Br}$$
$$\cdot \exp\left[-Kp^s R + (t - \tau R)p\right] dp. \quad (11)$$

Br is a Bromwich type contour; for simplicity we have dropped the indication that K, defined by equation (9), can depend upon the parameter s. Also for notational convenience we introduce the new time variable

$$t' = t - \tau R, \quad (12)$$

so that t' measures the time from a new origin associated with the arrival of the infinite frequency component of the pulse front at the point R. We also introduce the intermediate functions $F(p)$ and $f(p)$ in order that our notation cor-

responds with that used in the derivation of the method of steepest descent in Appendix A of Ewing et al (1957). However, we shall continue to use p rather than the ζ of Ewing et al as the independent variable.

$$F(p) = \frac{1}{2\pi jR} \quad (13)$$

and

$$Rf(p) = -KRp^s + t'p. \quad (14)$$

R is, of course, a large positive number. Equation (11) becomes

$$P(R, t') = \int_{Br} F(p) \exp\left[Rf(p)\right] dp. \quad (15)$$

A single saddle point occurs at

$$p = p_0 \equiv \left[\frac{sKR}{t'}\right]^{1/(1-s)}, \quad (16)$$

where $Rf(p)$ and its second derivative have the values

$$Rf(p_0) = -\left(\frac{1-s}{s}\right)t'\left(\frac{sKR}{t'}\right)^{1/(1-s)} \quad (17)$$

and

$$\left| R\frac{d^2f}{dp^2}\right|_{p=p_0} = (1-s)t'\left[\frac{t'}{sKR}\right]^{1/(1-s)}. \quad (18)$$

Upon noting that the contour Br crosses the real axis at an angle $\chi = \pi/2$, we can insert these expressions into the general form

$$P(R, t') \doteq \frac{\sqrt{2\pi}F(p_0) \exp\left[j\chi + Rf(p_0)\right]}{\left[\left| R\frac{d^2f}{dp^2}\right|_{p=p_0}\right]^{1/2}} \quad (19)$$

and obtain

$P(R, t')$

$$\doteq \frac{\exp\left\{-\left(\frac{1-s}{s}\right)t'\left[\frac{sKR}{t'}\right]^{1/(1-s)}\right\}}{R\left\{2\pi(1-s)t'\left[\frac{t'}{sKR}\right]^{1/(1-s)}\right\}^{1/2}}. \quad (20)$$

Simple time differentiation shows that this response has a peak at

$$t'_{peak} = \left(\frac{1-s}{1-\frac{s}{2}} \right)^{(1-s)/s} [sKR]^{1/s}, \qquad (21)$$

which yields a response peak magnitude of

$$P(R, t'_{peak})$$

$$= \frac{\left(1 - \frac{s}{2}\right)^{(2-s)/2s} \exp\left(\frac{s-2}{2s}\right)}{\sqrt{2\pi}[s(1-s)K]^{1/s}R^{1+1/s}}. \qquad (22)$$

In order to simulate crudely the AGC behavior of a seismic amplifier, we normalize the peak pressure response of equation (20) to unity by dividing (22) into it. In this way we find

$$P_{Norm}(R, t')$$

$$= \left(\frac{1-s}{1-\frac{s}{2}} \right)^{(2-s)/2s} \left[\frac{(sKR)^{1/s}}{t'} \right]^{(2-s)/(2-2s)}$$

$$\qquad\qquad\qquad (23)$$

$$\cdot \exp\left\{ \frac{2-s}{2s} - \left(\frac{1-s}{s}\right) \right.$$

$$\left. \cdot t'\left[\frac{sKR}{t'}\right]^{1/(1-s)} \right\}.$$

Before entering into computations using this steepest descent approximation to $P(R,t')$, we need to obtain some estimate as to just how good the approximation really is. The usual procedure in the error evaluation of an asymptotic approximation is to evaluate the next term in the asymptotic expansion. Although it is easy to set up the analytic expression for the next term, evaluation involves numerical integration of rather involved integrands containing third as well as second derivatives of $Rf(p)$ in equation (14). Even then we will only know a bound on the error as determined from this second term. Furthermore, the computational effort to evaluate this second term is very much greater than that required to evaluate the first term response given by equation (23). For practical expediency, we have avoided such laborious evaluations and have used the following two arguments to support our contention that the first term, equation (23), of the asymptotic expansion for $P(R,t')$ is an extremely good approximation in the range of large distances R that we shall encounter in this paper.

First we note that, for the particular choice of the parameter $s=0.5$, equation (11) has the form of the Laplace inversion of the function $R^{-1}\exp[-KR\sqrt{p}]$, which is known to yield precisely

$$P(R, t')\big|_{s=1/2} = \frac{K \exp\left[-\frac{K^2R^2}{4t'} \right]}{2R\sqrt{\pi}(t')^{3/2}}.$$

But this is exactly the expression that we obtain from our first term, unnormalized, saddle point solution (equation 20) for $s=0.5$. If our first term approximation is exact for $s=0.5$, it should, for the smooth waveforms that we shall encounter, not be too bad an approximation for s between 0.5 and 1.

We should note that the case $s=0.5$ has been studied extensively by Lamb (1962). Although Lamb makes some waveform computations of the type we shall make in this paper, his response results from a source excitation function which has the form of a turned-on, decaying exponential and contains a decay parameter which is taken implicitly to depend inversely upon the square of the distance R and thus does not exhibit the behavior of a realistic source. Our response, on the other hand, is that due to a delta-function source excitation, a choice which avoids the need to specify any source parameters at this time.

The second argument that our first term saddle point solution is sufficiently accurate for our needs is made by taking a numerical Fourier analysis of our saddle point solution back into the frequency plane in order to see just how well we can recover our original complex propagation function. In Appendix A we have followed the transform route for the attenuation function, $\alpha(f)$, using the waveforms that will be plotted later as Figure 3. Since the results of Appendix A show that we do indeed recover the attenuation function to within a couple of decibels for all frequencies where it is necessary that we do so, we conclude that our first term saddle point approximation has the accuracy required to give us confidence in our solution.

We are now in a position to study the waveforms predicted at large distances from a source for a wave traveling in a PL solid. Using the parameters from Table 1 for plexiglas, we see that, according to equation (9), $s=0.9227$ and $k_o=2.034\times10^{-5}$ imply $K=3.11\times10^{-6}$. The elas-

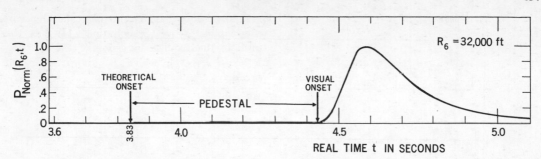

FIG. 2. Predicted normalized response at $R=R_6=32{,}000$ ft in a PL solid having the parameters of plate waves in plexiglas sheet. The parameters are $s=0.9227$, $k_0=2.034\times10^{-5}$ and $\tau=120\times10^{-6}$.

tic parameter τ is equal to 120×10^{-6} sec/ft for plexiglas and does not appear explicitly in our response functions except within the time variable t', a variable that is important when it is necessary to plot our waveforms in real time t with the aid of equation (12).

As Figure 2, using the above parameters for plexiglas, we show a real time plot of the normalized response, equation (23), predicted at $R=R_6=32{,}000$ ft. The most general observation is that the delta-function excitation pulse at $R=0$ has spread into a pulse having a width of roughly 0.2 sec without developing any additional zero crossings. The R_6 pulse appears to arrive at about 4.400 sec with an onset that does not seem to be associated with a discontinuity in the slope. Thus, even though our source excitation function contains all frequencies at constant amplitude, the PL solid selectively attenuates the high frequencies in such a manner that the pulse detected at a distance from the source does not have a well defined visual onset. However, we can make use of the preciseness of our computations from equation (23) to say more about the properties of the dispersed pulse. Referring to Figure 2, we note that during an initial portion of the response of duration $\tau R_6=3.830$ sec the amplitude of the pulse is, as a consequence of equation (12), precisely zero. Then at $t=\tau R_6$, which we associate with the theoretical onset, energy arrives that travels with the velocity of the projected infinite frequency component of the pulse. From the theoretical onset until a time $t=4.430$ sec, when the response is down 60 db from its peak value of unity, the response is finite but with amplitude more than 60 db down. This duration of about 0.600 sec, which deceptively appears to be part of the elastic arrival time, actually is part of the inelastic behavior.

We have chosen to call the response during the initial period a "pedestal." Generally, because of the small amplitude of the pedestal, scattering and other high frequency effects would frustrate any attempt to make a precise determination of the theoretical onset. The duration, but not the start, of the pedestal is arbitrarily set by our choice of the 60 db down point as the apparent (visual) onset; if we had, for example, chosen the 40 db down point to terminate the pedestal, the pedestal for a mean loss solid would have been lengthened by about 2.5 ms.

Although the lack of uniqueness of the duration of the pedestal results in its definition being vague, this in no way obscures the fact that the pedestal effect does exist.

At times after the termination of the pedestal, the broadened pulse has the form we would have expected from the computations of Lamb (1962) and Futterman (1962). The response monotonously increases until it reaches the peak at 4.594 sec. From the theoretical onset until the peak the response is controlled primarily by the exponential part of equation (23). Beyond the peak, the power-law factor in t' that multiplies the exponential is paramount.

Since our original aim was to be able to predict the variation of waveform distortion with increasing travel distance R, we computed waveforms (Figure 3) from equation (23) for 6 different values of $R=R_n$, including that for R_6 shown in Figure 2. Here we see that not only does the waveform spread roughly linearly with the travel distance R but that the pedestal duration seems to increase at essentially the same rate. Since we should expect our steepest descent approximation to become a better approximation with increasing R, we are strengthened in our conclusion that the pedestal effect is real and not just a

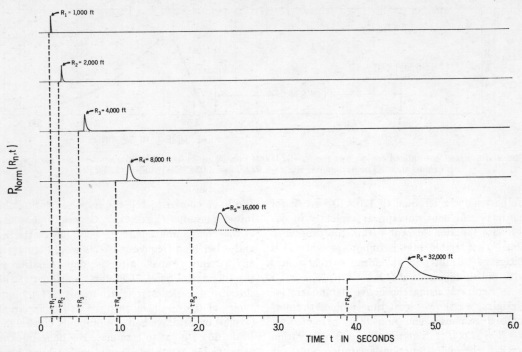

FIG. 3. Predicted normalized response for 6 different values of R after travelling through the PL solid of Figure 2.

manifestation of our asymptotic method of approximation.

At this point the reader may have a nagging suspicion that somewhere in our parameter evaluation we made a significant error in arriving at a value of the elastic parameter τ that is much too small and if we had arrived at say $\tau = 4.400/32,000 = 137.5 \times 10^{-6}$ sec/ft instead of 120×10^{-6} sec/ft, the pedestal would just disappear. However, a moment's reflection on this point will make it clear that the existence of the pedestal effect is a direct consequence of the assumptions of the PL model and its impulse response relation, equation (23). This latter expression is independent explicitly of the parameter τ; the only effect of changing the τ, according to equation (12), is to shift the inelastic part of the response (which equation (23) includes the pedestal) along the real time axis. Changing τ not only does not get rid of the pedestal but it leaves us with a completely incorrect set of real time values. The phase velocity, $v(\infty)$, at infinite frequency is, according to equation (7b) with $\theta(\omega) = \omega/v(\omega)$, just the reciprocal of τ. Assuming the larger value of τ would lead to a phase velocity at in-

finite frequency of 7270 ft/sec, which is well *below* the experimental phase velocity curve of Wuenschel (1965) that forms the basis of our parameter ·determination. To show this, in Figures 4a and 4b we reproduce Wuenschel's attenuation and phase velocity curves for plexiglas. We see that an extrapolation of the phase velocity curve in Figure 4b to infinite frequency could easily correspond to the 8330 ft/sec found from our deduced value for τ, but we would be hard put to envision just how this monotonously increasing curve could drop down to the 7270 ft/sec at infinite frequency that corresponds to a forced larger value for τ.

If this pedestal effect is real, why has it not been discussed by other investigators? One reason is that all computations, except that of Lamb (1962), have been based upon anelastic models that were subjected to straight numerical integration to get response waveforms and the region of the onset was not properly investigated. Furthermore, often the complex propagation function representing the anelastic behavior [such as the one used by Futterman (1962)] did not contain a true all-pass term which corre-

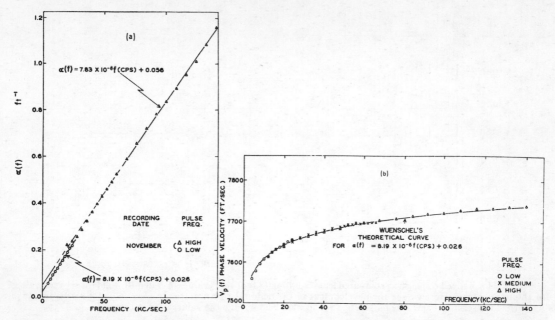

Fɪɢ. 4. (a) Wuenschel's attenuation data for plate waves in plexiglas sheet and his linear fittings to segments of this curve. (b) Wuenschel's corresponding dispersion curve after his removal of plate dispersion so that the curve will refer to a solid that is of infinite extent in three dimensions.

sponded to the arrival of the infinite frequency component. Lamb's work is of special interest to us not only because it is an exact solution and a special case of our PL response, equation (23) with $s = 0.5$, but because his waveform computations do not exhibit our pedestal effect. The resolution of this dilemma was most surprising. In Figure 5 we show a composite of response curves, according to equation (23), for 6 different values of s increasing from Lamb's value of $s = 0.5$ to $s = 0.9$. Here the parameter k_o as it appears in equation (8) was held constant at its plexiglas value in Table 1. Composite means that the response waveforms for all of the 6 values of R_n for each fixed value of s are displayed upon the same horizontal real time axis. For example, all of Figure 3 appears (ignoring a small difference in the value of s) in the single bottom trace of Figure 5. The lines extending from the pulses toward the left are the pedestals. It is quite evident from Figure 5 that, in agreement with Lamb's work, the top trace for $s = 0.5$ does not contain a significant pedestal as R increases from 1,000 ft to 32,000 ft. On the other hand, as s increases, the pedestal begins to appear and gradually takes on the duration at $s = 0.9$ that we

observed earlier. We should mention that as s approaches unity (a limit in which our causal solution is not valid), the duration of the pedestal will become infinite if we retain the assumption of our 1967 paper that the parameter k_o of equation (8) is independent of the parameter s. If, on the other hand, we require that K rather than k_o be independent of s, then the cosine $(s\pi/2)$ singularity in equation (9) does not yield an infinite value for K and, consequently, the phase lag $\theta(\omega)$ of equation (7b) remains finite. Although the assumption of an s-independent parameter K seems to be more realistic than the assumption of an s-independent parameter k_o, it is important to bear in mind that the parameter K can depend upon s in a nonsingular manner. Confirmation of a weak dependence of K on s can come only from further experimental work with other PL solids.

We have just seen that the pedestal effect becomes more pronounced as the parameter s approaches unity. A value of s near unity is more realistic than Lamb's value of 0.5. Consequently, we could clinch the argument for the existence of the pedestal if we were to demonstrate that it is the essentially linear behavior of the attenuation function rather than our particular choice of a

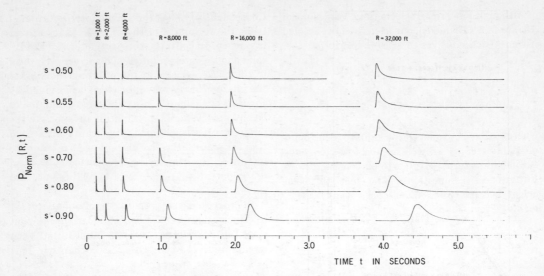

FIG. 5. Normalized response as computed from equation (23) at the 6 values of R in a PL solid for selected values of the parameter s. Here we see the development of the pedestal duration from a zero value for $s=0.5$ to one of more than 0.4 sec for $s=0.9$. The parameters k_0 and τ are held fixed to the values used in Figure 2.

power-law attenuation function with s nearly unity that is responsible for the pedestal. Since any essentially linear frequency attenuation function not identical to the one we chose is sufficient to prove this point, we shall take a deeper look into Futterman's (1962) truncated linear frequency model. Futterman assumes $\alpha(f)$ to vary precisely with the first power of the frequency but, as this assumption would violate causality if extended to infinite frequency, he truncates $\alpha(f)$ so that it is zero beyond some ill-defined high frequency that he somehow relates to the physical dimensions of the molecules in the solid. For our purposes, we need only require that the cutoff frequency have a value large enough to ensure that, at the 6 travel distances of computation, a noncausal precursor due to nonrealistic high frequency behavior is not confused with the pedestal. We shall redevelop Futterman's linear frequency model in a more general form than presented by Futterman in order that the ambiguities in some of the parameters of his formulation are easier to examine. We shall find it convenient to restate his results in terms of physically measured data rather than in terms of his low frequency cutoff (which we find conceptually and practically unnecessary). However, our effort will be directed not towards developing a response function from Futterman's model that is an alternative to our PL function

but rather toward supporting our PL model construction in its prediction of the pedestal effect.

THE TRUNCATED LINEAR-FREQUENCY ATTENUATION, CONSTANT-Q (TLF) SOLID

Futterman's model will be referred to in the following discussion as the TLF model. Futterman's terminology is based upon the complex index of refraction as used in electromagnetic theory rather than the complex propagation function used in our work. Similarly, he makes use of the K-K dispersion relations of electromagnetic theory rather than the Hilbert transform relations we have used. Since our preference is for the Hilbert transform formulation which is directly interpretable in terms of the minimum-phase property of the inelastic behavior, we shall rederive Futterman's phase velocity and Q relations in this way.

The minimum-phase requirement as stated in equation (4) is that $\hat{\alpha}(\omega)$ is the Hilbert transform of $\alpha(\omega)$. Mathematically, equation (4) simply states that $\hat{\alpha}(\omega)$ is the convolution of $\alpha(\omega)$ with $-1/(\pi\omega)$. The evenness of $\alpha(\omega)$ allows us to express this convolution in the form (Papoulis, 1962, p. 206),

$$\hat{\alpha}(\omega) = \frac{2\omega}{\pi} \int_0^\infty \frac{\alpha(\Omega)}{\Omega^2 - \omega^2} d\Omega. \quad (24)$$

Although Futterman selects three variations of $\alpha(\omega)$ to fit a TLF model, our needs will be satisfied if we restrict ourselves to his simplest form:

$$\alpha(\omega) = \begin{cases} 0 \text{ for } \omega < \omega_0 \text{ and } \omega > \omega_h \\ \dfrac{\pi B}{2} \, |\omega| \text{ for } \omega_0 \le \omega \le \omega_h \end{cases}, \quad (25)$$

where we have introduced the high frequency cutoff parameter ω_h as well as the low frequency cutoff parameter ω_0, in order to be consistent with Futterman's assumptions. The inelastic coefficient B is related to Wuenschel's b and Futterman's $Q_0 c$ by

$$B \equiv \frac{2b}{\pi} \equiv \frac{1}{\pi Q_0 c}. \quad (26)$$

Upon inserting $\alpha(\omega)$ from (25) into (24), using ordinary tables of integrals, and then constructing the complete phase lag function, $\theta(\omega)$, according to equation (4), we find

$$\theta(\omega) = \omega\tau_0 - \frac{B\omega}{2} \cdot \ln\left[\left|\frac{\omega_h^2}{\omega_h^2 - \omega^2}\right| \left|\frac{\omega^2 - \omega_0^2}{\omega_h^2}\right|\right]. \quad (27)$$

Note that although $\alpha(\omega)$ in (25) is truncated in ω, no such truncation appears in $\theta(\omega)$ as given by equation (27). This means $\theta(\omega)$ can take on the values $\theta(o)$ and $\theta(\infty)$. However, in order to interpret the parameter τ_0 in (27) which originated from the all-pass contribution, it is desirable to re-express (27) in terms of the phase velocity, $v(\omega)$

$$\frac{1}{v(\omega)} \equiv \frac{\theta(\omega)}{\omega} = \tau_0 - \frac{B}{2} \cdot \ln\left[\left|\frac{\omega_h^2}{\omega_h^2 - \omega^2}\right| \left|\frac{\omega^2 - \omega_0^2}{\omega_0^2}\right|\right]. \quad (28)$$

Equation (28) has the particular values

$$\frac{1}{v(0)} = \tau_0 \quad (29a)$$

and

$$\frac{1}{v(\infty)} = \tau_0 - B \ln\left|\frac{\omega_h}{\omega_0}\right|. \quad (29b)$$

Consequently, τ_0 has the simple interpretation of being the reciprocal of the phase velocity at *zero* frequency. In contrast, the parameter τ of the PL solid had the interpretation of being the reciprocal of the phase velocity at infinite frequency. Since the phase velocity curve for plexiglas as given by Figure 4 is a monotonously *increasing* function of the frequency, we find it difficult to associate the parameter τ_0 with a physically meaningful arrival time. Because the ln term becomes positive infinite at the truncation frequencies ω_0 and ω_h, it's apparent that $v(\omega_0)$ and $v(\omega_h)$ become zero there.

In order to obtain a reduction of equation (28) to Futterman's corresponding relation, we first take the parameter ω_h to be infinite and then require that ω be much greater than ω_0. This process yields, from (28).

$$\frac{1}{v(\omega)} \approx \tau_0 - B \ln \frac{\omega}{\omega_0}. \quad (30)$$

Upon replacing B by $1/(\pi Q_0 c)$ according to equation (26), we obtain Futterman's equation (25). However, as is evident from equation (30), we can no longer retain the interpretation of τ_0 as the reciprocal of the phase velocity at zero frequency, since τ_0 is now the reciprocal of the phase velocity at $\omega = \omega_0$. If we form $\theta(\omega)$ from (30) and use this with $(\alpha\omega)$ from (25) to construct the spatial Q according to (5), we find

$$Q(\omega) = \frac{1}{\pi}\left[\frac{\tau_0}{B} - \ln\frac{\omega}{\omega_0}\right], \quad (31)$$

which is also identical with Futterman's equation (26) if we use our equation (26) and identify τ_0 with $1/c$.

Although we have rederived Futterman's results for the phase velocity and spatial Q of a TLF solid, we are not satisfied with the requirement that a nonzero low frequency cutoff, $\omega = \omega_0$, was required to obtain his results. This low frequency cutoff can be avoided by the simple procedure of subtracting equation (29b) from equation (28) before making any assumptions on the parameters ω_0 and ω_h. In this way we find

$$\frac{1}{v(\omega)} = \frac{1}{v(\infty)} + \frac{B}{2} \ln\left[\frac{\omega_h^2 - \omega^2}{\omega^2 - \omega_0^2}\right]. \quad (32)$$

The first term on the right is just the parameter

τ of our PL solid model. Also, there is no difficulty here in taking the parameter ω_o to be zero. For ω_o set equal to zero,

$$\frac{1}{v(\omega)} \approx \tau + \frac{B}{2} \ln \left[\left(\frac{\omega_h}{\omega} \right)^2 - 1 \right]. \quad (33)$$

If we further take ω_h to be sufficiently large that $\omega << \omega_h$, we obtain

$$\frac{1}{v(\omega)} \approx \tau_h + B \ln \frac{\omega_h}{\omega}. \quad (34)$$

Note that we have appended the subscript h to the parameter τ, for it is evident from equation (34) that the limiting approximation requires that τ_h be interpreted as the reciprocal of the phase velocity at $\omega = \omega_h$ rather than at infinite frequency.

We find the form (34) for $1/v(\omega)$ to be superior to (30) obtained by Futterman because τ_h is smaller than τ_o and the approximation to the true all-pass behavior of the infinite frequency component is more realistic.

Since our purpose here is only to obtain the expression for the pedestal in a TLF solid, we shall not determine the $Q(\omega)$ to go with expression (34) nor further develop other TLF alternatives to Futterman's formulation. In order to obtain an expression for the pedestal, we must first construct the complex propagation function $\gamma(p)$. The deduction of $\gamma(p)$ is somewhat tricky and we shall proceed by presenting the derived form and then working backwards to show that it indeed yields the TLF forms of $\alpha(\omega)$ and $v(\omega)$.

We state that the TLF form for the complex propagation function has the simple form

$$\gamma(p) = p\tau_x - Bp \ln \frac{p}{\omega_x}, \quad (35)$$

where ω_x refers to either ω_o or ω_h and τ_x refers to either of τ_o or τ_h, respectively. The attenuation function $\alpha(f)$ is obtained from the real part of (35) when p approaches $j\omega$. The first term on the right becomes pure imaginary and does not contribute to $\alpha(f)$. We are left with

$$\alpha(\omega) = - B \text{ Re} \left[j\omega \ln \left\{ e^{j(\pi/2)} \frac{\omega}{\omega_x} \right\} \right], \quad (36)$$

where we have made use of the identity $j = \exp(j\pi/2)$. Then,

$$\alpha(\omega) = - B \text{ Re} \left[j\omega \ln \exp \left(j \frac{\pi}{2} \right) \right. $$
$$\left. + j\omega \ln \frac{\omega}{\omega_x} \right]. \quad (37)$$

The first term within the brackets in (37) is simply $-\omega\pi/2$, while the second term is pure imaginary and thus cannot contribute to $\alpha(f)$. The result is just (31) for $\alpha(f)$. In a similar way we have for the imaginary part of (35),

$$\theta(\omega) = \omega\tau_x - B \text{ Im} \left[j\omega \ln \exp \left(j \frac{\pi}{2} \right) \right. $$
$$\left. + j\omega \ln \frac{\omega}{\omega_x} \right]. \quad (38)$$

Here, the first term in the brackets, as we saw above, is real and thus cannot contribute to $\theta(f)$. The second term is pure imaginary and we have

$$\theta(\omega) = \omega\tau_x - B\omega \ln \frac{\omega}{\omega_x}. \quad (39)$$

With $1/v(\omega) = \theta(\omega)/\omega$, we arrive at (30) and (34) and our assumption that the complex propagation function is given by equation (35) is proven.

DISTORTION OF A PULSE PROPAGATING IN A TLF SOLID

The asymptotic mathematical development for obtaining the distortion response to a propagation function of the form of equation (35) follows almost exactly that for the PL solid. We shall only outline the corresponding steps. We wish to evaluate the integral, equation (10), subject to (35) rather than (6). This leads to the general form given by equation (15), where now

$$t' = t - R\tau_x \quad (40)$$

instead of (12) appears as the new time variable. Similarly,

$$Rf(p) = - R\gamma(p) + tp$$
$$= BRp \ln \left(\frac{p}{\omega_x} \right) + t'p \quad (41)$$

replaces (14). $F(p)$ is identical to (13). Note that the first term on the right side of the equality, i.e. $BRp \ln(p/\omega_x)$, is negative for $p < \omega_x$ and thus agrees in sign with $-Kp^s$ in equation (14) for the PL solid. However for $p > \omega_x$ the sign of BRp

$ln(p/\omega_x)$ changes and becomes opposite to that for the PL solid. Thus, we should now be prepared to expect difficulties at high frequencies even though difficulties were not indicated by the TLF form of $\alpha(\omega)$ which is truncated at a frequency ω_h well above ω (i.e. the high frequency form) or at a frequency ω_o well below ω (i.e. the low frequency form). Equation (41) leads to a saddle point at

$$p = p_0 = \omega_x \exp\left[-\left(1 + \frac{t'}{BR}\right)\right], \quad (42)$$

so that

$$Rf(p_0) = -BR\omega_x \exp\left[-\left(1 + \frac{t'}{BR}\right)\right] \quad (43)$$

and

$$\left| R \frac{d^2f}{dp^2} \right|_{p=p_0} = \frac{BR}{\omega_x} \exp\left[1 + \frac{t'}{BR}\right]. \quad (44)$$

The first term in the steepest descent relation, equation (19), then yields

$$P(R, t') \doteqdot \left[\frac{\omega_x}{2\pi BR^3}\right]^{1/2}$$
$$\cdot \exp\left\{-\frac{1}{2} - \frac{t'}{2BR} - BR\omega_x\right.$$
$$\left.\cdot \exp\left[-\left(1 + \frac{t'}{BR}\right)\right]\right\}. \quad (45)$$

This response has a peak at

$$t'_{\text{peak}} = BR[\ln(2BR\omega_x) - 1] \quad (46)$$

of magnitude

$$P(R, t'_{\text{peak}}) \doteqdot \left[\frac{\omega_x}{2\pi BR^3}\right]^{1/2}$$
$$\cdot \exp\left\{-\tfrac{1}{2} - \tfrac{1}{2}\ln(2BR\omega_x)\right\}. \quad (47)$$

The normalized response, which is the ratio of (45) to (47), becomes

$$P_{\text{Norm}}(R, t') \doteqdot \exp\left\{\frac{1}{2}\ln(2BR\omega_x)\right.$$
$$\left. - \frac{t'}{2BR} - BR\omega_x \exp\left(-1 - \frac{t'}{BR}\right)\right\}, \quad (48)$$

with t' defined by (40). The parameter ω_x (unlike the parameter τ_x) appears explicitly in the response (48). This result can be avoided if one returns to our expression (41) for $Rf(p)$ and re-expresses it in the form

$$Rf(p) = BRp \ln p + T'p, \quad (49)$$

where now

$$T' \equiv t' - BR \ln \omega_x$$
$$= t - R(\tau_x + B \ln \omega_x). \quad (50)$$

Repeating the algebra, we arrive at the alternate forms

$$P(R, T') = [2\pi BR^3]^{-1/2}$$
$$\cdot \exp\left\{-\frac{1}{2} - \frac{T'}{2BR} - BR\right.$$
$$\left.\cdot \exp\left(-1 - \frac{T'}{BR}\right)\right\} \quad (51)$$

and

$$P_{\text{Norm}}(R, T') = \exp\left\{\frac{1}{2}\ln(2BR) - \frac{T'}{2BR}\right.$$
$$\left. - BR \exp\left(-1 - \frac{T'}{BR}\right)\right\} \quad (52)$$

to equations (45) and (48), respectively. These responses have a peak at time

$$T'_{\text{peak}} = BR[\ln(2BR) - 1]. \quad (53)$$

In this report we are interested primarily in the normalized response; i.e., equation (52) plotted in real time t with the use of equation (50). In Figure 6b we have plotted $P_{\text{Norm}}(r,t)$ using the parameters of plexiglas at $R = 32,000$ ft. and have directly compared it with the similar response for a PL solid (Figure 6a). The most obvious difference is that the TLF response, computed according to equation (52), must be computed for negative T' in order to obtain the same result we obtain (Figure 6c) when we use equation (48). Although we do not need to specify the parameter ω_x when equation (52) is used, we do have to know ω_x when equation (50) is used to plot (52) in real time as we did in Figure 6b. There is a conceptual difficulty associated with the use of negative T'; we shall avoid this problem by stay-

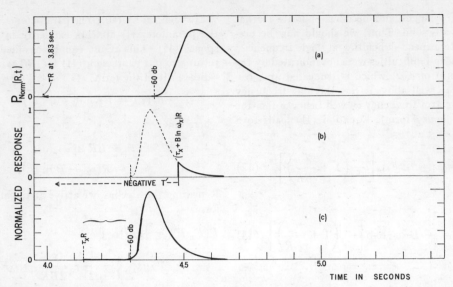

FIG. 6. (a) Normalized response for a PL solid as in Figure 2. (b) Normalized response for a TLF solid using equation (52). The dashed portion is for negative T'. (c) Same as (b) but computed according to equation (48). The parameters are $B=0.8\times10^{-6}$, $\omega_h=0.880\times10^6$, $\tau_h=129.2\times10^{-6}$ and $R=32{,}000$ ft.

ing with equation (48), which (Figure 6c) does seem to be properly behaved.

If we now compare the TLF predicted waveform of Figure 6c with that of the PL predicted waveform of Figure 6a, we see, aside from rather unimportant differences such as those in the precise onset times and the pulse-widths which could be due to the use of slightly different parameters, the two waveforms are very similar in their important features. The feature that concerns us the most is, of course, the existence of the pedestal. We see that for the waveform of Figure 6c there is indeed an indication of the existence of a pedestal in that the apparent onset is one of upward curvature followed by an ill-defined inflection point. As for the PL waveform, there is no point in time where the arrival slope is discontinuous. However, the duration of the pedestal is even more ill-defined than it was for the PL solid, because we have no argument to let us extend the negative T' used to plot Figure 6b back to any unique point in real time or, in the case of Figure 6c, back to a unique $\tau_x R$. There are other difficulties such as the behavior of equation (34) for frequencies above $\omega_h \exp(\tau_h/B)$. This behavior could affect the nature of the pedestal. Rather than further discuss the conceptual difficulties of the TLF model, we shall terminate the discussion by following Futterman's example and

making a computation for a very low Q solid so as to better bring out the effects of absorption that do exist.

As in Futterman's Figure 3, we assume the parameters $Q=5$, $R=327$ km, and $c=4.92$ km/sec at a low frequency cutoff of $\omega_o=0.001$. Since Q is actually a function of ω, we have from equation (31)

$$Q(\omega) = \frac{1}{\pi}\left[\frac{\tau_0}{B} - \ln\left(\frac{\omega}{\omega_0}\right)\right], \qquad (54)$$

from which we can solve for the parameter B.

$$B = \tau_0\left[\pi Q(\omega) - \ln\left|\frac{\omega_0}{\omega_*}\right|\right]^{-1}. \qquad (55)$$

At $\omega=\omega_o$, we have simply

$$B = \frac{\tau_0}{\pi Q(\omega_0)}, \qquad (56)$$

so that with $c=1/\tau_o$, we obtain $B=4.0\times10^{-6}$ in our fps system of units. We now have the parameters necessary to plot the response relation (48). For the computation of $P_{\text{Norm}}(R, t')$ we use the parameter pair ω_o and τ_o in place of ω_x and τ_x. The result is plotted as Figure 7b, where it is compared with Futterman's numerical integration shown as Figure 7a. Although our waveform

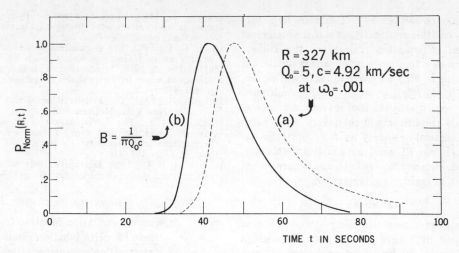

FIG. 7. (a) Futterman's numerical integration of his absorption-dispersion pair $(A3,D3)$ for a low-Q solid. (b) Our response for a TLF solid as computed from our analytic solution, equation (52) for his $(A1,D1)$ pair. Note that the shape of our response curve depends only upon the product BR.

peak comes in somewhat earlier, the general features are visually identical. It is, of course, the visual onset region that is of primary interest to us. Futterman's curve does indeed have the upward curvature that is characteristic of our PL response but his onset seems to show a well-defined break in slope in the response that we do not find in our solution of his TLF response. To obtain a better view of the visual onset region, we show the onset region of Figure 7b in much expanded detail as Figure 8. In spite of the ambiguity in arrival time of the theoretical onset for a TLF solid, we must conclude that the behavior

we obtain supports our contention from the PL solid discussion that a pedestal effect exists.

As a final remark on the TLF absorption model we should mention that the peak of the unnormalized response obtained by inserting equation (46) in (45) will show, after some algebraic manipulation, that the peak amplitude varies as the inverse square of the distance R traveled. This compares well with variation as the inverse to the power $1+1/s$ that we found for the PL solid. Our analysis of the TLF solid also quickly answers a question often raised as to how we handle the constant term in the linear fit to the attenua-

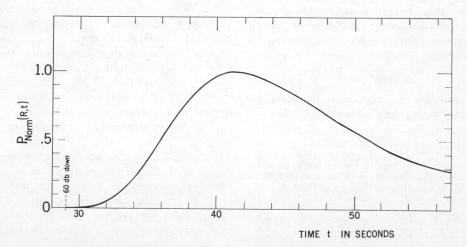

FIG. 8. Detail on the response in the TLF solid of Figure 7b illustrating the same lack of discontinuity in the slope of the response at the apparent onset that we observed in the PL solid.

tion data such as the one that appears in Figure 4a. If we call this constant α_o, it is $\alpha_o R$ which must be added to equation (25). Since the Hilbert transform of a frequency independent function is zero for all ω, $\theta(\omega)$ as given by (27) will be unaltered. Thus $F(p)$, equation (13), is multiplied by $\exp(-\alpha_o R)$, as is the final waveform response. The peak amplitude will then vary as $R^{-2}\exp(-\alpha_o R)$ instead of simply as R^{-2}. Since the parameter s in our PL solid must be less than unity, the two models are in even better agreement when the linear-fit constant α_o is retained.

SUMMARY

It has been shown that the pedestal effect, as exemplified in Figure 2, is a real effect which cannot be removed by redefining the arrival time. It has also been shown that it is not a by-product of the method of integration (steepest descent) nor of a particular parametric form of fitting the "essentially" linear attenuation function. If the high frequencies are truncated by passing the response through a low-pass filter, the early part of the pedestal will be removed; but the region of the apparent onset which has upward curvature will remain. Thus we conclude that the effect of absorption is to round off the apparent (visual) onset as well as the peak of the impulse response. Although the peak is still a unique point on the waveform and can be used to calculate precise velocities in the time domain, the visual onset has no such unique arrival time. Furthermore, when one considers refraction instead of reflection arrivals, the well-known time-integral effect on the onset of a refraction arrival for a purely elastic interface must be applied to our absorption response, a process further enhancing the pedestal effect.

ACKNOWLEDGMENT

This research was supported by grant GA 10840 from the National Science Foundation.

REFERENCES

Attewell, P. B., and Ramana, Y. V., 1966, Wave attenuation and internal friction as functions of frequency in rocks: Geophysics, v. 31, p. 1049–1056.
Ewing, W. M., Jardetsky, W. S., and Press, F., 1957, Elastic waves in layered media: New York, McGraw-Hill.
Futterman, W., 1962, Dispersive body waves, J. Geophys. Res., v. 67, p. 5279–5291.
Guillemin, E. A., 1963, Theory of linear physical systems: New York, Wiley & Sons, Inc.
Knopoff, L., 1964, Q: Rev. Geophysics, v. 2, p. 625–660.
Kolsky, H., 1960, Viscoelastic waves: International symposium on stress wave propagation in materials: New York, Interscience Publ. Inc.
Lamb, G. L., 1962, The attenuation of waves in dispersive media: J. Geophys. Res., v. 67, p. 5273–5277.
McDonal et al, 1958, Attenuation of shear and compressional waves in Pierre shale: Geophysics, v. 22, p. 421–439.
Papoulis, A., 1962, The Fourier integral and its application: New York, McGraw-Hill.
Strick, E., 1967, The determination of Q, dynamic viscosity, and transient creep curves from wave propagation measurements: Geophys. J. R. Astr. Soc., v. 13, p. 197–218.
Wuenschel, P. C., 1965, Dispersive body waves—an experimental study: Geophysics, v. 30, p. 539–557.

NOTATION

a_o = source excitation function (equal to zero for delta function excitation).

b = attenuation parameter after Wuenschel $= \pi B/2$.

B = attenuation parameter for TLF solid defined by equation (26).

c = reference phase velocity after Futterman.

f = frequency.

$j = \sqrt{-1}$.

k, k_o = attenuation parameters for PL solid defined by equation (8).

$K, K(s)$ = attenuation parameter for PL solid defined by equation (9).

$K(\omega) = -j\gamma(\omega)$ after Futterman.

$n(\omega)$ = complex index of refraction after Futterman $= Re\ n(\omega) + jIm\ n(\omega)$.

$p = j2\pi f$.

$P(R, t)$ = response in real time t after traveling distance R.

$Q(\omega)$ = spatial mechanical $Q = \theta(\omega)/2\alpha(\omega)$.

R, R_n = continuous and discrete values of travel distance.

s = attenuation parameter for PL solid defined by equation (8).

t = real time.

t', T' = delayed time according to equations (12) and (50).

$v(\omega)$ = phase velocity at angular frequency ω.

$\alpha(\omega)$ = attenuation per unit travel distance.

$\hat{\alpha}(\omega)$ = Hilbert transform of $\alpha(\omega)$.

$\gamma(\omega)$ = complex propagation function $= \alpha(\omega) + j\theta(\omega)$.

$\theta(\omega)$ = phase lag function.

τ, τ_∞ = reciprocal of phase velocity at infinite frequency.

τ_o, τ_x, τ_h = reciprocal of phase velocities at low, intermediate, and high frequencies, respectively.

$\omega = 2\pi f$.

$\omega_c, \omega_x, \omega_h$ = low, intermediate, and high values of ω, respectively.

Ω = integration variable in ω.

APPENDIX A
A DISCUSSION OF THE ACCURACY OF THE FIRST TERM STEEPEST DESCENT APPROXIMATION

The usual procedure for estimating the error term in an asymptotic expansion is rarely, if ever, used in the method of steepest descent, because the resulting error term can be evaluated only approximately. We have found it much more convenient and convincing to test the accuracy of our asymptotic expansions by carrying out a numerical Fourier analysis of the waveforms of Figure 3 [which were computed from equation (23)] to see just how well we can recover the original attenuation function, equation (8). The results are shown in Figure 9 for 5 of the 6 values of R_n. Although the original data were given only to about 150 hz, we have carried out the return Fourier analysis to about 250 hz. The original attenuations as obtained from equation (8) are shown as dashed lines and the Fourier recovered attenuations, as solid lines. It is quite apparent that the recovery is well within a couple of decibels for all attenuations less than, say, 70 db. At the higher frequencies, where the attenuations can exceed 70 db, we obtain computation error when the dynamic range of the IBM 7094 has apparently been exceeded. It is certainly true on intui-

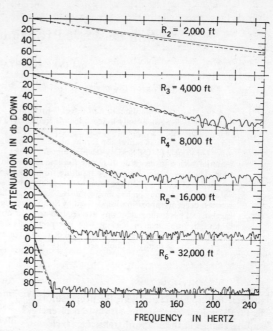

FIG. 9. Comparison at 5 values of R_n of the Fourier transform of $P_{Norm}(R_n, t)$ [determined from equation (23) and which was in turn derived from $\alpha(f)$ of equation (8)] with $\alpha(f)$. Here we have again used the parameters of Figure 2. The original $\alpha(f)$ are the dashed lines and the Fourier recovered are the solid lines.

tive grounds that the exact evaluation of the Fourier transform of such a simple and smooth expression as equation (23) could never behave in this irregular manner. Therefore we accept the analysis of Figure 9 as a verification that the first term of our steepest descent expansion is a sufficiently accurate approximation.

Reprinted from the Journal of Geophysical Research, v. 84, p. 4737–4748.

Constant Q-Wave Propagation and Attenuation

Einar Kjartansson

Rock Physics Project, Department of Geophysics, Stanford University, Stanford, California 94305

A linear model for attenuation of waves is presented, with Q, or the portion of energy lost during each cycle or wavelength, exactly independent of frequency. The wave propagation is completely specified by two parameters, e.g., Q and c_0, a phase velocity at an arbitrary reference frequency ω_0. A simple exact derivation leads to an expression for the phase velocity c as a function of frequency: $c/c_0 = (\omega/\omega_0)^\gamma$, where $\gamma = (1/\pi) \tan^{-1} (1/Q)$. Scaling relationships for pulse propagation are derived and it is shown that for a material with a given value of Q, the risetime or the width of the pulse is exactly proportional to travel time. The travel time for a pulse resulting from a delta function source at $x = 0$ is proportional to x^β, where $\beta = 1/(1 - \gamma)$. On the basis of this relation it is suggested that the velocity dispersion associated with anelasticity may be less ambiguously observed in the time domain than in the frequency domain. A steepest descent approximation derived by Strick gives a good time domain representation for the impulse response. The scaling relations are applied to field observations from the Pierre shale formation in Colorado, published by Ricker, who interpreted his data in terms of a Voigt solid with Q inversely proportional to frequency, and McDonal et al., who interpreted their data in terms of nonlinear friction. The constant Q theory fits both sets of data.

Introduction

A fundamental feature associated with the propagation of stress waves in all real materials is the absorption of energy and the resulting change in the shape of transient waveforms. Although a large number of papers have been written on the absorption of seismic waves in rocks, little, if any, general agreement exists about even the most fundamental properties of the processes involved. Table 1 shows a summary of the basic features of some of the different attenuation theories.

Early laboratory work on the absorption in rocks showed the loss per cycle or wavelength to be essentially independent of frequency. Since at that time no known linear theory could fit this observation, *Born* [1941] proposed that the loss was due to rate independent friction of the same kind as observed when two surfaces slide against each other. *Kolsky* [1956] and *Lomnitz* [1957] gave linear descriptions of the absorption that could account for the observed frequency independence and were also consistent with other independent observations of the transient creep in rocks and the change in shape of pulses propagating through thin rods. Despite this, and the fact that a satisfactory nonlinear friction model for attenuation has never been developed to the point where meaningful predictions could be made about the propagation of waves, nonlinear friction is commonly assumed to be the dominant attenuation mechanism, especially in crustal rocks [*McDonal et al.*, 1958; *Knopoff*, 1964; *White*, 1965; *Gordon and Davis*, 1968; *Lockner et al.*, 1977; *Johnston and Toksoz*, 1977].

A different type of theory for attenuation has been advocated by *Ricker* [1953, 1977]. In his model the absorption is described by adding a single term to the wave equation. Because of this simplicity, the theory of the propagation of transient waves has been further developed than for the other theories. For this reason, wavelets based on the Ricker theory have been commonly used in the computation of synthetic seismograms [*Boore et al.*, 1971; *Munasinghe and Farnell*, 1973], although the model contradicts practically all experimental observations of the frequency dependence of Q that is implied by the model. In this paper, we will discuss some of the data Ricker interpreted as in support of his theory.

Recently, there has been renewed interest in the effects of anelasticity on the wave propagation in rocks. *Liu et al.* [1976]

found that the change in the elastic moduli implied by the attenuation over the frequency range covered by the seismic body waves and free oscillations, was about an order of magnitude greater than the uncertainty in the measurements. The models used by *Liu et al.* [1976], as well as all of the other nearly constant Q (NCQ) models, have included at least one parameter that is in some way related to the range of frequencies over which the model gives Q nearly independent of frequency. How this cutoff is chosen appears to be quite arbitrary and the physical implications of the cutoff parameters are different between the models of *Lomnitz* [1957], *Futterman* [1962], *Strick* [1967], and *Liu et al.* [1976].

In this paper a linear description of attenuation is given that features Q exactly independent of frequency, without any cutoffs. The constant Q (CQ) model is mathematically much simpler than any of the NCQ models; it is completely specified by two parameters, i.e., phase velocity at an arbitrary reference frequency, and Q.

Most of the NCQ papers have described the wave phenomena in the frequency domain and have restricted their analysis to cases where Q is large ($Q > 30$). In contrast, the simplicity of the CQ description allows the derivation of exact analytical expressions for the various frequency domain properties such as the complex modulus, phase velocity, and the attenuation coefficient, that are valid over any range of frequencies and for any positive value of Q. In this paper more emphasis will be placed on the time domain description of transient phenomena, and exact expressions for the creep and relaxation functions and scaling relations for the transient wave pulse will be given. In addition, approximate expressions will be given for the impulse response, as a function of time, that results from a delta function excitation.

We will also show that when the frequency range is restricted and the losses are small, the results obtained from the various NCQ theories approach the same limit as those obtained from the CQ theory.

Definitions and Background

Seismic attenuation is commonly characterized by the quality parameter Q. It is most often defined in terms of the maximum energy stored during a cycle, divided by the energy lost during the cycle. When the loss is large this definition

Paper number 9B0272.
0148-0227/79/009B-0272$01.00

TABLE 1. Comparison of Attenuation Theories

Property	Theory			
	Friction	Voigt-Ricker	NCQ Band-Limited Near-Constant Q	CQ Linear Constant Q
Linearity	Nonlinear, velocity and Q depend on amplitude	Linear	Linear	Linear
Frequency dependence of Q	Independent	$1/Q \propto \omega$	Nearly independent in a frequency band	Independent
Frequency dependence of phase velocity	Independent	Independent at low frequencies	$C/C_0 \approx 1 + (1/\pi Q) \ln (\omega/\omega_0)$	$C/C_0 = (\omega/\omega_0)^\gamma$ $1/Q = \tan (\pi\gamma)$
Transient creep	None	$\Psi(t) \propto e^{-at}$	$\Psi(t) \approx (1/M_0)[1 + (2/\pi Q) \ln (1 + at)]$	$\Psi(t) \propto t^{2\gamma}$
Pulse broadening	Distorted or acausal	$\tau \propto T^{1/2}$	$\tau \overset{\sim}{\propto} T$	$\tau \propto T$
References	Born [1941] Knopoff [1964] White [1966] Walsh [1966] Lockner et al. [1977] Johnston and Toksöz [1977] Gordon and Davis [1968]	Voigt [1892] Ricker [1953, 1977] Collins [1960] Clark and Rupert [1960] Jaramillo and Colvin [1970] Balch and Smolka [1970]	Kolsky [1956] Lomnitz [1957] Futterman [1962] Azimi et al. [1968] Strick [1967, 1970] Liu et al. [1976] Kanamori and Anderson [1977] Minster [1978a]	Bland [1960] Strick [1967] This paper

becomes impractical; *O'Connell and Budiansky* [1978] suggested a definition in terms of the mean stored energy W and the energy loss ΔW, during a single cycle of sinusoidal deformation:

$$Q = \frac{4\pi W}{\Delta W} \qquad (1)$$

When this definition is used, Q is related to the phase angle between stress and strain, δ, according to

$$\frac{1}{Q} = \tan \delta \qquad (2)$$

The fact that amplitude dependence of the propagation velocity and Q at strains less than 10^{-6} has not been observed, strongly suggests that at these amplitudes the material response is dominated by linear effects, or in other words, the strain that results from a superposition of two stress functions is equal to the sum of the strains that result from the application of each stress function separately. When two effects are linearly related, the relationship may be expressed through a convolution. Thus the relationship between stress and strain in a linear material may be expressed as

$$\sigma(t) = m(t) * \epsilon(t) \qquad (3)$$

$$\epsilon(t) = s(t) * \sigma(t) \qquad (4)$$

where $\sigma(t)$ is the stress as a function of time, $\epsilon(t)$ is the strain, and $m(t)$ and $s(t)$ are real functions that vanish for negative time. The convolution operator $*$ is defined by

$$f(t) * g(t) = \int_{-\infty}^{\infty} f(t - t')g(t') \, dt' \qquad (5)$$

The relationship between stress and strain given in (3) and (4) was first given by *Boltzmann* [1876]. Our notation differs from Boltzmann's original notation only in that the functions $m(t)$ and $s(t)$ may include generalized functions such as the Dirac delta function or its derivatives. Combination of (3) and (4) implies that $m(t)$ and $s(t)$ must satisfy the condition

$$\delta(t) = m(t) * s(t) \qquad (6)$$

where $\delta(t)$ is the Dirac delta function.

Manipulations involving convolutions are usually facilitated by the use of the Fourier transform. We will use lower case letters to designate functions of time and capital letters for their Fourier transforms according to the definition

$$F(\omega) = \int_{-\infty}^{\infty} f(t)e^{-i\omega t}dt \qquad (7)$$

The inverse Fourier transform is then given by

$$f(t) = \frac{1}{2\pi} \int_{-\infty}^{\infty} F(\omega)e^{i\omega t} \, d\omega \qquad (8)$$

Bracewell [1965] gives a discussion of the formalism required for the extension to generalized functions.

Using the convolution theorem [*Bracewell*, 1965, p. 108], (3), (4), and (6) may be rewritten:

$$\Sigma(\omega) = M(\omega)E(\omega) \qquad (9)$$

$$E(\omega) = S(\omega)\Sigma(\omega) \qquad (10)$$

$$1 = M(\omega)S(\omega) \qquad (11)$$

where $\Sigma(\omega)$ is the Fourier transform of the stress, $E(\omega)$ is the Fourier transform of the strain, and $M(\omega)$ and $S(\omega)$ are the Fourier transforms of $m(t)$ and $s(t)$. Thus the stress and the strain are in the frequency domain related through a multiplication by a modulus $M(\omega)$ or compliance $S(\omega)$ just as in the purely elastic case, the only difference being that the modulus may be complex and frequency dependent. This relationship is commonly referred to as the correspondence principle. By a substitution of a unit step function into (3) and (4), it is easily shown that $m(t)$ and $s(t)$ are the first time derivatives of the relaxation and creep functions, where the relaxation function, $\bar{\Psi}(t)$, is the stress that results from a unit step in strain, and the creep function, $\Psi(t)$, is the strain that results from a unit step in stress.

When the stress-strain relations are combined with the equilibrium equation, the resulting one-dimensional wave equation has a solution that may be written in a form analogous to the classical case:

$$U(t, x) = e^{i(\omega t - kx)} \qquad (12)$$

where

$$k = \omega\left(\frac{\rho}{M(\omega)}\right)^{1/2} \qquad (13)$$

and ρ is the density of the material.

THE CONSTANT Q MODEL

The development so far has been completely general; no assumptions other than linearity and causality have been made about the properties of the material. We will now examine a particular form for the stress-strain relationships and show that it leads to a Q that is independent of frequency. Frequency independent Q implies that the loss per cycle is independent of the time scale of oscillation; therefore it might seem reasonable to try a material that has a creep function that plots as a straight line on a log-log plot, or

$$\Psi(t) \propto t^b$$

For the sake of convenience in subsequent manipulations, we will use a creep function of the form

$$\Psi(t) = \frac{1}{M_0\Gamma(1 + 2\gamma)}\left(\frac{t}{t_0}\right)^{2\gamma} \qquad t > 0$$
$$\Psi(t) = 0 \qquad\qquad\qquad t < 0 \qquad (14)$$

Γ is the gamma function which in all cases of interest to us has a value close to unity and t_0 is an arbitrary reference time introduced so that when t has the dimension of time, M_0 will have the dimension of modulus. Some of the properties of a material that has this creep function are discussed by *Bland* [1960, p. 54]. Response functions of this form have also been used to model dielectric losses in solids [*Jonscher*, 1977]. Differentiation of the expression in (14) yields

$$s(t) = \frac{2\gamma}{M_0\Gamma(1 + 2\gamma)}\left(\frac{t}{t_0}\right)^{2\gamma}\frac{1}{t} \qquad t > 0$$
$$s(t) = 0 \qquad\qquad\qquad\qquad t < 0 \qquad (15)$$

Taking the Fourier transform we get

$$S(\omega) = \frac{1}{M_0}\left(\frac{i\omega}{\omega_0}\right)^{-2\gamma} \qquad (16)$$

where

$$\omega_0 = \frac{1}{t_0} \qquad (17)$$

Using (11) we get

$$M(\omega) = M_0\left(\frac{i\omega}{\omega_0}\right)^{2\gamma} = M_0\left|\frac{\omega}{\omega_0}\right|^{2\gamma}e^{i\pi\gamma\,\mathrm{sgn}\,(\omega)} \qquad (18)$$

where

$$\mathrm{sgn}\,(\omega) = 1 \qquad \omega > 0$$
$$\mathrm{sgn}\,(\omega) = -1 \qquad \omega < 0 \qquad (19)$$

Taking the inverse Fourier transform of $M(\omega)$ and integrating, we get the relaxation function

$$\bar{\Psi}(t) = \frac{M_0}{\Gamma(1 - 2\gamma)}\left(\frac{t}{t_0}\right)^{-2\gamma} \qquad t > 0$$
$$\bar{\Psi}(t) = 0 \qquad\qquad\qquad\qquad t < 0 \qquad (20)$$

Figure 1 shows a plot of the constant Q creep function (14), and Figure 2, of the relaxation function (20), for several values of Q. Equation (18) shows that the argument of the modulus and thus the phase angle between the stress and the strain, is independent of frequency; therefore, it follows from the definition of Q (2) that Q is independent of frequency:

$$\frac{1}{Q} = \tan(\pi\gamma) \qquad (21)$$

or

$$\gamma = \frac{1}{\pi}\tan^{-1}\left(\frac{1}{Q}\right) \approx \frac{1}{\pi Q} \qquad (22)$$

Since both the creep and relaxation functions vanish for negative time, no strain can precede applied stress, nor can any stress precede applied strain; the material is causal.

To investigate the propagation of waves in the constant Q material, the modulus given by (18) may be substituted into the solution to the one dimensional wave equation, given by (12) and (13); the result may be written as

$$U(t, x) = e^{-\alpha x}e^{i\omega(t - x/c)} \qquad (23)$$

where

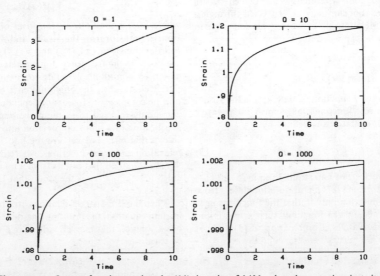

Fig. 1.　The constant Q creep function as given by (14), in units of $1/M_0$, plotted versus time in units of t_0.

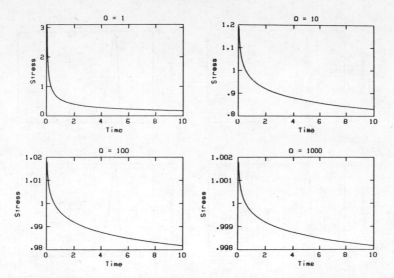

Fig. 2. The constant Q relaxation function as given by (20), in units of M_0, plotted versus time in units of t_0.

$$c = c_0 \left| \frac{\omega}{\omega_0} \right|^{\gamma} \tag{24}$$

$$\alpha = \tan\left(\frac{\pi\gamma}{2}\right) \text{sgn}(\omega)\frac{\omega}{c} \tag{25}$$

$$c_0 = \left(\frac{M_0}{\rho}\right)^{1/2} \Big/ \cos\left(\frac{\pi\gamma}{2}\right) \tag{26}$$

Since c is slightly dependent on frequency, constant Q is not exactly equivalent to assuming that α is proportional to frequency, as is often assumed in the literature. It is clear from (24), that c_0 is simply the phase velocity at the arbitrary reference frequency ω_0. In the final section of the paper, we discuss the low and high frequency limits for the phase velocity and the modulus, and the short and long term behavior of the creep function.

An alternative to (23) is to write the solution to the wave equation as

$$U(x, t) = \exp\left[i\omega\left(t - \frac{x}{c_s(i\omega)^{\gamma}}\right)\right] \tag{27}$$

where c_s is a constant related to M_0 by

$$c_s = \left(\frac{M_0}{\rho}\right)^{1/2} \omega_0^{-\gamma} \tag{28}$$

Use of the complex velocity notation, as in (27), often simplifies the algebra, e.g., in the derivation of reflection coefficients or when modeling wave propagation in two or three dimensions.

As most wave phenomena encountered in seismology are transient in nature, a time domain description of wave propagation is often more useful for modeling or comparison with data than a frequency domain description. The waveform that results from a delta function source, the impulse response, is particularly useful since the waveform that results from an arbitrary source is obtained by simply convolving the source with the impulse response. The Fourier transform of the impulse response, $b(t)$ is obtained by omitting the $i\omega t$ term in (12) or (23):

$$B(\omega) = e^{-\alpha x}e^{-i\,\omega x/c} \tag{29}$$

By substitution of (24) and (25) into (29), we get

$$B(\omega) = \exp\left\{-\frac{x\omega_0}{c_0}\left|\frac{\omega}{\omega_0}\right|^{1-\gamma}\left[\tan\left(\frac{\pi\gamma}{2}\right) + i\,\text{sgn}(\omega)\right]\right\} \tag{30}$$

The impulse response may be obtained by taking the inverse Fourier transform of $B(\omega)$ given by (30). Although we do not have an analytical expression for $b(t)$, we will present a useful approximate relation and some exact scaling relations. We will rewrite (30) as

$$B(\omega) = B_1(\omega_1) \tag{31}$$

where

$$\omega_1 = t_1\omega \tag{32}$$

$$t_1 = t_0\left(\frac{x\omega_0}{c_0}\right)^{\beta} \tag{33}$$

$$\beta = \frac{1}{1-\gamma} \approx 1 + \frac{1}{\pi Q} \tag{34}$$

and

$$B_1(\omega) = \exp\left\{-|\omega|^{1-\gamma}\left[\tan\left(\frac{\pi\gamma}{2}\right) + i\,\text{sgn}(\omega)\right]\right\} \tag{35}$$

It now follows from the similarity theorem [*Bracewell*, 1965; p. 101] that for any homogeneous material, the impulse response at any distance x from the source will be given by

$$b(t, x) = \frac{1}{t_1}b_1\left(\frac{t}{t_1}\right) \tag{36}$$

Equations (36) and (33) imply that in a given material, the travel time T, the pulse width τ and the pulse amplitude A are related according to

$$T \propto \tau \propto \frac{1}{A} \propto \left(\frac{x}{c_0}\right)^{\beta} \tag{37}$$

where any consistent operational definitions for the travel time and pulse width may be used. The proportionality between travel time and pulse width may be expressed as

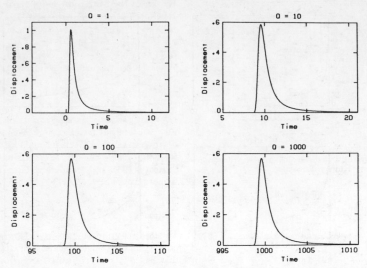

Fig. 3. The waveform $b(t)$, resulting from a unit impulse plane wave source at $x = Qc_0$, is plotted versus time in units of t_0. The waveform in computed using a numerical FFT algorithm.

$$\tau = C(Q)\frac{T}{Q} \qquad (38)$$

where $C(Q)$ is a function that depends only on Q. We will show that $C(Q)$ is nearly constant for $Q > 20$. Figure 3 shows a plot of the function $b(t)$, for several values of Q.

In order to illustrate the scaling relations, seismograms due to impulsive sources at several distances are plotted on a common set of axes in Figure 4. Figure 5 shows the same information but scaled according to distance, by dividing the time by the distance and multiplying the displacement by the distance. Velocity dispersion has the effect of delaying the pulses from the more distant sources more than would be expected for a constant propagation velocity. To further illustrate the dispersion, Figure 6 shows the results of the same kind of a numerical experiment as Figure 5, for a Q of 1000 and covering a larger range of distances. It may be concluded from Figures 5 and 6 that the dispersion due to the anelasticity is directly observable in the time domain when the travel time, in a homogeneous material, can be measured to within half a pulse width over a ratio of 10 in distance. This applies to high Q as well as low Q materials. To measure this effect in the earth would, however, require a careful control over the spatial variation in velocity.

The required control may be obtained when the wave travels the same path more than once. Waves reflected off the core-mantle interface may satisfy this condition for stations near the source. Assuming an average $Q = 160$ and a travel time of 936 s for one pass of ScS [Jordan and Sipkin, 1977], we obtain by a substitution into (34) a value for $\beta = 1.0020$. Equation (37) implies then that doubling the distance will result in a total traveltime of 1874.6 seconds for ScS_2, which is 2.6 s longer than would be expected if the dispersion were not present.

APPROXIMATIONS FOR TIME DOMAIN WAVELETS

So far we have made no assumptions about the value of Q (other than $Q > 0$), or the ranges of frequencies and travel times involved. Although we have been able to derive exact expressions for all frequency domain properties of the wave propagation, we do not have exact analytical expressions for

time domain wavelets or impulse responses. While modern computer techniques (e.g., the fast Fourier transform algorithm) make it relatively easy to transform data to the frequency domain and back, it is still useful to study the time domain wave form, especially since many earthquake data are still recorded in analog form. The need for a convenient time domain representation is demonstrated by the fact that wavelets based on the Voigt-Ricker model are often used by workers who do not accept the frequency dependence of Q implied by that model [e.g., Boore et. al., 1971; Munasinghe and Farnell, 1973].

Strick [1967] applied the causality requirement to the propagation of a wave pulse and found a form for the propagation function that satisfies this requirement. The constant Q transfer function (23), is a special case of Strick's function. Later Strick [1970] used the method of steepest descent to approximate the time domain impulse response. His expression has, in the notation used in this paper, the form

$$b_s(t, x) = \left\{ 2\pi\gamma t \left[\frac{(1 - \gamma)x}{c_s t} \right]^{-1/\gamma} \right\}^{-1/2}$$

$$\exp\left\{ -\frac{\gamma}{(1 - \gamma)} t \left[\frac{(1 - \gamma)x}{c_s t} \right]^{1/\gamma} \right\}$$

where $b_s(t, x)$ denotes Strick's approximation to the impulse

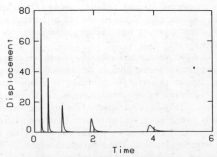

Figure 4. Seismograms resulting from sources at distances of 0.25, 0.5, 1, 2, and 4 times c_0. Q is 30.

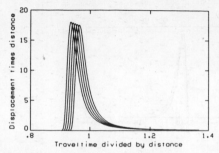

Fig. 5. Same seismograms as in Figure 4, but plotted as displacement times distance versus time divided by distance. The seismograms do not overlap due to velocity dispersion.

response, and c_s is defined by (28). By rearranging this expression, it may be written as

$$b_s(t, x) = \left(\frac{x}{c_s}\right)^{-\beta} t_s^{-(\gamma+1)/2\gamma} [2\pi\gamma(1-\gamma)^{-1/\gamma}]^{-1/2}$$

$$\exp\left[-\gamma(1-\gamma)^{(1-\gamma)/\gamma} t_s^{1-1/\gamma}\right] \qquad (39)$$

where

$$t_s = t\left(\frac{x}{c_s}\right)^{-\beta}$$

By differentiation we get the approximation for the differentiated impulse response, $b_{sv}(t, x)$:

$$b_{sv}(t, x) = \left(\frac{x}{c_s}\right)^{-\beta} b_s(t, x)\left[(1-\gamma)^{1/\gamma} t_s^{-1/\gamma} - \frac{\gamma+1}{2\gamma} t_s^{-1}\right]$$

$$(40)$$

It is evident from inspection of these expressions, that they do obey the correct scaling relations given by (37). Figures 7 and 8 show a comparison between the wave shapes computed by the fast Fourier transform method and those computed using the steepest descent approximation. They show an excellent agreement for the early part of the pulse, which includes most of the higher frequency information, while the steepest descent approximation underestimates the low frequency amplitudes in the later part of the pulse. This is not surprising since the assumptions involved in the steepest descent approximation break down at very low frequencies. This agreement contrasts with the result of *Minster* [1978a], who in his Figure 3 shows significant differences between arrivals computed using FFT methods and those computed using analytical expansions.

So far we have only considered the pulse propagation in homogeneous materials and given scaling relations applicable to materials with the same value of Q. As the wave shapes plotted in Figure 3 show a great deal of similarity for different values of Q, it should be possible to give scaling relations for different values of Q as well as for different distances.

When $Q^{-2} \ll 1$, the tangents in (22) and (25) may be replaced by their arguments. Thus (29) and (25) may be written as

$$B(\omega) \approx \exp\left\{-\frac{x|\omega|}{2Qc} - i\omega\frac{x}{c}\right\} \qquad (41)$$

where

$$c \approx c_0\left|\frac{\omega}{\omega_0}\right|^{1/\pi Q} \qquad (42)$$

By use of the Maclaurin series expansion of the exponential function, (42) may be written as

$$\frac{c}{c_0} = 1 + \frac{1}{\pi Q} \ln\left|\frac{\omega}{\omega_0}\right| + \frac{1}{2!}\left[\frac{1}{\pi Q} \ln\left|\frac{\omega}{\omega_0}\right|\right]^2 + \cdots \qquad (43)$$

When all the frequencies of interest satisfy the condition

$$\frac{1}{\pi Q} \ln\left|\frac{\omega}{\omega_0}\right| \ll 1 \qquad (44)$$

sufficient precision may be maintained by including only the first two terms of the expansion given in (43). The result is the dispersion relation given by many of the NCQ papers [e.g., *Kanamori and Anderson,* 1977]. Using the approximation indicated in (43), and dropping all terms involving the second or higher powers of $1/Q$, (41) becomes

$$B'(\omega) = \exp\left\{-\frac{x\omega}{c_0}\left[\frac{\text{sgn}(\omega)}{2Q} + i - \frac{i}{\pi Q} \ln\left|\frac{\omega}{\omega_0}\right|\right]\right\} \qquad (45)$$

The similarity and shift theorems [*Bracewell,* 1965, p. 101] may now be used to relate the approximate impulse response $b'(t)$ that has $B'(\omega)$ as its Fourier transform, as indicated by the following relations.

$$b'(t) = r b_1'(t') \qquad (46)$$

where

$$t' = rt - Q + \frac{1}{\pi} \ln\frac{r}{\omega_0} \qquad (47)$$

$$r = \frac{c_0 Q}{x} \qquad (48)$$

and $b_1'(t)$ is the inverse Fourier transform of

$$B_1'(\omega') = \exp\left\{-\omega'\left[\frac{1}{2} \text{sgn}(\omega') - \frac{i}{\pi} \ln|\omega'|\right]\right\} \qquad (49)$$

As long as the condition given by (44) holds, it is possible to obtain wave shapes for materials with different Q as well as different travel times by a combination of scaling and shifting of a single pulse shape. In particular, it follows from (46) and (48) that the amplitude of the pulse will be approximately proportional to Q. This result, combined with the exact scaling relations (37), implies that the function $C(Q)$, defined by (38) approaches a constant value as Q becomes large. In order to test the usefulness of (38) we have evaluated numerically the value of $C(Q)$. The results are plotted in Figure 9 for two pulse width definitions and three different travel time definitions. These curves show that the value of $C(Q)$ is practically inde-

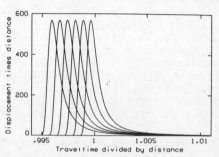

Fig. 6. Seismograms resulting form sources at distances of 0.01, 0.1, 1, 10, 100, and 1000 times c_0 plotted in the same manner as in Figure 5, for $Q = 1000$. This shows that the dispersion effect, relative to the pulse width, is independent of Q when $Q \gg 1$.

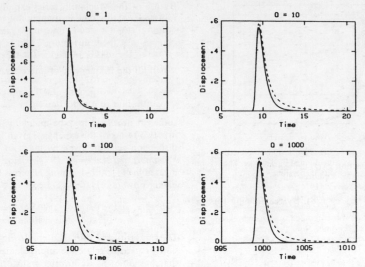

Fig. 7. Comparison of waveforms computed using (39) (solid line), to the waveforms from Figure 3 (dashed line).

pendent of Q, for Q greater than about 20. The similarity of the pulse shape for different values of Q implies that the pulse broadening along the wave path may be summed and (38) written as

$$\tau \approx \int C(Q) \frac{dT}{Q} \approx C \int \frac{dT}{Q} \qquad (50)$$

This relation may provide the basis for a practical method for inverting models for the anelastic properties of rocks in situ when the wave sources are sufficiently impulsive and the waves are recorded on broadband instruments. The ambiguities involved in using the pulse breadth in this manner are far less than those involved in the use of amplitudes in a narrow frequency band, since a number of purely elastic effects, such as focusing from curved interfaces, can have large effects on the amplitudes of seismic signals. It has the advantage over spectral methods that the measurement may be done on a

clearly defined phase of the wave form [*Gladwin and Stacey,* 1974]. It should be noted that (38) and (50) apply for other pulsewidth measures than risetime, but the value of $C(Q)$ will of course be different.

FIELD MEASUREMENTS OF ATTENUATION

There have been relatively few field studies of the propagation of transient wave pulses in rocks. *Gladwin and Stacey* [1974] found that the rise time τ, which they defined as the maximum amplitude divided by the maximum slope on the seismogram, could be fitted by an expression of the form

$$\tau = \tau_0 + C \frac{T}{Q} \qquad (51)$$

where τ_0 indicates the rise time of the source and C was a constant with a value of 0.53 ± 0.04. This value is in reason-

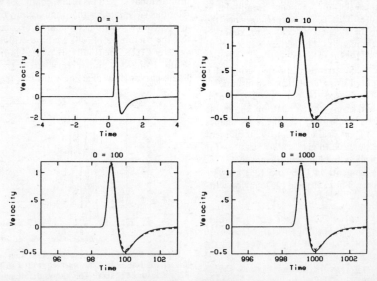

Fig. 8. Comparison of waveforms computed using (40) (solid line), to waveforms computed using numerical FFT algorithms (dashed line).

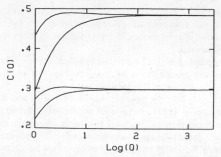

Fig. 9. Plot of the function $C(Q)$ defined by (39). Each pair of curves was computed using as a pulse width measure, the rise time definition of *Gladwin and Stacey* [1974], i.e., maximum amplitude divided by maximum slope. The top pair of curves applies to the impulse response $b(t)$, and the lower curve applies to its derivative. The lower curve in each pair was computed using as travel time T the arrival time of the peak of the pulse, and the upper was computed using the arrival time of maximum slope. The asymptotic values are 0.485 and 0.298.

ably good agreement with the value predicted on the basis of the CQ theory of 0.485 for large Q (Figure 9).

McDonal et al. [1958] performed experiments in wells drilled into the Pierre shale formation near Limon, Colorado. Fourier analysis of their data indicated that individual Fourier components of the wave forms decayed exponentially in amplitude with distance and that this decay was proportional to frequency. The attenuation per 1000 feet (305 m) was given in decibels as 0.12 times frequency. Substitution of this value into (41) and using a velocity of 7000 feet/s (2130 m/s) gives Q equal to 32. This result was obtained at depths of several hundred feet. Deep reflections indicated that the attenuation decreased with depth with the average attenuation down to a depth of 4000 feet (1220 m) corresponding to a Q of approximately 100. Their wave forms did not show a large amount of broadening over a ratio of 5 in travel times; this indicates that the sources were long compared to the impulse response of the wavepath so the assumption of a delta function source is not appropriate. However, if the rise times of the wave forms shown in Figures 3 and 6 of *McDonal et al.* [1958], are fitted to the expression (51), a reasonable fit may be obtained using $C = 0.5$ and $Q = 30$. This is consistent with the first part of the source being approximately a delta function in velocity or a step function in displacement.

Ricker [1953, 1977] described experiments done in 1948 in the same formation. Wave forms were recorded by three geophones at depths of 422, 622, and 822 feet (129, 190, and 251 m), for shots at depths less than 300 feet (92 m) in adjacent wells. Figure 10 shows a plot of pulse width versus travel time [*Ricker*, 1977, Figure 15.23]. Ricker fitted these data by a function of the form

$$\tau = at^{1/2} \qquad (52)$$

This relation is in direct conflict with (37), as well as the experimental result of *Gladwin and Stacey* [1974]. According to *Ricker* [1977, p. 198], this observation is the strongest, if not the only evidence supporting the applicability of his theory to seismic waves. By inspection of Figure 10 it appears that the data could just as well be fitted by a function of the form (51) used by *Gladwin and Stacey* [1974]. *McDonal et al.* [1958] criticized Ricker's experiment on the basis that each shot was recorded by no more than three geophones, and that wave

forms from different shots were not comparable because 'One can not shoot a second time in the same hole because the same hole is not there any more.' This is probably the reason for some of the scatter in Ricker's data, particularly from the 300 foot shots. This error can be reduced, however by adjusting the parameter τ_0 in (51) for each shot, provided that it is recorded by at least two geophones. Thus we have fitted the wavelet breadth data to a model given by

$$\tau = \tau_{0i} + C \int \frac{dT}{Q} \qquad (53)$$

In order to facilitate the integration, the travel time data were fitted to the form

$$T = a(x_g - x_s) + b(x_g{}^2 - x_s{}^2) \qquad (54)$$

where x_g is the depth to the geophone and x_s is the depth to shot. This expression implies that the velocity as function of depth will be given by

$$V = \frac{1}{a + 2bx} \qquad (55)$$

As Ricker did not specify which of the data points were obtained from the same shot, it was only possible to determine the source widths for each shot depth. For the pulse width measure used by Ricker, the value of the parameter C in (53) is approximately unity. Figure 11 shows a plot of the data from Figure 10, with the source width subtracted, compared to a straight line with a slope of $1/Q = 1/32$. The data points for the geophone at 622 feet tend to be above the curve; this can be explained by attenuation decreasing with depth. This result implies that both Ricker's data and the data of McDonal et. al. are consistent with the linear constant Q model, and both give the same value for Q. This is particularly significant in light of the fact that they interpreted their data very differently, and that neither of them considered a constant or near constant linear attenuation in the interpretation of their data. The apparent conflict between the observations of *Ricker* [1953] and *McDonal et. al.* [1958] has been noted by many authors including *Gladwin and Stacey* [1974], *Reiter and Monfort* [1977], and *Bless and Ahrens* [1977].

COMPARISON WITH NEARLY CONSTANT Q THEORIES

Lomnitz [1956] investigated the transient creep in rocks at low stress levels. He found that the shear strain resulting from

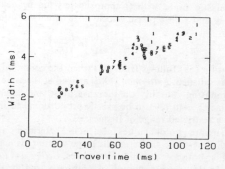

Fig. 10. Pulse width as a function of travel time in Pierre shale. Data from Figure 15.23 in *Ricker* [1977]. Geophones are at depths of 422, 622, and 822 feet (129, 190, and 251 m). Sources are at 25 foot (7.6 m) intervals at depths from 100 to 300 feet (30.5 to 91.5 m). Numbers indicate sources, 1 for 100 feet (30.5 m), to 9 for 300 feet (91.5 m).

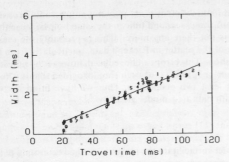

Fig. 11. The data in Figure 10, after subtraction of the initial pulse widths, compared with predicted pulse widths for $Q = 32$. Both Q and the source widths were determined by simultaneous least-square inversion.

a step in applied stress could be described to within the experimental error with a creep function of the form

$$\Psi(t) = \frac{1}{M_0} [1 + q \ln (1 + at)] \qquad (56)$$

where a is a frequency much greater than the sample rate or the time resolution of the experiment. He found that the fit to the data was insensitive to the value of a, as long as it was large. For Q greater than about 20, (56) is approximately equal to the CQ creep function (14). By using the first two terms from the MacLaurin series expansion of the exponential function, (14) may be rewritten

$$\Psi(t) = \frac{1}{M_0} e^{2\gamma \ln (t/t_0)} \approx \frac{1}{M_0}\left[1 + \frac{2}{\pi Q} \ln \left(\frac{t}{t_0}\right)\right]$$

when $t_0 \ll t$, this is approximately equal to

$$\Psi(t) = \frac{1}{M_0}\left[1 + \frac{2}{\pi Q} \ln \left(1 + \frac{t}{t_0}\right)\right] \qquad (57)$$

Later, *Lomnitz* [1957, 1962] used his creep law and the superposition principle to derive a model for wave attenuation with Q approximately independent of frequency for large Q. *Pandit and Savage* [1973] measured Q for several rock samples with Q ranging from 30 to 300 and found good agreement between values determined at sonic frequencies and those derived from transient creep measurements over several tens of seconds.

Kolsky [1956] did experiments on the propagation of ultrasonic pulses in polymers and found the pulse width to be proportional to travel time. To model his data, he used a viscoelastic model with Q approximately independent of frequency and with a phase velocity that varied according to

$$\frac{c}{c_0} = 1 + \frac{1}{\pi Q} \ln \left(\frac{\omega}{\omega_0}\right) \qquad (58)$$

Equation (58) follows from (43) when the condition given in (44) is satisfied. *Futterman* [1962] arrived at the same formula by imposing causality on the wave pulse and assuming the parameter α in (29) to be exactly proportional to frequency over a restricted range of frequencies.

There are two difficulties inherent in Futterman's approach, which necessitate limits on the range where Q is nearly constant, at both low and high frequencies. *Collins and Lee* [1956] showed that the assumption of a nonzero limit for the phase velocity as frequency approaches zero, implies that Q must approach infinity at zero frequency. Futterman's formulation was based on a finite value of the refractive index at zero frequency and is thus incompatible with constant Q, where the

phase velocity has no nonzero limit as frequency approaches zero. It can also be shown [e.g., *Azimi et. al.,* 1968], that α proportional to frequency at high frequencies leads to a violation of causality.

It appears that these limitations, which are peculiar to Futterman's approach, have led many workers to assume that a physically realizable formulation with Q exactly independent of frequency was not possible. *Liu et. al.* [1976] and *Kanamori and Anderson* [1977] have used viscoelastic distributions to derive dispersion relations of the form shown in (58). Viscoelastic density functions are discussed in the appendix, and it is shown how the constant Q model can be derived from distributions of dashpots and springs.

DISCUSSION

Of the two assumptions that provide the basis for the constant Q model, linearity is the more fundamental, and it has also been more frequently questioned in the literature than the frequency independence of Q. Nonlinear, rate independent friction was originally proposed [e.g., *Born,* 1941] to explain the frequency independence of Q, since at that time no simple linear models were available that could account for this observation. As summarized in Table 1, all of the nonlinear friction mechanisms that have been proposed have several features in common. These include the dependence of the effective elastic moduli on strain amplitude, proportionality of $1/Q$ to strain at low amplitudes, frequency independence of both Q and the moduli, distortion of waveforms and cusped stress-strain loops, and the absence of any transient creep or relaxation. *Mindlin and Deresiewicz* [1953] analyzed the losses due to friction between spheres in contact, and found the attenuation to be proportional to amplitude at low amplitudes. *White* [1966] claimed that the introduction of static friction into this model had the effect of making Q independent of amplitude. This claim cannot be correct since it may be shown [*Mavko,* 1979], that static friction cannot increase the loss. *Walsh* [1966] considered the sliding across barely closed elliptical cracks and found the loss for closed cracks with zero normal force to be independent of amplitude. However this model cannot, as shown by *Savage* [1969], explain loss independent of amplitude for the whole rock. The required distributions of elliptical cracks would imply that the effective elastic moduli of the rock, as functions of confining pressure, are discontinuous at all values of confining pressure. *Mavko* [1979] has considered a more general case of nonelliptical cracks and found the attenuation to depend on amplitude in much the same manner as in the contact sphere model of Mindlin and Deresiewicz. All of the above models feature a decrease in the effective moduli with strain amplitude due to the increase in area of the sliding surfaces. Decrease of both velocity and Q, similar to what would be expected on the basis of the above models, has been observed in laboratory studies of rocks, [*Gordon and Davis,* 1968; *Winkler et al.,* 1979], but only at strains greater than about 10^{-6} to 10^{-5}. At lower strains both Q and wave velocities are found to be independent of amplitude.

The dependence of the wave velocity on frequency is such that it is difficult to separate it from the effects of spatial heterogeneities. There is however an increasing amount of evidence in support of the frequency dependence of the elastic moduli. Seismic models for the whole earth show much improved agreement with the free oscillation data when the frequency dependence of the elastic moduli is taken into account [*Anderson et. al.,* 1977]. It is also well established that for many rocks the elastic moduli derived from ultrasonic pulse measurements are significantly greater than the moduli derived from low frequency deformation experiments [*Simmons and Brace*, 1965]. This difference is generally larger for lossy materials. *Gretener* [1961] analyzed well logging data from several oil wells in Canada and found statistically significant differences between observed travel times from surface sources to geophones in wells and travel times predicted on the basis of high frequency continuous velocity logs. *Strick* [1971] showed that these differences could be explained by the dispersion associated with linear attenuation with Q nearly independent of frequency.

Brennan and Stacey [1977] measured both Q and elastic moduli in low frequency deformation experiments, at strains of 10^{-6}, and found the moduli to vary with frequency as predicted by linearity. The stress-strain loops were elliptical although earlier experiments at larger amplitudes showed cusped stress-strain loops [*McKavanagh and Stacey*, 1974].

Because the principle of superposition does not apply to the nonlinear solid friction models, it is difficult to predict their effects on the propagation of transient stress pulses. *Walsh* [1966] pointed out that the losses due to friction cannot be described through the use of complex moduli although this is frequently attempted [e.g., *Johnston and Toksoz*, 1977]. It is easily shown [e.g., *Gladwin and Stacey*, 1974], that the use of complex frequency independent moduli leads to acausal waveforms that arrive before they are excited. *Savage and Hasegawa* [1967] used the stress-strain hysteresis loops implied by several different friction models, to model wave propagation. The results showed significant amounts of distortion which has never been observed experimentally.

From these observations it may be concluded that at strain amplitudes of interest in seismology, the propagation and attenuation of waves are dominated by linear effects, with some nonlinear effects showing up at strains of 10^{-5} or greater. This amplitude corresponds to a stress amplitude of 5 bars, since the ambient seismic noise level is on the order of 10^{-11} in strain, and studies of earthquake source mechanisms indicate stress changes of 1–100 bars [*Hanks*, 1977]; it is evident that nonlinear effects can only be significant very near the source.

While a good case can be made for the linearity of the absorption of seismic energy at low amplitudes, no such simple answer can be given to the question of the frequency dependence of the attenuation. Theoretical models of specific attenuation mechanisms are often formulated in terms of relaxation times, each of which implies a creep function that is a decaying exponential. A model that has a single relaxation time is often referred to as the standard linear solid and has Q proportional and inversely proportional to frequency at high and low frequencies, respectively. Cases where inertial effects may play a role, such as in the flow of low viscosity fluids [*Mavko and Nur*, 1979], feature even stronger variation of attenuation with frequency. It may be shown [*Kjartansson*, 1978] that, in materials with sharply defined heterogeneities (e.g., grain boundaries or pores), absorption due to processes controlled by diffusion, such as phase transformations or thermal relaxation, leads to Q proportional to $\omega^{1/2}$ and $\omega^{-1/2}$ at high and low frequencies, even for uniform distributions of identical pores or crystals.

For these types of mechanisms, the approximate frequency independence of Q that is observed indicates distributions of time constants associated with the individual absorbing elements. It may be shown, for example, that the frequency at which maximum absorption occurs for mechanisms involving the diffusion of heat, is inversely proportional to the square of the minimum dimension of the inhomogenieties involved. The empirical observation that Q, in solids, varies much slower than even the square root of frequency, is thus an expression of the statistical nature of the inhomogenieties. It is interesting that dielectric losses in solids show the same type of frequency dependence as do the energy losses in stress waves [*Jonscher*, 1977].

While Q is probably not strictly independent of frequency, there is no reason to believe that any of the band-limited near-constant Q theories better approximate the wave propagation in real materials than the constant Q model. Therefore nothing is gained in return for the mathematical complexity and potential inconsistency in using, for example, the absorption band model of *Liu et al.* [1976].

Strick [1967] obtained a transfer function for wave propagation of which the constant Q is a special case. He rejected the CQ case, however, on the basis that the lack of an upper bound for the phase velocity was in violation of causality. Strick's three-parameter model is equivalent to the CQ model, with an additional time delay applied to the waveform. *Strick* [1970] computed waveforms for his models, and found that the detectable onset of the signal always arrived significantly later than the applied time shift. He termed this delay 'pedestal' and attributed to it significance that has been subject to some controversy. For the CQ case, the pedestal arrives when the source is excited. *Minster* [1978b] argued that the presence of the pedestal was an indication of the need for a high frequency cutoff of the type built into the model of Liu et al. This pedestal controversy points to a limitation shared by all of continuum mechanics; no continuum model, including the CQ model, can have any significance at wavelengths shorter than the molecular separation nor at periods longer than the age of the universe. This covers approximately 32 orders of magnitude in frequency, which for a Q of 100 implies a change in velocity of about 26%. The possibility that some 'calculable' energy might arrive 26% earlier than any detectable energy, is hardly a sufficient reason to introduce a high frequency cutoff. Calculable values of physical parameters outside the observable range are common in other fields, such as in solutions to the diffusion equation and in statistics. *Minster* [1978b] and *Lundquist* [1977] suggest that the cutoff should be at periods between 0.1 and 1 s for the mantle. Such cutoffs have never been observed for any of the rocks that have been studied in the laboratory, where the range of frequencies extends up to about 1 MHz.

Lomnitz's [1957] attenuation model has often been criticized [*Kogan*, 1966; *Liu et al.*, 1976; *Kanamori and Anderson*, 1977] on the basis that the lack of an upper bound for the transient creep would not permit mountains or large scale gravity anomalies to last through geologic time. Since the Lomnitz creep function is practically equivalent to the constant Q creep function for large values of time and Q, this criticism applies equally to the constant Q model. However, it does not pass the test of substituting numbers into the expressions (56) or (14). For example, for a material with a Q of 100, the strain that results from the application of a unit stress is only about 33% larger over a period of one billion years, than for the first millisecond of applied stress. Thus neither the constant Q theory, nor any of the NCQ theories can explain the large strains required by plate tectonics. The fact that brittle deformation only takes place in the uppermost part of the crust, with exception of localized areas of unusually rapid tectonic activity, may indicate that over geologic time most of the earth deforms as a viscous fluid with Q for shear near zero. The assumption, implicit in the band-limited NCQ model of *Liu et al.* [1976], that Q approaches infinity outside the range of observations, is thus particularly inappropriate for low frequency shear deformations in the mantle.

CONCLUSIONS

Contrary to what has often been assumed in the past, it is possible to formulate a description of wave propagation and attenuation with Q exactly independent of frequency, that is both linear and causal. The wave propagation properties of materials can be completely specified by only two parameters, for example, Q and phase velocity at an arbitrary reference frequency. This simplicity makes it practical to derive exact expressions describing, in the frequency domain, the wave propagation for any positive value of Q. The dispersion that accompanies any linear energy absorption leads to a propagation velocity of any transient disturbance that is not only a function of the material, but also of the past history of the wave. Review of available data indicates that the assumption of linearity is well justified for seismic waves, but it is likely that Q is weakly dependent on frequency. There is, however, no indication that any of the NCQ theories that we have discussed provide a better description of the attenuation in actual rocks than the constant Q theory does.

APPENDIX: VISCOELASTIC MODELS

In the literature on viscoelasticity, it is common to describe the behavior of materials through networks of springs and dashpots, often characterized by either relaxation or retardation spectra. It has been claimed that only attenuation models given in terms of such networks are physically realizable, and models derived by other means have been termed 'ad hoc' [e.g., *Minster, 1978a*].

While it is possible to give physical models for attenuation that can not be modeled by spring-dashpot networks, [e.g., *Mavko and Nur, 1979*], the formulation of viscoelastic models in terms of relaxation spectra is often useful. *Gross* [1953] has summarized the relationships between the various functions that have been used to characterize viscoelastic materials. In his notation the retardation frequency density function $N(s)$, is related to the creep function according to

$$\Psi(t) = - \int_0^\infty N(s)e^{-ts}\,ds \qquad (A1)$$

and the relaxation frequency density function $\bar{N}(s)$, is related to the relaxation function according to

$$\bar{\Psi}(t) = \int_0^\infty \bar{N}(s)e^{-ts}\,ds \qquad (A2)$$

Kanamori and Anderson [1977] used a relaxation function of the form

$$\bar{N}(s) = As^{-1} \qquad s_1 < s < s_2$$
$$\bar{N}(s) = 0 \qquad \text{elsewhere} \qquad (A3)$$

to derive an absorption band NCQ model. The constant Q model may be specified by

$$\bar{N}(s) = \frac{M_0 \sin(2\pi\gamma)}{\pi}(st_0)^{2\gamma}s^{-1} \qquad (A4)$$

Using the definition of the gamma function and the identity

$$\Gamma(z)\Gamma(1-z) = \frac{\pi}{\sin(\pi z)} \qquad (A5)$$

the constant Q relaxation function (20), is readily obtained. Since the constant Q model is mathematically a special case of the power law models of *Strick* [1967] and *Azimi et al.* [1968], it follows that those models do also have spring-dashpot representations.

Acknowledgments. This work has been greatly stimulated by numerous discussions with Amos Nur and Michael Gladwin. The author has enjoyed discussions with many individuals about particular aspects of this theory, including David M. Boore, Jon Claerbout, George Moeckel, Bernard Minster, Henry Swanger and Kenneth Winkler. This work was supported by National Science Foundation grant EAR76-22501 and by the U.S. Department of Energy, Office of Basic Energy Sciences contract EY-76-03-0325. The author was supported by a Cecil and Ida Green Fellowship in Geophysics during part of this work.

REFERENCES

Anderson, D. L., H. Kanamori, R. S. Hart, and H.-P. Liu, The earth as a seismic absorption band, *Science, 196,* 1104–1106, 1977.

Azimi, S. A., A. V. Kalinin, V. V. Kalinin, and B. L. Pivovarov, Impulse and transient characteristics of media with linear and quadratic absorption laws, *Phys. Solid Earth, 1968,* 88–93, 1968.

Balch, A. H., and F. R. Smolka, Plane and spherical transient Voight waves, *Geophysics, 35,* 745–761, 1970.

Bland, D. R., *The Theory of Linear Viscoelasticity,* Pergamon, New York, 1960.

Bless, S. J., and T. J. Ahrens, Measurements of the longitudinal modulus of Pierre clay shale at varying strain rates, *Geophysics, 42,* 34–40, 1977.

Boltzmann, L., Zur Theorie der elastiche Nachwirkung, *Ann. Phys. Chem. Erqanzung, 7,* 624–654, 1876.

Boore, D. M., K. L. Larner, and K. Aki, Comparison of two independent methods for the solution of wave-scattering problems: Response of a sedimentary basin to vertically incident SH waves, *J. Geophys. Res., 76,* 558–569, 1971.

Born, W. T., The attenuation constant of earth materials, *Geophysics, 6,* 132–148, 1941.

Bracewell, R., *The Fourier Transform and Its Applications,* 381 pp., McGraw-Hill, New York, 1965.

Brennan, B. J., and F. D. Stacey, Frequency dependence of elasticity of rock—Test of seismic velocity dispersion, *Nature, 268,* 220–222, 1977.

Clark, G. B., and G. B. Rupert, Plane and spherical waves in a Voigt medium, *J. Geophys. Res., 71,* 2047–2053, 1966.

Collins, F., Plane compressional Voigt waves, *Geophysics, 25,* 483–504, 1960.

Collins, F., and C. C. Lee, Seismic wave attenuation characteristics from pulse experiments, *Geophysics, 21,* 16–40, 1956.

Futterman, W. I., Dispersive body waves, *J. Geophys. Res., 67,* 5279–5291, 1962.

Gladwin, M. T., and F. D. Stacey, Anelastic degradation of acoustic pulses in rock, *Phys. Earth Planet. Interiors, 8,* 332–336, 1974.

Gordon, R. B., and L. A. Davis, Velocity and attenuation of seismic waves in imperfectly elastic rock, *J. Geophys. Res., 73,* 3917–3935, 1968.

Gretener, P. E. F., An analysis of the observed time discrepancies between continuous and conventional well velocity surveys, *Geophysics, 26,* 1–11, 1961.

Gross, B., *Mathematical Structure of the Theories of Viscoelasticity,* 74 pp., Hermann, Paris, 1953.

Hanks, T. C., Earthquake stress drops, ambient tectonic stresses and stresses that drive plate motions, *Pure Appl. Geophys., 115,* 441–458, 1977.

Jaramillo, E. E., and J. D. Colvin, Transient waves in a Voigt medium, *J. Geophys. Res., 75,* 5767–5773, 1970.

Johnston, D. H., and N. Toksoz, Attenuation of seismic waves in dry and saturated rocks (abstract), *Geophysics, 42* 1511, 1977.

Jonscher, A. K., The "universal" dielectric response, *Nature, 267,* 673–679, 1977.

Jordan, T. H., and S. A. Sipkin, Estimation of the attenuation operator for multiple ScS waves, *Geophys. Res. Lett., 4,* 167–170, 1977.

Kanamori, H., and D. L. Anderson, Importance of physical dispersion in surface wave and free oscillation problems: Review, *Rev. Geophys. Space. Phys., 15,* 105–112, 1977.

Kjartansson, E., Thermal relaxation, an attenuation mechanism for porous rocks (abstract), *Eos Trans. AGU, 59,* 324, 1978.

Knopoff, L., Q, *Rev. Geophys. Space Phys., 2,* 625–660, 1964.

Kogan, S. Y., A brief review of seismic wave absorption theories, II, *Phys. Solid Earth, 1966,* 678–683, 1966.

Kolsky, H., The propagation of stress pulses in viscoelastic solids, *Phys. Mag., 1,* 693–710, 1956.

Liu, H.-P., D. L. Anderson, and H. Kanamori, Velocity dispersion due to anelasticity; Implications for seismology and mantle composition, *Geophys. J. Roy. Astron. Soc., 47,* 41–58, 1976.

Lockner, D. A., J. B. Walsh, and J. D. Byerlee, Changes in seismic velocity and attenuation during deformation of grainite, *J. Geophys. Res., 82,* 5374–5378, 1977.

Lomnitz, C., Creep measurements in igneous rocks, *J. Geol., 64,* 473–479, 1956.

Lomnitz, C., Linear dissipation in solids, *J. Appl. Phys., 28,* 201–205, 1957.

Lomnitz, C., Application of the logarithmic creep law to stress wave attenuation in the solid earth, *J. Geophys. Res., 67,* 365–368, 1962.

Lundquist, G., Evidence for a frequency dependent Q (abstract), *Eos Trans. AGU, 58,* 1182, 1977.

Mavko, G. M., Frictional attenuation: An inherent amplitude dependence, *J. Geophys. Res., 84,* in press, 1979.

Mavko, G. M., and A. Nur, Wave attenuation in partially saturated rocks, *Geophysics, 44,* 161–178, 1979.

McDonal, F. J., F. A. Angona, R. L. Mills, R. L. Sengbush, R. G. van Nostrand, and J. E. White, Attenuation of shear and compressional waves in Pierre shale, *Geophysics, 23,* 421–439, 1958.

McKavanagh, B. M., and F. D. Stacey, Mechanical hysteresis in rocks at low strain amplitudes and seismic frequencies, *Phys. Earth Planet. Interiors, 8,* 246–250, 1974.

Mindlin, R. D., and H. Deresiewicz, Elastic spheres in contact under varying oblique forces, *J. Appl. Mech., 20,* 327–344, 1953.

Minster, J. B., Transient and impulse responses of a one-dimensional linearly attenuating medium, I, Analytical results, *Geophys. J. Roy. Astron. Soc., 52,* 479–501, 1978a.

Minster, J. B., Transient and impulse responses of a one-dimensional linearly attenuating medium, II, A parametric study, *Geophys. J. Roy. Astron. Soc., 52,* 503–524, 1978b.

Munasinghe, M., and G. W. Farnell, Finite difference analysis of Rayleigh wave scattering at vertical discontinuities, *J. Geophys. Res., 78,* 2454–2466, 1973.

O'Connell, R. J., and B. Budiansky, Measures of dissipation in viscoelastic media, *Geophys. Res. Lett., 5,* 5–8, 1978.

Pandit, B. I., and J. C. Savage, An experimental test of Lomnitz's theory of internal friction in rocks, *J. Geophys. Res., 78,* 6097–6099, 1973.

Reiter, L., and M. E. Monfort, Variations in initial pulse width as a function of anelastic properties and surface geology in central California, *Bull. Seismol. Soc. Amer., 67,* 1319–1338, 1977.

Ricker, N., The form and laws of propagation of seismic wavelets, *Geophysics, 18,* 10–40, 1953.

Ricker, N., *Transient Waves in Visco-Elastic Media,* 278 pp., Elsevier, Amsterdam, 1977.

Savage, J. C., Comments on 'Velocity and attenuation of seismic waves in imperfectly elastic rock' by R. B. Gordan and L. A. Davis, *J. Geophys. Res., 74,* 726–728, 1969.

Savage, J. C., and H. S. Hasegawa, Evidence for a linear attenuation mechanism, *Geophysics, 32,* 1003–1014, 1967.

Simmons, G., and W. F. Brace, Comparison of static and dynamic measurements of compressibility of rocks, *J. Geophys. Res., 70,* 5649–5656, 1965.

Strick, E., The determination of Q, dynamic viscosity and creep curves from wave propagation measurements, *Geophys. J. Roy. Astron. Soc., 13,* 197–218, 1967.

Strick, E., A predicted pedestal effect for pulse propagation in constant-Q solids, *Geophysics, 35,* 387–403, 1970.

Strick, E., An explanation of observed time discrepancies between continuous and conventional well velocity surveys, *Geophysics, 36,* 285–295, 1971.

Voight, W., Uber innere Ribung fester Korper, insbesondere der Metalle, *Ann. Phys. Chem. Neue Folge, 47,* 671–693, 1892.

Walsh, J. B., Seismic wave attenuation in rock due to friction, *J. Geophys. Res., 71,* 2591–2599, 1966.

White, J. E., Static friction as a source of seismic attenuation, *Geophysics, 31,* 333–339, 1966.

Winkler, K., A. Nur, and M. Gladwin, Friction and seismic attenuation in rocks, *Nature, 277,* 528–531, 1979.

(Received June 8, 1978;
revised January 8, 1979;
accepted February 6, 1979.)